Volume 70

Epigenetics and Cancer, Part A

Advances in Genetics, Volume 70

Serial Editors

Theodore Friedmann
University of California at San Diego, School of Medicine, USA

Jay C. Dunlap
Dartmouth Medical School, Hanover, NH, USA

Stephen F. Goodwin
University of Oxford, Oxford, UK

Volume 70

Epigenetics and Cancer, Part A

Edited by

Zdenko Herceg

Epigenetics Group
International Agency for Research on Cancer (IARC)
Lyon, France

Toshikazu Ushijima

Carcinogenesis Division
National Cancer Centre Research Institute
Tokyo, Japan

AMSTERDAM • BOSTON • HEIDELBERG • LONDON
NEW YORK • OXFORD • PARIS • SAN DIEGO
SAN FRANCISCO • SINGAPORE • SYDNEY • TOKYO

ELSEVIER

Academic Press is an imprint of Elsevier

Academic Press is an imprint of Elsevier

525 B Street, Suite 1900, San Diego, CA 92101-4495, USA
30 Corporate Drive, Suite 400, Burlington, MA 01803, USA
32 Jamestown Road, London, NW1 7BY, UK
Radarweg 29, PObox 211, 1000 AE Amsterdam, The Netherlands

First edition 2010

ISBN: 978-0-12-380866-0
ISSN: 0065-2660

For information on all Academic Press publications
visit our website at elsevierdirect.com

Printed and bound in USA

10 11 12   10 9 8 7 6 5 4 3 2 1

Contents

SECTION II   EPIGENETIC EVENTS UNDERLYING BIOLOGICAL PHENOMENA   143

SECTION III CANCER EPIGENOME   245

9 Cancer Epigenome   247
Matthias Lechner, Chris Boshoff, and Stephan Beck

10 Identification of Driver and Passenger DNA Methylation in Cancer by Epigenomic Analysis   277
Satish Kalari and Gerd P. Pfeifer

SECTION IV EPIGENETIC THERAPY AND EPIGENETIC DRUGS 325

Contributors

Numbers in parentheses indicate the pages on which the authors' contributions begin.

Stephan Beck (245) UCL Cancer Institute, University College London, London, United Kingdom

Chiara Biancotto (341) Department of Experimental Oncology, European Institute of Oncology, Via Adamello 16, Milan, Italy

Chris Boshoff (245) UCL Cancer Institute, University College London, London, United Kingdom

George A. Calin (87) Department of Experimental Therapeutics and Cancer Genetics, University of Texas, M.D. Anderson Cancer Center, Houston, Texas, USA

Manel Esteller (25) Cancer Epigenetics and Biology Program (PEBC), The Bellvitge Institute for Biomedical Research (IDIBELL), Hospital Duran i Reynals, Avinguda Gran Via de L'Hospitalet 199-203, E-08907 L'Hospitalet de Llobregat, Barcelona, Catalonia, Spain

Muller Fabbri (87) Department of Molecular Virology, Immunology, and Medical Genetics and Comprehensive Cancer Center, Ohio State University, Biomedical Research Tower, Columbus, Ohio, USA

Gianmaria Frigè (341) Department of Experimental Oncology, European Institute of Oncology, Via Adamello 16, Milan, Italy

Zdenko Herceg (1, 57, 325) Epigenetics Group, International Agency for Research on Cancer (IARC), 69008 Lyon, France

Megan P. Hitchins (201) Adult Cancer Program, Lowy Cancer Research Centre, Prince of Wales Clinical School, University of New South Wales, Sydney, Australia

Satish Kalari (277) Department of Cancer Biology, Beckman Research Institute of the City of Hope, Duarte, California, USA

Marta Kulis (25) Cancer Epigenetics and Biology Program (PEBC), The Bellvitge Institute for Biomedical Research (IDIBELL), Hospital Duran i Reynals, Avinguda Gran Via de L'Hospitalet 199-203, E-08907 L'Hospitalet de Llobregat, Barcelona, Catalonia, Spain

Matthias Lechner (245) UCL Cancer Institute, University College London, London, United Kingdom

Derek Hock Kiat Lim (145) Department of Medical & Molecular Genetics, School of Clinical and Experimental Medicine, University of

Birmingham College of Medical and Dental Sciences; and West Midlands Region Genetics Service, Birmingham Women's Hospital, Edgbaston, Birmingham, United Kingdom

Eamonn Richard Maher (145) Department of Medical & Molecular Genetics, School of Clinical and Experimental Medicine, University of Birmingham College of Medical and Dental Sciences; and West Midlands Region Genetics Service, Birmingham Women's Hospital, Edgbaston, Birmingham, United Kingdom

Samson Mani (325) Epigenetics Group, International Agency for Research on Cancer (IARC), Lyon, France

Saverio Minucci (341) Department of Experimental Oncology, European Institute of Oncology, Via Adamello 16; and Department of Biomolecular Sciences and Biotechnologies, University of Milan, Via Celoria 26, Milan, Italy

Rabih Murr (101) Friedrich Miescher Institute for Biomedical Research, Maulbeerstrasse 66, 4058 Basel, Switzerland

Gerd P. Pfeifer (277) Department of Cancer Biology, Beckman Research Institute of the City of Hope, Duarte, California, USA

Carla Sawan (57) Epigenetics Group, International Agency for Research on Cancer (IARC), 69008 Lyon, France

Hiromu Suzuki (309) Department of Biochemistry; and First Department of Internal Medicine, Sapporo Medical University, Sapporo, Japan

Minoru Toyota (309) Department of Biochemistry, Sapporo Medical University, Sapporo, Japan

Toshikazu Ushijima (1) Carcinogenesis Division, National Cancer Center Research Institute, Tokyo, Japan

Akira Watanabe (177) Center for iPS Cell Research and Application (CiRA), Kyoto University, Shogoin, Sakyo-ku, Kyoto, Japan

Yasuhiro Yamada (177) Center for iPS Cell Research and Application (CiRA), Kyoto University, Shogoin, Sakyo-ku, Kyoto, Japan

Preface

Epigenetics is a fascinating and rapidly expanding field of modern biology. Over the past decade, the field has witnessed a remarkable improvement in our knowledge of the importance of epigenetic events in the control of both normal cellular processes and abnormal events associated with disease. Both the scientific and medical communities now recognize that epigenetic changes lie at the heart of many complex diseases, most notably cancer.

Epigenetic events have been shown to be associated with virtually every step of tumor development and progression. They are also likely to occur very early in tumor development. The advent and rapid development of new technologies for epigenomics has started to unravel molecular features of cancer cells responsible for cancer development and progression, and to identify novel targets for diagnostics and therapeutics. Accurate measurement of various epigenetic modifications allowed us to evaluate the contribution of environmental, dietary, and lifestyle factors to human cancers. These advances have turned academic, medical, and public attention to the application of epigenetics to cancer prevention, diagnosis, and treatment.

We felt that it is important to deliver many conceptual breakthroughs and technological advances that are likely to revolutionize the traditional concept of cancer and cancer research to a broad scientific and medical community. For this book, we have invited many leading scientists, who have made important contributions to epigenetics and epigenomics and shaped the current trends in the field. We have attempted to "cover" the state-of-the-art in cancer epigenetics and cutting-edge technologies in epigenomics. Our aim was to discuss the state of science and future research needs covering the most recent advances, both conceptual and technological, and to provide novel opportunities for cancer prevention, diagnosis, and treatment. Although this book is intended primarily for academic and professional readers (from basic science to clinical researchers and epidemiologists), we believe that it will appeal to and be used by a wider audience among healthcare workers.

We thank all the authors for their valuable contribution and for making this book what it is. We are much obliged to the reviewers, who spend their precious time reviewing the manuscripts; their constructive criticism and candid opinions significantly improved both the scope and quality of the chapters. Special thanks are due to Drs. Andrea Baccareli, Amir Eden, Robert Dante, Aleksandra Fučić, Koraljka Gall-Trošelj, Anastas Gospodinov, Hector Hernandez-Vargas, Barry Iacopetta, Atsushi Kaneda, Yutaka Kondo, Vladimir Krutovskikh, Saadi Khochbin,

Heinz Linhart, Joel Mason, John Mattick, Kent Nephew, Magali Olivier, Anupam Paliwal, Gerd Pfeifer, Christoph Plass, Hidenobu Soejima, Hiromu Suzuki, Minoru Toyota, Thomas Vaissière, Paolo Vineis, André Verdel, Joseph Wiemels, Nick Wong, and Daniel Worthley. We are also grateful to Sandrine Montigny (from IARC, Lyon) for her excellent secretarial help and final formatting of the chapters. Thanks are also due to Zoe Kruze and Narmada Thangavelu (from Elsevier) for their understanding and patience during (often lengthy) process of preparation and review of the manuscripts. We thank all our colleagues in our respective laboratories (in Lyon and Tokyo) for their understanding, enthusiasm, and support during the preparation of this book.

Zdenko Herceg
Lyon, France
Toshikazu Ushijima
Tokyo, Japan

1

Introduction: Epigenetics and Cancer

Zdenko Herceg* and Toshikazu Ushijima†

*Epigenetics Group, International Agency for Research on Cancer (IARC), Lyon, France
†Carcinogenesis Division, National Cancer Center Research Institute, Tokyo, Japan

I. Introduction
II. Basic Epigenetic Mechanisms—Ghosts Above the Genes
III. Epigenetic Events and Biological Phenomena—Gene Wiring to Instruct Inheritance
  A. Genomic imprinting and cancer
  B. Epigenetic codes in stem cells and cancer stem cells
  C. Constitutional epimutations and cancer susceptibility
IV. Cancer Epigenome—Zooming in on Genome-Wide Scale
  A. Cancer epigenome
  B. Identification of driver DNA methylation changes
  C. Epigenetic drivers and genetic passengers in cancer
V. Epigenetics and Environmental Factors—Where Your Genes Meet the Environment
  A. Epigenetic alterations induced by dietary and other environmental factors
  B. Induction of epigenetic changes by chronic inflammation
  C. Maternal diet, early life exposure, and epigenetic processes
  D. Folate, one carbon metabolism, and DNA methylation in cancer
VI. Epigenetic Changes in Cancer—Markers in Tracking Cancer Cells
  A. Epigenetic biomarkers for cancer pathophysiology

Advances in Genetics, Vol. 70
0065-2660/10 $35.00
DOI: 10.1016/S0065-2660(10)70001-8

ABSTRACT

The field of epigenetics has witnessed a recent explosion in our knowledge on the importance of epigenetic events in the control of both normal cellular processes and abnormal events associated with diseases, moving this field to the forefront of biomedical research. Advances in the field of cancer epigenetics and epigenomics have turned academic, medical, and public attention to the potential application of epigenetics in cancer control. A tremendous pace of discovery in this field requires that these recent conceptual breakthroughs and technological state-of-the-art in epigenetics and epigenomics are updated and summarized in one book with cancer focus. This book is primarily intended to academic and professional audience; however, an attempt has been made to make it understandable by and appealing to a wider audience among healthcare workers. The main aim of this book is to produce an authoritative and comprehensive reference source in print and online, covering all critical aspects of epigenetics and epigenomics and their implications in cancer research. This book discusses the state of science and determines the future research needs, covering most recent advances, both conceptual and technological, and their implication for better understanding of molecular mechanisms of cancer development and progression, early detection, risk assessment, and prevention of cancer. In this chapter, we describe the main aim and scope of this book and provide a brief emphasis of each of 22 chapters regrouped into eight major parts. © 2010, Elsevier Inc.

I. INTRODUCTION

Epigenetics represents a new frontier in cancer research owing to the fact that epigenetic events have emerged as key mechanisms in regulation of critical biological processes and in development of human diseases. Although nongenetic phenomena have long been considered critical for our understanding of the mechanisms and causes of complex diseases such as cancer, it was not until recently that epigenetics has attracted considerable attention of scientific and medical communities. The field of epigenetics has witnessed a recent explosion in our knowledge on the importance of epigenetic events in the control of both normal cellular processes and abnormal events associated with diseases. Historically, the "epigenetics" was used to describe all biological phenomena that do not follow normal genetic rules. The term "epigenetics" was coined by Conrad Waddington in 1942 to describe the discipline in biology that studies "the interactions of genes with their environment that bring the phenotype into being." The field of epigenetics is considered as one of the most rapidly expanding fields of modern biology that has enormous impact on our thinking and understanding of biological phenomena and complex diseases, notably cancer. Since the introduction of the term "epigenetics," a number of biological events that are not coded in DNA sequence itself have been considered as epigenetic phenomena. These include imprinting, position-effect variegation in the fruit fly, paramutations, and X-chromosome inactivation. Recent discoveries in epigenetics and epigenomics revealed that different epigenetic events may share common molecular mechanisms.

In a broader sense, epigenetics can be considered as an interface between genotype and phenotype. Epigenetics encompasses mechanisms that modify the final outcome of the genetic code without altering the underlying DNA sequence. The importance of epigenetic principle is highlighted by the fact that all cells in any given organism share an identical genome with other cell types, yet they exhibit striking morphological and functional properties. Therefore, it is obvious that epigenetic events define the identity and proliferation potential of different cells in the body, the features that are typically deregulated in cancer. Nowadays, the term "epigenetics" may be defined as the study of all changes that are stably transmitted over many rounds of cell divisions, but that do not alter the nucleotide sequence (genetic code). Epigenetic inheritance includes DNA methylation, histone modifications, and RNA-mediated gene silencing, all of which are essential mechanisms that allow the stable propagation of gene activity states from one generation of cells to the next. Consistent with the importance of epigenetic mechanisms, deregulation of epigenetic states is intimately linked to human diseases, most notably cancer (Feinberg and Tycko, 2004; Herceg, 2007; Jones and Baylin, 2007).

Recent discoveries in the field of cancer epigenetics have turned academic, medical, and public attention to the potential application of epigenetic and epigenomics in cancer control. A spectacular pace of discoveries in this field requires that these recent conceptual breakthroughs and technological state-of-the-art in epigenetics and epigenomics are updated and summarized in one book with focus on cancer. Although several books on "epigenetics" are available, these are either becoming increasingly out of date (owing to the fact that the field of epigenetics is rapidly expanding), not focused on cancer epigenetics, or too technical. This book is primarily intended to academic and professional audience; however, an attempt has been made to make it understandable by and appealing to a wider audience among healthcare workers. The main aim of this book is to produce an authoritative and comprehensive reference source in print and online, covering all critical aspects of epigenetics and epigenomics and their implications in cancer research and cancer control. This book discusses the state of science and future research needs, covering most recent advances, both conceptual and technological, and their implication for better understanding of molecular mechanisms of cancer development and progression, early detection, risk assessment, and prevention of cancer. Technological advances in epigenomics for cancer research and molecular epidemiology are also discussed. The content of the book is organized into 22 chapters which can be regrouped into 8 major parts. Particular emphasis is given to: (i) basic epigenetic mechanisms in the regulation of critical cellular processes, (ii) epigenetic events underlying biological phenomena, (iii) cancer epigenome, (iv) epigenetic changes induced by environmental and dietary/lifestyle factors, (v) epigenetic biomarkers, (vi) epigenetic therapy and epigenetic drugs, (vii) application of epigenetics in molecular epidemiology and epigenetic cancer prevention, and (viii) epigenetic databases.

II. BASIC EPIGENETIC MECHANISMS—GHOSTS ABOVE THE GENES

Major epigenetic mechanisms include DNA methylation, covalent posttranslational modifications of histone proteins, and RNA-mediated gene silencing. Different types of epigenetic modifications are intimately linked and often act in self-reinforcing manner in regulation of different cellular processes (Fig. 1.1). Epigenetic mechanisms are essential for embryonic development, cell differentiation, protection against viral genomes, and are likely to be important for the integration of endogenous and environmental signals during the life of an organism (Feinberg et al., 2006; Herceg, 2007; Jaenisch and Bird, 2003). By analogy, deregulation of epigenetic mechanisms has been associated with a variety of human diseases, most notably cancer (Egger et al., 2004; Feinberg and Tycko, 2004; Feinberg et al., 2006; Jones and Baylin, 2002; Ushijima, 2005).

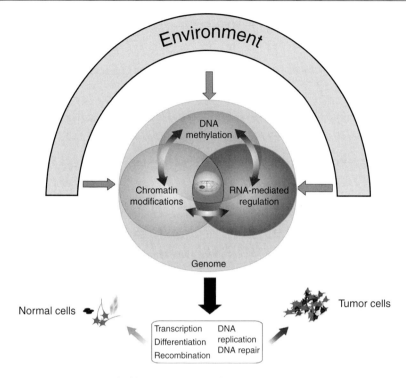

Figure 1.1. An intimate and self-reinforcing cross talk between different types of epigenetic information. Epigenetic mechanisms regulate many cellular processes directly or indirectly and play critical roles in cellular responses to environmental and endogenous stimuli. There is an intimate and self-reinforcing cross talk between different types of epigenetic information. This is proposed to constitute the "epigenetic code" that modulates genetic code in response to endogenous and environmental cues. Epigenetic code is important to maintain gene expression profiles and chromatin structure in a heritable manner over many cell generations, and may dictate cellular outcomes by regulating cellular processes such as gene transcription, proliferation, and DNA repair. Deregulation of epigenetic mechanisms may promote the development of abnormal phenotypes and diseases including cancer. (*Adapted from* Elsevier Ltd. and *Carla Sawan, Thomas Vaissière, Rabih Murr,* and *Zdenko Herceg:* epigenetic drivers and genetic passengers on the road to cancer. Mutation Research/Fundamental and Molecular Mechanisms of Mutagenesis 2008 642(1–2) © 2008 Elsevier B.V.)

DNA methylation and histone acetylation and methylation are the major epigenetic modifications that are most intensively studied in the context of gene transcription and abnormal events that lead to oncogenic process. Once the patterns of these epigenetic marks have been set up, they are propagated autonomously over many cell generations. Wealth of evidence suggests that these marks are dynamically linked in the epigenetic control of gene expression and that their deregulation play an important role in tumorigenesis (Esteller, 2007; Feinberg *et al.*, 2006; Jones and Baylin, 2002, 2007). Disruption of one of these two epigenetic mechanisms inevitably affects the other. For example, hypermethylation of CpG island in gene promoters triggers deacetylation of local histones, whereas lower levels of histone acetylation (hypoacetylation) seem to predispose to targeted DNA methylation. Therefore, there is an intimate communication between histone acetylation and DNA methylation. Recent studies have begun to reveal the mechanisms underlying these events, although it remains unclear who initiates the talk upon a given environmental or endogenous signal and what is the hierarchical order of epigenetic events during unscheduled gene silencing in tumor cells.

The best-studied epigenetic mechanism is DNA methylation. The methylation of DNA refers to the covalent addition of a methyl group to the 5-carbon (C^5) position of cytosine bases that are located 5′ to a guanosine base. This is a very small chemical modification mark on DNA molecule that does not alter DNA code; however, it may have major regulatory consequences. Aberrant DNA methylation is tightly connected to a wide variety of human malignancies. Two forms of aberrant DNA methylation are found in human cancer: the overall loss of 5-methyl-cytosine (global hypomethylation) and gene promoter-associated (CpG island-specific) hypermethylation (Feinberg and Tycko, 2004; Jones and Baylin, 2007). While the precise consequences of genome-wide hypomethylation are still debated (activation of cellular proto-oncogenes, induction of chromosome instability), hypermethylation of gene promoters is in turn associated with gene inactivation. When hypermethylated, gene promoters become unable to bind the factors that are responsible for gene expression. The gene thus becomes inactivated. A large number of studies indicated that the silencing of tumor suppressor genes and other cancer-related genes may occur through hypermethylation of their promoters. Unscheduled hypermethylation of gene promoters represents an attractive target for early diagnosis, risk assessment, and cancer prevention. For example, the genes that are the target of DNA hypermethylation early in the tumor development, in a high percentage of cases, and in the cancer type-specific manner are of particular interest. In Chapter 2, Manel Esteller and Marta Kulis introduce basic biology of DNA methylation and illustrate the roles of regional hypermethylation and global hypomethylation in carcinogenesis.

Many recent studies have also implicated aberrant patterns of histone modifications and chromatin remodeling in human malignancies. The histone modifications usually occur at the N-terminal "tails" of histones protruding from nucleosomes. These posttranslational modifications include acetylation, methylation, phosphorylation, and ubiquitination. The different histone modifications appear to act in a combinatorial and consistent fashion in regulation of several key cellular processes. This led to the concept known as the "histone code."

The importance of histone modifications is demonstrated by the fact that mechanisms involving these modifications are essential during development and that their deregulation can lead to human diseases. Histone proteins have thus emerged as key carriers of epigenetic information, constituting a fundamental and critical regulatory system that extends beyond the genetic information. Interest in histone modifications has further grown over the last few years with the discovery and characterization of a large number of histone-modifying molecules and protein complexes. Alterations in these chromatin-based processes could lead to mutations in oncogenes, tumor suppressor genes, or DNA repair genes, resulting in genomic instability, oncogenic transformation, and the development of cancer. Importantly, aberrant activity of histone-modifying factors may promote cancer development by misregulating chromatin structure and activity, an example of which is frequently found in human leukemia. Growing evidence suggests that aberrant epigenetic regulation of key cellular processes, most notably gene transcription and DNA repair, may be involved in oncogenesis. Chapter 3 (by Zdenko Herceg and Carla Sawan) introduces basic biology of histone modifications and histone code, and discusses how chromatin modifications may regulate critical cellular process as well as how different forms of epigenetic information in chromatin are disrupted in cancer.

More recently, the role of noncoding RNAs in the control of cellular processes is beginning to emerge. The microRNAs (miRNAs) are small (20–22 nucleotides) RNAs located either within the introns or exons of protein-coding genes (70%) or in intergenic regions (30%), and able to regulate the expression of about 30% of human genes (Bartel, 2004; Calin and Croce, 2006). This regulation is mediated by the direct interaction between miRNAs and messenger RNAs (mRNAs) and it can be accomplished by two mechanisms, translational repression or mRNA degradation (Croce and Calin, 2005; Molnar et al., 2008). It has been proposed that a single miRNA may have several targets as well as a single mRNA may be targeted by different miRNAs. However, the function of most of the miRNAs remains largely unknown. Accumulating evidence indicates that deregulation of miRNAs is linked to several steps of cancer initiation and progression. Interestingly, miRNAs appear to be able to act as either tumor suppressors or oncogenes by affecting distinct genes of gene families involved in critical biological processes such as proliferation and differentiation. Many miRNA genes are located in the genomic loci known as fragile sites and are therefore susceptible to either loss or amplification. Several recent studies indicated that miRNA profiles differ significantly between cancer and normal tissues and also between different tumors. Interestingly, miRNA profiling revealed distinct patterns that may classify cancers according to the developmental lineage and differentiation status, lending miRNAs useful tools in cancer diagnostics and prognosis (Calin and Croce, 2006). Although we are just beginning to understand the complexity of mechanisms regulated by miRNAs, this most recent type of epigenetic inheritance is likely to provide important information to the overall knowledge in cancer biology. While numerous studies revealed the distinct miRNA profiles in a variety of human malignancies, much less is known about the profiles and function of long noncoding RNAs in cancer (Mercer et al., 2009). In Chapter 4, George Calin and colleagues discuss the basic epigenetic mechanisms involving miRNAs in the regulation of gene expression and their significance in cancer development and progression.

Histone modifications, DNA methylation, and RNA-mediated silencing are epigenetic modifications whose patterns can be regarded as heritable marks that ensure accurate transmission of the chromatin states and gene expression profiles over many cell generations. Accumulating evidence suggests that an epigenetic cross talk, such as interplay between DNA methylation and histone modifications, may be involved in the process of gene transcription and aberrant gene silencing in tumors (Jaenisch and Bird, 2003; Murr et al., 2007; Vaissiere et al., 2008). Although the molecular mechanism of gene activation is relatively well understood, the hierarchical order of events and dependencies in the course of gene silencing during cancer development and progression remains largely unknown. While several studies suggested that DNA methylation patterns guide histone modifications (including histone acetylation and methylation) during gene silencing, other set of studies argues that DNA methylation evolved to take its cues primarily from histone modification states (Jaenisch and Bird, 2003; Thomson et al., 2010). In Chapter 5, Rabih Murr summarizes current knowledge on the interplay between DNA methylation, histone modifications, and miRNAs during gene silencing and its importance in the integration of exogenous and intrinsic stimuli in the control of key cellular processes. Implication of this epigenetic interplay for cancer therapy and prevention is also discussed.

III. EPIGENETIC EVENTS AND BIOLOGICAL PHENOMENA—GENE WIRING TO INSTRUCT INHERITANCE

A. Genomic imprinting and cancer

Genomic imprinting refers to the conditioning of parental genomes mediated by epigenetic mechanism during gametogenesis ensuring that a specific locus is exclusively expressed from either maternal or paternal genome in the offspring. Around 80 genes have so far been found imprinted in humans and mice, although a recent estimation suggested that as many as 600 genes are potentially imprinted. Imprinted genes play critical roles in developmental and cellular processes; therefore, loss of imprinting (LOI) due to epigenetic alterations leads to abnormal biallelic expression, resulting in several human syndromes. Importantly, pathological biallelic expression of several genes caused by LOI is associated with human cancer. In general, the potential significance of epigenetic dysfunction in human malignancies is illustrated by the fact that the LOI and loss of X-chromosome inactivation occur at much higher frequency compared to genetic mutations. With the exception of the IGF2/H19, an extensively imprinted locus whose deregulation is implicated in human malignancies, much remains unknown. The IGF2/H19 locus has been studied in childhood tumors such as Wilms tumor and overgrowth syndromes such as the Beckwith–Wiedemann syndrome. The strong support for a gatekeeper role for LOI of IGF2 also in Wilms tumor has come from the studies showing that Beckwith–Wiedemann syndrome, a prenatal overgrowth disorder, predisposes to various embryonal tumors including Wilms tumor. Consistent with an important role of LOI of IGF2 in cancers of adults, recent studies have demonstrated that epimutation of IGF2/H19 locus is a common epigenetic event in adults and is associated with fivefold increased incidence of colorectal cancer. These studies argue that LOI may be a frequent mechanism by which epigenetic alterations predisposes to the development of cancer. In Chapter 6, Eamonn Maher and Derek Lim discuss recent advances on imprinting and LOI and mechanistic insights into developmental syndromes and human malignancies.

B. Epigenetic codes in stem cells and cancer stem cells

Stem cells constitute a minority of cell population in adult tissues, yet they play the key role in the development and tissue homeostasis. The main properties of stem cells are the self-renewal, essential for maintenance of the stem cell pool, and the ability to differentiate in different lineages required for the integrity and function of tissues. Given their special properties, stem cells are tightly regulated by multiple genes and gene networks. This control prevents the shift in the balance between self-renewal and differentiation. Deregulation of epigenetic information (encoded in DNA methylation, histone modification patterns, and noncoding RNAs) in cells with pluripotent potential may alter defining properties of stem cells, self-renewal, and differentiation potential, leading to cancer initiation and progression (Feinberg et al., 2006; Shukla et al., 2008). Histone modifications appear to play the key role in establishing and maintaining distinct gene expression patterns and consequently pluripotent state and differentiation fates of stem cells. Given that all cell types and many cancers are derived from pluripotent cells, a better understanding of epigenetic mechanisms controlling pluripotency and reprograming will greatly impact on many areas of modern biology and will help to tailor efficient therapeutic strategies (Feinberg et al., 2006). In Chapter 7, Yasuhiro Yamada and Akira Watanabe review current knowledge on stem cells and cancer stem cells, discuss epigenetic mechanisms that regulate "stemness" and pluripotency, and speculate how these new findings impact cancer therapeutics.

C. Constitutional epimutations and cancer susceptibility

In addition to the germ line mutations that represent a high risk of developing specific cancer, accumulating evidence argues that epigenetic changes can also occur constitutionally to confer a risk of developing particular types of human cancer. These epigenetic aberrations are known as "constitutional epimutations" that alter gene activity to confer a phenotype without change within the DNA code of the affected gene (Cubas et al., 1999). Constitutional epimutations are characterized by promoter methylation and transcriptional silencing of a single allele of the gene in normal somatic tissues. Example of constitutional epimutations is particularly well characterized in Lynch syndrome, an autosomal dominant cancer susceptibility syndrome characterized by the early development of several human malignancies, including cancer of colorectum and uterus (Lynch, 1999). Recent studies have indicated that epimutations are associated with distinct patterns of inheritance depending on the underlying mechanisms (Hitchins and Ward, 2007; Hitchins et al., 2007). In Chapter 8, Megan Hitchins describes "the journey of discovery of epimutations" and reviews different types of epimutations, with a focus of human disease phenotypes, notably cancer. The author also discusses potential mechanisms underlying epimutations and describes how these correlate with the observed patterns of inheritance.

IV. CANCER EPIGENOME—ZOOMING IN ON GENOME-WIDE SCALE

Tumor development is a complex multistep process and human cancers are characterized by profound abnormalities in the genome and the epigenome. Such abnormalities include genetic changes (mutations, chromosomal rearrangements) and epigenetic changes, including aberrant DNA methylation and histone modifications, both of which are believed to be triggered by exposure to environmental, dietary, and lifestyle factors. Therefore, in addition to genetic analysis, a comprehensive epigenetic profiling of cancer genomes is essential in identifying causative changes involved in cancer development and progression, regardless whether these changes are inherited or acquired during the life.

A. Cancer epigenome

In comparison with the genome, the epigenome is believed to be more complex and highly dynamic, which may underlay and/or reflect different functional states in time and space. This dynamics is governed by reversible modifications of genomic DNA (CpG methylation) and core histone proteins (histone modifications). Emerging technologies for detection of epigenetic changes and recent progress in the field of epigenomics promise to rapidly advance the capacity to address important issues in cancer research. Such tools are already in use to characterize tumor samples in high-throughput settings. While the analysis of cancer genomes is well underway as part of multiple national and international efforts, the analysis of cancer epigenomes is still at an early stage (American Association for Cancer Research Human Epigenome Task Force, 2008; Jones and Martienssen, 2005). However, recent technological advances that now allow cancer epigenetics to be studied genome-wide have already begun to provide both biological insight in the process of cancer development and critical information for new avenues in translational research (Lister *et al.*, 2009). In Chapter 9, Stephan Beck and colleagues review recent progress in high-resolution and genome-wide analyses on DNA methylation and its implications for cancer research.

B. Identification of driver DNA methylation changes

Human cancer genomes exhibit widespread epigenetic alterations, among which changes in DNA methylation patterns are particularly well documented. This includes global loss of methyl-cytosine (caused by DNA hypomethylation of repetitive sequences) and unscheduled hypermethylation of CpG islands in the promoter of a wide range of genes. In recent years, the analysis of DNA methylation patterns in cancer has progressed from single gene studies, which focused on potentially important candidate genes, to a more global analysis in which large number or nearly all gene promoters are analyzed (Hernandez-Vargas *et al.*, 2010; Martinez *et al.*, 2009; Rauch *et al.*, 2008). In Chapter 10, Gerd Pfeifer and Satish Kalari give an overview of these genome-scale methylation-profiling techniques and summarize the key information obtained with these approaches. The current knowledge on the specificity of methylation aberrations in cancer at a genome-wide level is covered. The authors also discuss how to identify those DNA methylation changes that are important for the processes of cancer initiation, progression, or metastasis (driver methylation changes) as well as challenges associated with distinguishing these from methylation changes that are merely passenger events during cancer development and progression.

C. Epigenetic drivers and genetic passengers in cancer

Genetic changes and aneuploidy are associated with alterations in DNA sequence, and they are a hallmark of the malignant process. Similarly, epigenetic alterations are universally present in human cancer and result in heritable changes in gene expression over many cell generations, leading to functional consequences equivalent to those induced by genetic alterations. Intriguingly, accumulating evidence suggests that epigenetic changes may precede and provoke genetic changes (Sawan *et al.*, 2008). Epigenetic alterations may occur early in tumor development and may trigger a spectrum of genetic alterations such as mutations and chromosomal aberrations. There are different pathways by which disruption of epigenetic states (DNA methylation, histone modifications, and noncoding RNAs) either individually or in combination may trigger genetic changes. In this scenario, epigenetic events are primary events while genetic changes (such as mutations) may simply be a consequence of disrupted epigenetic states. Aberrant epigenetic events affect multiple genes and cellular pathways in a nonrandom fashion and this can predispose to induction and accumulation of genetic changes in the course of tumor initiation and progression. Chapter 11 (by Minoru Toyota and Hiromu Suzuki) addresses how these considerations may be important for better understanding of the process of tumorigenesis and molecular events underlying the acquisition of drug resistance, as well as the development of novel strategies for cancer therapy.

V. EPIGENETICS AND ENVIRONMENTAL FACTORS—WHERE YOUR GENES MEET THE ENVIRONMENT

A. Epigenetic alterations induced by dietary and other environmental factors

An important role of dietary, lifestyle, and environmental factors in the development of a wide variety of cancers is well supported by both epidemiological and laboratory-based studies (Herceg, 2007). Environmental and dietary factors known to play important roles in the etiology of human cancer include chemical carcinogens, dietary toxins (such as aflatoxin B1), and physical carcinogens (UV and ionizing radiation), whereas tobacco smoking, alcohol abuse, and excess exposure to sunlight are lifestyle factors known to contribute to human cancer. Epigenetic mechanisms are believed to play a critical role in response to both endogenous stimuli and exogenous (environmental) factors. These mechanisms are thus physiological tools used by cells to establish and maintain gene expression patterns that are appropriate for specific environmental cues. It is believed that epigenetic mechanisms in animals also play important roles in the adaptation and response to environmental exposures. In many instances, however, clear-cut causal relationship between epigenetic states and environmental factors proved to be difficult to establish. This stems from the fact that environmental factors are likely to induce subtle changes, which are often cumulative, thus quantitative manifestation of phenotypic traits may occur after repetitive exposure over a long period of time. For these reasons, epidemiology proved to be incapable of identifying complex environmental factors and dietary regimes that induce and/or promote tumor development by triggering epigenetic changes.

This was in part due to the lack of epidemiological and laboratory-based studies that addressed the role of epigenetic changes induced by environment and nutrition with sufficient statistical power. In Chapter 1, Vol. 71, John Mathers, Gordon Strathdee, and Caroline Relton review recent studies implicating epigenetic changes induced by environment, diet, and lifestyle in human cancer. The authors also argue that epigenetic marks represent attractive candidates for the development of surrogate endpoints that could be used in dietary or lifestyle intervention studies for cancer prevention.

B. Induction of epigenetic changes by chronic inflammation

Infectious agents including viruses, such as human papillomavirus (HPV), Epstein–Barr virus (EBV), and human hepatitis virus (HBV), and bacteria such as *Helicobacter pylori* may also alter expression of host genes via an epigenetic strategy. Epigenetic mechanisms including DNA methylation and chromatin modifications are known to regulate viral gene expression. It has been shown that methylation of integrated HPV-associated primary cancers and cervical cancer cell lines inhibits the transcription of most viral genes. Interestingly, CpG methylation appears to correlate with HPV pathogenesis, suggesting that methylation of HPV DNA is implicated in the development and progression of cervical cancer. EBV genomes are also subject to host cell-dependent epigenetic modifications including DNA methylation, binding of regulatory proteins and histone modifications, and different EBV latency types are associated with distinct viral epigenotypes. Associations between EBV and HBV infection and promoter hypermethylation of several genes were commonly found in several cancers including hepatocellular, gastric, and nasopharyngeal carcinoma. These studies provide strong evidence that infectious agents may employ epigenetic strategies to deregulate cellular processes and promote tumorigenesis; however, underlying mechanisms are poorly understood. In Chapter 2, Vol. 71, Toshikazu Ushijima reviews current knowledge on inducers of epigenetic changes with a focus on chronic inflammation induced by bacterial and viral infections. The authors also discuss molecular mechanisms underlying aberrant epigenetic states leading to the development of cancer.

C. Maternal diet, early life exposure, and epigenetic processes

Several studies have reported that diet may influence epigenetic patterns and this could explain many diet-associated disorders (Gluckman *et al.*, 2008). During early development, both paternal and maternal genomes undergo a striking epigenetic reprograming, most notably through DNA demethylation immediately after fertilization. After implantation, methylation patterns are reestablished via *de novo* methylation (Reik, 2007). This epigenetic reprograming during early development must be a well-tuned process since it is an attempt to establish a configuration of the genome that can respond to changing needs of the early life development. How maternal diet may affect the phenotype of the offspring by epigenetic mechanisms and how this early life exposure may modulate susceptibility to cancer and other diseases in childhood and later life are discussed by Kent Thornburg and collaborators (Chapter 3, Vol. 71).

D. Folate, one carbon metabolism, and DNA methylation in cancer

Folate is a methyl donor that plays an essential role in DNA synthesis and biological methylation reactions, including DNA methylation. Folate deficiency may be implicated in the development of genomic DNA hypomethylation, which is an early epigenetic event found in many cancers. Numerous studies employing *in vitro* systems, animal models, and human interventional studies have tested this hypothesis. While numerous technical challenges remain in this important field of research, changes in folate intake appear to be capable of modulating DNA methylation levels in the human colonic mucosa and this may potentially alter colorectal cancer (CRC) risk. In Chapter 4, Vol. 71 (by Robyn Ward and Jia Liu), the impact of folate intake and one carbon metabolism on cancer susceptibility is discussed. The authors also discuss existing evidence on folate and its relationship to DNA methylation using CRC as an example. The evidence from animal, human, and *in vitro* studies on the effects of folate deficiency and supplementation on epigenetic states including DNA methylation and histone modifications is provided.

VI. EPIGENETIC CHANGES IN CANCER—MARKERS IN TRACKING CANCER CELLS

Epigenetic changes (e.g., promoter-specific hypermethylation) occur early and at high frequency in different human malignancies; this feature combined with high sensitivity and specificity of detection may be exploited in the area of molecular diagnostics and cancer risk assessment. As a result, the research on epigenetic changes as potential biomarkers is in full swing. Chapters 5–7, Vol. 71 review advantages of epigenetic biomarkers for cancer pathophysiology, such as response to therapeutics, prognosis, and occurrence of metastasis, over those involving genomic alterations and gene expression changes, and describe current examples.

A. Epigenetic biomarkers for cancer pathophysiology

In recent years, the development of high-throughput and genome-wide analytic methods has opened the possibility of identifying simultaneously multiple changes in gene expression as well as epigenetic alterations affecting the epigenome of cancer cells. The main question raised by such studies is to determine which alterations, or combinations thereof, can be interpreted as reliable biomarkers for providing information about the carcinogenesis process. This assessment should be done from the viewpoint of the suspected, primary role of such alterations in the initial steps of tumorigenesis. For example, the molecular events that occur in early stage of cancers or in precursor lesions are more likely to have a direct influence on cancer occurrence and progression than those that accumulate at later stage of cancer development. Among the latter, many alterations may be considered as "passengers" that represent a mere consequence of the highly disturbed genomic and epigenomic instability that accompanies the progression of many cancers. Chapter 5, Vol. 71 (by Dajun Deng and colleagues) discusses the challenge of incorporating current knowledge in epigenetics and epigenomics to identify new biomarkers that may be useful in the early detection and treatment of cancer.

B. Detection of epigenetic changes in body fluids

Tumor-derived cell-free circulating DNA isolated from the plasma and serum of individuals with cancer has been shown to contain cancer-associated alterations. While the origin and possible function of this free circulating DNA is not fully understood, it represents an attractive target for biomarker discovery. In addition to genetic changes (mutations, microsatellite alterations), plasma DNA from individuals with tumors was shown to harbor epigenetic changes, namely alterations in DNA methylation at CpG sites in the promoter regions of a wide range of tumor suppressor genes and other cancer-associated genes. Epigenetic changes are tumor-specific and thus have the potential to serve as highly specific biomarkers. In addition, DNA methylation changes appear early in tumor development and can be found in virtually every type of human cancer, thus they can provide particularly attractive markers with broad application in diagnostics and risk assessment. The development of epigenetic markers for cancer-bearing individuals could similarly enhance the management of their disease. In Chapter 6, Vol. 71, Triantafillos Liloglou, and John Fields summarize recent progress in the development and application of new assays for the detection of DNA methylation changes in body fluids with both time- and cost-effectiveness.

C. Epigenetic biomarkers for cancer risk assessment

Epigenetic events are shown to influence virtually each steps in tumor development; therefore, understanding epigenetic changes associated with cancer onset, progression, and metastasis are fundamental to improving our abilities to successfully treat and prevent cancer. Epigenetic alterations in comparison with genetic changes are typically acquired in a gradual manner. These features offer an enormous potential for prevention strategies. Based on quantitative estimates over two-thirds of the cancer incidence accounted for by environmental and dietary factors, therefore the majority of cancers are potentially avoidable.

VII. EPIGENETIC DRUGS ON THE RISE—WAKING UP SLEEPING BEAUTY

The interest in the epigenetics of cancer is strongly augmented by the recent realization that epigenetic changes can be exploited as a powerful tool in the clinic and as a novel approach in cancer treatment. A distinguishing feature of epigenetic changes in comparison with genetic changes is that they are reversible; therefore, aberrant DNA methylation, histone acetylation, and methylation are attractive targets for therapeutic intervention. Many efforts and resources have been mobilized in the development of different therapeutic approaches that are known as "epigenetic therapies." The ubiquity of epigenetic changes in many malignancies and other significant human diseases has triggered an impressive quest for the development of "epigenetic drugs" and epigenetic therapies. A number of agents have been subjected to an intensive investigation, many of which have been found capable of altering epigenetic states including DNA methylation patterns and histone modification states. Chapters 12 and 13, Vol. 70, give an overview of recent development and opportunities in the field of epigenetic therapy of cancer.

A. DNA demethylating and coupling therapies

Different approaches are directed to modify DNA methylation states in cancer cells and are based on specific properties of various chemical agents affecting the activity of the enzymes involved in the establishment and maintenance of DNA methylation. Among these agents, demethylating agents (inhibitors of DNA methyltransferases) are the most extensively studied epigenetic agents. These include 5-azacytidine (5-aza-CR) and 5-aza-2-deoxycytidine (5-aza-CdR), both of which efficiently inhibit DNMTs and lower DNA methylation levels in a variety of cancer cell lines, leading to reactivation of gene expression (Egger et al., 2004). 5-aza-CR and 5-aza-CdR are nucleoside analogs of cytosine and the mechanism by which they inhibit DNA methylation involves incorporation of these molecules at the position of cytosine during DNA replication. This event results in trapping and inactivation of DNA methyltransferase; therefore, event transient treatment of cells with demethylating agents can lead to a long lasting demethylation effect (Egger et al., 2004). In a similar manner, pseudoisocytosine (also known as zebularine) can induce efficient demethylation (Marquez et al., 2005). Numerous studies showed that these agents can efficiently reactivate the expression of aberrantly silenced genes in a variety of cancer cells. In Chapter 12, Vol. 70, Zdenko Herceg gives a brief overview of epigenetic therapy, and detailed explanation on demethylation therapy, such as mode of action, clinical indication, dose adjustment, and assessment of adverse effects as well as the principles of combinatorial therapies that couple DNA methylation inhibitors with HDAC inhibitors as well as coupling therapies.

B. Histone modification therapy of cancer

Similar to DNA methylation changes, aberrant histone acetylation and methylation are attractive targets for the epigenetic therapy. Indeed, a number of drugs that are capable of altering levels or patterns of histone modifications have been discovered, and many of these drugs are now in clinical trials. Inhibitors of histone deacetylases (HDACs) turned out to be effective against specific human cancers. For example, HDAC inhibitor vorinostat has already been approved by FDA. Other HDAC inhibitors of various chemical structures are currently in clinical trials. Drugs targeting histone methylation are also under development. In Chapter 13, Severio Minucci and colleagues summarize how HDAC inhibitors work, and describe clinical effects of vorinostat and other drugs for which clinical effects are available. The status of development of other drugs targeting different histone modifications is also covered.

VIII. EPIGENETICS AND CANCER PREVENTION—PROTECT YOUR EPIGENOME

A. Epigenetics in molecular epidemiology

Recent progress in epigenomics and emergence of powerful technologies for the detection of epigenetic changes in high-throughput settings holds promise to advance our capacity to evaluate the contribution of epigenetic changes induced by the environmental epimutagens to human cancer. These powerful tools are beginning to be applied to large population-based and case-control studies which offer some of the most exciting opportunities to study the contribution of epigenetic events to specific human cancers. Chapter 7, Vol. 71 (by Yasuhito Yuasa) discusses the application of epigenetics to molecular epidemiology, such as assessment of exposure to environmental epimutagens by epigenetic markers. It also covers different considerations relevant to molecular epidemiology studies, regarding the adoption of stringent criteria for the design, conduct, and evaluation of studies in which epigenetic markers are applied.

B. Epigenetic cancer prevention

A plethora of studies points to a fundamental role of epigenetic changes in cancer development and progression. Accumulation of aberrant epigenetic changes is observed even in histologically normal predisposed tissues, indicating that induction of epigenetic alterations occurs in very early stages of human carcinogenesis. In animal models, genomic hypomethylation suppresses intestinal tumorigenesis and treatment with 5-aza-dC delays androgen-independent disease in prostate cancer model. These observations led to the realization that epigenetic changes have tremendous potential in the prevention of cancer. Reversibility and gradual acquisition of epigenetic alterations are key features that offer an enormous potential for prevention strategies. Based on quantitative estimates over two-thirds of the cancer incidence accounted for by environmental and dietary factors, therefore the majority of cancers are potentially avoidable. In Chapter 8, Vol. 71, Jia Chen reviews current knowledge on the application of epigenetics in the development of novel strategies for cancer prevention.

IX. EPIGENETICS DATABASES AND COMPUTATIONAL METHODOLOGIES—RESOURCE AND TOOLBOX FOR COMMUNITY

The field of epigenetics has seen surge of interest with the recent technical advances that allow robust, quantitative, and genome-wide studies of epigenetic alterations. This resulted in an explosion of different types of epigenetic data. The list of genes altered by epigenetic mechanisms is rapidly expanding (several hundreds of genes have been reported to be modified by epigenetic mechanisms to date) and with the Human Epigenome Project in preparation, a more comprehensive epigenetic landscape of the cancer epigenome will be available. Despite the fact that progress in identification of different types of epigenetic changes and the genes altered through epigenetic deregulation has been remarkably rapid, much remains unknown. Even for the insiders, the comparison of experimental data and the extraction of trends represent a difficult task due to the fact that experimental strategies, techniques used, and data processing differ considerably between the studies. The data on epigenetic changes are heterogeneous and often range from the global measurement of specific modification to the exact and quantitative detection of epigenetic change or pattern. Furthermore, publication standards for data processing and presentation of epigenetic alterations have not yet been established. In his Chapter 9, Vol. 71, Maté Ongenaert overviews different tools including databases available in the field of cancer epigenetics. How the combined expertise of researchers in different fields may be applied to provide better tools and resources for the scientific community in the field of cancer epigenetics is also discussed.

X. CLOSING REMARKS

Both the scientific and medical communities now recognize that epigenetic changes lie in the heart of several important human diseases, most notably cancer. Epigenetic events have been associated with virtually every step of tumor development and progression, and epigenetic alterations are believed to occur early in tumor development and may precede the malignant process (Belinsky, 2004; Laird, 2003; Nephew and Huang, 2003). Therefore, epigenetic deregulation can be exploited as a powerful tool in the clinic and as novel approach to early diagnosis, prediction of clinical outcome, and risk assessment (Belinsky, 2004; Egger et al., 2004; Fraga et al., 2005; Laird, 2003; Seligson et al., 2005). An important distinction between epigenetic and genetic alterations is intrinsic reversibility of the former, making cancer-associated changes in DNA methylation, histone modifications, and expression of noncoding RNAs particularly attractive targets for the epigenetic therapy (Egger et al., 2004; Feinberg and Tycko, 2004; Jones and Baylin, 2007). Another distinguishing feature of epigenetic changes is that they arise in a gradual manner, leading to a progressive silencing of specific genes. This represents an exciting opportunity that can also be exploited in the development of novel strategies for the modulation of the susceptibility to diseases and prevention.

The race is on to find efficient drugs and therapeutic strategies that can reverse epigenetic changes and unscheduled gene silencing. A number of drugs that are capable of altering levels or patterns of DNA methylation or histone modifications have been discovered, and many of these drugs are now in clinical trials. However, despite the promise of epigenetic therapy, several concerns need to be addressed before it can be fully exploited in clinics. There is a need to develop target-specific DNMT inhibitors and isotype-selective HDAC inhibitors to minimize the toxicity associated with these drugs. Early clinical trials with demethylating agents showed relatively strong cytotoxic effect and were not well tolerated. However, these drugs were used at relatively high doses, and more recent studies with significantly lower doses showed encouraging results with relatively mild cytotoxicity. Therefore, combinatorial therapies that target different epigenetic mechanisms such as DNA methylation inhibitors with HDAC inhibitors may prove particularly efficient. Recent studies including clinical trials should give an answer on important questions regarding dosing schedules, routes of administration, and combination regimens.

The intrinsic reversibility of epigenetic alterations represents an exciting opportunity not only for cancer therapy but also for the development of novel strategies for cancer prevention. It is hoped that it may become feasible to select appropriate combinations of epigenetic drugs to revert or block the functional consequences of these alterations in early pre-neoplastic lesions. By targeting specifically and simultaneously multiple pathways based on epigenetic signatures, epigenetic intervention may confer a greater therapeutic or preventive efficacy, while having less side effects than conventional cytotoxic drugs. Reversibility of epigenetic events offers a unique opportunity for chemoprevention or diet-based intervention that targets critical epigenetic pathways. However, because diet-induced epigenetic modulation varies during different developmental periods and early life, it is critical to consider the issue of "window of vulnerability" that may directly impact the timing of effective strategies for cancer prevention. For example, dietary effects may be the greatest during embryogenesis and early development (Waterland and Jirtle, 2003). Therefore, further studies are needed to test whether dietary intervention in adult population could result in sufficient and sustaining restoration of epigenetic patterns in target tissues. Most of the studies aiming to investigate the role of diet on epigenetic states have focused on one carbon metabolism intermediates and have been carried out in animal models. Therefore, more comprehensive studies need to be carried out, and should focus on other dietary components and effects of dietary regimes in humans. A particular attention should be given to early life exposure and epigenetic reprograming during "the window of vulnerability" and their effect on the susceptibility to diseases in later life. Intriguingly, epigenetic changes seem to affect not only the health of the exposed individual but also the future generations. Therefore, further studies are required to substantiate the observations that human germ line could "capture" information about the ancestral environment through epigenetic states and pass it to the next generation.

The first human methylome at single base resolution for human embryonic stem cells and fetal fibroblasts was reported in 2009 (Lister et al., 2009), and this study highlighted that comprehensive methylome analysis is highly informative and holds considerable significance for the proposed stem cell origin of human cancer. Despite tremendous progress in recent years, the field of cancer epigenomics is still at an early stage and no comprehensive analysis of a cancer epigenome has been performed. However, several studies aiming to profile entire cancer methylome at single base resolution are well underway. Therefore, near future is likely to bring comprehensive epigenetic landscapes in different cancer types that will be the great leap forward in our understanding of mechanisms of cancer development and progression and pave the way for comprehensive functional studies.

Remarkable advances in epigenomics and emergence of powerful technologies allows the detection of epigenetic changes in high-throughput and genome-wide settings. The advent and rapid development of massively parallel sequencing technologies (next generation and next–next generation sequencing) has dramatically accelerated cancer research and opened up new perspectives. This holds promise to advance our capacity to elucidate mechanisms underlying tumor development and progression and to evaluate the contribution of epigenetic changes induced by the environmental, dietary, and lifestyle epimutagens to human cancer (Fig. 1.1). In particular, it will also help in elucidating the role of epigenetic changes induced by bacterial and viral agents and chronic inflammation in several common human cancers. These powerful tools are beginning to be applied to large population-based and case-control studies. Large cohort and case-control studies offer some of the most exciting opportunities to study the contribution of epigenetic events induced by the diet and environment to human cancer. Such examples are the European Prospective Investigation into Cancer and Nutrition (EPIC), a large prospective cohort study designated to investigate the relationship between diet, various lifestyles and the incidence of cancer in a number of European countries (Riboli and Kaaks, 1997), and the International Childhood Cancer Cohort Consortium (I4C) (Brown et al., 2007). However, the application of epigenetic markers to epidemiological studies requires careful considerations in the design of such studies. Sample size should be estimated based on power calculations which depend on the background rate of the marker (e.g., among controls) and the expected strength of the association (e.g., difference in the markers between cases—or exposed—and controls). Different considerations relevant to molecular epidemiology studies, regarding the adoption of stringent criteria for the design, conduct, and analysis of studies in which epigenetic markers are applied, should be taken into account.

While deregulation of epigenetic mechanisms is primarily studied in the context of cancer biology which is the focus of this book, epigenetic changes have been also implicated in several developmental syndromes (Egger et al., 2004; Feinberg and Tycko, 2004; Jiang et al., 2004), diabetes, obesity, cardiovascular diseases (Maier and Olek, 2002; McKinsey and Olson, 2004), and neurological disorders (Urdinguio et al., 2009). Therefore, basic epigenetic concepts and mechanisms reviewed in this book will be of interest for those working on other complex human diseases and in related fields of modern biology. In conclusion, the epigenetic research is a field in full swing and the near future is likely to bring long-awaited answers that will help in better understanding how tumors develop and progress, and provide important information for the development of novel and efficient strategies for cancer control, a major public health priority in the 21st century. We believe that this book will spread awareness of opportunities and challenges that epigenetics may offer.

Acknowledgment

Thanks to Dr. Hector Hernandez-Vargas for reading and useful suggestions, and Thomas Vaissiere for the illustrations.

References

Bartel, D. P. (2004). MicroRNAs: Genomics, biogenesis, mechanism, and function. *Cell* **116**, 281–297.

Belinsky, S. A. (2004). Gene-promoter hypermethylation as a biomarker in lung cancer. *Nat. Rev. Cancer* **4**, 707–717.

Brown, R. C., Dwyer, T., Kasten, C., Krotoski, D., Li, Z., Linet, M. S., Olsen, J., Scheidt, P., and Winn, D. M. (2007). Cohort profile: The International Childhood Cancer Cohort Consortium (I4C). *Int. J. Epidemiol.* **36**, 724–730.

Calin, G. A., and Croce, C. M. (2006). MicroRNA signatures in human cancers. *Nat. Rev. Cancer* **6**, 857–866.

Croce, C. M., and Calin, G. A. (2005). miRNAs, cancer, and stem cell division. *Cell* **122**, 6–7.

Cubas, P., Vincent, C., and Coen, E. (1999). An epigenetic mutation responsible for natural variation in floral symmetry. *Nature* **401**, 157–161.

Egger, G., Liang, G., Aparicio, A., and Jones, P. A. (2004). Epigenetics in human disease and prospects for epigenetic therapy. *Nature* **429**, 457–463.

Esteller, M. (2007). Cancer epigenomics: DNA methylomes and histone-modification maps. *Nat. Rev. Genet.* **8**, 286–298.

Feinberg, A. P., and Tycko, B. (2004). The history of cancer epigenetics. *Nat. Rev. Cancer* **4**, 143–153.

Feinberg, A. P., Ohlsson, R., and Henikoff, S. (2006). The epigenetic progenitor origin of human cancer. *Nat. Rev. Genet.* **7**, 21–33.

Fraga, M. F., Ballestar, E., Villar-Garea, A., Boix-Chornet, M., Espada, J., Schotta, G., Bonaldi, T., Haydon, C., Ropero, S., Petrie, K., Iyer, N. G., Perez-Rosado, A., *et al.* (2005). Loss of acetylation at Lys16 and trimethylation at Lys20 of histone H4 is a common hallmark of human cancer. *Nat. Genet.* **37**, 391–400.

Gluckman, P. D., Hanson, M. A., Cooper, C., and Thornburg, K. L. (2008). Effect of in utero and early-life conditions on adult health and disease. *N. Engl. J. Med.* **359**, 61–73.

Herceg, Z. (2007). Epigenetics and cancer: Towards an evaluation of the impact of environmental and dietary factors. *Mutagenesis* **22**, 91–103.

Hernandez-Vargas, H., Lambert, M. P., Le Calvez-Kelm, F., Gouysse, G., McKay-Chopin, S., Tavtigian, S. V., Scoazec, J. Y., and Herceg, Z. (2010). Hepatocellular carcinoma displays distinct DNA methylation signatures with potential as clinical predictors. *PLoS One* **5**, e9749.

Hitchins, M. P., and Ward, R. L. (2007). Erasure of MLH1 methylation in spermatozoa-implications for epigenetic inheritance. *Nat. Genet.* **39**, 1289.

Hitchins, M. P., Wong, J. J., Suthers, G., Suter, C. M., Martin, D. I., Hawkins, N. J., and Ward, R. L. (2007). Inheritance of a cancer-associated MLH1 germ-line epimutation. *N. Engl. J. Med.* **356**, 697–705.

American Association for Cancer Research Human Epigenome Task Force; European Union, Network of Excellence, Scientific Advisory Board. (2008). Moving AHEAD with an International Human Epigenome Project. *Nature* **454**, 711–715.

Jaenisch, R., and Bird, A. (2003). Epigenetic regulation of gene expression: How the genome integrates intrinsic and environmental signals. *Nat. Genet.* **33**(Suppl.), 245–254.

Jiang, Y. H., Bressler, J., and Beaudet, A. L. (2004). Epigenetics and human disease. *Annu. Rev. Genomics Hum. Genet.* **5**, 479–510.

Jones, P. A., and Baylin, S. B. (2002). The fundamental role of epigenetic events in cancer. *Nat. Rev. Genet.* **3**, 415–428.

Jones, P. A., and Baylin, S. B. (2007). The epigenomics of cancer. *Cell* **128**, 683–692.

Jones, P. A., and Martienssen, R. (2005). A blueprint for a Human Epigenome Project: The AACR Human Epigenome Workshop. *Cancer Res.* **65**, 11241–11246.

Laird, P. W. (2003). The power and the promise of DNA methylation markers. *Nat. Rev. Cancer* **3**, 253–266.

Lister, R., Pelizzola, M., Dowen, R. H., Hawkins, R. D., Hon, G., Tonti-Filippini, J., Nery, J. R., Lee, L., Ye, Z., Ngo, Q. M., Edsall, L., Antosiewicz-Bourget, J., *et al.* (2009). Human DNA methylomes at base resolution show widespread epigenomic differences. *Nature* **462**, 315–322.

Lynch, H. T. (1999). Hereditary nonpolyposis colorectal cancer (HNPCC). *Cytogenet. Cell Genet.* **86**, 130–135.

Maier, S., and Olek, A. (2002). Diabetes: A candidate disease for efficient DNA methylation profiling. *J. Nutr.* **132**, 2440S–2443S.

Marquez, V. E., Kelley, J. A., Agbaria, R., Ben-Kasus, T., Cheng, J. C., Yoo, C. B., and Jones, P. A. (2005). Zebularine: A unique molecule for an epigenetically based strategy in cancer chemotherapy. *Ann. N. Y. Acad. Sci.* **1058**, 246–254.

Martinez, R., Martin-Subero, J. I., Rohde, V., Kirsch, M., Alaminos, M., Fernandez, A. F., Ropero, S., Schackert, G., and Esteller, M. (2009). A microarray-based DNA methylation study of glioblastoma multiforme. *Epigenetics* **4**, 255–264.

McKinsey, T. A., and Olson, E. N. (2004). Cardiac histone acetylation—therapeutic opportunities abound. *Trends Genet.* **20**, 206–213.

Mercer, T. R., Dinger, M. E., and Mattick, J. S. (2009). Long non-coding RNAs: Insights into functions. *Nat. Rev. Genet.* **10**, 155–159.

Molnar, V., Tamasi, V., Bakos, B., Wiener, Z., and Falus, A. (2008). Changes in miRNA expression in solid tumors: An miRNA profiling in melanomas. *Semin. Cancer Biol.* **18**, 111–122.

Murr, R., Vaissiere, T., Sawan, C., Shukla, V., and Herceg, Z. (2007). Orchestration of chromatin-based processes: Mind the TRRAP. *Oncogene* **26**, 5358–5372.

Nephew, K. P., and Huang, T. H. (2003). Epigenetic gene silencing in cancer initiation and progression. *Cancer Lett.* **190**, 125–133.

Rauch, T. A., Zhong, X., Wu, X., Wang, M., Kernstine, K. H., Wang, Z., Riggs, A. D., and Pfeifer, G. P. (2008). High-resolution mapping of DNA hypermethylation and hypomethylation in lung cancer. *Proc. Natl. Acad. Sci. USA* **105**, 252–257.

Reik, W. (2007). Stability and flexibility of epigenetic gene regulation in mammalian development. *Nature* **447**, 425–432.

Riboli, E., and Kaaks, R. (1997). The EPIC Project: Rationale and study design. European Prospective Investigation into Cancer and Nutrition. *Int. J. Epidemiol.* **26**(Suppl. 1), S6–S14.

Sawan, C., Vaissiere, T., Murr, R., and Herceg, Z. (2008). Epigenetic drivers and genetic passengers on the road to cancer. *Mutat. Res.* **642**, 1–13.

Seligson, D. B., Horvath, S., Shi, T., Yu, H., Tze, S., Grunstein, M., and Kurdistani, S. K. (2005). Global histone modification patterns predict risk of prostate cancer recurrence. *Nature* **435**, 1262–1266.

Shukla, V., Vaissiere, T., and Herceg, Z. (2008). Histone acetylation and chromatin signature in stem cell identity and cancer. *Mutat. Res.* **637**, 1–15.

Thomson, J. P., Skene, P. J., Selfridge, J., Clouaire, T., Guy, J., Webb, S., Kerr, A. R., Deaton, A., Andrews, R., James, K. D., Turner, D. J., Illingworth, R., *et al.* (2010). CpG islands influence chromatin structure via the CpG-binding protein Cfp1. *Nature* **464**, 1082–1086.

Urdinguio, R. G., Sanchez-Mut, J. V., and Esteller, M. (2009). Epigenetic mechanisms in neurological diseases: Genes, syndromes, and therapies. *Lancet Neurol.* **8**, 1056–1072.

Ushijima, T. (2005). Detection and interpretation of altered methylation patterns in cancer cells. *Nat. Rev. Cancer* **5**, 223–231.

Vaissiere, T., Sawan, C., and Herceg, Z. (2008). Epigenetic interplay between histone modifications and DNA methylation in gene silencing. *Mutat. Res.* **659**, 40–48.

Waterland, R. A., and Jirtle, R. L. (2003). Transposable elements: Targets for early nutritional effects on epigenetic gene regulation. *Mol. Cell. Biol.* **23**, 5293–5300.

Basic Epigenetic Mechanisms in the Regulation of Critical Cellular Processes

2

DNA Methylation and Cancer

Marta Kulis and Manel Esteller
Cancer Epigenetics and Biology Program (PEBC), The Bellvitge Institute for
Biomedical Research (IDIBELL), Hospital Duran i Reynals, Avinguda Gran
Via de L'Hospitalet 199-203, E-08907 L'Hospitalet de Llobregat, Barcelona,
Catalonia, Spain

ABSTRACT

DNA methylation is one of the most intensely studied epigenetic modifications
in mammals. In normal cells, it assures the proper regulation of gene expression
and stable gene silencing. DNA methylation is associated with histone modifications and the interplay of these epigenetic modifications is crucial to regulate the
functioning of the genome by changing chromatin architecture. The covalent
addition of a methyl group occurs generally in cytosine within CpG dinucleotides which are concentrated in large clusters called CpG islands. DNA methyltransferases are responsible for establishing and maintenance of methylation
pattern. It is commonly known that inactivation of certain tumor-suppressor

Advances in Genetics, Vol. 70
Copyright 2010, Elsevier Inc. All rights reserved.

0065-2660/10 $35.00
DOI: 10.1016/S0065-2660(10)70002-X

genes occurs as a consequence of hypermethylation within the promoter regions and a numerous studies have demonstrated a broad range of genes silenced by DNA methylation in different cancer types. On the other hand, global hypomethylation, inducing genomic instability, also contributes to cell transformation. Apart from DNA methylation alterations in promoter regions and repetitive DNA sequences, this phenomenon is associated also with regulation of expression of noncoding RNAs such as microRNAs that may play role in tumor suppression. DNA methylation seems to be promising in putative translational use in patients and hypermethylated promoters may serve as biomarkers. Moreover, unlike genetic alterations, DNA methylation is reversible what makes it extremely interesting for therapy approaches. The importance of DNA methylation alterations in tumorigenesis encourages us to decode the human epigenome. Different DNA methylome mapping techniques are indispensable to realize this project in the future. © 2010, Elsevier Inc.

I. INTRODUCTION

To establish an organism, classical genetic processes are not sufficient. For proper development and cell functioning, the epigenetic phenomena are absolutely required, controlling gene expression. Epigenetic could be defined as study of heritable changes in gene expression that occur independently of changes in primary DNA sequence. Distinct epigenomes may explain differences in cell state during development when zygote transforms to somatic tissue, even if the DNA stays the same. It seems also logic that epigenetic correlates with several diseases, that has been proved in monozygotic twins that are sharing exactly the same genetic material but not necessarily the same epigenome (Fraga *et al.*, 2005).

Epigenetic regulation of gene expression is mediated by mechanisms such as methylation of DNA, modifications of histones, and positioning of nucleosome along the DNA. The interplay between epigenetic components guarantees proper balance between transcriptional activity and repression by changing chromatin architecture. Thus, regulation of packaging of DNA ensures maintenance of correct chromosome replication, gene expression, and stable gene silencing (Esteller, 2007). DNA methylation is one of the most intensely studied epigenetic modifications in mammals and it has an important impact on normal cell physiology. As this DNA modification seems to be a critical player in the transcriptional regulation, it is not surprising that defects in this mechanism may lead to various diseases, including cancer (Esteller, 2008). Indeed, in 1983 Feinberg and Vogelstein observed reduction of DNA methylation of specific genes in human colon cancer cells, comparing with normal tissues (Feinberg and Vogelstein, 1983). In the same year, Gama-Sosa *et al.* described a global reduction of 5-methylcytosine content of DNA from tumor samples (Gama-Sosa et al., 1983). Since these findings, being the first proof of molecular epigenetic abnormalities in cancer, we have significantly broadened our knowledge in that field (Esteller, 2008). Even if there is still far less known about epigenetic inheritance system and mechanisms of action than traditional genetics, the importance of DNA methylation in carcinogenesis is beyond any doubt.

In this chapter, we will try to summarize our actual state of knowledge concerning DNA methylation alteration in tumor progression.

II. THE MOLECULAR BASIS OF DNA METHYLATION

The propagation of DNA methylation is well understood at the biochemical level. It consists of the covalent addition of a methyl group ($-CH_3$) that occurs exclusively at the 5 position of the cytosine moiety (Fig. 2.1A). Intriguingly, methylated cytosine could be converted into 5-hydroxymethylcytosine (hmC), structurally similar to its unmodified counterpart (Tahiliani et al., 2009). This kind of modification was found in cerebellar Purkinje neurons and in embryonic stem cells but the biological function of hmC and its possible impact on DNA silencing machinery is still poorly understood (Kriaucionis and Heintz, 2009).

5-Methylcytosines are observed within CpG dinucleotides. Non-CpG methylation is uncommon although 5-methylcytosines have been found within CpA and, to a smaller extent, in CpT restrictively in embryonic stem cells, but not in somatic tissues (Ramsahoye et al., 2000). Moreover, recent studies of 5-methylcytosine distribution across mammalian genome, from both human embryonic stem cells and fetal fibroblasts, at single-base-resolution revealed that nearly 25% of all DNA methylation identified in embryonic stem cells was in a non-CpG context (Lister et al., 2009). Cytosine methylation in CHG and CHH (where H = A, T, or C) was found to be exclusive for stem cells and it seems that non-CpG methylation is lost while cells become differentiated. Thus, non-CpG methylation might be related to the origin and maintenance of pluripotent state (Lister et al., 2009).

Contrary to expectation, the whole genome is characterized by rather low overall CpG content. Typically, these dinucleotides are concentrated in large clusters, called CpG islands (CGIs) that are enriched mostly in the promoter and/or the first exon region. According to the calculation of CpG prevalence, nearly 60% of human promoters are characterized by high-CpG content (Saxonov et al., 2006). Nevertheless, CpG density itself does not influence gene expression. This regulation is dependent solely on the DNA

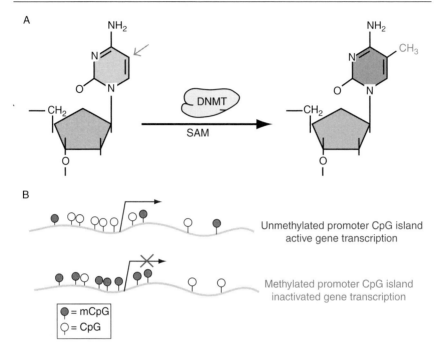

Figure 2.1. Cytosine methylation. (A) Methylation at 5 position of cytosine moiety, catalyzed by DNMT, in the presence of S-adenosyl-methionine (SAM). (B) Unmethylated CpGs within the promoter regulation region (dark brown) do not abolish transcription, gene is expressed. When the DNA methylation of CpGs within the regulatory region occurs, gene becomes silenced.

methylation process. Normally, CGIs are unmethylated in transcriptionally active genes, whereas silenced genes are characterized by methylation within promoter region (e.g., tissue-specific or developmental genes; Fig. 2.1B).

The methylation reaction of cytosines is mediated by a class of enzymes called DNA methyltransferases (DNMTs) that catalyze the transfer of the methyl group from S-adenosyl-methionine onto cytosine. Five members of the DNMT family have been identified in mammals: DNMT1, DNMT2, DNMT3a, DNMT3b, and DNMT3L. However, as far as we know, only DNMT1, DNMT3a, and DNMT3b interplay to produce the global cytosine methylation pattern. These independently encoded proteins are classified as de novo enzymes (DNMT3a and DNMT3b) or as a maintenance enzyme (DNMT1). DNMT2 and DNMT3L are not thought to function as cytosine methyltransferases. However, DNMT3L was shown to stimulate de novo DNA methylation by DNMT3a and to mediate transcriptional repression through interaction with histone deacetylase 1 (Chedin et al., 2002; Deplus et al., 2002).

DNMT1 appears to be involved in restoring the parental DNA methylation pattern in the newly synthesized DNA daughter strand, thereby ensuring the methylation status of CGIs through multiple cell generations. DNMT1 exhibits a preference for hemimethylated substrates and it possesses a domain targeting to replication foci. As a confirmation of the important role of this enzyme in proper cell functioning and development, it should be mentioned that the loss of DNMT1 function results in embryonic lethality in mice (Li *et al.*, 1992).

De novo DNA methylation during embryogenesis and germ cell development are carried out by DNMT3 family (DNMT3a and DNMT3b). Inactivation of each of these genes leads to severe phenotypes (Okano *et al.*, 1999). *DNMT3a* knock-out mice die shortly after birth and embryonic lethality is observed in case of the absence of *DNMT3b*. Thus, DNMT3a seems to be responsible for the methylation of sequences critical for late developmental or after birth, whereas DNMT3b may be more important for early developmental stages (Okano *et al.*, 1999). Moreover, DNMT3b is thought to be specialized for DNA methylation of particular regions of the genome, as has been shown by the studies of the Immunodeficiency, Centromere instability and facial abnormalities (ICF) syndrome, a disease caused by specific mutation in *DNMT3b* (Jin *et al.*, 2008). In ICF, the DNA hypomethylation of satellite II and III repeats, CGI on inactive chromosome X, and some sporadic repeats was demonstrated, which therefore implies that these sequences are plausible targets for DNMT3b.

How DNA methylation contributes to the inhibition of expression still remains unclear and various hypotheses have been proposed. First, for some transcription factors, for example, AP-2, c-myc, CREB/ATF, E2F, and NF-kB, DNA methylation was thought to create a physical barrier, abolishing access to promoter binding sites. This might be true, but only for a subset of transcription factors. However, further research showed that transcription factors often do not show a preference for unmethylated sequences (e.g., CTF (CCAAT-binding transcription factor)). Nevertheless, in the well-documented transcriptional regulation by CTCF (CCCTC-binding factor), DNA methylation of target sequences is crucial. CTCF binds to imprinting control regions (ICRs) of imprinted genes and is essential for transcription of the INK/ARF locus (Rodriguez *et al.*, 2010). DNA methylation of corresponding CTCF-binding sites abolishes its association and therefore contributes to the permanent silencing of several genes at the respective loci.

Another model of gene inactivation mediated by DNA methylation is related to methyl-CpG binding domain proteins (MBDs). In general, DNA methylation is considered to be an initiation step for establishing the inactive chromatin state. It is followed by an MBDs association that, in turn, recruits histone deacetylases known as repressive epigenetic modification enzymes (Fig. 2.2). The chromatin compacts and gene silencing is achieved.

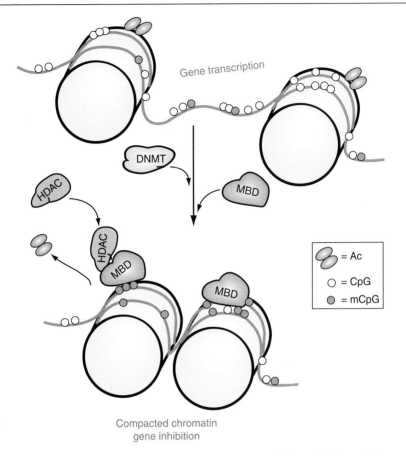

Figure 2.2. Mechanism of gene silencing mediated by DNA methylation. DNA methylation is followed by methyl-CpG-binding proteins (MBD) association that, in turn, recruits histone deacetylases (HDAC). As a consequence, chromatin becomes compacted and gene expression is inhibited.

Nevertheless, in some processes, other mechanisms act before DNA methylation occurs. For example, in *Neurospora crassa* DNA methylation depends on the methylation of histone 3 at lysine 9 (H3K9) and only this modification can trigger epigenetic gene repression (Tamaru and Selker, 2001).

Nevertheless, MBDs are not the only class of proteins capable of acting as HDAC-dependent transcriptional repressors by association with methylated DNA sequences. Kaiso-like family of proteins binds to methylated DNA by zinc-finger motifs and has been reported to be involved in gene silencing (Filion *et al.*, 2006; Prokhortchouk *et al.*, 2001). Unlike MBDs, Kaiso recognizes unmethylated sequences also but its exact mechanism of action remains still unclear (Daniel *et al.*, 2002). In addition, proteins containing SET and RING finger-associated domain, structurally unrelated to MBDs, can also associate with methyl-CpG and were shown to bind to various tumor-suppressor genes in cancer cells (Unoki *et al.*, 2004).

Recently, a growing number of studies show the involvement of poly-comb group proteins (PcG) in establishing the DNA methylation pattern. It has been conjectured that DNMT1 and DNMT3B play distinct roles in the func-tioning of Polycomb-repressive complexes 1 and 2 (PRC1 and PRC2). DNMT1 seems to be essential for recruiting BMI1, a PRC1 component, to PcG bodies, while DNMT3B is possibly needed for maintaining the monoubiquitination of H2AK119, a PRC1-mediated mark (Hernandez-Munoz *et al.*, 2005; Jin *et al.*, 2009). On the other hand, EZH2, a PRC2 component, is thought to mediate recruitment of DNMT3B to target sequences of some genes, particularly those involved in developmental regulation (Vire *et al.*, 2006). It was postulated that first the target genes are marked by H3K27 methylation and then *de novo* DNA methylation of sequences of concern takes place (Ohm *et al.*, 2007; Widschwendter *et al.*, 2007). Moreover, it was also reported that in cancer cells up to 5% of promoters containing CpGs were silenced by H3K27 trimethy-lation which was independent of DNA methylation (Kondo *et al.*, 2008). As the exact links between PcG regulation and DNA methylation are still unclear, this finding adds a novel layer of complexity to the epigenetic gene silencing.

In summary, even the above short explanation of the DNA methyla-tion-mediated gene silencing clearly illustrates how all epigenetic components interact in a complex manner to regulate gene expression.

III. DNA METHYLATION ABNORMALITIES IN CANCER

Appropriate DNA methylation is essential for development and proper cell functioning, thus any abnormalities in this process may lead to various diseases, including cancer. Indeed, tumor cells are characterized by a different methylome from that of normal cells. Interestingly, both hypo- and hypermethylation events can be observed in cancer. Generally, a global decrease in methylated CpG content is observed. This phenomenon contributes to genomic instability and, less frequently, to activation of silenced oncogenes. On the other hand, CGI hypermethylation in promoters of specific genes has been shown as a critical hallmark in many cancer cells (Paz *et al.*, 2003). An increasing number of genes have been reported to be inactivated by a DNA methylation mechanism during tumorigenesis that mainly act as tumor suppressors in normal tissues.

It should be mentioned that some of the current hypotheses highlight the role of epigenetic modification in early stages of tumor development and even in cancer predisposition. It has been proposed that epigenetic disruptions are the initiating events leading to the occurrence of "cancer progenitor cells" (Feinberg *et al.*, 2006). Furthermore, both genetic and epigenetic alterations may lead to tumor progression. In this context, the existence of DNA methylation abnormalities that appear before mutations and that are involved in tumorigenesis is strong evidence in supporting this theory.

A. DNA hypomethylation

In a normal cell, pericentromeric heterochromatin is highly methylated. Satellite sequences, repetitive genomic sequences (such as LINE, SINE, IAP, and Alu elements) are silenced, thereby ensuring genomic integrity and stability. In a wide variety of tumors, however, this mechanism is disrupted and loss of DNA methylation of normally inactivated regions occurs. As a consequence, there is a greater chance of undesired mitotic recombination. Transposable elements are then reactivated and can integrate at random sites in the genome, leading to mutagenesis and genomic instability. This may be related to tumor development and could be associated with distinct stages of the disease. For example, loss of DNA methylation of Sat2 (juxtacentromeric satellite 2) and Satα (centromeric satellite α) in breast cancer is linked to early tumor development while in ovarian cancer they contribute to tumor progression and was proposed as being a highly significant marker of poor prognosis (Costa *et al.*, 2006; Widschwendter *et al.*, 2004a,b).

Loss of DNA methylation may also activate latent viral sequences incorporated in the genome, what could possibly conduct to tumor progression. For instance, genomes of genital human papillomaviruses (HPVs) are efficiently repressed by CpG methylation after infection. DNA hypomethylation phenomena allow HPV reexpression that may be accompanied by cervical cancer progression (Badal *et al.*, 2003). The latency of Epstein–Barr virus, another known virus related to different lymphomas, also appears to be controlled by epigenetic modifications. DNA methylation was shown to repress the expression of nuclear antigens, which are responsible for the oncogenic properties of some alternative promoters (Takacs *et al.*, 2010).

DNA hypomethylation of individual genes in cancer cells is rather uncommon. The majority of the promoters affected by loss of DNA methylation belong to tissue-specific genes. For instance, a family of cancer/testis (CT) antigens, which are restricted to adult testicular germ cells under normal conditions, were reported to be aberrantly activated in various types of human cancer (Caballero and Chen, 2009). For example, in melanoma and colorectal cancer, melanoma-associated CT antigens *MAGE* are reactivated by DNA hypomethylation of promoter region (Kim *et al.*, 2006; Weber *et al.*, 1994). DNA methylation in the Synuclein-γ gene (SNCG) was reported to be cell-type specific; it was reported to be hypomethylated in breast, ovarian and gastric cancer but remains hypermethylated in urothelial carcinomas (Dokun et al., 2008; Gupta et al., 2003; Yanagawa et al., 2004).

Depending on the functions of the genes affected by hypomethylation, this phenomenon may be related to distinct stages of tumorigenesis. Loss of DNA methylation in the CDH3 gene promoter results in overexpression of P-cadherin in colorectal and breast carcinomas. As a consequence, decreased cell polarity promotes cell invasion, motility, and migration, associated with worse patient survival (Milicic et al., 2008; Paredes et al., 2005; Ribeiro et al., 2010). According to recent studies, DNA hypomethylation may be involved in anticancer drug resistance. This is the case of the multidrug resistance 1 gene (MDR1) whose promoter appears to be hypomethylated in breast cancer cells. Concomitant overexpression in tumor and serum of breast cancer patients of P-glycoprotein involved in drug resistance is correlated with advanced tumor stages and it has been proposed to possibly affect the effectiveness of chemotherapy (Baker et al., 2005; Chekhun et al., 2006; Sharma et al., 2010).

Despite the loss of DNA methylation in cancer, no actively demethylating enzyme has been discovered. Moreover, there is no clear evidence of any reduction in DNMTs activity in cancer. On the contrary, overexpression of these enzymes was reported (Ahluwalia et al., 2001; Lin et al., 2007; Roll et al., 2008). The matter of the mechanism of hypomethylation phenomena in cancer arises. Recent hypotheses take into consideration the possible role of catalytically inactive variants of DNMT3B in this process. More than 20 variants of this enzyme that encode truncated proteins lacking catalytic domain have been identified in several cancer cell lines (Ostler et al., 2007). It was proposed that inactive variants of DNMT3B negatively regulate DNA methyltrasferase activity. They may abolish the DNA methylation mechanism either by titrating the DNMT3B binding partners, thus interfering with the DNA methylation machinery or by competing with the active form to bind to the target DNA sequences. Indeed, overexpression of one of the inactive variants of DNMT3B (DNMT3B7) changes the methylation pattern of 293 cells (Ostler et al., 2007). Nevertheless, to date, the involvement of DNMT inactive variants has been demonstrated only in leukemia and hepatocellular carcinoma (Roman-Gomez et al., 2005; Saito et al., 2002). Further research is therefore needed to elucidate the basis of global hypomethylation in different cancer types.

B. DNA hypermethylation

In cancer, despite global hypomethylation, certain genes undergo inactivation as a consequence of hypermethylation of CGIs in regulatory regions, which are unmethylated in nonmalignant tissues. This kind of epigenetic silencing was first demonstrated in the studies of retinoblastoma patients in which the hypermethylation in the promoter of the retinoblastoma tumor-suppressor gene (*RB1*) was discovered (Greger *et al.*, 1989). Since then, a large number of tumor-suppressor genes silenced by DNA hypermethylation in tumor tissues have been identified. It is beyond doubt that DNA methylation-associated silencing plays a crucial role in tumorigenesis and is a hallmark of all types of human cancer (Esteller, 2008).

Genes that acquire hypermethylation in regulatory regions are involved in a variety of important cellular pathways. For instance, two cell cycle-related genes, $p16^{INK4a}$ (*CDKN2A*) and $p15^{INK4a}$ (*CDKN2B*), undergo DNA methylation-mediated silencing in different types of cancer. They are both involved in the control of G1 progression, acting as cyclin-dependent kinase inhibitors and are important tumor suppressors (Drexler, 1998). Various genes associated with DNA repair processes are also hypermethylated in tumor tissues, thus confirming the fact that epigenetic events may promote classical genetic alteration such as mutations (Drexler, 1998, 2007). This is the case for MGMT (O-6-methylguanine-DNMT), which in normal tissues averts the negative effect of DNA alkylation and was shown to be silenced in many types of carcinomas (Esteller *et al.*, 1999; Weller *et al.*, 2010). Defective DNA mismatch repair is observed in a significant portion of gastric cancers and is associated with hypermethylation of *MLH1* (Fleisher *et al.*, 2001). The methylation pattern of *BRCA1*, which is involved in DNA repair of double-stranded breaks, transcription, and recombination, has also been well studied in breast and ovarian cancer (Catteau and Morris, 2002).

Silencing by DNA hypermethylation of genes involved in cell adhesion, for example, CDH1 (E-cadherin) and CDH13 (H-cadherin) may lead to invasion and/or metastasis, and thereby tumor progression (Katoh, 2005; Kim *et al.*, 2005). Some of the genes hypermethylated in cancer are connected with cancer cell survival as they have pro-apoptotic functions, for example, death-associated protein kinase 1 (*DAPK1*), mediating interferon-γ-induced apoptosis or TMS1, activating apoptotic signaling pathways (Gordian *et al.*, 2009; Michie *et al.*, 2010).

DNA hypermethylation affects also the genes encoding upstream regulators of many cellular processes. For instance, transcription factors (e.g., GATA4 and GATA5) or genes involved in different receptor-mediated signaling pathways (e.g., ESR1, RARB2, and CRBP1) are often hypermethylated in cancer (Akiyama *et al.*, 2003; Esteller *et al.*, 2002; Widschwendter *et al.*, 2004a,b).

Taken together, the aforementioned examples confirm that aberrant CGI methylation contributes to cancer development and progression by affecting the genes involved in crucial cellular processes. It should be mentioned that although there is a subset of genes that are hypermethylated in a many tumor types (e.g., $p16^{INK4a}$), overall, recent studies have shown that the methylation of the promoters is cancer-cell-specific. Each cancer type can be defined by a specific "methylome." Moreover, differences in DNA methylation patterns may reflect the stage of tumorigenesis or even distinguish between the particular histological features. Some examples of genes silenced by CGI hypermethylation in particular tumor are listed in Table 2.1.

Recently, a group of small noncoding RNAs, called microRNAs (miR-NAs), have been proposed to play an important role in tumorigenesis because of their deregulated levels of expression in normal and cancer cells (Lujambio and Esteller, 2009). It has also been shown that in a subset of human tumors, miRNA processing machinery may be disrupted, leading to the defects in the miRNAs expression (Melo et al., 2009). miRNAs are short, 20- to 30-nucleotide-long, regulatory RNAs (Cai et al., 2009). They act at the posttranscriptional level by matching the 3′-untranslated region of a target gene. If the seed sequence of miRNAs presents perfect or near-perfect complementarity to the recognized sequence, target mRNA is degraded. If the binding remains imperfect, the target gene is inactivated by translational inhibition. miRNAs are involved in different cellular mechanisms, functioning as tumor suppressor, oncogenes, negative or positive regulators of distinct processes, depending on the gene they are targeting. Since miRNAs may be upregulated or downregulated in cancer cells, the possibility that they might be regulated by DNA methylation events presented itself (Lu et al., 2005a,b) (Fig. 2.3). Indeed, several studies confirmed that miRNAs acting as tumor-suppressor genes are silenced by hypermethylation of CGIs in their promoter regions (Davalos and Esteller, 2010) (Fig. 2.3). For instance, DNA methylation-mediated inactivation of miR-124a in colon cancer modulates the activity of its target—the CDK6 oncogene. As a consequence, RB protein, a tumor-suppressor gene, crucial in cell-cycle regulation, may be inactivated by CDK6-mediated phosphorylation (Lujambio et al., 2007). Similarly, DNA hypermethylation of miRNA-127 in bladder cancer results in aberrant expression of the oncogenic factor BCL-6 (Saito et al., 2006). Thus, DNA methylation events that not only occur in protein-coding genes but also are related to noncoding RNAs expression seem to be important in tumor formation and progression (Fig. 2.3).

According to most reports, the methylation of those hypermethylated genes that are silenced in cancer occurs in the CGI of the promoter and regulatory regions. Intriguingly, some studies claim that inhibition of gene expression may take place in non-CGI regions. That is the case for MASPIN and TIMP1, which are regulated by methylation of CpGs separated one from another even by 300–500 bp (Futscher et al., 2002; Veerla et al., 2008). Studies of induction mechanism of mammalian primary response genes by Toll-like receptors revealed the possible role of nucleosome density at different types of promoters (containing CGI or not) (Ramirez-Carrozzi et al., 2009).

Table 2.1. Examples of Genes Silenced by CpG Island Promoter Hypermethylation in Most Frequently Reported Cancer Types

Gene	Function	Breast cancer	Lung cancer	Prostate cancer	Leukemia/ lymphomas	Colon cancer
APC	Antagonist of the Wnt signaling pathway involved in cell migration and adhesion	X	X	X	–	X
BMAL1	Core component of the circadian clock	–	–	–	X	–
BRCA1	DNA repair double-stranded breaks, transcription	X	–	–	–	X
CDH1	E-cadherin, cell adhesion	X	X	X	X	X
CDH13	H-cadherin, cell adhesion	X	X	–	X[a]	X
CDKN2A (p16^{INK4a})	CDK4 inhibitor, control of cell-cycle G1 progression	X	X	X	X[b]	X
CDKN2B (p15^{INK4b})	CDK4 and CDK6 inhibitor, control of cell-cycle G1 progression	–	–	–	–	–
p14^{ARF}	Control of cell-cycle G1 progression, stabilizer of the tumor-suppressor protein p53	–	–	–	–	X
COX2	Cyclooxygenase, prostaglandin biosynthesis	–	–	–	–	X
CRBP1	Transport of retinol necessary for growth or differentiation of epithelial tissues	–	X	–	X	X
DAPK1	Positive mediator of γ-interferon induced programmed cell death	–	X	–	X	X
ESR1	Estrogen receptor, regulation of gene expression	X	X	X	X	X
GATA4	GATA family of zinc-finger transcription factors	–	–	–	–	X
GATA5	GATA family of zinc-finger transcription factors	–	–	–	–	X
GSTP1	Metabolism, detoxification, and elimination of genotoxic foreign compounds	–	X	X	–	–
HIC1	Transcription factor	X	X	–	X	X
IGFBP3	Insulin-like growth factor-binding protein	–	X	–	–	–
MGMT	DNA repair	–	X	–	X[a]	X
MLH1	DNA mismatch repair, DNA damage signaling	–	–	–	–	X
NORE1A	Ras effector homolog	–	X	–	–	–
PYCARD	TMS1/ASC, apoptotic signaling pathways	X	X	–	–	–
RARB2	Retinoic acid receptor, limits growth of many cell types by regulating gene expression	–	X	X	X	X
RASSF1A	Inhibit the accumulation of cyclin D1, involved in cell cycle arrest at G1/S phase transition and DNA repair	X	X	X	X	X
TLE1	Groucho homolog	–	–	–	X	–
TP73	p53 family of transcription factors, apoptotic response to DNA damage	–	–	–	X	–

X, reported frequent hypermethylation.
[a] DNA methylation only in lymphomas.
[b] DNA methylation only in leukemia.

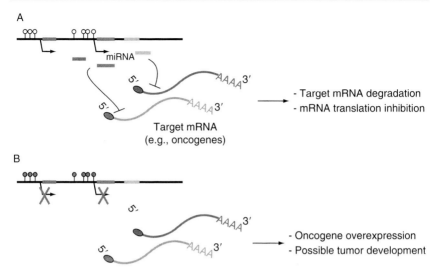

Figure 2.3. Regulation of miRNAs expression by DNA methylation. (A) When regulatory regions within miRNAs are unmethylated, miRNAs are expressed and may inactivate target genes by either degradation or translational inhibition. miRNAs act as tumor-suppressors if they target oncogenes. (B) miRNAs, expression is abolished by their promoters methylation, target genes are not silenced. Oncogenes, silenced in normal cells by miRNAs, can now be expressed, thus contribute to tumor development.

Although the number of identified genes silenced by DNA methylation is growing rapidly, it is still not clear how they acquire hypermethylation during tumorigenesis. Current studies incline toward the hypothesis of nonrandom DNA methylation process. First, it has been shown that every cancer type has not only a unique methylation profile with several common hypermethylated genes but also numerous cancer-cell-specific ones. Therefore, there would be a little probability for aberrant DNA methylation to occur completely by chance. Second, the proposed interplay between PcG proteins and DNA methylation events may lead to the designation of a target sequence for methylation (Simon and Lange, 2008). EZH2, one of PRC2 components, was shown to interact with DNMTs at least at some gene promoters (e.g., *TBX3* in breast cancer, *HOXD10*, *SIX3*, *KCNA1*, and *MYT1* in osteosarcoma) (Vire *et al.*, 2006). These genes appeared to be enriched for trimethylation of H3K27, a specific mark for EZH2 activity. Thus, PcG proteins might mark some specific sequences that become further methylated during cancer development (Schlesinger *et al.*, 2007).

IV. DNA METHYLATION MARKERS

Despite our constantly growing understanding of carcinogenesis, there is still an eager needs to design novel, powerful tools that can be applied as part of clinical practice. Tumor biomarkers are indispensable not only for early diagnosis of cancer but also for prognosis, prediction of therapeutic response, monitoring therapy, or assessment of risk of recurrence after curative surgery. Many different molecules, including nucleic acids, proteins, or even small metabolites, could be used as the indicators of tumorigenesis. Nevertheless, to be applied in clinical practice, they should present some crucial characteristics such as easy detection and measurement, preferentially in noninvasive manner, selectivity, and specificity. The use of DNA markers in this context has various advantages over, for example, proteins. DNA is stable, easy to isolate from different kinds of material and, unlike proteins, relatively small amounts of material are needed to perform the assays. Some genetic alterations have already been proposed as being valuable in diagnosis (Chin and Gray, 2008), but it has also been suggested that changes in the DNA methylation pattern may also aid in following cancer progression (Caballero and Chen, 2009).

DNA hypermethylation events have several advantages with respect to their use as tumor biomarkers (Mulero-Navarro and Esteller, 2008). As mentioned above, particular genes may be methylated at different stages of tumor development. They can serve to detect both early and premalignant stages of a disease or to assess the risk of progression to malignancy. This information could be crucial to the choice of the correct treatment for patients and to avoid the needless use of more aggressive therapy. DNA methylation abnormalities may occur in a very precocious, initial stage of neoplasia or may even affect the genes that are not directly related to tumorigenesis but which confer a "mutator phenotype" (e.g., MGMT).

Considering that some of the genes acquire tissue-specific DNA methylation in cancer (e.g., BRCA1 is hypermethylated exclusively in breast and ovarian carcinomas), the identification of different cancer types by DNA methylation analysis may also be possible. Furthermore, this kind of studies could yield information about the primary origin of metastatic tumors.

It should be mentioned that genes affected by DNA methylation do not necessarily contribute explicitly to the tumorigenesis in such a way that they could be considered putative biomarkers. For diagnostic purposes, it would be sufficient to prove the correlation between DNA methylation pattern and tumor development. For example, this is the case for Vimentin which at first does not seem to be involved in cancer formation because although it is methylated in colon cancer and unmethylated in normal colorectal epithelial crypt cells, it is unexpressed in both cases (Chen et al., 2005). Nevertheless, a significant association of Vimentin methylation with Dukes' stage and preferences for developing metastasis and peritoneal dissemination in colorectal carcinomas, detected in colon cancer patients, has encouraged the consideration of the hypermethylation of this gene to be a putative biomarker (Shirahata et al., 2009).

DNA methylation markers are conjectured to play a role as predictive factors. They may be useful for determining not only the outcome of the disease but also the individual response to chemotherapy. Furthermore, monitoring DNA methylation profiles after treatment or curative surgery may yield evidence concerning the efficiency of applied therapy or detect disease recurrence.

One of the biggest advantages of DNA methylation biomarkers is the ease with which they can be detected and analyzed. The regions which may be subject to DNA methylation events are well defined. Hence, in theory, not extremely sophisticated methods may be sufficient to detect whether a gene of interest is methylated or not. In comparison, mutations inactivating tumor-suppressor genes are normally distributed over many sites throughout the sequence, requiring more complex analyses to check their occurrence. In addition, arising from the fact that DNA can be released from tumor tissues, methylated genes can be examined from biological fluids (Wong, 2001). Identification of DNA methylation-based biomarkers in specimens like sputum, plasma, serum, or urine obtained in a noninvasive manner could be promising sources for early detection of carcinomas (Table 2.2). However, the choice of the appropriate technique for the routine clinical detection of gene hypermethylation is a problem that remains to be resolved. Even if there were a wide range of different methods available to determinate the DNA methylation patterns (reviewed below), their optimization and validation would need to be applicable in quotidian usage.

Table 2.2. Examples of Potential DNA Methylation Biomarkers in Different Types of Cancer and Biological Material in Which the Genes of Interest Could be Detected (Mulero-Navarro and Esteller, 2008)

Gene	Cancer type	Detection material
BRCA1	Ovary	Serum
CDKN2A ($p16^{INK4a}$)	Lung	Sputum
	Colon	Stool
CDKN2B ($p15^{INK4b}$)	Leukemia	Blood
GSTP1	Prostate	Urine, ejaculate
MGMT	Colon	Stool
	Lung	Sputum
RASSF1A	Lung	Sputum
	Prostate	Urine
	Breast	Serum
SFRP2	Colon	Stool

It should be stressed that the majority of methylated genes postulated to be satisfactory candidates for biomarkers have been checked only in a small-scale studies. For the required clinical validation of putative epigenetic biomarkers, prospective trials or meta/pooled analyses are needed. Moreover, the assessment of a marker's ability to diagnose preclinical disease as well as false-positive and false-negative rate should be established. Our hope is that DNA methylation markers, together with transcriptome and proteome biomarkers, will be a crucial and potent tool in future medicine.

V. DETERMINING THE CANCER DNA METHYLOME

The importance of DNA methylation alterations in tumorigenesis encourages us to decode the human epigenome. Nowadays, various methods are available to assess the DNA methylation pattern of particular DNA regions. Important progress has also been made toward completing a large-scale analysis of DNA methylation.

5-Methylcytosines can be detected by three possible methods based on (1) bisulfite conversion of DNA, (2) methyl-sensitive restriction enzymes, or (3) chromatin immunoprecipitation (ChIP) assays (Table 2.3).

Bisulfite treatment takes advantage of the fact that methylated cytosine becomes less reactive. Thus, bisulfite treatment of DNA converts only unmethylated cytosines to uracil, whereas methylated nucleotides stay unchanged (Fig. 2.4). Afterward, the methylation pattern of defined CpGs could be detected by the methylation-specific PCR (MSP) with specific primers complementary either to the originally methylated or to the unmethylated product (Herman and Baylin, 2001). MSP is the easiest and quickest method that can be used in everyday laboratory practice to analyze DNA methylation in large number of samples. A disadvantage of this method is that only a few CpGs can be checked. To assess the DNA methylation status of multiple CpGs (e.g., within promoter regions), sequencing of amplified fragment could be performed after bisulfite conversion of DNA (bisulfite sequencing PCR [BSP]). The combination of bisulfite treatment with quantitative methylation-specific PCR (known also as MethyLight) or pyrosequencing provides information about the minimal aberrant DNA methylation (Eads et al., 2000; Uhlmann et al., 2002).

Although bisulfite treatment is a reliable method to distinguish between methylated and unmethylated cytosine, recent discoveries of hmC in embryonic stem cells and Purkinje neurons call for more careful usage of this technique. hmC is structurally similar to mC and both nucleotides are not converted to uracil upon bisulfite treatment, implying that a proportion of genomic loci identified as methylated may actually be hydroxymethylated (Huang et al., 2010).

Table 2.3. Selected Methods for Screening for Methylation Alterations

Method	Short description	Advantages	Disadvantages
Based on bisulfite conversion of DNA			
MSP	Methylation specific PCR; specially designed two pairs of primers to detect methylated or unmethylated CpGs	Easy to perform, quick analysis of large number of samples, low cost	Only few CpGs are checked, not quantitative
BSP	Bisulfite sequencing PCR; sequencing of amplified fragment to establish the methylation status of multiple CpGs	Several CpGs of a particular CGI are checked (10–30)	Quite costly and laborious, time consuming
MethyLight	Quantitative methylation-specific PCR	Sensitive, quantitative, high throughput, able to detect minimal amounts of aberrant DNA methylation	Quite costly, do not quantify methylation at the individual nucleotide level
Based on methylation-sensitive restriction enzymes			
RLGS	Restriction landmark genomic scanning; digested DNA fragments are separated on a 2D gel. Detection based on radioactive labeling of unmethylated digestion sites	Genome-wide screening, quantitative (spots signal reflect the copy number), limited sensitivity	Detects only CpGs located within restriction site, large amounts of DNA needed, lack of signal for methylated sites
MS-RDA	Methylation-sensitive representational difference analysis; digested DNA fragments ligated with universal adaptors. After amplification, libraries of products with unmethylated sites are obtained	Genome-wide screening, limited sensitivity	Detects only CpGs located within restriction site, lack of signal for methylated sites
HELP assay	*Hpa*II tiny fragments enrichment by ligation-mediated PCR; DNA samples digested with methylation-sensitive *Hpa*I as well as its methylation-insensitive isoschizomer. Digested fragments analyzed on two parallel DNA microarrays	Genome-wide screening, positive representation of both hypo- and hypermethylation status	Detects only CpGs located within restriction site
CHARM	Comprehensive high-throughput array-based relative methylation analysis; digested DNA fragments (e.g., by *McrBc*) analyzed on a specially design array, smoothing the data of neighboring genomic location increases sensitivity and specificity	Genome-wide screening, regions of lower CpG density taken into the consideration, highly quantitative	Enzymatic digestion limitations
Based on affinity assays			
MeDIP	Methylated DNA immunoprecipitation; immunoprecipitation of methylated DNA fragments using antibody against 5-methylcytosine, following by hybridization to DNA microarray	High-resolution maps of methylome, global distribution of mCpGs studied, genome-wide screening	Large amounts of DNA needed, restricted to the antibody specificity, low resolution of detection

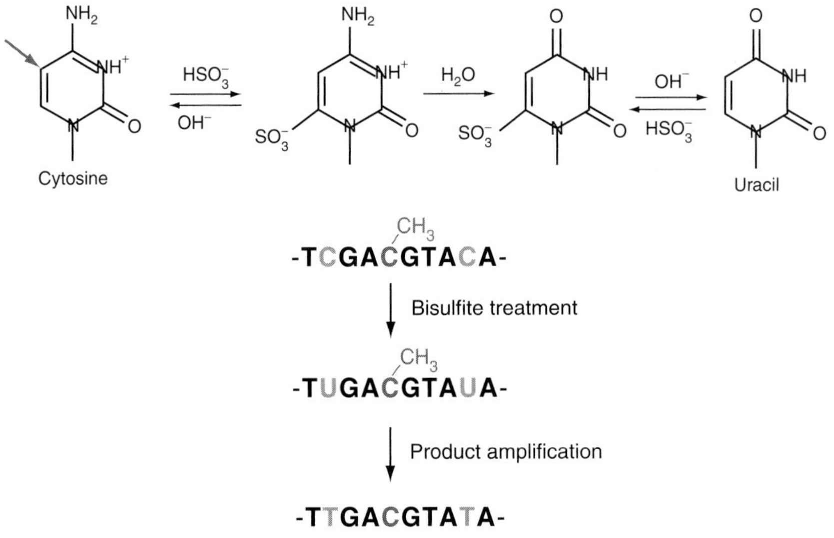

$$-\text{TCGAC}\overset{\overset{\displaystyle CH_3}{|}}{\text{C}}\text{GTACA-}$$

↓ Bisulfite treatment

$$-\text{TUGAC}\overset{\overset{\displaystyle CH_3}{|}}{\text{C}}\text{GTAUA-}$$

↓ Product amplification

$$-\text{TTGACGTATA-}$$

Figure 2.4. Bisulfite conversion of DNA. (A) Bisulfite treatment reaction converts cytosines to uracil. Methylation at 5 position of cytosine ring (marked with red arrow) protects whole compound from conversion. (B) After bisulfite treatment, followed by product amplification, unmethylated cytosines are changed to thymidine, whereas methylated nucleotides stay unaffected.

Another approach to detecting methylated cytosines is based on methylation-sensitive restriction enzymes that can digest the target sequence only if it is not methylated. *Hpa*II, *Sac*II, *Not*I, and *Bst*UI are the enzymes commonly used for this type of assay. The analyses where methylation-sensitive enzymes cleavage is involved can often be successfully used to evaluate global DNA methylation status evaluation. However, one of the significant disadvantages of this kind of strategy is its limitation to recognition only those CpGs that are localized within the enzyme restriction site. Moreover, the CpG density can hinder proper analysis.

One of the first methods applied for genome-wide methylation studies was restriction landmark genomic scanning (RLGS) (reviewed in Rush and Plass, 2002), in which DNA samples of, for example, normal and tumor tissue are digested by methylation-sensitive enzyme (e.g., *Not*I). DNA fragments are labeled at the restriction site by radionucleotides. Then, electrophoresis is performed, followed by a second digestion, this time with a frequently cutting enzyme. The smaller DNA fragments obtained in this way are separated by electrophoresis in the second direction. As a result, the image of labeled spots corresponding to the unmethylated sites is revealed. The analysis of differences in spot patterns between the two samples provides information about changes in DNA methylation pattern. Furthermore, the spot signal is correlated with quantification of gene copy number.

The first step of another method, methylation-sensitive representational difference analysis, also requires digestion by methyl-sensitive enzymes of DNA samples from the two sources that are to be compared (Kaneda et al., 2003). Unlike with RLGS, the amplicons are not labeled but ligated with short DNA fragments (known as universal adaptors) containing an overhang complementary to the restriction site. Products are then amplified and the libraries of products with unmethylated sites are obtained for screening.

The development of microarray systems has facilitated the identification of differentially methylated regions. In differential methylation hybridization, which combines restriction enzyme assays with microarrays, the first step consists of digesting DNA samples with a frequent cutter and further ligation of the derived smaller fragments to the special linkers (Huang et al., 1999). Second, linked fragments are cleaved by methylation-sensitive restriction enzymes such as BstUI and HpaII. A linker-PCR is then performed, which amplifies only methylated sequences that were not cleaved by BstUI or HpaII. In the final step, the probes are identified after hybridization to the microarray.

Frequently, a disadvantage of the sensitive enzyme digestion approaches is a lack of signal for negative results. In RLGS for instance, methylation of a particular CpG site in a cancer sample is demonstrated only by the lack of the spot signal in the corresponding normal tissue. One method for tackling this problem is HpaII tiny fragment enrichment by ligation-mediated PCR (HELP assay) (Khulan et al., 2006). The main advantage of this approach is its use of two restriction enzymes, HpaII and its methylation-insensitive isoschizomer, MspI. The analysis of digested fragments on genomic DNA microarrays gives a positive representation of both hypomethylation and hypermethylation status in the genome.

The third mode for distinguishing between methylated and unmethylated cytosines is related to ChIP. For example, the use of specific antibodies against MBD proteins (methyl-CpG-binding proteins which specifically recognize 5-methylcytosines) allows to immunoprecipitate the methylated fragments of DNA that can be subjected to further analysis to identify them (Ballestar et al., 2003; Lopez-Serra et al., 2006). It could be also mentioned that 5-methylcytosines recognition of MBDs serves to select methylated DNA fragments using affinity columns or beads (Brock et al., 2001).

The combination of the ChIP method with hybridization to the DNA microarray, known as the ChIP-on-chip assay, provides information about large-scale methylation status. Moreover, the development of the antibodies against 5-methylcytosine presents the opportunity to identify multiple distinct CpG sites across the genome, when combined with DNA microarray analyses (Me-DIP strategy) (Weber et al., 2005).

Microarray technology has improved in recent years and is now an important tool that can be used for global DNA methylation analyses. The use of commercial tilling microarray platforms containing overlapping oligonucleotides that span broad regions of genome reveals important information about DNA methylation patterns within large fragments of sequences. An interesting array-based approach has recently been proposed, known as comprehensive high-throughput array-based relative methylation analysis (CHARM) (Irizarry et al., 2008). Unlike the majority of similar approaches, the measurement of methylation content by this method is not restricted to the CGIs but takes into the consideration regions of lower CpG density. Furthermore, CHARM is also highly quantitative, which is a great advantage in the study of DNA methylation profiles in cancer.

Nowadays, as a result of the technological advances, massive sequencing methods, instead of microarray assays, can be used to evaluate large-scale methylation patterns. Various sequencing platforms are available commercially, such as SOLiD, Illumina GoldenGate, and Illumina Infinium. Measurement of genome-scale DNA methylation status involves bisulfite conversion of the samples as a first step. As mentioned above, this kind of treatment transforms only unmethylated cytosine, resulting in the occurrence of a pseudopolymorphism: cytosines are transformed to thymidines in the final product, whereas 5-methylcytosines continue to be detected as cytosines. Therefore, the performed analysis is similar to that for SNPs detection. The massive sequencing technologies are accurate means of detecting DNA methylation profiles. Nevertheless, their high cost and currently relatively poor coverage of the genome limit their application in everyday laboratory and clinical practice.

Although there has been notable progress in the analysis of genome-wide DNA methylation patterns, there is still a great demand to reveal the human epigenome (Esteller, 2006). This idea has captured the interest of several public and private organizations. For instance, the Human Epigenome Consortium was founded with the aim of identifying the methylation-variable positions within the human genome. This kind of detailed DNA methylation map could provide with a great deal of interesting information that would significantly advance our knowledge of epigenetic events in cancer.

VI. THERAPY PERSPECTIVES

Unlike genetic alterations, epigenetic events, including DNA methylation, are reversible, which makes it extremely interesting from the point of view of developing new approaches to therapy. The agents capable of restoring the normal cell DNA methylation pattern have been developed in the course of several studies. The strategies use special drugs to inactivate DNMTs. As an expected result of this approach, DNMT inhibitors should trigger the loss of the overall level that is thought to randomly reactivate several methylation-silenced genes (reviewed in Yoo and Jones, 2006). Generally, one of the main advantages of DNMT inhibitors is that, unlike other chemotherapeutic drugs, they do not target cells for immediate death. Possible death and/or inhibition of tumor progression are believed to be a consequence of reactivation of some important cellular pathways. Nevertheless, potentially, a specific DNA methylation pattern could be restored in cancer cells after the removal of the demethylating agent. According to some research, the use of DNMT inhibitors could be unsafe because global hypomethylation may contribute to tumor progression (Gius et al., 2004). Although various studies confirm the tumorigenesis inhibition, the possible side effects related to loss of methylation should be borne in mind.

Compounds currently known to act as DNMT inhibitors can be divided into two groups: nucleoside and nonnucleoside analogs. Members of the first group are characterized by different modifications in the cytosine ring. They contribute to the loss of methylation by incorporation into newly synthesized DNA. Modifications within the nucleoside structure abolish the normally observed elimination of DNMTs during the methylation reaction. Enzymes are "caught" by the formation of a covalent complex with the drugs, which thereby inhibits their activity. Intriguingly, DNMT inhibitors may induce the proteosomal degradation of targeted enzymes (at least in the case of DNMT1) (Ghoshal et al., 2005).

By now, the most studied and promising demethylation agents, commonly used in everyday laboratory practice, are 5-azacytidine (5-Aza-CR, vidaza) and 5-aza-2'-deoxycytidine (5-Aza-CdR, decitabine) (Christman, 2002). These two main drugs with anticancer properties have already been approved by the US Food and Drug Administration (FDA) for the treatment of myelodysplastic syndromes (MDS) and chronic myelomonocytic leukemia (CML) (Kaminskas et al., 2005; Steensma, 2009). They have several advantages that are important in clinical practice. They can be used in low doses, which reduce toxicity (Issa et al., 2005). It is also believed that because they act by incorporation into synthesized DNA they preferentially affect rapidly dividing cells, including cancerogenous cells (Liang et al., 2002). The most common side effects of these drugs are myelosuppressions as well as nausea and infections. Generally, 5-Aza-CR and 5-Aza-CdR are well tolerated, with notable clinical activity in MDS and CML patients. However, this kind of treatment seems to fail in solid tumors (Ghoshal and Bai, 2007). 5-Aza-CR and 5-Aza-CdR require intravenous or subcutaneous injection that may produce inconvenience in their use. In addition, both drugs seem to be relatively unstable in aqueous solution and furthermore are subjected to deamination enzymes that inactivate them (Chabot et al., 1983). The development of zebularine, another nucleoside analog with DNMT-inhibiting properties, is believed to resolve this problem (Zhou et al., 2002), since it is more stable in aqueous solution and is not affected by deaminases because it acts as their inhibitor (Carlow and Wolfenden,

1998). The mode of action of zebularine is similar to that of 5-Aza analogs but it seems that a much higher dose is required for efficient treatment. However, zebularine can be administrated orally and its effects seem to be less toxic (Cheng *et al.*, 2004). In addition, the combination of zebularine and 5-Aza-CdR treatment serves to enhance the antitumoral activity of 5-Aza-CdR (Lemaire *et al.*, 2009).

 The risk of using nucleoside analogs as DNMT inhibitors is associated with their incorporation into the DNA, which may lead to toxicity and undesired side effects because they could have a potential to form mutagenic lesion. For this reason, a wide range of nonnucleoside analogs, which bind to the catalytic or cofactor site of DNMTs without becoming incorporated into the DNA, are currently under investigation. Over the years, several compounds have been proposed as being DNMT inhibitors, including SGI-1027 (Datta *et al.*, 2009), RG108 (Brueckner *et al.*, 2005), hydralazine (Deng *et al.*, 2003), procainamide (Lu *et al.*, 2005a,b), or psammaplin (Pina *et al.*, 2003). However, none of them has so far been successfully introduced into clinical practice. This state of affairs should prompt further research in that field. Another strategy consists of developing antisense oligonucleotides that can inhibit the DNMT at the translational level. MG98 is representative of this class of inhibitors and its ability to repress DNA methylation of tumor-suppressor genes has been investigated in preclinical studies and clinical phase I/II trials (Amato, 2007). By now, in the phase I studies of patients with myelodysplastic syndromes and acute myeloid leukemia althought reduction of DNMT1 expression following MG98 administration was confirmed in selected patients, no clinical activity was observed in the dose level tested (Klisovic *et al.*, 2008). The results from the treatment of patients with advanced solid malignances seem to be more optimistic. In nearly 80% of cases, the mechanism of DNMT1 suppression by MG98 was proved. Moreover, the drug was well tolerated and some early evidences of clinical activity were reported (Plummer *et al.*, 2009).

 In conclusion, it should be remembered that the epigenetic regulation of a gene is a result of the interplay between several epigenetic components. Thus, for efficient therapy the simultaneous use of different agents is highly recommended. For instance, the combination of the treatment with 5-Aza-CdR and histone deacetylase inhibitor was reported to give better therapeutic effects. It is beyond doubt that epigenetic therapy is an interesting and potentially powerful tool for tackling cancer, but much work is still required to produce a treatment that is also satisfactory for solid tumors.

VII. CONCLUSIONS

Since its discovery, DNA methylation has captured a lot of interest and its fundamental role in proper cell functioning is broadly recognized. Numerous studies have shown the key role of DNA methylation in tumor development and progression. Particularly, promoter hypermethylation resulting in gene silencing is one of the central players in cancer. In addition to classical genetic abnormalities, DNA methylation aberrations are considered as a second layer of tumorigenesis complexicity. Moreover, as it seems that DNA hypermethylation is not a random process and it can accurately characterize type, stage or histology of specific tumor, the usage of DNA methylation markers holds a promise to be valuable in everyday clinical practice. Many trials are ongoing with hope to provide marker panels that allow early diagnosis, assessment of risk of development and recurrence or prognosis by noninvasive manner. Finally, what makes the findings of DNA hypermethylation events in cancer extremely interesting is the reversibility of that process. The chance to return to the normal cell phenotype by means of DNA methylation inhibitors is an immense advantage that the epigenetic processes have above classical genetic events. The approval of two drugs, 5-azacytidine and 5-aza-2'-deoxycytidine in myelodysplastic syndromes and chronic myelomonocytic leukemia treatment, stimulates us even more to make progress in that field.

The crucial role of DNA methylation and its promising applications in cancer therapies underline the growing need to investigate this phenomenon at a global scale. Although among the last years, an amazing progress in screening technologies has been made, the decoding of human epigenome is still ahead. There is a great hope that the discovery of global patterns of epigenetic modifications enables us to better understand this process and will be crucial in successful strategies of cancer treatment. To achieve this aim, the international collaboration between many laboratories, public, and private institutions, is required to be continued. Moreover, the multidisciplinary investigation, combining different branches of science, from biology thought biophysics and informatics, will be welcome to deeper our knowledge about epigenome which is potentially as important in tumorigenesis as genome itself. In our biggest hope, personal epigenome analysis will be feasible soon and may open the way to novel putative translational application in patients.

References

Ahluwalia, A., Hurteau, J. A., Bigsby, R. M., and Nephew, K. P. (2001). DNA methylation in ovarian cancer. II. Expression of DNA methyltransferases in ovarian cancer cell lines and normal ovarian epithelial cells. *Gynecol. Oncol.* **82,** 299–304.

Akiyama, Y., Watkins, N., Suzuki, H., Jair, K. W., van Engeland, M., Esteller, M., Sakai, H., Ren, C. Y., Yuasa, Y., Herman, J. G., et al. (2003). GATA-4 and GATA-5 transcription factor genes and potential downstream antitumor target genes are epigenetically silenced in colorectal and gastric cancer. *Mol. Cell. Biol.* **23,** 8429–8439.

Amato, R. J. (2007). Inhibition of DNA methylation by antisense oligonucleotide MG98 as cancer therapy. *Clin. Genitourin. Cancer* **5,** 422–426.

Badal, V., Chuang, L. S., Tan, E. H., Badal, S., Villa, L. L., Wheeler, C. M., Li, B. F., and Bernard, H. U. (2003). CpG methylation of human papillomavirus type 16 DNA in cervical cancer cell lines and in clinical specimens: Genomic hypomethylation correlates with carcinogenic progression. *J. Virol.* **77**, 6227–6234.

Baker, E. K., Johnstone, R. W., Zalcberg, J. R., and El-Osta, A. (2005). Epigenetic changes to the MDR1 locus in response to chemotherapeutic drugs. *Oncogene* **24**, 8061–8075.

Ballestar, E., Paz, M. F., Valle, L., Wei, S., Fraga, M. F., Espada, J., Cigudosa, J. C., Huang, T. H., and Esteller, M. (2003). Methyl-CpG binding proteins identify novel sites of epigenetic inactivation in human cancer. *EMBO J.* **22**, 6335–6345.

Brock, G. J., Huang, T. H., Chen, C. M., and Johnson, K. J. (2001). A novel technique for the identification of CpG islands exhibiting altered methylation patterns (ICEAMP). *Nucleic Acids Res.* **29**, E123.

Brueckner, B., Garcia Boy, R., Siedlecki, P., Musch, T., Kliem, H. C., Zielenkiewicz, P., Suhai, S., Wiessler, M., and Lyko, F. (2005). Epigenetic reactivation of tumor suppressor genes by a novel small-molecule inhibitor of human DNA methyltransferases. *Cancer Res.* **65**, 6305–6311.

Caballero, O. L., and Chen, Y. T. (2009). Cancer/testis (CT) antigens: Potential targets for immunotherapy. *Cancer Sci.* **100**, 2014–2021.

Cai, Y., Yu, X., Hu, S., and Yu, J. (2009). A brief review on the mechanisms of miRNA regulation. *Genomics Proteomics Bioinformatics* **7**, 147–154.

Carlow, D., and Wolfenden, R. (1998). Substrate connectivity effects in the transition state for cytidine deaminase. *Biochemistry* **37**, 11873–11878.

Catteau, A., and Morris, J. R. (2002). BRCA1 methylation: A significant role in tumour development? *Semin. Cancer Biol.* **12**, 359–371.

Chabot, G. G., Bouchard, J., and Momparler, R. L. (1983). Kinetics of deamination of 5-aza-2′-deoxycytidine and cytosine arabinoside by human liver cytidine deaminase and its inhibition by 3-deazauridine, thymidine or uracil arabinoside. *Biochem. Pharmacol.* **32**, 1327–1328.

Chedin, F., Lieber, M. R., and Hsieh, C. L. (2002). The DNA methyltransferase-like protein DNMT3L stimulates de novo methylation by Dnmt3a. *Proc. Natl. Acad. Sci. USA* **99**, 16916–16921.

Chekhun, V. F., Kulik, G. I., Yurchenko, O. V., Tryndyak, V. P., Todor, I. N., Luniv, L. S., Tregubova, N. A., Pryzimirska, T. V., Montgomery, B., Rusetskaya, N. V., *et al.* (2006). Role of DNA hypomethylation in the development of the resistance to doxorubicin in human MCF-7 breast adenocarcinoma cells. *Cancer Lett.* **231**, 87–93.

Chen, W. D., Han, Z. J., Skoletsky, J., Olson, J., Sah, J., Myeroff, L., Platzer, P., Lu, S., Dawson, D., Willis, J., *et al.* (2005). Detection in fecal DNA of colon cancer-specific methylation of the nonexpressed vimentin gene. *J. Natl. Cancer Inst.* **97**, 1124–1132.

Cheng, J. C., Yoo, C. B., Weisenberger, D. J., Chuang, J., Wozniak, C., Liang, G., Marquez, V. E., Greer, S., Orntoft, T. F., Thykjaer, T., *et al.* (2004). Preferential response of cancer cells to zebularine. *Cancer Cell* **6**, 151–158.

Chin, L., and Gray, J. W. (2008). Translating insights from the cancer genome into clinical practice. *Nature* **452**, 553–563.

Christman, J. K. (2002). 5-Azacytidine and 5-aza-2′-deoxycytidine as inhibitors of DNA methylation: Mechanistic studies and their implications for cancer therapy. *Oncogene* **21**, 5483–5495.

Costa, F. F., Paixao, V. A., Cavalher, F. P., Ribeiro, K. B., Cunha, I. W., Rinck, J. A., Jr., O'Hare, M., Mackay, A., Soares, F. A., Brentani, R. R., *et al.* (2006). SATR-1 hypomethylation is a common and early event in breast cancer. *Cancer Genet. Cytogenet.* **165**, 135–143.

Daniel, J. M., Spring, C. M., Crawford, H. C., Reynolds, A. B., and Baig, A. (2002). The p120(ctn)-binding partner Kaiso is a bi-modal DNA-binding protein that recognizes both a sequence-specific consensus and methylated CpG dinucleotides. *Nucleic Acids Res.* **30**, 2911–2919.

Datta, J., Ghoshal, K., Denny, W. A., Gamage, S. A., Brooke, D. G., Phiasivongsa, P., Redkar, S., and Jacob, S. T. (2009). A new class of quinoline-based DNA hypomethylating agents reactivates tumor suppressor genes by blocking DNA methyltransferase 1 activity and inducing its degradation. *Cancer Res.* **69**, 4277–4285.

Davalos, V., and Esteller, M. (2010). MicroRNAs and cancer epigenetics: A macrorevolution. *Curr. Opin. Oncol.* **22**, 35–45.

Deng, C., Lu, Q., Zhang, Z., Rao, T., Attwood, J., Yung, R., and Richardson, B. (2003). Hydralazine may induce autoimmunity by inhibiting extracellular signal-regulated kinase pathway signaling. *Arthritis Rheum.* **48**, 746–756.

Deplus, R., Brenner, C., Burgers, W. A., Putmans, P., Kouzarides, T., de Launoit, Y., and Fuks, F. (2002). Dnmt3L is a transcriptional repressor that recruits histone deacetylase. *Nucleic Acids Res.* **30**, 3831–3838.

Dokun, O. Y., Florl, A. R., Seifert, H. H., Wolff, I., and Schulz, W. A. (2008). Relationship of SNCG, S100A4, S100A9 and LCN2 gene expression and DNA methylation in bladder cancer. *Int. J. Cancer* **123**, 2798–2807.

Drexler, H. G. (1998). Review of alterations of the cyclin-dependent kinase inhibitor INK4 family genes p15, p16, p18 and p19 in human leukemia-lymphoma cells. *Leukemia* **12**, 845–859.

Eads, C. A., Danenberg, K. D., Kawakami, K., Saltz, L. B., Blake, C., Shibata, D., Danenberg, P. V., and Laird, P. W. (2000). MethyLight: A high-throughput assay to measure DNA methylation. *Nucleic Acids Res.* **28**, E32.

Esteller, M. (2006). The necessity of a human epigenome project. *Carcinogenesis* **27**, 1121–1125.

Esteller, M. (2007). Cancer epigenomics: DNA methylomes and histone-modification maps. *Nat. Rev. Genet.* **8**, 286–298.

Esteller, M. (2008). Epigenetics in cancer. *N. Engl. J. Med.* **358**, 1148–1159.

Esteller, M., Hamilton, S. R., Burger, P. C., Baylin, S. B., and Herman, J. G. (1999). Inactivation of the DNA repair gene O6-methylguanine-DNA methyltransferase by promoter hypermethylation is a common event in primary human neoplasia. *Cancer Res.* **59**, 793–797.

Esteller, M., Guo, M., Moreno, V., Peinado, M. A., Capella, G., Galm, O., Baylin, S. B., and Herman, J. G. (2002). Hypermethylation-associated inactivation of the cellular retinol-binding-protein 1 gene in human cancer. *Cancer Res.* **62**, 5902–5905.

Feinberg, A. P., and Vogelstein, B. (1983). Hypomethylation distinguishes genes of some human cancers from their normal counterparts. *Nature* **301**, 89–92.

Feinberg, A. P., Ohlsson, R., and Henikoff, S. (2006). The epigenetic progenitor origin of human cancer. *Nat. Rev. Genet.* **7**, 21–33.

Filion, G. J., Zhenilo, S., Salozhin, S., Yamada, D., Prokhortchouk, E., and Defossez, P. A. (2006). A family of human zinc finger proteins that bind methylated DNA and repress transcription. *Mol. Cell. Biol.* **26**, 169–181.

Fleisher, A. S., Esteller, M., Tamura, G., Rashid, A., Stine, O. C., Yin, J., Zou, T. T., Abraham, J. M., Kong, D., Nishizuka, S., *et al.* (2001). Hypermethylation of the hMLH1 gene promoter is associated with microsatellite instability in early human gastric neoplasia. *Oncogene* **20**, 329–335.

Fraga, M. F., Ballestar, E., Paz, M. F., Ropero, S., Setien, F., Ballestar, M. L., Heine-Suner, D., Cigudosa, J. C., Urioste, M., Benitez, J., *et al.* (2005). Epigenetic differences arise during the lifetime of monozygotic twins. *Proc. Natl. Acad. Sci. USA* **102**, 10604–10609.

Futscher, B. W., Oshiro, M. M., Wozniak, R. J., Holtan, N., Hanigan, C. L., Duan, H., and Domann, F. E. (2002). Role for DNA methylation in the control of cell type specific maspin expression. *Nat. Genet.* **31**, 175–179.

Gama-Sosa, M. A., Slagel, V. A., Trewyn, R. W., Oxenhandler, R., Kuo, K. C., Gehrke, C. W., and Ehrlich, M. (1983). The 5-methylcytosine content of DNA from human tumors. *Nucleic Acids Res.* **11**, 6883–6894.

Ghoshal, K., and Bai, S. (2007). DNA methyltransferases as targets for cancer therapy. *Drugs Today (Barc.)* **43**, 395–422.

Ghoshal, K., Datta, J., Majumder, S., Bai, S., Kutay, H., Motiwala, T., and Jacob, S. T. (2005). 5-Aza-deoxycytidine induces selective degradation of DNA methyltransferase 1 by a proteasomal pathway that requires the KEN box, bromo-adjacent homology domain, and nuclear localization signal. *Mol. Cell. Biol.* **25**, 4727–4741.

Gius, D., Cui, H., Bradbury, C. M., Cook, J., Smart, D. K., Zhao, S., Young, L., Brandenburg, S. A., Hu, Y., Bisht, K. S., *et al.* (2004). Distinct effects on gene expression of chemical and genetic manipulation of the cancer epigenome revealed by a multimodality approach. *Cancer Cell* **6**, 361–371.

Gordian, E., Ramachandran, K., and Singal, R. (2009). Methylation mediated silencing of TMS1 in breast cancer and its potential contribution to docetaxel cytotoxicity. *Anticancer Res.* **29**, 3207–3210.

Greger, V., Passarge, E., Hopping, W., Messmer, E., and Horsthemke, B. (1989). Epigenetic changes may contribute to the formation and spontaneous regression of retinoblastoma. *Hum. Genet.* **83**, 155–158.

Gupta, A., Godwin, A. K., Vanderveer, L., Lu, A., and Liu, J. (2003). Hypomethylation of the synuclein gamma gene CpG island promotes its aberrant expression in breast carcinoma and ovarian carcinoma. *Cancer Res.* **63**, 664–673.

Herman, J. G., and Baylin, S. B. (2001). Methylation-specific PCR. *Curr. Protoc. Hum. Genet.* Chapter 10, Unit 10 6.

Hernandez-Munoz, I., Taghavi, P., Kuijl, C., Neefjes, J., and van Lohuizen, M. (2005). Association of BMI1 with polycomb bodies is dynamic and requires PRC2/EZH2 and the maintenance DNA methyltransferase DNMT1. *Mol. Cell. Biol.* **25**, 11047–11058.

Huang, T. H., Perry, M. R., and Laux, D. E. (1999). Methylation profiling of CpG islands in human breast cancer cells. *Hum. Mol. Genet.* **8**, 459–470.

Huang, Y., Pastor, W. A., Shen, Y., Tahiliani, M., Liu, D. R., and Rao, A. (2010). The behaviour of 5-hydroxymethylcytosine in bisulfite sequencing. *PLoS One* **5**, e8888.

Irizarry, R. A., Ladd-Acosta, C., Carvalho, B., Wu, H., Brandenburg, S. A., Jeddeloh, J. A., Wen, B., and Feinberg, A. P. (2008). Comprehensive high-throughput arrays for relative methylation (CHARM). *Genome Res.* **18**, 780–790.

Issa, J. P., Gharibyan, V., Cortes, J., Jelinek, J., Morris, G., Verstovsek, S., Talpaz, M., Garcia-Manero, G., and Kantarjian, H. M. (2005). Phase II study of low-dose decitabine in patients with chronic myelogenous leukemia resistant to imatinib mesylate. *J. Clin. Oncol.* **23**, 3948–3956.

Jacinto, F. V., and Esteller, M. (2007). MGMT hypermethylation: A prognostic foe, a predictive friend. *DNA Repair (Amst.)* **6**, 1155–1160.

Jin, B., Tao, Q., Peng, J., Soo, H. M., Wu, W., Ying, J., Fields, C. R., Delmas, A. L., Liu, X., Qiu, J., *et al.* (2008). DNA methyltransferase 3B (DNMT3B) mutations in ICF syndrome lead to altered epigenetic modifications and aberrant expression of genes regulating development, neurogenesis and immune function. *Hum. Mol. Genet.* **17**, 690–709.

Jin, B., Yao, B., Li, J. L., Fields, C. R., Delmas, A. L., Liu, C., and Robertson, K. D. (2009). DNMT1 and DNMT3B modulate distinct polycomb-mediated histone modifications in colon cancer. *Cancer Res.* **69**, 7412–7421.

Kaminskas, E., Farrell, A., Abraham, S., Baird, A., Hsieh, L. S., Lee, S. L., Leighton, J. K., Patel, H., Rahman, A., Sridhara, R., *et al.* (2005). Approval summary: Azacitidine for treatment of myelo-dysplastic syndrome subtypes. *Clin. Cancer Res.* **11**, 3604–3608.

Kaneda, A., Takai, D., Kaminishi, M., Okochi, E., and Ushijima, T. (2003). Methylation-sensitive representational difference analysis and its application to cancer research. *Ann. N. Y. Acad. Sci.* **983**, 131–141.

Katoh, M. (2005). Epithelial-mesenchymal transition in gastric cancer (Review). *Int. J. Oncol.* **27,** 1677–1683.

Khulan, B., Thompson, R. F., Ye, K., Fazzari, M. J., Suzuki, M., Stasiek, E., Figueroa, M. E., Glass, J. L., Chen, Q., Montagna, C., *et al.* (2006). Comparative isoschizomer profiling of cytosine methylation: The HELP assay. *Genome Res.* **16,** 1046–1055.

Kim, J. S., Han, J., Shim, Y. M., Park, J., and Kim, D. H. (2005). Aberrant methylation of H-cadherin (CDH13) promoter is associated with tumor progression in primary nonsmall cell lung carcinoma. *Cancer* **104,** 1825–1833.

Kim, K. H., Choi, J. S., Kim, I. J., Ku, J. L., and Park, J. G. (2006). Promoter hypomethylation and reactivation of MAGE-A1 and MAGE-A3 genes in colorectal cancer cell lines and cancer tissues. *World J. Gastroenterol.* **12,** 5651–5657.

Klisovic, R. B., Stock, W., Cataland, S., Klisovic, M. I., Liu, S., Blum, W., Green, M., Odenike, O., Godley, L., Burgt, J. V., *et al.* (2008). A phase I biological study of MG98, an oligodeoxynucleotide antisense to DNA methyltransferase, in patients with high-risk myelodysplasia and acute myeloid leukemia. *Clin. Cancer Res.* **14,** 2444–2449.

Kondo, Y., Shen, L., Cheng, A. S., Ahmed, S., Boumber, Y., Charo, C., Yamochi, T., Urano, T., Furukawa, K., Kwabi-Addo, B., *et al.* (2008). Gene silencing in cancer by histone H3 lysine 27 trimethylation independent of promoter DNA methylation. *Nat. Genet.* **40,** 741–750.

Kriaucionis, S., and Heintz, N. (2009). The nuclear DNA base 5-hydroxymethylcytosine is present in Purkinje neurons and the brain. *Science* **324,** 929–930.

Lemaire, M., Momparler, L. F., Raynal, N. J., Bernstein, M. L., and Momparler, R. L. (2009). Inhibition of cytidine deaminase by zebularine enhances the antineoplastic action of 5-aza-2′-deoxycytidine. *Cancer Chemother. Pharmacol.* **63,** 411–416.

Li, E., Bestor, T. H., and Jaenisch, R. (1992). Targeted mutation of the DNA methyltransferase gene results in embryonic lethality. *Cell* **69,** 915–926.

Liang, G., Gonzales, F. A., Jones, P. A., Orntoft, T. F., and Thykjaer, T. (2002). Analysis of gene induction in human fibroblasts and bladder cancer cells exposed to the methylation inhibitor 5-aza-2′-deoxycytidine. *Cancer Res.* **62,** 961–966.

Lin, R. K., Hsu, H. S., Chang, J. W., Chen, C. Y., Chen, J. T., and Wang, Y. C. (2007). Alteration of DNA methyltransferases contributes to 5′CpG methylation and poor prognosis in lung cancer. *Lung Cancer* **55,** 205–213.

Lister, R., Pelizzola, M., Dowen, R. H., Hawkins, R. D., Hon, G., Tonti-Filippini, J., Nery, J. R., Lee, L., Ye, Z., Ngo, Q. M., *et al.* (2009). Human DNA methylomes at base resolution show widespread epigenomic differences. *Nature* **462,** 315–322.

Lopez-Serra, L., Ballestar, E., Fraga, M. F., Alaminos, M., Setien, F., and Esteller, M. (2006). A profile of methyl-CpG binding domain protein occupancy of hypermethylated promoter CpG islands of tumor suppressor genes in human cancer. *Cancer Res.* **66,** 8342–8346.

Lu, J., Getz, G., Miska, E. A., Alvarez-Saavedra, E., Lamb, J., Peck, D., Sweet-Cordero, A., Ebert, B. L., Mak, R. H., Ferrando, A. A., *et al.* (2005a). MicroRNA expression profiles classify human cancers. *Nature* **435,** 834–838.

Lu, Q., Wu, A., and Richardson, B. C. (2005b). Demethylation of the same promoter sequence increases CD70 expression in lupus T cells and T cells treated with lupus-inducing drugs. *J. Immunol.* **174,** 6212–6219.

Lujambio, A., and Esteller, M. (2009). How epigenetics can explain human metastasis: A new role for microRNAs. *Cell Cycle* **8,** 377–382.

Lujambio, A., Ropero, S., Ballestar, E., Fraga, M. F., Cerrato, C., Setien, F., Casado, S., Suarez-Gauthier, A., Sanchez-Cespedes, M., Git, A., *et al.* (2007). Genetic unmasking of an epigenetically silenced microRNA in human cancer cells. *Cancer Res.* **67,** 1424–1429.

Melo, S. A., Ropero, S., Moutinho, C., Aaltonen, L. A., Yamamoto, H., Calin, G. A., Rossi, S., Fernandez, A. F., Carneiro, F., Oliveira, C., et al. (2009). A TARBP2 mutation in human cancer impairs microRNA processing and DICER1 function. Nat. Genet. 41, 365–370.

Michie, A. M., McCaig, A. M., Nakagawa, R., and Vukovic, M. (2010). Death-associated protein kinase (DAPK) and signal transduction: Regulation in cancer. FEBS J. 277, 74–80.

Milicic, A., Harrison, L. A., Goodlad, R. A., Hardy, R. G., Nicholson, A. M., Presz, M., Sieber, O., Santander, S., Pringle, J. H., Mandir, N., et al. (2008). Ectopic expression of P-cadherin correlates with promoter hypomethylation early in colorectal carcinogenesis and enhanced intestinal crypt fission in vivo. Cancer Res. 68, 7760–7768.

Mulero-Navarro, S., and Esteller, M. (2008). Epigenetic biomarkers for human cancer: The time is now. Crit. Rev. Oncol. Hematol. 68, 1–11.

Ohm, J. E., McGarvey, K. M., Yu, X., Cheng, L., Schuebel, K. E., Cope, L., Mohammad, H. P., Chen, W., Daniel, V. C., Yu, W., et al. (2007). A stem cell-like chromatin pattern may predispose tumor suppressor genes to DNA hypermethylation and heritable silencing. Nat. Genet. 39, 237–242.

Okano, M., Bell, D. W., Haber, D. A., and Li, E. (1999). DNA methyltransferases Dnmt3a and Dnmt3b are essential for de novo methylation and mammalian development. Cell 99, 247–257.

Ostler, K. R., Davis, E. M., Payne, S. L., Gosalia, B. B., Exposito-Cespedes, J., Le Beau, M. M., and Godley, L. A. (2007). Cancer cells express aberrant DNMT3B transcripts encoding truncated proteins. Oncogene 26, 5553–5563.

Paredes, J., Albergaria, A., Oliveira, J. T., Jeronimo, C., Milanezi, F., and Schmitt, F. C. (2005). P-cadherin overexpression is an indicator of clinical outcome in invasive breast carcinomas and is associated with CDH3 promoter hypomethylation. Clin. Cancer Res. 11, 5869–5877.

Paz, M. F., Fraga, M. F., Avila, S., Guo, M., Pollan, M., Herman, J. G., and Esteller, M. (2003). A systematic profile of DNA methylation in human cancer cell lines. Cancer Res. 63, 1114–1121.

Pina, I. C., Gautschi, J. T., Wang, G. Y., Sanders, M. L., Schmitz, F. J., France, D., Cornell-Kennon, S., Sambucetti, L. C., Remiszewski, S. W., Perez, L. B., et al. (2003). Psammaplins from the sponge Pseudoceratina purpurea: Inhibition of both histone deacetylase and DNA methyltransferase. J. Org. Chem. 68, 3866–3873.

Plummer, R., Vidal, L., Griffin, M., Lesley, M., de Bono, J., Coulthard, S., Sludden, J., Siu, L. L., Chen, E. X., Oza, A. M., et al. (2009). Phase I study of MG98, an oligonucleotide antisense inhibitor of human DNA methyltransferase 1, given as a 7-day infusion in patients with advanced solid tumors. Clin. Cancer Res. 15, 3177–3183.

Prokhortchouk, A., Hendrich, B., Jorgensen, H., Ruzov, A., Wilm, M., Georgiev, G., Bird, A., and Prokhortchouk, E. (2001). The p120 catenin partner Kaiso is a DNA methylation-dependent transcriptional repressor. Genes Dev. 15, 1613–1618.

Ramirez-Carrozzi, V. R., Braas, D., Bhatt, D. M., Cheng, C. S., Hong, C., Doty, K. R., Black, J. C., Hoffmann, A., Carey, M., and Smale, S. T. (2009). A unifying model for the selective regulation of inducible transcription by CpG islands and nucleosome remodeling. Cell 138, 114–128.

Ramsahoye, B. H., Biniszkiewicz, D., Lyko, F., Clark, V., Bird, A. P., and Jaenisch, R. (2000). Non-CpG methylation is prevalent in embryonic stem cells and may be mediated by DNA methyltransferase 3a. Proc. Natl. Acad. Sci. USA 97, 5237–5242.

Ribeiro, A. S., Albergaria, A., Sousa, B., Correia, A. L., Bracke, M., Seruca, R., Schmitt, F. C., and Paredes, J. (2010). Extracellular cleavage and shedding of P-cadherin: A mechanism underlying the invasive behaviour of breast cancer cells. Oncogene 29, 392–402.

Rodriguez, C., Borgel, J., Court, F., Cathala, G., Forne, T., and Piette, J. (2010). CTCF is a DNA methylation-sensitive positive regulator of the INK/ARF locus. Biochem. Biophys. Res. Commun. 392, 129–134.

Roll, J. D., Rivenbark, A. G., Jones, W. D., and Coleman, W. B. (2008). DNMT3b overexpression contributes to a hypermethylator phenotype in human breast cancer cell lines. Mol. Cancer 7, 15.

Roman-Gomez, J., Jimenez-Velasco, A., Agirre, X., Cervantes, F., Sanchez, J., Garate, L., Barrios, M., Castillejo, J. A., Navarro, G., Colomer, D., *et al.* (2005). Promoter hypomethylation of the LINE-1 retrotransposable elements activates sense/antisense transcription and marks the progression of chronic myeloid leukemia. *Oncogene* **24,** 7213–7223.

Rush, L. J., and Plass, C. (2002). Restriction landmark genomic scanning for DNA methylation in cancer: Past, present, and future applications. *Anal. Biochem.* **307,** 191–201.

Saito, Y., Kanai, Y., Sakamoto, M., Saito, H., Ishii, H., and Hirohashi, S. (2002). Overexpression of a splice variant of DNA methyltransferase 3b, DNMT3b4, associated with DNA hypomethylation on pericentromeric satellite regions during human hepatocarcinogenesis. *Proc. Natl. Acad. Sci. USA* **99,** 10060–10065.

Saito, Y., Liang, G., Egger, G., Friedman, J. M., Chuang, J. C., Coetzee, G. A., and Jones, P. A. (2006). Specific activation of microRNA-127 with downregulation of the proto-oncogene BCL6 by chromatin-modifying drugs in human cancer cells. *Cancer Cell* **9,** 435–443.

Saxonov, S., Berg, P., and Brutlag, D. L. (2006). A genome-wide analysis of CpG dinucleotides in the human genome distinguishes two distinct classes of promoters. *Proc. Natl. Acad. Sci. USA* **103,** 1412–1417.

Schlesinger, Y., Straussman, R., Keshet, I., Farkash, S., Hecht, M., Zimmerman, J., Eden, E., Yakhini, Z., Ben-Shushan, E., Reubinoff, B. E., *et al.* (2007). Polycomb-mediated methylation on Lys27 of histone H3 pre-marks genes for de novo methylation in cancer. *Nat. Genet.* **39,** 232–236.

Sharma, G., Mirza, S., Parshad, R., Srivastava, A., Datta Gupta, S., Pandya, P., and Ralhan, R. (2010). CpG hypomethylation of MDR1 gene in tumor and serum of invasive ductal breast carcinoma patients. *Clin. Biochem.* **43,** 373–379.

Shirahata, A., Sakata, M., Sakuraba, K., Goto, T., Mizukami, H., Saito, M., Ishibashi, K., Kigawa, G., Nemoto, H., Sanada, Y., *et al.* (2009). Vimentin methylation as a marker for advanced colorectal carcinoma. *Anticancer Res.* **29,** 279–281.

Simon, J. A., and Lange, C. A. (2008). Roles of the EZH2 histone methyltransferase in cancer epigenetics. *Mutat. Res.* **647,** 21–29.

Steensma, D. P. (2009). Decitabine treatment of patients with higher-risk myelodysplastic syndromes. *Leuk. Res.* **33**(Suppl. 2), S12–S17.

Tahiliani, M., Koh, K. P., Shen, Y., Pastor, W. A., Bandukwala, H., Brudno, Y., Agarwal, S., Iyer, L. M., Liu, D. R., Aravind, L., *et al.* (2009). Conversion of 5-methylcytosine to 5-hydroxymethylcytosine in mammalian DNA by MLL partner TET1. *Science* **324,** 930–935.

Takacs, M., Banati, F., Koroknai, A., Segesdi, J., Salamon, D., Wolf, H., Niller, H. H., and Minarovits, J. (2010). Epigenetic regulation of latent Epstein-Barr virus promoters. *Biochim. Biophys. Acta* **1799,** 228–235.

Tamaru, H., and Selker, E. U. (2001). A histone H3 methyltransferase controls DNA methylation in *Neurospora crassa*. *Nature* **414,** 277–283.

Uhlmann, K., Brinckmann, A., Toliat, M. R., Ritter, H., and Nurnberg, P. (2002). Evaluation of a potential epigenetic biomarker by quantitative methyl-single nucleotide polymorphism analysis. *Electrophoresis* **23,** 4072–4079.

Unoki, M., Nishidate, T., and Nakamura, Y. (2004). ICBP90, an E2F-1 target, recruits HDAC1 and binds to methyl-CpG through its SRA domain. *Oncogene* **23,** 7601–7610.

Veerla, S., Panagopoulos, I., Jin, Y., Lindgren, D., and Hoglund, M. (2008). Promoter analysis of epigenetically controlled genes in bladder cancer. *Genes Chromosomes Cancer* **47,** 368–378.

Vire, E., Brenner, C., Deplus, R., Blanchon, L., Fraga, M., Didelot, C., Morey, L., Van Eynde, A., Bernard, D., Vanderwinden, J. M., *et al.* (2006). The Polycomb group protein EZH2 directly controls DNA methylation. *Nature* **439,** 871–874.

Weber, J., Salgaller, M., Samid, D., Johnson, B., Herlyn, M., Lassam, N., Treisman, J., and Rosenberg, S. A. (1994). Expression of the MAGE-1 tumor antigen is up-regulated by the demethylating agent 5-aza-2'-deoxycytidine. *Cancer Res.* **54,** 1766–1771.

Weber, M., Davies, J. J., Wittig, D., Oakeley, E. J., Haase, M., Lam, W. L., and Schubeler, D. (2005). Chromosome-wide and promoter-specific analyses identify sites of differential DNA methylation in normal and transformed human cells. *Nat. Genet.* **37,** 853–862.

Weller, M., Stupp, R., Reifenberger, G., Brandes, A. A., van den Bent, M. J., Wick, W., and Hegi, M. E. (2010). MGMT promoter methylation in malignant gliomas: Ready for personalized medicine? *Nat. Rev. Neurol.* **6,** 39–51.

Widschwendter, M., Jiang, G., Woods, C., Muller, H. M., Fiegl, H., Goebel, G., Marth, C., Muller-Holzner, E., Zeimet, A. G., Laird, P. W., *et al.* (2004a). DNA hypomethylation and ovarian cancer biology. *Cancer Res.* **64,** 4472–4480.

Widschwendter, M., Siegmund, K. D., Muller, H. M., Fiegl, H., Marth, C., Muller-Holzner, E., Jones, P. A., and Laird, P. W. (2004b). Association of breast cancer DNA methylation profiles with hormone receptor status and response to tamoxifen. *Cancer Res.* **64,** 3807–3813.

Widschwendter, M., Fiegl, H., Egle, D., Mueller-Holzner, E., Spizzo, G., Marth, C., Weisenberger, D. J., Campan, M., Young, J., Jacobs, I., *et al.* (2007). Epigenetic stem cell signature in cancer. *Nat. Genet.* **39,** 157–158.

Wong, I. H. (2001). Methylation profiling of human cancers in blood: Molecular monitoring and prognostication (review). *Int. J. Oncol.* **19,** 1319–1324.

Yanagawa, N., Tamura, G., Honda, T., Endoh, M., Nishizuka, S., and Motoyama, T. (2004). Demethylation of the synuclein gamma gene CpG island in primary gastric cancers and gastric cancer cell lines. *Clin. Cancer Res.* **10,** 2447–2451.

Yoo, C. B., and Jones, P. A. (2006). Epigenetic therapy of cancer: Past, present and future. *Nat. Rev. Drug Discov.* **5,** 37–50.

Zhou, L., Cheng, X., Connolly, B. A., Dickman, M. J., Hurd, P. J., and Hornby, D. P. (2002). Zebularine: A novel DNA methylation inhibitor that forms a covalent complex with DNA methyltransferases. *J. Mol. Biol.* **321,** 591–599.

3

Histone Modifications and Cancer

Carla Sawan and Zdenko Herceg

Epigenetics Group, International Agency for Research on Cancer (IARC), 69008 Lyon, France

Advances in Genetics, Vol. 70
Copyright 2010, Elsevier Inc. All rights reserved.

0065-2660/10 $35.00
DOI: 10.1016/S0065-2660(10)70003-1

ABSTRACT

It is now widely recognized that epigenetic events are important mechanisms underlying cancer development and progression. Epigenetic information in chromatin includes covalent modifications (such as acetylation, methylation, phosphorylation, and ubiquitination) of core nucleosomal proteins (histones). A recent progress in the field of histone modifications and chromatin research has tremendously enhanced our understanding of the mechanisms underlying the control of key physiological and pathological processes. Histone modifications and other epigenetic mechanisms appear to work together in establishing and maintaining gene activity states, thus regulating a wide range of cellular processes. Different histone modifications themselves act in a coordinated and orderly fashion to regulate cellular processes such as gene transcription, DNA replication, and DNA repair. Interest in histone modifications has further grown over the last decade with the discovery and characterization of a large number of histone-modifying molecules and protein complexes. Alterations in the function of histone-modifying complexes are believed to disrupt the pattern and levels of histone marks and consequently deregulate the control of chromatin-based processes, ultimately leading to oncogenic transformation and the development of cancer. Consistent with this notion, aberrant patterns of histone modifications have been associated with a large number of human malignancies. In this chapter, we discuss recent advances in our understanding of the mechanisms controlling the establishment and maintenance of histone marks and how disruptions of these chromatin-based mechanisms contribute to tumorigenesis. We also suggest how these advances may facilitate the development of novel strategies to prevent, diagnose, and treat human malignancies. © 2010, Elsevier Inc.

I. INTRODUCTION

Genomic DNA is packaged into a highly compacted DNA–protein complex, called chromatin, which imposes constraints on cellular processes (such as gene transcription, DNA replication, and DNA repair) that use chromatin DNA as a template. To deal with an inaccessible chromatin structure, eukaryotic cells have evolved mechanisms known as chromatin modifications (histone markings) that are believed to facilitate opening of chromatin through reversible modifications of core histones (Grant, 2001). Covalent modification of conserved residues in core histones by acetylation, phosphorylation, methylation, and ubiquitination is a reversible posttranslational modification thought to be an important mechanism by which cells regulate chromatin accessibility and function of chromatin DNA.

It is now accepted that histone modifications and histone-modifying complexes play important roles in critical cellular processes and cell fates. Histone modifications act in a coordinated and orderly fashion and directly affect chromatin structure and function. Consequently, they influence chromatin-based processes, including gene transcription, DNA repair, and DNA replication (Kouzarides, 2007). Therefore, a remarkable progress in the field of histone modifications and chromatin research has enhanced our understanding of epigenetic mechanisms in the control of cellular processes.

Moreover, patterns of histone marks are profoundly altered in cancer cells (Sharma *et al.*, 2010). Epigenetic deregulation involving histone-modifying complexes and histone marks may be an important mechanism underlying the development and progression of the disease. These mechanisms may contribute to oncogenesis through deregulation of gene transcription and DNA repair. Furthermore, accumulating evidence argue that different types and subtypes of cancer may have specific patterns of histone modifications (histone modification signatures), a phenomenon that can be exploited in biomarker discovery. Finally, since histone modifications are reversible, they represent potential targets for cancer therapy and prevention.

The aim of this chapter is to review the recent progress in the field of histone modifications and chromatin research that has enhanced our understanding of epigenetic mechanisms in the control of cellular processes and abnormal events involved in tumorigenesis. We also discuss the implications of these advances on the development of new approaches for diagnosis, treatment, and prevention of cancer.

II. CHROMATIN ORGANIZATION

In the nucleus of eukaryotic cells, DNA is highly compacted and organized into a condensed structure, called chromatin, by both histone and nonhistone proteins. Chromatin is often divided into two distinct functional forms: heterochromatin, a condensed and highly repressive form and euchromatin, a more "relaxed" form that can be generally permissive for transcription. The basic unit of the chromatin is the nucleosome which is formed of 147 base pairs of DNA wrapped 1.6 times around an octamer composed of two H3–H4 histone dimers bridged together as a stable tetramer that is flanked by two separate H2A–H2B dimers (Fig. 3.1). The addition of other factors, such as linker histone H1 and nonhistone chromatin proteins, results in higher order chromatin organization and compaction (Fig. 3.1; Clausell *et al.*, 2009; Eickbush and Moudrianakis, 1978; Luger *et al.*, 1997). These core histone molecules are among the most evolutionarily conserved proteins highlighting the likelihood of the critical functions of these small proteins. Histone proteins are composed of a structured globular

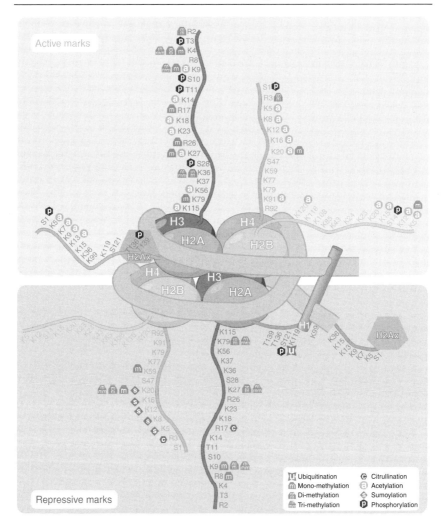

Figure 3.1. Schematic representation of the nucleosome. DNA is wrapped around histone octamer of the four core histones H2A, H2B, H3, and H4. Histone H1, the linker protein, is bound to DNA between nucleosomes. Different amino acids constituting histone tails are represented along with the different covalent modification specific of each residue. Active marks are represented in the upper part of the figure and repressive marks are represented in the lower part of the figure. Lysine (K), arginine (R), serine (S), and threonine (T).

central domain by which they interact together and that is in close contact to DNA and a less structured amino-terminal domain (Eickbush and Moudrianakis, 1978; Luger *et al.*, 1997). Each of the four core histones contains 20–30 amino acids in their N-terminal domain, called histone "tail." Histone H2A is unique in having an additional 37 amino acids carboxy-terminal domain that protrudes from the nucleosome. Histone tails extend from the surface of the nucleosome and include the majority of sites for posttranslational modifications (Fig. 3.1). Different histone modifications have been identified, including acetylation, phosphorylation, methylation, ubiquitilation, and sumoylation. Combination of different histone modifications in the globular and N-terminal domains dictates the status of a gene (active or repressed transcription) through dynamic chromatin structure (Goll and Bestor, 2002; Grant, 2001), thus regulating different cellular processes such as transcription, DNA replication, and DNA repair.

III. HISTONE MODIFICATIONS

Histone modifications are posttranslational covalent additions to the N-terminal and C-terminal of histone tails, although examples of modifications within the globular domain have been identified. Histones are currently known to be subjected to nine different types of posttranslational modifications including acetylation, methylation, phosphorylation, ubiquitination, sumoylation, ADP-ribosylation, deimination, proline isomerization, and the newly identified propionylation (Kouzarides, 2007; Liu et al., 2009). Histone modifications are further enriched by the fact that lysine residues can accept mono-, di-, and tri-methyl groups and that arginine residues can be mono- or di-methylated. There are more than 60 residues of histones where modifications have been detected (Kouzarides, 2007) and given the number of new modification sites that are identified every year; this list is likely to further grow in years to come. Histone modifications regulate and control chromatin structure with different modifications yielding distinct functional impact on a variety of DNA-based processes (Table 3.1). Moreover, numerous recent reports have shown the presence of site-specific combinations and interdependence of different histone modifications form the so-called "histone code". Different combinations of histone modifications may result in different biological outcomes. The role of histone modifications and their crosstalk in different cellular processes will be discussed in the following sections.

A. Acetylation

Acetylation is a reaction based on the introduction of an acetyl functional group into a chemical compound. It is a reaction involving the hydrogen atom of a hydroxyl group with an acetyl group (CH_3CO). Lysine acetylation refers to the

Table 3.1. Different Histone Modifications on Different Histone Residues

Modification	Histone	Residue	Oganism	Enzyme	Possible role
Acetylation	H2A	K4	Y	Esa1	Ta
		K5	M	Tip60, Hat1, P300/CBP	Ta, R
		K7	Y	Hat1, Esa1	Ta
	H2B	K5	M	ATF2	Ta
		K11	Y	Gcn5	Ta
		K12	M	ATF2, P300/CBP	Ta
		K15	M	ATF2, P300/CBP	
		K16	Y	Gcn5, Esa1	Ta
		K20	M	P300	Ta
	H3	K4	Y	Esa1, Hpa2	
		K9	M/Y	Gcn5, SRC-1	Ta, R
		K14	M/Y	Gcn5, PCAF, Tip60, SRC-1, hTFIIIC90, TAF1, p300/ Gcn5, Esa1, Elp3, Hpa2, TAF1, Sas2, Sas3	Ta, R, Rep
		K18	M/Y	P300, CBP/Gcn5 (SAGA)	Ta, R
		K23	M/Y	P300, CBP/Gcn5 (SAGA), Sas3	Ta, R
		K27	M	Gcn5	Ta, R
	H4	K5	M/Y	Hat1, Tip60, ATF2, p300/Hat1, Esa1, Hpa2	Ta, R, Rep
		K8	M/Y	Gcn5, PCAF, Tip60, ATF2, p300/Esa1, Elp3	Ta, R, Rep
		K12	M/Y	Hat1, Tip60/Hat1, Esa1, Hpa2	Ta, R, Rep
		K16	M/Y	MOF, Gcn5, Tip60, ATF2/Gcn5, Esa1, Sas2	Ta, R
Methylation	H1	K26	M	EZH2	Tr
	H3	R2	M	CARM1	Ta
		K4	M/Y	MLL4, SET1, MLL1, SET7/9, MYD3/Set1	Ta
		R8	M	PMRT5	Tr
		K9	M/Y	SUV39h1, SUV39h2, ESET, G9A, EZH2, Eu-HMTase1/ Clr4, S.p. Clr4	Ta, Tr
		R17	M	CARM1	Ta
		R26	M	CARM1	Ta
		K27	M	EZH2, G9A	Ta, Tr
		K36	M/Y	HYPB, NSD1/Set2, S.c.	Ta
		K79	M/Y	DOT1L/S.c. Dot-1	Ta, Tr, R
	H4	R3	M	PRMT1, PRMT5	Ta
		K20	M/Y	PR-SET7, SUV4-20/SET9	Ta, Tr, R

Table 3.1. (*Continued*)

Modification	Histone	Residue	Oganism	Enzyme	Possible role
Phosphorylation	H2AX	S129	Y	S.c Mec1,Tel1	R
		S139	M	ATM, ATR, DNA-PK	R
	H2A	T119	M	NHK-1	Tr, R
	H2B	S10	Y	Ste20	R
		S14	M	Mst1	R
	H3	S10	M/Y	TG2, Aurora B, MSK1, MSK2/Snf1	Ta
		T11	M	Dlk/ZIP	
		S28	M	MSK1, MSK2	
	H4	S1	Y	CK1	R
Ubiquitination	H2A	K119	M	Ring 1b	Ta, R
	H2B	K120	M	RNF 20/40	Ta, R
		K123	Y	Rad6	Ta, R

Lysine (K), arginine (R), serine (S), threonine (T) in mammalian cells (M), and yeast (Y) carried out by different enzymes and regulating different cellular processes such as transcriptional activation (Ta), transcriptional repression (Tr), DNA repair (R), and DNA replication (Rep).

ε-amino group of a lysine residue. It is a reversible modification referred to as N^{ε}-acetylation that is different from the N^{α}-acetylation, an acetylation of the N-terminal α-amine of proteins, a widespread modification in eukaryotes.

In 1964, Allfrey first discovered that histone proteins exist in an acetylated form (Allfrey *et al.*, 1964). Few years later, Gershey defined histone acetylation as an N^{ε}-acetylation. Lysine acetylation is controlled by two types of enzymes, histone acetyltransferases (HATs) which use acetyl-CoA, that is specifically recognized and bound by the Arg/Gln-X-X-Gly-X-Gly/Ala segment of HATs, to transfer an acetyl group to the ε-amino groups on the N-terminal tails of histones (Loidl, 1994), and histone deacetylases (HDACs) which reverse this modification (Brownell and Allis, 1995; Rundlett *et al.*, 1996; Walkinshaw *et al.*, 2008). Soon after the discovery of the first HATs and HDACs in the mid-1990s, identification of a large number of HATs followed, resulting in a renowned interest in histone acetylation. Most HATs come in the form of large multiprotein complexes. Different components of HAT complexes ensure locus targeting and chromosomal-domain and substrate-specificity. Based on sequence similarity, HATs can be organized into families, which seem to display different mechanisms of histone substrate binding and catalysis (Lee and Workman, 2007). The dynamic equilibrium of lysine acetylation *in vivo* is governed by the opposing actions of HATs and HDACs. Similar to acetyltransferases, the HDACs are also part of large multiprotein complexes (Ekwall, 2005).

Within a histone octamer, positively charged histone tails protrude out from the central domain of the nucleosome and are believed to bind the negatively charged DNA through charge interactions or mediate interactions between nucleosomes contributing in chromatin compaction (Fletcher and Hansen, 1995; Luger et al., 1997). Lysine acetylation is believed to neutralize the positive charge of histone tail, weakening histone–DNA (Horiuchi et al., 1995; Steger and Workman, 1996) or nucleosome–nucleosome interactions (Fletcher and Hansen, 1996; Luger and Richmond, 1998), inducing a conformational change (Norton et al., 1989). This results in destabilizing nucleosome and chromatin structure, thus facilitating access to the DNA for different nuclear factors, such as the transcription complex. In agreement with this model, hyperacetylation of histones is now considered as a hallmark of transcriptionally active chromatin. Deacetylation of histones by HDACs results in a decrease in the space between the nucleosome and the DNA, leading to a closed (heterochromatin-like) chromatin conformation that diminishes accessibility for transcription factors.

B. Methylation

Protein methylation is a covalent posttranslational modification commonly occurring on carboxyl groups of glutamate, leucine, and isoprenylated cysteine or on the side-chain nitrogen atoms of lysine, arginine, and histidine residues (Clarke, 1993). Described in 1964, histones have long been known to be substrates for methylation (Murray, 1964). Histone methylation occurs on the side-chain nitrogen atoms of lysine and arginine of histones. The most heavily methylated histone is histone H3, followed by histone H4.

Arginine can be either mono- or di-methylated, with the latter in symmetric or asymmetric configuration (Gary and Clarke, 1998). Protein arginine methyltransferases (PRMTs) are the enzymes that catalyze arginine methylation. PRMTs share a conserved catalytic core but are very different on the N- and C- terminal regions which are likely to determine the substrate specificity (Weiss et al., 2000). There are two types of PRMTs: the type I enzymes that catalyze mono- and asymmetric di-methylation of arginine and the type II enzymes that catalyze mono- and symmetric di-methylation of arginine (Gary and Clarke, 1998). Several studies suggested that arginine methylation may play a role in expression of the genes involved in tumor suppression (Pal et al., 2004; Singh et al., 2004).

Similar to arginine methylation, lysine methylation can occur in mono- (me), di- (2me), and tri- (3me) methylated forms. Some of the lysine residues that are methylated in histones H3 and H4 are also found to be substrates for acetylation. The enzymes that catalyze methylation on lysine residues have been grouped into two classes: the lysine-specific SET domain-containing histone methyltransferases (HMTs) that share a strong homology in a 140-amino acid catalytic domain known as the SET (Su(var), Enhancer of Zeste, and Trithorax) domain and the non-SET containing lysine HMTs. It is important to note that not all SET domain-containing proteins are HMTs nor are all the HMT activities are mediated by SET domains (Feng et al., 2002; Ng et al., 2002).

Lysine methylation is extremely diverse in its consequences. Depending upon a particular lysine, methylation may serve as a marker of transcriptionally active euchromatin or transcriptionally repressed heterochromatin (Table 3.1; Sims et al., 2003). For instance, histone H3 K9, H4 K20, and H3 K27 methylation are mainly involved in formation of heterochromatin. On the other hand, histone H3 K4, H3 K36, and H3 K79 methylation correlate with euchromatin (Fig. 3.1). Interestingly, a strict equilibrium is needed between histone H3 K4 and H3 K27 methylation, which are respectively an activator and a repressor of transcription, for the maintenance of "stemness" transcription patterns to maintain pluripotency of embryonic stem cells (Bernstein et al., 2006). Moreover, it seems that histone H3 clipping, a mechanism of cleavage of the 21 amino acids of histone tails following induction of gene transcription and histone eviction, occurs on histone tails that carry repressive histone marks (Santos-Rosa et al., 2009).

Until very recently, the dogma was that methylation is an irreversible process. With the identification of the first lysine demethylase LSD1 (lysine-specific demethylase 1) in 2004, our view of histone methylation regulation changed to a much more dynamic view of histone methylation, opening the way for the identification of numerous histone demethylases (Shi et al., 2004). LSD1 demethylates both mono- and di-methylated K4 on histone H3 (Shi et al., 2004). In 2006, the first jumonji-domain-containing demethylase, JHDM1A, was identified as a mono- and di-methyl histone H3 K36 (Tsukada et al., 2006). The jumonji (JmjC)-domain-containing proteins belong to the dioxygenase superfamily and use a demethylation mechanism distinct from that of LSD1/KDM1 (Anand and Marmorstein, 2007). These enzymes can demethylate tri-methylated lysine residues. The JMJD2/KDM4 demethylases are a tri-methyl demethylase family that were reported soon after the first JHDM1A was discovered (Whetstine et al., 2006). Over the past few years, a series of studies identified additional jumonji-domain-containing families that have me K4, me K9, me K27, and me K36 as their substrate. Moreover, one histone arginine demethylase, namely JMJD6, was recently identified and found to belong to JMD2 family (Chang et al., 2007). Despite the tremendous and exciting progress in the last few years, the field of histone demethylases is still in its early days and we only have sporadic knowledge of the biological role of these enzymes.

C. Phosphorylation

Phosphorylation is the addition of a phosphate (PO_4) group to a protein or other organic molecule. It is a reversible posttranslational modification that usually occurs on serine, threonine, and tyrosine residues in eukaryotic proteins. Histone phosphorylation was first observed in the late-1960s (Gutierrez and Hnilica, 1967; Kleinsmith *et al.*, 1966), and the first kinase responsible for phosphorylating histone tail was shown to be an AMP-dependent kinase (Langan, 1968). On the other hand, phosphatases mediate the removal of the phosphate group. Phosphorylation adds a single negative charge to the histone tail, thus altering chromatin structure and influencing transcription by facilitating interaction between transcription factors and other chromatin components. Histones H1, H2A, H2B, H3, and H4 have all been identified as being susceptible to phosphorylation at serine or threonine residues although histone H3 phosphorylation is the most studied and today there are several well-characterized and conserved posphorylated H3 residues: T3, S10, T11, S28, and T45 (Ahn *et al.*, 2005; Baker *et al.*, 2010; Foster and Downs, 2005; Krishnamoorthy *et al.*, 2006; Nowak and Corces, 2004; Thiriet and Hayes, 2009). The distinct phosphorylation pattern of histones has been linked to different cellular processes such as transcription, mitosis, DNA repair, apoptosis, and chromosome condensation (Cheung *et al.*, 2000a). Several distinct kinases are required for the phosphorylation of histones on different residues (Table 3.1). Phosphorylation of histone H2AX is induced by a DNA damage signaling pathway, and this modification is dependent on phosphatidylinositol (PI) 3-kinase-related kinase (PIKK), such as Mec1 in yeast and ATM, ATR, and DNA-PK in mammals (Foster and Downs, 2005). Histone H2B S14 phosphorylation plays a role in apoptosis and is catalyzed by the Mst1 (mammalian sterile-20-like kinase) (Ahn *et al.*, 2005). Phosphorylation of histones H3 S10 and H3 S28 is known to be involved in transcriptional activation during mitosis and is regulated by the Aurora kinases, which are highly conserved from yeast to human (Nowak and Corces, 2004). Other kinases in the MSK/RSK/Jil-1 family can also mediate phosphorylation of histone H3 S10 to regulate gene expression (Nowak and Corces, 2004). Recently, phosphorylation of H3 T45 mediated by the S phase kinase CdC7-Dbf4 in *Saccharomyces cerevisiae* has been identified as a replication-associated histone modification in budding yeast (Baker *et al.*, 2010). On the other hand, protein phosphatase 2 A (PP2A) is a phosphatase that dephosphorylates phosphorylated H3 S10 and results in transcription inhibition (Nowak and Corces, 2004).

While acetylation and methylation are histone marks that are recognized by proteins that have a binding domain, like chromo- and bromodomains that recognize, respectively, acetyl and methyl marks, the domains recognizing the phospho-mark are still unknown. Therefore, much remains to be done to establish signaling pathways and events that are downstream of histone phosphorylation.

D. Ubiquitination

Ubiquitin is a 76 amino acids protein that is highly conserved in eukaryotes. Ubiquitination is a posttranslational modification that involves the attachment of ubiquitin molecule(s) to proteins through formation of an isopeptide bound between the ubiquitin C-terminus and the side chain lysine of target proteins. The protein ubiquitination reaction involves three sequential separate enzymatic activities: E1 (activating), E2 (conjugating), and E3 (ligase) enzymes (Pickart and Eddins, 2004). Substrates can be mono- or poly-ubiquitinated, whereas poly-ubiquitination targets proteins for degradation via the 26S proteasome, mono-ubiquitination generally acts as a tag that marks the substrate protein for a particular function. In vivo, histones H1, H2A, H2B, and H3 can all be ubiquitinated at lysine residues; however, H2A and H2B ubiquitination is most common (Zhang, 2003). Histone H2A ubiquitination was mapped to the highly conserved residue Lys 119 and affects between 5% and 15% of the available H2A; it is a relatively abundant modification, whereas ubiquitination of H2B is less abundant (1–2%) (Goldknopf and Busch, 1975). Typically, histone ubiquitination is associated with active and "open" chromatin; however, histone ubiquitination has been linked with both transcriptional activation and silencing (Zhang, 2003). Ubiquitinated histones were believed to affect gene transcription by altering higher order chromatin structure and exposing DNA due to their size. Histones marked by ubiquitin are also known to behave as key signals for subsequent histone modifications by recruiting additional regulatory molecules (Briggs et al., 2002). The main function of histone ubiquitination seems to be transcriptional control. Ubiquitinated H2A plays an important role in transcriptional activation and several active genes have been shown to contain a high percentage of ubiquitinated H2A (Barsoum and Varshavsky, 1985; Levinger and Varshavsky, 1982; Nickel et al., 1989). Surprisingly, ubiquitinated H2A was also linked to the inhibition of transcription (Dawson et al., 1991; Nickel et al., 1989; Parlow et al., 1990). Ubiquitination of histone H2B was also linked to activation and inhibition of transcription (Zhang, 2003). Consequently, it appears that histone ubiquitination regulates transcription in a positive or a negative manner depending on genomic context. Histone ubiquitination is also believed to affect other histone modifications, especially acetylation and methylation. For example, HDAC6 was shown to bind ubiquitin through its zinc-finger domain. Histone H3 K4 and H3 K79 methylation were shown to be dependent on Rad6-mediated H2B K123 ubiquitination (Briggs et al., 2002). The effect of ubiquitination on histone acetylation and methylation can explain its role in both activation and inhibition of transcription. For instance, it has been proposed that ubiquitination of H2B occurs mostly in euchromatin leading to histone H3 K4 and H3 K79 methylation, which would prevent Sir proteins from association with active euchromatic regions, thereby restricting Sir proteins to heterochromatic regions to mediate silencing. At the same time in euchromatin, the ubiquitination would activate the transcription by methylating histone H3 K4 and by facilitating the transcriptional elongation (Ng et al., 2003).

Ubiquitin proteases are the enzymes that mediate histone deubiquitination, thus controlling transcriptional regulation. Interestingly, sequential ubiquitination and deubiquitination are both involved in gene activation (Henry et al.,

2003). However, following Rad6-mediated H2B K123 monoubiquitination and the subsequent methylation of histone H3, Ubp8 is recruited to histone H2B to then remove the ubiquitin moiety, a necessary step prior to transcriptional initiation (Henry et al., 2003). Furthermore, Ubp8 mutations result in increased levels of H2B ubiquitination and decreased levels of SAGA-regulated transcription (Henry et al., 2003).

E. Histone code

All histone modifications described above are considered as heritable epigenetic marks since they are maintained with high fidelity through cell division. Over the past few years, the field of epigenetics provided numerous evidence arguing that histone modifications act in a combinatorial and consistent manner, leading to the concept of "histone code" (Jenuwein and Allis, 2001; Strahl and Allis, 2000). Different histone modifications present on histone tails generate a "code" read by different cellular machineries and dictating different cellular outcomes (Jenuwein and Allis, 2001). Therefore, the combinatorial nature of different histone marks adds a layer of complexity in recruiting epigenetic modifiers and regulating cellular processes (Fig. 3.2). Moreover, the histone code predicts a chronology in establishment of a specific modification pattern. For instance, Msk1/2-mediated H3 S10 phosphorylation enhances binding of GCN5 leading to acetylation of histone H3 K14 and methylation of histone H3 K4, and further inhibits histone H3 K9 methylation resulting in open chromatin conformation (Lo et al., 2000; Rea et al., 2000). Moreover, phosphorylation of H3 S10 favors H3 K9 acetylation since Aurora-B kinase can only bind unmodified or acetylated histone H3 K9, thus preventing SUV39H1 binding and histone H3 K9 methylation (Hirota et al., 2005). On the other hand, histone H3 K9 methylation inhibits H3 S10 phosphorylation and represses gene transcription (Rea et al., 2000). Recently, phosphorylation of histone H3 T6 by protein kinase C beta I was shown to be a major event in preventing LSD1 from demethylating histone H3 K4 during androgen receptor dependent gene activation (Metzger et al., 2010).

　　　Histone modification crosstalk can occur between different histones. Histone H3 K4 and H3 K79 methylation that are involved in transcriptional activation depends on and are regulated by histone H2B K123 ubiquitination (Sun and Allis, 2002). The combination between a specific histone modifications is due to the targeting of histone-modifying enzymes to a specific residue and the specificity of these enzymes for their target substrate. Indeed, epigenetic marks on histone tails provide binding sites for specific domains of effector proteins (Jenuwein and Allis, 2001). For instance, bromodomains target proteins to acetylated residues, whereas chromodomains target proteins for methylation

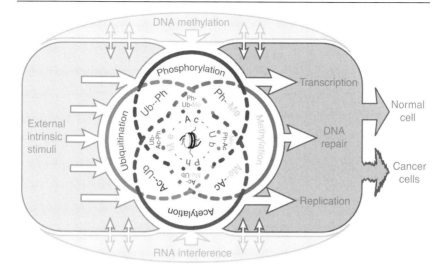

Figure 3.2. External and intrinsic stimuli influence different histone modifications: acetylation (Ac), methylation (Me), phosphorylation (Ph), and ubiquitination (Ub), and the crosstalk between each other or with other epigenetic mechanisms such as DNA methylation and RNA interference. These epigenetic modifications tightly regulate chromatin-based cellular processes such as transcription, DNA repair, and DNA replication to safeguard cells from external and internal stress. Any deregulation in these histone modifications or in their crosstalk will lead to abnormal regulation of chromatin-based processes which may promote oncogenic transformation and cancer development.

marks (Bannister *et al.*, 2001; Owen *et al.*, 2000). Together, the combinatorial and sequential modification of histone tail are a promising field of research that should give a better understanding of different cellular processes.

IV. ROLE OF HISTONE MODIFICATIONS IN CELLULAR PROCESSES

Covalent modifications on histone tails are now established as key regulators of chromatin-based processes. This section deals with the role of different histone modifications in the regulation and coordination of transcription, DNA repair, and DNA replication.

A. Transcription

In response to different stimuli, the regulation of gene expression in eukaryotes requires chromatin modifiers and specific histone modifications that bring about an "open" permissive chromatin. These epigenetic modifiers facilitate and open the way for transcription factors to the DNA and activate a cascade of events resulting in gene transcription. On the other hand, histone modifications can also promote transcriptional repression by inducing a condensed chromatin state and closing DNA accessibility to transcription factors. Thus, histone modifications dictate whether the chromatin state is permissive for transcription (Fig. 3.2).

Histone lysine acetylation has long been correlated with transcriptional activation. Indeed, through decreasing histone charge, acetylation is believed to weaken histone–DNA interaction, thus relaxing the chromatin structure and opening the way for transcription machinery (Workman and Kingston, 1998). Moreover, acetylated histones may serve as docking sites for recruitment of other transcription regulators influencing other histone modifications (de la Cruz et al., 2005; Hassan et al., 2002; Jenuwein and Allis, 2001). Histone acetylation influences gene transcription at two levels, while global histone acetylation correlates with general transcriptional activity (Gottesfeld and Forbes, 1997), specific promoter acetylation controls the activity of corresponding genes. However, specific promoter acetylation occurs in a context of global acetylation and deacetylation that regulates basal transcription levels to facilitate a rapid transcriptional repression (Bulger, 2005; Vogelauer et al., 2000). Reversal of acetylation correlates with transcriptional repression and the enzymes that carry out histone deacetylation are present in numerous repressive complexes (Vogelauer et al., 2000).

Histone methylation plays two different roles in gene transcription depending on the residue and histone modified: arginine methylation of histones H3 and H4 and methylation of histone H3 K4 or H3 K36 (Huang et al., 2005; Lee et al., 2005; Shilatifard, 2006). COMPASS-catalyzed histone H3K4 methylation is associated with RNA polymerase II at its initiating form. Methylated histone H3 K36 is found at $3'$ end of active genes in combination with the elongating form of RNA polymerase II (Shilatifard, 2006). Similar to the above mechanisms, methylated histone H3 K4 may serve as docking sites for downstream effectors that are involved with transcriptional activation, thus affecting gene expression (Sims and Reinberg, 2006). On the other hand, methylation of histones H3 K9, H3 K20, and H3 K27 exhibits a repressive effect on gene transcription. Similar to above mechanisms, chromatin modifiers that mediate these methylation marks function by inducing repressive chromatin state and recruiting repressive complexes to transcription sites (Hansen et al., 2008; Sims and Reinberg, 2006). Conversely, lysine demethylation is associated with transcriptional activation. For instance, LSD1 in a complex with the androgen receptor demethylates histone H3 K9 and results in its transcriptional activation (Metzger et al., 2005).

Histone phosphorylation also plays a role in several cellular processes. Phosphorylation of histone H3 S10 has been closely linked to gene activation. MSK1/2 has been linked to transcriptional activation of mitogen-stimulated immediate-early response genes, such as c-fos and c-jun (Mahadevan et al., 1991). Moreover, crosstalk between phosphorylation of histone H3 S10 and other histone modifications has been extensively studied and it is now well established that phosphorylation influences several histone modifications giving different transcriptional outcomes. Phosphorylation of histone H3 S10 can enhance acetylation of histone H3 K14 (Cheung et al., 2000b; Lo et al., 2000), abolish acetylation of histone H3 K9 (Edmondson et al., 2002), and inhibit methylation of histone H3 K9 (Rea et al., 2000), thus favoring an active transcriptional state.

Ubiquitination of histones has several transcriptional outcomes depending on the substrate and the histone proteins that are targets of this modification. Transcriptional repression is mediated by histone H2A K119 ubiquitination (Wang et al., 2006), whereas transcriptional activation is mediated by histone H2B K120 ubiquitination (Zhu et al., 2005). Consequently, enzymes responsible for deubiquitination promote either gene transcription (e.g., Ubp8) or transcriptional repression (e.g., Ubp10) depending on the histone and the substrate deubiquitinated as well as the chromatin-modifying and transcriptional complexes that lie downstream in the signaling cascade (Gardner et al., 2005).

B. DNA repair

Eukaryote cells continuously face numerous endogenous and exogenous genotoxic stresses that can cause deleterious DNA lesions, including DNA double-strand breaks (DSBs). To deal with these threats, cells have evolved mechanisms of DNA damage repair that maintain the genomic stability and prevent oncogenic transformation and development of diseases (Lindahl and Wood, 1999; Mills et al., 2003). Two major pathways are involved in DSB repair: homologous recombination and nonhomologous end joining (Jackson, 2002; Valerie and Povirk, 2003). Compacted chromatin can be a major obstacle to the orchestration of DNA repair and other chromatin-based processes. After the induction of DNA damage, chromatin must first be relaxed to give access to repair proteins to the site of breaks. Biochemical and molecular studies have revealed the link between different histone modifications and DNA repair highlighting the major role of chromatin remodeling enzymes in repair mechanisms (Table 3.1; Peterson and Cote, 2004; van Attikum and Gasser, 2005). For efficient repair, the chromatin structure needs to be altered and access to the sites of break needs to be ensured, both of which require posttranslational histone modifications, ATP-dependent nucleosome mobilization, and exchange of histone variants. In this section, we focus on the role of posttranslational histone modifications in DNA repair.

One of the earliest events in DSB signaling is the phosphorylation of histone H2AX, a variant of histone H2A. This phosphorylation is carried out by the phosphatase inositol-3 family of kinase (PI3K) and is spread over kilobases (in yeast) and megabases (in mammal cells) from the break site (Redon et al., 2002; Shroff et al., 2004). This modification is required for retention and accumulation of repair proteins to the damaged sites (Downs et al., 2004; Stucki et al., 2005). Moreover, it has been shown that histone H2AX phosphorylation is required for the recruitment of HATs to the sites of breaks. The recruitment of HATs is mediated by Arp4 and leads to acetylation of the chromatin surrounding the breaks, thereby relaxing the chromatin and facilitating access to repair proteins (Downs et al., 2004). Binding of NuA4 HAT complexes and subsequent acetylation of histone H4 is concomitant with H2AX phosphorylation (Downs and Cote, 2005). Moreover, defects in histone H3 acetylation result in sensitivity to DNA damaging agents, consistent with its importance in DNA repair (Masumoto et al., 2005). In agreement with an important role of histone acetylation in DNA repair, recent study provided evidence that TRRAP/TIP60 is essential for recruitment and loading of repair proteins to the site of breaks (Loizou et al., 2006; Murr et al., 2006, 2007).

More recently, the role of histone methylation in DNA repair received considerable attention. Methylation of histone H4 K20 in fission yeast was shown to be essential for the recruitment of Crb2, a checkpoint adaptor protein with homology to 53BP1, to sites of breaks and insure proper checkpoint activation in response to DNA damage breaks (Botuyan et al., 2006). In human cells, 53BP1 may function in a very similar manner (Sanders et al., 2004). Interestingly, TIP60 binds to the heterochromatic histone mark 3me H3 K9, triggering acetylation and activation of DNA DSB repair. Although 3me H3 K9 is not required for the recruitment of Tip60 to sites of DNA damage, the interaction of TIP60/ATM with the MRN complex is sufficient for chromatin localization. However, the interaction with 3me H3 K9 is essential for TIP60 HAT stimulation and the initiation of downstream repair events (Sun et al., 2009). In vitro studies showed a direct role of histone H4 K79 methylation in the formation of ionizing radiation-induced 53BP1 foci; however, the underlying mechanism remains unknown (Huyen et al., 2004).

Another modification that has been recently linked to DNA damage is histone ubiquitination. Ubiquitination of histones H3 and H4 by the CUL4-DDB-Roc1 complex seems important for recruitment of the XPC repair proteins in response to ultraviolet (UV)-induced DNA damage (Wang et al., 2006). Moreover, evidence of the role of mono-ubiquitination of histone H2A in UV-induced repair coincides with H2AX phosphorylation (Bergink et al., 2006). Phosphorylation of serine 1 of histone H4 coincides with a decrease in histone acetylation. This suggests that serine-1 phosphorylation occurs after H4 deacetylation and that it may regulate restoration of chromatin configuration after repair is completed (Utley et al., 2005). Together, these studies highlight DNA repair as dynamic and complex processes that necessitate a highly concerted actions of various chromatin modifiers and remodelers.

C. DNA replication

DNA replication occurs in the S phase of the cell cycle and it is initiated at discrete sites on the chromosome called origins of replication. DNA replication represents a delicate process for the cells since it requires a high fidelity during the duplication of DNA sequence and maintenance and propagation of chromatin states. This cellular process involves several critical steps: access of the replication machinery to DNA, disruption of the parental nucleosomes ahead of the replication fork, nucleosome assembly on the daughter duplex DNA, and propagation of the epigenetic state. All these events are regulated by a network of histone-modifying complexes that control the access to DNA and nucleosomal organization. Although the role of chromatin modifications in DNA replication remains poorly understood, several studies have provided evidence that histone posttranslational modifications can control efficiency and timing of the replication origin activity (Azuara et al., 2006; Kurdistani and Grunstein, 2003; Lin et al., 2003; Perry et al., 2004; Vogelauer et al., 2002; Zhang et al., 2002). As compacted chromatin can limit and prevent access for replication machinery to the DNA, it can be hypothesized that histone modifications play a critical role in setting the chromatin status for early and late origins of DNA replication. In agreement with this idea, it has been shown that in yeast, histone acetylation in the vicinity of replication origins affects replication timing. Indeed, higher levels of histone acetylation coincide with earlier firing of an origin (Vogelauer et al., 2002). In human, acetylation of histone tails is also found to correlate with replication timing (Cimbora et al., 2000). Consistent with the idea that acetylation opens the way to DNA replication machinery, several studies showed that the HAT HBO1 is associated with replication factor MCM2 and with the origin recognition complex 1 subunit of the human initiator protein. These findings suggest that targeting of histone acetylation at the origin of replication establishes a chromatin structure that is favorable for DNA replication (Burke et al., 2001; Iizuka and Stillman, 1999). Interestingly, ING5 containing HBO1 HAT complex associates with MCM2-7 helicase and appears to be essential for DNA replication in humans, consistent with the finding that depletion of either ING5 or HBO1 impairs S phase progression (Doyon et al., 2006). Recent studies in S. cerevisiae showed a dynamic regulation of acetylation of histones H3 and H4 around an origin of replication (Unnikrishnan et al., 2010). Further studies are needed to examine the exact mechanism and the implication of other histone modifications such as methylation, phosphorylation, and ubiquination. It is likely that histone phosphorylation, similar to acetylation, could also play a role in making DNA accessible to DNA replication machinery and in restoring compacted chromatin configuration after DNA replication is completed.

V. DEREGULATION OF HISTONE MODIFICATIONS AS A MECHANISM UNDERLYING CANCER DEVELOPMENT

Different critical cellular phenomena including DNA repair, transcription, DNA replication, and cell cycle are controlled by histone modifications. Therefore, histone modifications are tightly controlled processes that require a fair amount of flexibility and coordination in response to intrinsic or external stimuli. Deregulation in any of these histone modifications may shift the balance of gene expression leading to alterations in critical cellular processes such as transcription, proliferation, apoptosis, and DNA repair, ultimately resulting in cellular transformation and malignant outgrowth (Fig. 3.2). Over the last few years, a number of studies reported aberrant epigenetic patterns resulting in unscheduled silencing of tumor-suppressor genes and other cancer-associated genes in a variety of human cancers (Feinberg and Tycko, 2004; Jones and Baylin, 2002). Moreover, deregulation of histone modifications may trigger genetic changes through aberrant functions of cellular processes such as DNA repair, gene transcription, and DNA replication (Sawan et al., 2008).

An aberrant activity of several histone-modifying enzymes has been found in cancer cells (Esteller, 2006). Missense and truncating mutation as well as the loss of heterozygosity at the locus of p300 HAT gene was associated with hypoacetylation in different human cancers (Gayther et al., 2000). Fusion proteins from MLL gene translocation with p300 or CBP acetyltransferases have been linked to leukemia (Yang, 2004). Impaired activity of HDACs has been suggested to contribute to deregulation of epigenetic code leading to cancer. For instance, HDAC6 deregulation has been linked to breast cancer, whereas aberrant activity of HDAC2 was associated with several sporadic cancers (Ozdag et al., 2006; Ropero et al., 2006; Zhang et al., 2004). Furthermore, aberrations in histone methyltransferases are believed to contribute to cancer development and progression. Overexpression of the polycomb group protein Ezh2 has been implicated in metastatic prostate cancer and associated with poor prognosis (Varambally et al., 2002). LSD1 overexpression has been linked to increasing grade in prostate cancer, a feature that could be exploited for prognostic purposes (Kahl et al., 2006). LSD1 has been found strongly expressed in poorly differentiated neuroblastoma and in estrogen receptor (ER)-negative breast cancers (Schulte et al., 2009).

Epigenetic alterations in a large number of genes have been frequently found in human cancers; however, with the advent of high-throughput and genome-wide profiling of epigenetic changes, this list is likely to rapidly grow in coming years. These alterations have been linked to abnormal cellular proliferation, invasiveness, metastatic progression, and therapy resistance (Gronbaek et al., 2007; Gupta and Massague, 2006). Deregulation of chromatin modifiers and remodelers alters cellular processes, most notably gene transcription and DNA repair (Fig. 3.2). Indeed, HATs are found mutated in several cancer types. P300 and CBP have been characterized as tumor suppressors and act as transcriptional cofactors for a variety of oncoproteins such as p53, pRB, Myb, Jun, and Fos (Yang, 2004). Moreover, HDAC1 is known to interact with tumor-suppressor protein Rb and the E2F transcription factor leading to E2F gene repression. Histone H3 kinases seem to play a role in tumorigenesis since MLTK-α plays a significant role in neoplastic transformation and cancer development by regulating oncogenes such as c-jun and c-myc (Cho et al., 2004). Therefore, it is not surprising that alteration in expression and function of chromatin modifiers and remodelers would induce altered transcription of oncoproteins leading to carcinogenesis (Brehm et al., 1998). On the other hand, the HAT TIP60 mutations abrogating its acetyltransferase activity may impair DNA break repair mechanisms resulting in mutations and genomic instability (Ikura et al., 2000). It has been shown that mutations in a subunit of Tip60, Yng2 result in a repair defect following replication fork stall (Choy and Kron, 2002). Together, these studies show the link between epigenetic deregulation and aberrant cellular functions that may be the major causes of cancer development.

Recent studies also suggest that epigenetic disruption in stem cells populations may be the origin of cancer. The idea is that these epigenetic aberrations occur in normal stem cell progenitors and that they are among earliest events in cancer development (Feinberg et al., 2006). This is supported by the discovery of altered progenitors in normal tissues of cancer patients and is consistent with the fact that tumors have heterogeneous population of cells with diverse tumorigenic properties (Al-Hajj et al., 2003). Since epigenetic mechanisms are essential for maintenance of the equilibrium between self-renewal and differentiation in stem cells (Niwa, 2007), it is evident that any alteration in epigenetic status of normal progenitor cells may result in expansion of aberrant progenitor populations and set the stage for accumulation of further epigenetic and genetic aberrations leading to carcinogenesis. Together, these studies underscore the importance that the deregulated function of molecules and complexes responsible for different histone-modifying activities plays in key cellular processes and in a variety of human cancers (Fig. 3.2).

VI. HISTONE MODIFICATIONS IN CANCER AS POTENT BIOMARKERS

Cancer cells exhibit an altered epigenome compared to normal counterparts. Global DNA hypomethylation of the genome combined with gene promoter-specific hypermethylation has long been associated with cancer (Feinberg and

Tycko, 2004; Sawan et al., 2008; Vaissiere et al., 2009b). Changes in histone modifications have also been detected in different cancer types. Indeed, Fraga et al. (2005) demonstrated that loss of histone H4 K16 acetylation and H4 K20 trimethylation is a hallmark of human cancers. me H3 K4, 2me H3 K9, 3me H3 K9, Ac H3, and Ac H4 were all found to be significantly reduced in prostate cancer compared to nonmalignant prostate tissue (Ellinger et al., 2010). Decrease in 2me H3 K4, 2me H3 K9, or Ac H3 K18 levels were shown to be markers of pancreatic adenocarcinomas, lung, and kidney cancers (Manuyakorn et al., 2010; Seligson et al., 2009). The studies aiming to identify biomarkers of non-small cell lung carcinoma revealed an aberrant pattern of histone H4 modifications with hyperacetylation of histones H4 K5 and H4 K8 in cancer cells compared to normal lung epithelium as well as hypoacetylation of histones H4 K12 and H4 K16 (Van Den Broeck et al., 2008). Loss of 3me H4 K20 was found to be more frequent in squamous cell carcinoma than in adenocarcinoma, although it was associated with poor survival of lung cancer patients with adenocarcinoma subtype (Van Den Broeck et al., 2008). These data suggest that the deregulation of histone H4 modifications may play an important role in lung carcinogenesis and identify 3me H4 K20 as potential epigenetic biomarkers for early detection of squamous cell carcinoma. Moreover, Barlesi et al. (2007) demonstrated that patients with a pathologic tumor stage II of lung carcinoma exhibiting lower level of histone H2A K5 acetylation have significantly worse survival compared to those with high H2A K5 acetylation (Barlesi et al., 2007). Stage 1 patients with large cell or squamous carcinoma harboring high levels of histone H3 K4 dimethylation have favorable survival prognosis. Similarly, adenocarcinoma patients with lower histone H3 K9 acetylation have better survival than others (Barlesi et al., 2007). Taken together, these studies demonstrate that different histone modifications can be used as a biomarker for early diagnosis of cancer and powerful prognostic predictors.

Epigenetic modifications are considered as attractive biomarkers for early detection of cancer in part due to the fact that they can be assessed in body fluids of cancer-bearing individuals. A considerable effort has been put in developing new techniques for easier detection of epigenetic modifications in body fluids. Techniques for detection of DNA methylation changes in body fluids have already been developed (Vaissiere et al., 2009a). Since it has been shown that plasma contains, in addition to genomic DNA, free nucleosomes (Chan et al., 2003; Rumore and Steinman, 1990), several studies attempted to detect histone modifications including methylated histone H3 K9 in plasma using cell-free nucleosomes combined with ELISA and real-time PCR (Deligezer et al., 2008). Although these studies are in early stages, the presence of cancer-derived nucleosomes with specific patterns of histone modifications may prove a useful biomarker in predicting clinical outcomes.

VII. CONCLUSION AND FUTURE DIRECTIONS

The research on histone modifications is a new and rapidly expanding field that holds promises to advance our understanding of the key processes underlying tumor development and progression. Recent studies argue that histone modifications act in a coordinated and orderly fashion to regulate cellular processes such as gene transcription, DNA repair, and DNA replication. These studies have also suggested an existence of an intimate and self-reinforcing crosstalk and interdependence between histone marks including acetylation, methylation, and phosphorylation. Consistent with critical roles of histone modifications in key cellular processes, a large body of evidence has suggested that deregulation of histone marking is intimately linked to a number of human malignancies. Therefore, better understanding of the mechanisms controlling the establishment and maintenance of histone marks and how the disruption of these mechanisms contribute to tumorigenesis will facilitate the development of novel strategies to prevent, diagnose, and treat human malignancies. The fact that epigenetic alterations including histone modifications are, in contrast to genetic changes, in turn reversible, may provide the basis for development of new therapeutic and prevention approaches targeting aberrant histone modifications. Many epigenetic therapies targeting epigenetic modifiers have been recently developed. These include DNA methyltransferase inhibitors and HAT inhibitors, both of which aim to reactivate epigenetically silenced tumor suppressors, DNA repair genes, and other cancer-associated genes (Egger et al., 2004). A number of agents altering histone modifiers have been discovered and are currently in clinical trials (Egger et al., 2004). Although epigenetic therapy based on modulation of histone-modifying activities still faces a major challenge regarding toxic side effects, the agents capable of targeting histone modifiers have already entered clinical practices and are routinely used in the treatment of specific human malignancies. Finally, a striking reconfiguration of histone marks in cancer cells compared to its normal counterparts represents an attractive target for biomarker discovery. Although for successful and reliable application of biomarkers based on histone modifications in clinics or molecular epidemiology a number of criteria need to be satisfied, recent technological advances in epigenomics may prove particularly valuable in profiling large sample series from molecular epidemiology studies and clinical trials for early diagnostic and prognostic purposes as well as in monitoring the efficacy of epigenetics-based cancer therapies and preventive strategies. Therefore, future studies on histone modifications should greatly impact on many areas of modern biology and will help to tailor efficient strategies for cancer therapy and prevention.

Acknowledgments

C. S. is supported by a PhD fellowship from l'Association pour la Recherche contre le Cancer. The work in Epigenetics Group at IARC is supported by grants from the National Institutes of Health/ National Cancer Institute (NIH/NCI), United States; l'Association pour la Recherche sur le Cancer (ARC), France, la Ligue Nationale (Française) Contre le Cancer, France, and Agence Nationale de Recherhe Contre le Sida et Hépatites Virales (ANRS, France), and the Swiss Bridge Award (to Z. H.).

References

Ahn, S. H., Henderson, K. A., Keeney, S., and Allis, C. D. (2005). H2B (Ser10) phosphorylation is induced during apoptosis and meiosis in S. cerevisiae. Cell Cycle 4, 780–783.

Al-Hajj, M., Wicha, M. S., Benito-Hernandez, A., Morrison, S. J., and Clarke, M. F. (2003). Prospective identification of tumorigenic breast cancer cells. Proc. Natl. Acad. Sci. USA 100, 3983–3988.

Allfrey, V. G., Faulkner, R., and Mirsky, A. E. (1964). Acetylation and methylation of histones and their possible role in the regulation of RNA synthesis. Proc. Natl. Acad. Sci. USA 51, 786–794.

Anand, R., and Marmorstein, R. (2007). Structure and mechanism of lysine-specific demethylase enzymes. J. Biol.l Chem. 282, 35425–35429.

Azuara, V., Perry, P., Sauer, S., Spivakov, M., Jorgensen, H. F., John, R. M., Gouti, M., Casanova, M., Warnes, G., Merkenschlager, M., et al. (2006). Chromatin signatures of pluripotent cell lines. Nat. Cell Biol. 8, 532–538.

Baker, S. P., Phillips, J., Anderson, S., Qiu, Q., Shabanowitz, J., Smith, M. M., Yates, J. R., 3rd, Hunt, D. F., and Grant, P. A. (2010). Histone H3 Thr 45 phosphorylation is a replication-associated post-translational modification in S. cerevisiae. Nat. Cell Biol. 12, 294–298.

Bannister, A. J., Zegerman, P., Partridge, J. F., Miska, E. A., Thomas, J. O., Allshire, R. C., and Kouzarides, T. (2001). Selective recognition of methylated lysine 9 on histone H3 by the HP1 chromo domain. Nature 410, 120–124.

Barlesi, F., Giaccone, G., Gallegos-Ruiz, M. I., Loundou, A., Span, S. W., Lefesvre, P., Kruyt, F. A., and Rodriguez, J. A. (2007). Global histone modifications predict prognosis of resected non small-cell lung cancer. J. Clin. Oncol. 25, 4358–4364.

Barsoum, J., and Varshavsky, A. (1985). Preferential localization of variant nucleosomes near the 5'-end of the mouse dihydrofolate reductase gene. J. Biol. Chem. 260, 7688–7697.

Bergink, S., Salomons, F. A., Hoogstraten, D., Groothuis, T. A., de Waard, H., Wu, J., Yuan, L., Citterio, E., Houtsmuller, A. B., Neefjes, J., et al. (2006). DNA damage triggers nucleotide excision repair-dependent monoubiquitylation of histone H2A. Genes Dev. 20, 1343–1352.

Bernstein, B. E., Mikkelsen, T. S., Xie, X., Kamal, M., Huebert, D. J., Cuff, J., Fry, B., Meissner, A., Wernig, M., Plath, K., et al. (2006). A bivalent chromatin structure marks key developmental genes in embryonic stem cells. Cell 125, 315–326.

Botuyan, M. V., Lee, J., Ward, I. M., Kim, J. E., Thompson, J. R., Chen, J., and Mer, G. (2006). Structural basis for the methylation state-specific recognition of histone H4-K20 by 53BP1 and Crb2 in DNA repair. Cell 127, 1361–1373.

Brehm, A., Miska, E. A., McCance, D. J., Reid, J. L., Bannister, A. J., and Kouzarides, T. (1998). Retinoblastoma protein recruits histone deacetylase to repress transcription. Nature 391, 597–601.

Briggs, S. D., Xiao, T., Sun, Z. W., Caldwell, J. A., Shabanowitz, J., Hunt, D. F., Allis, C. D., and Strahl, B. D. (2002). Gene silencing: Trans-histone regulatory pathway in chromatin. Nature 418, 498.

Brownell, J. E., and Allis, C. D. (1995). An activity gel assay detects a single, catalytically active histone acetyltransferase subunit in Tetrahymena macronuclei. *Proc. Natl. Acad. Sci. USA* **92**, 6364–6368.

Bulger, M. (2005). Hyperacetylated chromatin domains: Lessons from heterochromatin. *J. Biol. Chem.* **280**, 21689–21692.

Burke, T. W., Cook, J. G., Asano, M., and Nevins, J. R. (2001). Replication factors MCM2 and ORC1 interact with the histone acetyltransferase HBO1. *J. Biol. Chem.* **276**, 15397–15408.

Chan, K. C., Zhang, J., Chan, A. T., Lei, K. I., Leung, S. F., Chan, L. Y., Chow, K. C., and Lo, Y. M. (2003). Molecular characterization of circulating EBV DNA in the plasma of nasopharyngeal carcinoma and lymphoma patients. *Cancer Res.* **63**, 2028–2032.

Chang, B., Chen, Y., Zhao, Y., and Bruick, R. K. (2007). JMJD6 is a histone arginine demethylase. *Science* **318**, 444–447.

Cheung, P., Allis, C. D., and Sassone-Corsi, P. (2000a). Signaling to chromatin through histone modifications. *Cell* **103**, 263–271.

Cheung, P., Tanner, K. G., Cheung, W. L., Sassone-Corsi, P., Denu, J. M., and Allis, C. D. (2000b). Synergistic coupling of histone H3 phosphorylation and acetylation in response to epidermal growth factor stimulation. *Mol. Cell* **5**, 905–915.

Cho, Y. Y., Bode, A. M., Mizuno, H., Choi, B. Y., Choi, H. S., and Dong, Z. (2004). A novel role for mixed-lineage kinase-like mitogen-activated protein triple kinase alpha in neoplastic cell transformation and tumor development. *Cancer Res.* **64**, 3855–3864.

Choy, J. S., and Kron, S. J. (2002). NuA4 subunit Yng2 function in intra-S-phase DNA damage response. *Mol. Cell. Biol.* **22**, 8215–8225.

Cimbora, D. M., Schubeler, D., Reik, A., Hamilton, J., Francastel, C., Epner, E. M., and Groudine, M. (2000). Long-distance control of origin choice and replication timing in the human beta-globin locus are independent of the locus control region. *Mol. Cell. Biol.* **20**, 5581–5591.

Clarke, S. (1993). Protein methylation. *Curr. Opin. Cell. Biol.* **5**, 977–983.

Clausell, J., Happel, N., Hale, T. K., Doenecke, D., and Beato, M. (2009). Histone H1 subtypes differentially modulate chromatin condensation without preventing ATP-dependent remodeling by SWI/SNF or NURF. *PLoS One* **4**, e0007243.

Dawson, B. A., Herman, T., Haas, A. L., and Lough, J. (1991). Affinity isolation of active murine erythroleukemia cell chromatin: Uniform distribution of ubiquitinated histone H2A between active and inactive fractions. *J. Cell. Biochem.* **46**, 166–173.

de la Cruz, X., Lois, S., Sanchez-Molina, S., and Martinez-Balbas, M. A. (2005). Do protein motifs read the histone code? *Bioessays* **27**, 164–175.

Deligezer, U., Akisik, E. E., Erten, N., and Dalay, N. (2008). Sequence-specific histone methylation is detectable on circulating nucleosomes in plasma. *Clin. Chem.* **54**, 1125–1131.

Downs, J. A., Allard, S., Jobin-Robitaille, O., Javaheri, A., Auger, A., Bouchard, N., Kron, S. J., Jackson, S. P., and Cote, J. (2004). Binding of chromatin-modifying activities to phosphorylated histone H2A at DNA damage sites. *Mol. Cell* **16**, 979–990.

Downs, J. A., and Cote, J. (2005). Dynamics of chromatin during the repair of DNA double-strand breaks. *Cell Cycle* **4**, 1373–1376.

Doyon, Y., Cayrou, C., Ullah, M., Landry, A. J., Cote, V., Selleck, W., Lane, W. S., Tan, S., Yang, X. J., and Cote, J. (2006). ING tumor suppressor proteins are critical regulators of chromatin acetylation required for genome expression and perpetuation. *Mol. Cell* **21**, 51–64.

Edmondson, D. G., Davie, J. K., Zhou, J., Mirnikjoo, B., Tatchell, K., and Dent, S. Y. (2002). Site-specific loss of acetylation upon phosphorylation of histone H3. *J. Biol. Chem.* **277**, 29496–29502.

Egger, G., Liang, G., Aparicio, A., and Jones, P. A. (2004). Epigenetics in human disease and prospects for epigenetic therapy. *Nature* **429**, 457–463.

Eickbush, T. H., and Moudrianakis, E. N. (1978). The histone core complex: An octamer assembled by two sets of protein–protein interactions. *Biochemistry* **17,** 4955–4964.

Ekwall, K. (2005). Genome-wide analysis of HDAC function. *Trends Genet.* **21,** 608–615.

Ellinger, J., Kahl, P., von der Gathen, J., Rogenhofer, S., Heukamp, L. C., Gutgemann, I., Walter, B., Hofstadter, F., Buttner, R., Muller, S. C., *et al.* (2010). Global levels of histone modifications predict prostate cancer recurrence. *Prostate* **70,** 61–69.

Esteller, M. (2006). Epigenetics provides a new generation of oncogenes and tumour-suppressor genes. *Br. J. Cancer* **94,** 179–183.

Feinberg, A. P., Ohlsson, R., and Henikoff, S. (2006). The epigenetic progenitor origin of human cancer. *Nat. Rev. Genet.* **7,** 21–33.

Feinberg, A. P., and Tycko, B. (2004). The history of cancer epigenetics. *Nat. Rev. Cancer* **4,** 143–153.

Feng, Q., Wang, H., Ng, H. H., Erdjument-Bromage, H., Tempst, P., Struhl, K., and Zhang, Y. (2002). Methylation of H3-lysine 79 is mediated by a new family of HMTases without a SET domain. *Curr. Biol.* **12,** 1052–1058.

Fletcher, T. M., and Hansen, J. C. (1995). Core histone tail domains mediate oligonucleosome folding and nucleosomal DNA organization through distinct molecular mechanisms. *J. Biol. Chem.* **270,** 25359–25362.

Fletcher, T. M., and Hansen, J. C. (1996). The nucleosomal array: Structure/function relationships. *Crit. Rev. Eukaryot. Gene Expr.* **6,** 149–188.

Foster, E. R., and Downs, J. A. (2005). Histone H2A phosphorylation in DNA double-strand break repair. *FEBS J.* **272,** 3231–3240.

Fraga, M. F., Ballestar, E., Villar-Garea, A., Boix-Chornet, M., Espada, J., Schotta, G., Bonaldi, T., Haydon, C., Ropero, S., Petrie, K., *et al.* (2005). Loss of acetylation at Lys16 and trimethylation at Lys20 of histone H4 is a common hallmark of human cancer. *Nat. Genet.* **37,** 391–400.

Gardner, R. G., Nelson, Z. W., and Gottschling, D. E. (2005). Ubp10/Dot4p regulates the persistence of ubiquitinated histone H2B: Distinct roles in telomeric silencing and general chromatin. *Mol. Cell. Biol.* **25,** 6123–6139.

Gary, J. D., and Clarke, S. (1998). RNA and protein interactions modulated by protein arginine methylation. *Prog. Nucleic Acid Res. Mol. Biol.* **61,** 65–131.

Gayther, S. A., Batley, S. J., Linger, L., Bannister, A., Thorpe, K., Chin, S. F., Daigo, Y., Russell, P., Wilson, A., Sowter, H. M., *et al.* (2000). Mutations truncating the EP300 acetylase in human cancers. *Nat. Genet.* **24,** 300–303.

Goldknopf, I. L., and Busch, H. (1975). Remarkable similarities of peptide fingerprints of histone 2A and nonhistone chromosomal protein A24. *Biochem. Biophys. Res. Commun.* **65,** 951–960.

Goll, M. G., and Bestor, T. H. (2002). Histone modification and replacement in chromatin activation. *Genes Dev.* **16,** 1739–1742.

Gottesfeld, J. M., and Forbes, D. J. (1997). Mitotic repression of the transcriptional machinery. *Trends Biochem. Sci.* **22,** 197–202.

Grant, P. A. (2001). A tale of histone modifications. *Genome Biol.* **2,** REVIEWS0003.

Gronbaek, K., Hother, C., and Jones, P. A. (2007). Epigenetic changes in cancer. *APMIS* **115,** 1039–1059.

Gupta, G. P., and Massague, J. (2006). Cancer metastasis: Building a framework. *Cell* **127,** 679–695.

Gutierrez, R. M., and Hnilica, L. S. (1967). Tissue specificity of histone phosphorylation. *Science* **157,** 1324–1325.

Hansen, K. H., Bracken, A. P., Pasini, D., Dietrich, N., Gehani, S. S., Monrad, A., Rappsilber, J., Lerdrup, M., and Helin, K. (2008). A model for transmission of the H3K27me3 epigenetic mark. *Nat. Cell Biol.* **10,** 1291–1300.

Hassan, A. H., Prochasson, P., Neely, K. E., Galasinski, S. C., Chandy, M., Carrozza, M. J., and Workman, J. L. (2002). Function and selectivity of bromodomains in anchoring chromatin-modifying complexes to promoter nucleosomes. *Cell* **111,** 369–379.

Henry, K. W., Wyce, A., Lo, W. S., Duggan, L. J., Emre, N. C., Kao, C. F., Pillus, L., Shilatifard, A., Osley, M. A., and Berger, S. L. (2003). Transcriptional activation via sequential histone H2B ubiquitylation and deubiquitylation, mediated by SAGA-associated Ubp8. *Genes Dev.* **17,** 2648–2663.

Hirota, T., Lipp, J. J., Toh, B. H., and Peters, J. M. (2005). Histone H3 serine 10 phosphorylation by Aurora B causes HP1 dissociation from heterochromatin. *Nature* **438,** 1176–1180.

Horiuchi, J., Silverman, N., Marcus, G. A., and Guarente, L. (1995). ADA3, a putative transcriptional adaptor, consists of two separable domains and interacts with ADA2 and GCN5 in a trimeric complex. *Mol. Cell. Biol.* **15,** 1203–1209.

Huang, S., Litt, M., and Felsenfeld, G. (2005). Methylation of histone H4 by arginine methyltransferase PRMT1 is essential in vivo for many subsequent histone modifications. *Genes Dev.* **19,** 1885–1893.

Huyen, Y., Zgheib, O., Ditullio, R. A., Jr., Gorgoulis, V. G., Zacharatos, P., Petty, T. J., Sheston, E. A., Mellert, H. S., Stavridi, E. S., and Halazonetis, T. D. (2004). Methylated lysine 79 of histone H3 targets 53BP1 to DNA double-strand breaks. *Nature* **432,** 406–411.

Iizuka, M., and Stillman, B. (1999). Histone acetyltransferase HBO1 interacts with the ORC1 subunit of the human initiator protein. *J. Biol. Chem.* **274,** 23027–23034.

Ikura, T., Ogryzko, V. V., Grigoriev, M., Groisman, R., Wang, J., Horikoshi, M., Scully, R., Qin, J., and Nakatani, Y. (2000). Involvement of the TIP60 histone acetylase complex in DNA repair and apoptosis. *Cell* **102,** 463–473.

Jackson, S. P. (2002). Sensing and repairing DNA double-strand breaks. *Carcinogenesis* **23,** 687–696.

Jenuwein, T., and Allis, C. D. (2001). Translating the histone code. *Science* **293,** 1074–1080.

Jones, P. A., and Baylin, S. B. (2002). The fundamental role of epigenetic events in cancer. *Nat. Rev. Genet.* **3,** 415–428.

Kahl, P., Gullotti, L., Heukamp, L. C., Wolf, S., Friedrichs, N., Vorreuther, R., Solleder, G., Bastian, P. J., Ellinger, J., Metzger, E., *et al.* (2006). Androgen receptor coactivators lysine-specific histone demethylase 1 and four and a half LIM domain protein 2 predict risk of prostate cancer recurrence. *Cancer Res.* **66,** 11341–11347.

Kleinsmith, L. J., Allfrey, V. G., and Mirsky, A. E. (1966). Phosphoprotein metabolism in isolated lymphocyte nuclei. *Proc. Natl. Acad. Sci. USA* **55,** 1182–1189.

Kouzarides, T. (2007). Chromatin modifications and their function. *Cell* **128,** 693–705.

Krishnamoorthy, T., Chen, X., Govin, J., Cheung, W. L., Dorsey, J., Schindler, K., Winter, E., Allis, C. D., Guacci, V., Khochbin, S., *et al.* (2006). Phosphorylation of histone H4 Ser1 regulates sporulation in yeast and is conserved in fly and mouse spermatogenesis. *Genes Dev.* **20,** 2580–2592.

Kurdistani, S. K., and Grunstein, M. (2003). Histone acetylation and deacetylation in yeast. *Nat. Rev.* **4,** 276–284.

Langan, T. A. (1968). Histone phosphorylation: Stimulation by adenosine $3',5'$-monophosphate. *Science* **162,** 579–580.

Lee, D. Y., Teyssier, C., Strahl, B. D., and Stallcup, M. R. (2005). Role of protein methylation in regulation of transcription. *Endocr. Rev.* **26,** 147–170.

Lee, K. K., and Workman, J. L. (2007). Histone acetyltransferase complexes: One size doesn't fit all. *Nat. Rev.* **8,** 284–295.

Levinger, L., and Varshavsky, A. (1982). Selective arrangement of ubiquitinated and D1 protein-containing nucleosomes within the *Drosophila* genome. *Cell* **28,** 375–385.

Lin, C. M., Fu, H., Martinovsky, M., Bouhassira, E., and Aladjem, M. I. (2003). Dynamic alterations of replication timing in mammalian cells. *Curr. Biol.* **13,** 1019–1028.

Lindahl, T., and Wood, R. D. (1999). Quality control by DNA repair. *Science* **286,** 1897–1905.

Liu, B., Lin, Y., Darwanto, A., Song, X., Xu, G., and Zhang, K. (2009). Identification and characterization of propionylation at histone H3 lysine 23 in mammalian cells. *J. Biol. Chem.* **284,** 32288–32295.

Lo, W. S., Trievel, R. C., Rojas, J. R., Duggan, L., Hsu, J. Y., Allis, C. D., Marmorstein, R., and Berger, S. L. (2000). Phosphorylation of serine 10 in histone H3 is functionally linked in vitro and in vivo to Gcn5-mediated acetylation at lysine 14. *Mol. Cell* **5,** 917–926.

Loidl, P. (1994). Histone acetylation: Facts and questions. *Chromosoma* **103,** 441–449.

Loizou, J. I., Murr, R., Finkbeiner, M. G., Sawan, C., Wang, Z. Q., and Herceg, Z. (2006). Epigenetic information in chromatin: The code of entry for DNA repair. *Cell Cycle* **5,** 696–701.

Luger, K., Mader, A. W., Richmond, R. K., Sargent, D. F., and Richmond, T. J. (1997). Crystal structure of the nucleosome core particle at 2.8 Å resolution. *Nature* **389,** 251–260.

Luger, K., and Richmond, T. J. (1998). The histone tails of the nucleosome. *Curr. Opin. Genet. Dev.* **8,** 140–146.

Mahadevan, L. C., Willis, A. C., and Barratt, M. J. (1991). Rapid histone H3 phosphorylation in response to growth factors, phorbol esters, okadaic acid, and protein synthesis inhibitors. *Cell* **65,** 775–783.

Manuyakorn, A., Paulus, R., Farrell, J., Dawson, N. A., Tze, S., Cheung-Lau, G., Hines, O. J., Reber, H., Seligson, D. B., Horvath, S., et al. (2010). Cellular histone modification patterns predict prognosis and treatment response in resectable pancreatic adenocarcinoma: Results from RTOG 9704. *J. Clin. Oncol.* **28,** 1358–1365.

Masumoto, H., Hawke, D., Kobayashi, R., and Verreault, A. (2005). A role for cell-cycle-regulated histone H3 lysine 56 acetylation in the DNA damage response. *Nature* **436,** 294–298.

Metzger, E., Imhof, A., Patel, D., Kahl, P., Hoffmeyer, K., Friedrichs, N., Muller, J. M., Greschik, H., Kirfel, J., Ji, S., et al. (2010). Phosphorylation of histone H3T6 by PKCbeta(I) controls demethylation at histone H3K4. *Nature* **464**(7289), 792–796.

Metzger, E., Wissmann, M., Yin, N., Muller, J. M., Schneider, R., Peters, A. H., Gunther, T., Buettner, R., and Schule, R. (2005). LSD1 demethylates repressive histone marks to promote androgen-receptor-dependent transcription. *Nature* **437,** 436–439.

Mills, K. D., Ferguson, D. O., and Alt, F. W. (2003). The role of DNA breaks in genomic instability and tumorigenesis. *Immunol. Rev.* **194,** 77–95.

Murr, R., Loizou, J. I., Yang, Y. G., Cuenin, C., Li, H., Wang, Z. Q., and Herceg, Z. (2006). Histone acetylation by Trrap-Tip60 modulates loading of repair proteins and repair of DNA double-strand breaks. *Nat. Cell Biol.* **8,** 91–99.

Murr, R., Vaissiere, T., Sawan, C., Shukla, V., and Herceg, Z. (2007). Orchestration of chromatin-based processes: Mind the TRRAP. *Oncogene* **26,** 5358–5372.

Murray, K. (1964). The occurrence of Epsilon-N-methyl lysine in histones. *Biochemistry* **3,** 10–15.

Ng, H. H., Feng, Q., Wang, H., Erdjument-Bromage, H., Tempst, P., Zhang, Y., and Struhl, K. (2002). Lysine methylation within the globular domain of histone H3 by Dot1 is important for telomeric silencing and Sir protein association. *Genes Dev.* **16,** 1518–1527.

Ng, H. H., Robert, F., Young, R. A., and Struhl, K. (2003). Targeted recruitment of Set1 histone methylase by elongating Pol II provides a localized mark and memory of recent transcriptional activity. *Mol. Cell* **11,** 709–719.

Nickel, B. E., Allis, C. D., and Davie, J. R. (1989). Ubiquitinated histone H2B is preferentially located in transcriptionally active chromatin. *Biochemistry* **28,** 958–963.

Niwa, H. (2007). How is pluripotency determined and maintained? *Development (Cambridge, England)* **134,** 635–646.

Norton, V. G., Imai, B. S., Yau, P., and Bradbury, E. M. (1989). Histone acetylation reduces nucleosome core particle linking number change. *Cell* **57,** 449–457.

Nowak, S. J., and Corces, V. G. (2004). Phosphorylation of histone H3: A balancing act between chromosome condensation and transcriptional activation. *Trends Genet.* 20, 214–220.

Owen, D. J., Ornaghi, P., Yang, J. C., Lowe, N., Evans, P. R., Ballario, P., Neuhaus, D., Filetici, P., and Travers, A. A. (2000). The structural basis for the recognition of acetylated histone H4 by the bromodomain of histone acetyltransferase gcn5p. *EMBO J.* 19, 6141–6149.

Ozdag, H., Teschendorff, A. E., Ahmed, A. A., Hyland, S. J., Blenkiron, C., Bobrow, L., Veerakumarasivam, A., Burtt, G., Subkhankulova, T., Arends, M. J., et al. (2006). Differential expression of selected histone modifier genes in human solid cancers. *BMC Genomics* 7, 90.

Pal, S., Vishwanath, S. N., Erdjument-Bromage, H., Tempst, P., and Sif, S. (2004). Human SWI/SNF-associated PRMT5 methylates histone H3 arginine 8 and negatively regulates expression of ST7 and NM23 tumor suppressor genes. *Mol. Cell. Biol.* 24, 9630–9645.

Parlow, M. H., Haas, A. L., and Lough, J. (1990). Enrichment of ubiquitinated histone H2A in a low salt extract of micrococcal nuclease-digested myotube nuclei. *J. Biol. Chem.* 265, 7507–7512.

Perry, P., Sauer, S., Billon, N., Richardson, W. D., Spivakov, M., Warnes, G., Livesey, F. J., Merkenschlager, M., Fisher, A. G., and Azuara, V. (2004). A dynamic switch in the replication timing of key regulator genes in embryonic stem cells upon neural induction. *Cell Cycle* 3, 1645–1650.

Peterson, C. L., and Cote, J. (2004). Cellular machineries for chromosomal DNA repair. *Genes Dev.* 18, 602–616.

Pickart, C. M., and Eddins, M. J. (2004). Ubiquitin: Structures, functions, mechanisms. *Biochim. Biophy. Acta* 1695, 55–72.

Rea, S., Eisenhaber, F., O'Carroll, D., Strahl, B. D., Sun, Z. W., Schmid, M., Opravil, S., Mechtler, K., Ponting, C. P., Allis, C. D., et al. (2000). Regulation of chromatin structure by site-specific histone H3 methyltransferases. *Nature* 406, 593–599.

Redon, C., Pilch, D., Rogakou, E., Sedelnikova, O., Newrock, K., and Bonner, W. (2002). Histone H2A variants H2AX and H2AZ. *Curr. Opin. Genet. Dev.* 12, 162–169.

Ropero, S., Fraga, M. F., Ballestar, E., Hamelin, R., Yamamoto, H., Boix-Chornet, M., Caballero, R., Alaminos, M., Setien, F., Paz, M. F., et al. (2006). A truncating mutation of HDAC2 in human cancers confers resistance to histone deacetylase inhibition. *Nat. Genet.* 38, 566–569.

Rumore, P. M., and Steinman, C. R. (1990). Endogenous circulating DNA in systemic lupus erythematosus. Occurrence as multimeric complexes bound to histone. *J. Clin. Invest.* 86, 69–74.

Rundlett, S. E., Carmen, A. A., Kobayashi, R., Bavykin, S., Turner, B. M., and Grunstein, M. (1996). HDA1 and RPD3 are members of distinct yeast histone deacetylase complexes that regulate silencing and transcription. *Proc. Natl. Acad. Sci. USA* 93, 14503–14508.

Sanders, S. L., Portoso, M., Mata, J., Bahler, J., Allshire, R. C., and Kouzarides, T. (2004). Methylation of histone H4 lysine 20 controls recruitment of Crb2 to sites of DNA damage. *Cell* 119, 603–614.

Santos-Rosa, H., Kirmizis, A., Nelson, C., Bartke, T., Saksouk, N., Cote, J., and Kouzarides, T. (2009). Histone H3 tail clipping regulates gene expression. *Nat. Struct. Mol. Biol.* 16, 17–22.

Sawan, C., Vaissiere, T., Murr, R., and Herceg, Z. (2008). Epigenetic drivers and genetic passengers on the road to cancer. *Mutat. Res.* 642, 1–13.

Schulte, J. H., Lim, S., Schramm, A., Friedrichs, N., Koster, J., Versteeg, R., Ora, I., Pajtler, K., Klein-Hitpass, L., Kuhfittig-Kulle, S., et al. (2009). Lysine-specific demethylase 1 is strongly expressed in poorly differentiated neuroblastoma: Implications for therapy. *Cancer Res.* 69, 2065–2071.

Seligson, D. B., Horvath, S., McBrian, M. A., Mah, V., Yu, H., Tze, S., Wang, Q., Chia, D., Goodglick, L., and Kurdistani, S. K. (2009). Global levels of histone modifications predict prognosis in different cancers. *Am. J. Pathol.* 174, 1619–1628.

Sharma, S., Kelly, T. K., and Jones, P. A. (2010). Epigenetics in cancer. *Carcinogenesis* 31, 27–36.

Shi, Y., Lan, F., Matson, C., Mulligan, P., Whetstine, J. R., Cole, P. A., Casero, R. A., and Shi, Y. (2004). Histone demethylation mediated by the nuclear amine oxidase homolog LSD1. *Cell.* **119**, 941–953.

Shilatifard, A. (2006). Chromatin modifications by methylation and ubiquitination: Implications in the regulation of gene expression. *Annu. Rev. Biochem.* **75**, 243–269.

Shroff, R., Arbel-Eden, A., Pilch, D., Ira, G., Bonner, W. M., Petrini, J. H., Haber, J. E., and Lichten, M. (2004). Distribution and dynamics of chromatin modification induced by a defined DNA double-strand break. *Curr. Biol.* **14**, 1703–1711.

Sims, R. J., 3rd, Nishioka, K., and Reinberg, D. (2003). Histone lysine methylation: A signature for chromatin function. *Trends Genet.* **19**, 629–639.

Sims, R. J., 3rd, and Reinberg, D. (2006). Histone H3 Lys 4 methylation: Caught in a bind? *Genes Dev.* **20**, 2779–2786.

Singh, V., Miranda, T. B., Jiang, W., Frankel, A., Roemer, M. E., Robb, V. A., Gutmann, D. H., Herschman, H. R., Clarke, S., and Newsham, I. F. (2004). DAL-1/4.1B tumor suppressor interacts with protein arginine N-methyltransferase 3 (PRMT3) and inhibits its ability to methylate substrates in vitro and in vivo. *Oncogene* **23**, 7761–7771.

Steger, D. J., and Workman, J. L. (1996). Remodeling chromatin structures for transcription: What happens to the histones? *Bioessays* **18**, 875–884.

Strahl, B. D., and Allis, C. D. (2000). The language of covalent histone modifications. *Nature* **403**, 41–45.

Stucki, M., Clapperton, J. A., Mohammad, D., Yaffe, M. B., Smerdon, S. J., and Jackson, S. P. (2005). MDC1 directly binds phosphorylated histone H2AX to regulate cellular responses to DNA double-strand breaks. *Cell* **123**, 1213–1226.

Sun, Y., Jiang, X., Xu, Y., Ayrapetov, M. K., Moreau, L. A., Whetstine, J. R., and Price, B. D. (2009). Histone H3 methylation links DNA damage detection to activation of the tumour suppressor Tip60. *Nat. Cell Biol.* **11**, 1376–1382.

Sun, Z. W., and Allis, C. D. (2002). Ubiquitination of histone H2B regulates H3 methylation and gene silencing in yeast. *Nature* **418**, 104 108.

Thiriet, C., and Hayes, J. J. (2009). Linker histone phosphorylation regulates global timing of replication origin firing. *J. Biol. Chem.* **284**, 2823–2829.

Tsukada, Y., Fang, J., Erdjument-Bromage, H., Warren, M. E., Borchers, C. H., Tempst, P., and Zhang, Y. (2006). Histone demethylation by a family of JmjC domain-containing proteins. *Nature* **439**, 811–816.

Unnikrishnan, A., Gafken, P. R., and Tsukiyama, T. (2010). Dynamic changes in histone acetylation regulate origins of DNA replication. *Nat. Struct. Mol. Biol.* **17**(4), 430–437.

Utley, R. T., Lacoste, N., Jobin-Robitaille, O., Allard, S., and Cote, J. (2005). Regulation of NuA4 histone acetyltransferase activity in transcription and DNA repair by phosphorylation of histone H4. *Mol. Cell. Biol.* **25**, 8179–8190.

Vaissiere, T., Cuenin, C., Paliwal, A., Vineis, P., Hoek, G., Krzyzanowski, M., Airoldi, L., Dunning, A., Garte, S., Hainaut, P., *et al.* (2009a). Quantitative analysis of DNA methylation after whole bisulfitome amplification of a minute amount of DNA from body fluids. *Epigenetics* **4**, 221–230.

Vaissiere, T., Hung, R. J., Zaridze, D., Moukeria, A., Cuenin, C., Fasolo, V., Ferro, G., Paliwal, A., Hainaut, P., Brennan, P., *et al.* (2009b). Quantitative analysis of DNA methylation profiles in lung cancer identifies aberrant DNA methylation of specific genes and its association with gender and cancer risk factors. *Cancer Res.* **69**, 243–252.

Valerie, K., and Povirk, L. F. (2003). Regulation and mechanisms of mammalian double-strand break repair. *Oncogene* **22**, 5792–5812.

van Attikum, H., and Gasser, S. M. (2005). The histone code at DNA breaks: A guide to repair? *Nat. Rev.* **6**, 757–765.

Van Den Broeck, A., Brambilla, E., Moro-Sibilot, D., Lantuejoul, S., Brambilla, C., Eymin, B., Khochbin, S., and Gazzeri, S. (2008). Loss of histone H4K20 trimethylation occurs in preneoplasia and influences prognosis of non-small cell lung cancer. *Clin. Cancer Res.* **14**, 7237–7245.

Varambally, S., Dhanasekaran, S. M., Zhou, M., Barrette, T. R., Kumar-Sinha, C., Sanda, M. G., Ghosh, D., Pienta, K. J., Sewalt, R. G., Otte, A. P., et al. (2002). The polycomb group protein EZH2 is involved in progression of prostate cancer. *Nature* **419**, 624–629.

Vogelauer, M., Rubbi, L., Lucas, I., Brewer, B. J., and Grunstein, M. (2002). Histone acetylation regulates the time of replication origin firing. *Mol. Cell* **10**, 1223–1233.

Vogelauer, M., Wu, J., Suka, N., and Grunstein, M. (2000). Global histone acetylation and deacetylation in yeast. *Nature* **408**, 495–498.

Walkinshaw, D. R., Tahmasebi, S., Bertos, N. R., and Yang, X. J. (2008). Histone deacetylases as transducers and targets of nuclear signaling. *J. Cell. Biochem.* **104**, 1541–1552.

Wang, H., Zhai, L., Xu, J., Joo, H. Y., Jackson, S., Erdjument-Bromage, H., Tempst, P., Xiong, Y., and Zhang, Y. (2006). Histone H3 and H4 ubiquitylation by the CUL4-DDB-ROC1 ubiquitin ligase facilitates cellular response to DNA damage. *Mol. Cell* **22**, 383–394.

Weiss, V. H., McBride, A. E., Soriano, M. A., Filman, D. J., Silver, P. A., and Hogle, J. M. (2000). The structure and oligomerization of the yeast arginine methyltransferase, Hmt1. *Nat. Struct. Biol.* **7**, 1165–1171.

Whetstine, J. R., Nottke, A., Lan, F., Huarte, M., Smolikov, S., Chen, Z., Spooner, E., Li, E., Zhang, G., Colaiacovo, M., et al. (2006). Reversal of histone lysine trimethylation by the JMJD2 family of histone demethylases. *Cell* **125**, 467–481.

Workman, J. L., and Kingston, R. E. (1998). Alteration of nucleosome structure as a mechanism of transcriptional regulation. *Annu. Rev. Biochem.* **67**, 545–579.

Yang, X. J. (2004). The diverse superfamily of lysine acetyltransferases and their roles in leukemia and other diseases. *Nucleic Acids Res.* **32**, 959–976.

Zhang, J., Xu, F., Hashimshony, T., Keshet, I., and Cedar, H. (2002). Establishment of transcriptional competence in early and late S phase. *Nature* **420**, 198–202.

Zhang, Y. (2003). Transcriptional regulation by histone ubiquitination and deubiquitination. *Genes Dev.* **17**, 2733–2740.

Zhang, Z., Yamashita, H., Toyama, T., Sugiura, H., Omoto, Y., Ando, Y., Mita, K., Hamaguchi, M., Hayashi, S., and Iwase, H. (2004). HDAC6 expression is correlated with better survival in breast cancer. *Clin. Cancer Res.* **10**, 6962–6968.

Zhu, B., Zheng, Y., Pham, A. D., Mandal, S. S., Erdjument-Bromage, H., Tempst, P., and Reinberg, D. (2005). Monoubiquitination of human histone H2B: The factors involved and their roles in HOX gene regulation. *Mol. Cell* **20**, 601–611.

4

Epigenetics and miRNAs in Human Cancer

Muller Fabbri* and George A. Calin[†]

*Department of Molecular Virology, Immunology, and Medical Genetics and Comprehensive Cancer Center, Ohio State University, Biomedical Research Tower, Columbus, Ohio, USA
[†]Department of Experimental Therapeutics and Cancer Genetics, University of Texas, M.D. Anderson Cancer Center, Houston, Texas, USA

ABSTRACT

Epigenetic factors and microRNAs (miRNAs) are regulators of gene expression. Their regulatory function is frequently aberrant in cancer. In this chapter, we show that a tight connection occurs between miRNAs and epigenetics. Epigenetic factors can be responsible for the aberrancies of the miRNome (defined as the full spectrum of miRNAs for a specific genome) observed in cancer. Indeed, miRNAs undergo the same epigenetic regulatory laws like any other protein-coding gene. Moreover, a specific group of miRNAs (defined as epi-miRNAs) can directly target effectors of the epigenetic machinery (such as DNA methyltransferases, histone deacetylases, and polycomb repressive complex genes) and

Advances in Genetics, Vol. 70
Copyright 2010, Elsevier Inc. All rights reserved.

0065-2660/10 $35.00
DOI: 10.1016/S0065-2660(10)70004-3

indirectly affect the expression of tumor suppressor genes, whose expression is controlled by epigenetic factors. The result of this epigenetic–miRNA interaction is a new layer of complexity in gene regulation, whose comprehension opens new avenues to understand human cancerogenesis and to achieve new cancer treatments. © 2010, Elsevier Inc.

I. INTRODUCTION

Epigenetics is defined as all heritable changes in gene expression not associated with concomitant alterations in the DNA sequence. Gene epigenetic regulation usually includes DNA methylation and histone modifications.

DNA methylation is a reversible process, which occurs in specific genomic areas called CpG islands (Herman and Baylin, 2003; Weber et al., 2007), and is catalyzed by three major DNA methyltransferases: DNMT1 (maintenance DNMT), which preserves the methylation patterns throughout each cell division (Li et al., 1992, 1993), and DNMT3a and DNMT3b (de novo DNMTs), which transfer a methyl group to previously unmethylated genomic regions (Okano et al., 1999). CpG islands are frequently located at the promoter region of a gene, and their hypermethylated status prevents gene expression.

Histones are the main protein components of chromatin and can undergo posttranslational modifications (such as methylation, acetylation, etc.) which can determine whether chromatin is in the accessible, and early replicating form (called euchromatin), or in the inaccessible, and late replicating form (called heterochromatin). These histone modifications responsible for the chromatin status constitute the so-called "histone code." For instance, while acetylation of histones H3 and H4 and methylation of lysine 4 of histone H3 (H3K4me) are found in euchromatin, di- or trimethylation of lysine 9 of histone 3 (H3K9me) is a characteristic mark of heterochromatin, highly conserved in fungi, plants, and animals (Bernstein et al., 2007). These histone modifications are catalyzed by several histone deacetylases (HDACs) and histone methyltransferases (HMTs). Finally, the polycomb repressive complex 2 (PRC2) can mediate epigenetic gene silencing by trimethylating histone H3 lysine 27 (H3K27me3), a mark of heterochromatin.

microRNAs (miRNAs) are noncoding RNAs (ncRNAs) which regulate gene expression. miRNAs are involved in a variety of biological processes, including development, differentiation, apoptosis, and cell proliferation (Ambros and Lee, 2004; Bartel, 2004; Carleton et al., 2007; He and Hannon, 2004; Pasquinelli et al., 2005; Plasterk, 2006).

The biogenesis of miRNAs is initiated by an RNA polymerase II, which initially transcribes the miRNA gene into a long, capped, and polyadenylated precursor, called pri-miRNA (Cai et al., 2004; Lee et al., 2004). By means of a double-stranded RNA-specific ribonuclease called Drosha, in conjunction with its binding partner DGCR8 (DiGeorge syndrome critical region gene 8 or Pasha), the pri-miRNA is processed into a hairpin RNA precursor (pre-miRNA), about 70–100 nucleotides (nt) long (Cullen, 2004). Subsequently, the pre-miRNA translocates from the nucleus to the cytoplasm, by means of Exportin 5/Ran GTP. Once in the cytoplasm, the precursor is cleaved into a 18–24 nt duplex by a ribonucleoproteic complex, composed of a ribonuclease III (Dicer), and TRBP (HIV-1 transactivating response RNA-binding protein). Finally, the duplex interacts with a large protein complex called RISC (RNA-induced silencing complex), which includes proteins of the Argonaute family (Ago1–4 in humans). One strand of the miRNA duplex remains stably associated with RISC and becomes the mature miRNA, which guides the RISC complex mainly (but not exclusively) to the 3′-untranslated region (3′-UTR) of the target mRNAs. According to the miRNA:mRNA degree of base pair complementarity, the target mRNA can be cleaved (in case of perfect Watsonian match) or its translation into protein can be prevented (in case of imperfect Watsonian match). Overall, the effect of miRNAs is to silence the expression of the target mRNAs either by mRNA cleavage or by translational repression. However, it has been recently shown that miRNAs can actually also increase the expression of a target mRNA (Vasudevan et al., 2007). Figure 4.1 summarizes the events that occur during miRNA biogenesis.

Each mRNA can be targeted by more than one miRNA (Vatolin et al., 2006), and each miRNA can target hundreds different transcripts. For instance, it has been demonstrated that a cluster of two miRNAs (namely miR-15a and miR-16) can affect the expression of about 14% of the human genome in a leukemic cell line (Calin et al., 2008).

miRNAs are frequently located in cancer-associated genomic regions (CAGRs), which include fragile sites (FRA) where tumor suppressor genes (TSGs) are located, and regions of frequent LOH, deletion, amplification, and translocation (Calin et al., 2004). Since this first report, several groups have identified aberrancies of the miRNome (defined as the full spectrum of miRNAs in a specific genome) in almost all human tumors (Croce, 2009; Fabbri et al., 2009; Garzon et al., 2006). Tumor-specific signatures of deregulated miRNAs have been identified for all tested human cancers, which sometimes harbor prognostic implications (Barbarotto et al., 2008; Calin and Croce, 2006; Khoshnaw et al., 2009; Lowery et al., 2009; Ortholan et al., 2009; Pekarsky et al., 2005).

In an attempt to decode which mechanisms underlie the abnormal miRNA expression observed in cancer, an increasing number of studies have investigated how miRNAs are regulated. It is now widely accepted that miRNAs undergo the same regulatory mechanisms of conventional protein-coding genes (PCG), including epigenetic regulation. Intriguingly, a subgroup of these

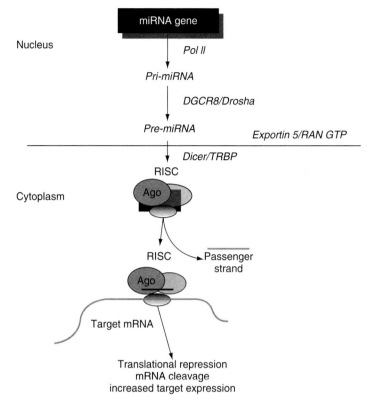

Figure 4.1. Biogenesis of miRNAs.

ncRNAs (epi-miRNAs) control, directly and indirectly, the expression of epigenetic effectors, such as DNMTs, HDACs, and polycomb genes, casting a new
light on miRNA functions as both genetic and epigenetic regulators. This
chapter focuses on two aspects: miRNAs as targets of epigenetic changes and
miRNAs as regulators of the epigenetic machinery (epi-miRNAs) and on their
implications for human cancerogenesis.

II. miRNAS AS TARGETS OF EPIGENETIC CHANGES IN CANCER

Paralleling what occurs for other PCGs, aberrant epigenetic regulation is responsible for abnormal miRNA expression in several malignancies.

By treating bladder cancer cells with both a DNA demethylating agent
(5-aza-2'-deoxycytidine, 5-AZA) and an HDAC inhibitor (4-phenylbutyric
acid), Saito et al. (2006) found that about 5% of all human miRNAs increased

their expression levels. Among the reexpressed miRNAs, miR-127 was upregulated about 49 times in treated versus untreated cells and reduced the expression of the oncogene BCL6, one of its direct target genes. This miRNA is embedded in a CpG island region and is epigenetically silenced by both promoter hypermethylation and histone modifications in cancer cells (Saito et al., 2006). Despite miR-127 belongs to a large miRNA cluster, which includes miR-136, -431, -432, and -433, it is the only member of the cluster whose expression increases upon treatment with the two epigenetic drugs (Saito et al., 2006). Moreover, when each drug was used alone, no variation in miR-127 expression was observed (Saito et al., 2006), indicating that both DNA methylation and histone modifications affect the epigenetic regulation of miR-127. This seminal work shows that indeed miRNAs undergo epigenetic regulation, that it is a complex epigenetic regulation (involving both methylation and histone modifications), and that there are differences among miRNAs which even belong to the same cluster. Lujambio et al. created a double knockout (DKO) for DNMT1 and DNMT3b in the colorectal cancer cell line HCT-116, and compared miRNA expression profile of DKO and wild-type cells. About 6% of the 320 analyzed miRNAs were upregulated in the DKO cells (Lujambio et al., 2007). Among the deregulated miRNAs, only miR-124a was embedded in a CpG island which is densely methylated in the cancer cell line, but not in the normal tissue. Upregulation of miR-124a reduces the levels of CDK6 and impacts on the phosphorylation status of CDK6-downstream effector Rb protein (Lujambio et al., 2007). More recently, Prosper's group has identified a signature of 13 miRNAs embedded in CpG islands, with high heterochromatic markers (such as high levels of K9H3me2 and/or low levels of K4H3me3) in acute lymphoblastic leukemia (ALL) patients (Agirre et al., 2009; Roman-Gomez et al., 2009). Among these, miR-124a was methylated in 59% of ALLs, and its promoter hypermethylation was associated with higher relapse rate and mortality rate versus nonhypermethylated cases, being miR-124a promoter methylation status an independent prognostic factor for disease-free and overall survival (Agirre et al., 2009). Finally, supporting Lujambio's results, also in ALL the impact of miR-124a in the CDK6-Rb pathway was demonstrated by showing that miR-124a directly silences CDK6 (Agirre et al., 2009). Hypermethylation of miR-124a promoter is also involved in the formation of epigenetic field defect, a gastric cancer predisposing condition characterized by accumulation of abnormal DNA methylation in normal-appearing gastric mucosa, mostly induced by Helicobacter pylori infection (Ando et al., 2009). These findings also suggest that miR-124a promoter hypermethylation is an early event in gastric carcinogenesis. In addition to miR-124a, also miR-107, another epigenetically controlled miRNA, targets CDK6 and impacts this oncogenic pathway in pancreatic cancer (Lee et al., 2009b). In HCT-116 cells deficient for DNMT1 and DNMT3B, Brueckner et al. (2007) showed increased expression of let-7a-3, a miRNA normally silenced by promoter hypermethylation in the wild-type cell line. In lung adenocarcinoma primary tumors, let-7a-3 promoter was found hypomethylated with respect to the normal counterpart (Brueckner *et al.*, 2007), whereas hypermethylation of let-7a-3 promoter was described in epithelial ovarian cancer, paralleled the low expression of insulin-like growth factor-II expression and was associated with a good prognosis (Lu *et al.*, 2007). Therefore, DNA methylation could act as protective mechanism by silencing miRNA with oncogenic function. Also, the miRNA-200 family participates in the maintenance of an epithelial phenotype and loss of its expression can result in epithelial to mesenchymal transition (EMT). Furthermore, the loss of expression of miR-200 family members is linked to an aggressive cancer phenotype. Vrba *et al.* found that aberrant DNA methylation of the miR-200c/141 CpG island is closely linked to their inappropriate silencing in cancer cells. The epigenetic regulation of this miRNA cluster appears evolutionarily conserved, since similar results were obtained in mouse (Vrba *et al.*, 2010). Interestingly, no variation in miRNA expression was observed in lung cancer cells treated with demethylating agents or HDAC inhibitors or their combination (Yanaihara *et al.*, 2006). Another miRNA which is under epigenetic control is miR-1. In hepatocarcinoma, miR-1 is frequently silenced by promoter hypermethylation (Datta *et al.*, 2008). However, in DNMT1 null HCT-116 cells (but not in DNMT3B null cells), hypomethylation and reexpression of miR-1-1 were observed (Datta *et al.*, 2008), revealing a key role for the *maintenance* DNMT in the regulation of this miRNA. These studies bring to the question: which factors determine the degree of epigenetic control on miRNA genes? Han *et al.* (2007) observed that neither 5-AZA nor DNMT1 deletion alone can recapitulate miRNA expression profile of DKO DNMT1/DNMT3B HCT-116 cells. Also, Lehmann *et al.* (2008) found that in breast cancer cell lines 5-AZA reactivates miR-9-1 (hypermethylated in up to 86% of primary tumors), but not miR-124a-3, miR-148, miR-152, and miR-663 (hypermethylated as well) (Lehmann *et al.*, 2008). Previously, Meng *et al.* (2008) observed that in malignant, but not in normal cholangiocytes, 5-AZA induces reexpression of miR-370. Overall, these results indicate that the epigenetic control of miRNAs is both cancer- and miRNA-specific.

A. Epigenetics can regulate miRNA expression in metastatic tumors

Lujambio *et al.* (2008) treated three lymph-node metastatic cell lines with 5-AZA and identified three miRNAs which showed cancer-specific CpG island hypermethylation: miR-148a, miR-34b/c, and miR-9. The reintroduction of miR-148a and miR-34b/c in cancer cells with epigenetic inactivation inhibited cell motility and their metastatic potential in xenograft models, and was associated with downregulation of miRNA oncogenic target genes, such as c-MYC, E2F3, CDK6, and TGIF2 (Lujambio *et al.*, 2008). Finally, promoter hypermethylation of these three miRNAs was significantly associated with metastasis formation also in human malignancies (Lujambio *et al.*, 2008). miR-34b/c cluster is epigenetically regulated also in colorectal cancer (promoter hypermethylation in 90% of primary colorectal cancer tumors vs. normal colon mucosa) (Toyota *et al.*, 2008), whereas epigenetic silencing of miR-9 and miR-148a (together with miR-152, -124a, and -663) was described also in breast cancer (Lehmann *et al.*, 2008).

The relationship between miRNA and cognate host gene epigenetic regulation was addressed by Grady *et al.* (2008) by studying miR-342, located in an intron of the EVL (Ena/Vasp-like) gene. EVL promoter hypermethylation occurs in 86% of colorectal cancers and is already present in 67% of adenomas, suggesting that it is an early event in colon cancerogenesis. The combined treatment with 5-AZA and the HDAC inhibitor trichostatin A restores the synchronized expression of EVL and miR-342. The EGFL7 gene, frequently downregulated in several cancer cell lines and in primary bladder and prostate tumors, hosts miR-126 in one of its introns. While the mature miR-126 can be encoded by three different transcripts of the cognate host gene, each of them with its own promoter, miR-126 is concomitantly upregulated with one of EGFL7 transcripts which has a CpG island promoter, when cancer cell lines are treated with inhibitors of DNA methylation and histone deacetylation, indicating that silencing of intronic miRNAs in cancer may occur by means of epigenetic changes of cognate host genes (Saito *et al.*, 2009). In summary, miRNAs are encoded by either noncoding RNA genes, which have their own promoters, or noncoding sequences in introns of PCGs. In the latter case, the expression of miRNAs is driven usually by the same promoters as for the corresponding proteins.

Finally, Fazi *et al.* showed that transcription factors can recruit epigenetic effectors at miRNA promoter regions and contribute to the regulation of their expression. The AML1/ETO fusion oncoprotein is the aberrant product of t(8;21) translocation in acute myeloid leukemia (AML) and can bind to the pre-miR-223 region. The oncoprotein recruits epigenetic effectors (i.e., DNMTs, HDAC1, and MeCP2), leading to aberrant hypermethylation of the CpG in close proximity to the AML1/ETO-binding site, and H3–H4 deacetylation of the same chromatin region (Fazi *et al.*, 2007).

In SkBr3 breast cancer cell line, Scott *et al.* (2006) were able to demonstrate that 27 miRNA expression levels are rapidly modified (5 up- and 22 downregulated) by a treatment with the HDAC inhibitor LAQ824, indicating that some miRNAs are mainly silenced by histone modifications. In A549 lung cancer cell line, the HDAC inhibitor SAHA deregulates 64 miRNA (more than twofold change) targeting genes involved in angiogenesis, apoptosis, chromatin modification, cell proliferation, and differentiation (Lee *et al.*, 2009a).

In summary, there is clear experimental evidence that miRNAs undergo epigenetic regulation like any other PCG. These epigenetic factors are ultimately responsible for the expression/silencing of miRNAs and, indirectly, affect their target gene levels, contributing to the normal or malignant phenotype.

III. epi-miRNAS: REGULATORS OF THE EPIGENETIC MACHINERY

With the term epi-miRNAs, we define those miRNAs which target, directly or indirectly, effectors of the epigenetic machinery. The first evidence of the existence of epi-miRNAs was obtained in lung cancer cell lines, where a family of miRNAs (miR-29a, -29b, and -29c) directly targets DNMT3a and DNMT3b (Fabbri et al., 2007). Reexpression of miR-29s induces disruption of de novo DNA methylation and leads to a global DNA hypomethylation of cancer cells. Interestingly, TSGs, epigenetically silenced by promoter hypermethylation (such as FHIT and WWOX in lung cancer), were reexpressed as a consequence of their promoter CpG island demethylation occurring after miR-29s treatment, and led to cancer cell apoptosis and inhibition of lung cancer growth in xenograft models (Fabbri et al., 2007). This discovery identified a previously unknown mechanism: miRNAs can indirectly regulate gene expression by affecting their epigenetic regulation. In mouse embryonic stem cells (ES), two independent groups have shown that members of miR-290 cluster directly target RBL2, an inhibitor of DNMT3 genes (Benetti et al., 2008; Sinkkonen et al., 2008). ES Dicer null cells are characterized by no expression of the miR-290 cluster, overexpression of RBL2, and disruption of de novo methylation pathway, leading to increased telomere recombination and aberrant telomere elongation. Restoration of the miRNA cluster reverted this phenotype (Benetti et al., 2008; Sinkkonen et al., 2008). Interestingly, the regulatory effect of miR-290 cluster on de novo DNMTs was not observed in human embryonic kidney 293 cells following Dicer knockdown, suggesting that miR-290 targeting effect on DNMT3s might be cell- and/or species-specific (Sinkkonen et al., 2008).

In an AML model, Garzon et al. showed that in addition to the de novo DNMTs, miR-29b is also able to indirectly silence DNMT1, by directly targeting SP1, a transactivator of the DNMT1 gene (Garzon et al., 2009). DNMT3b expression is also under the control of miR-148a and miR-148b. By binding to an unusual site in the coding region (not the 3'-UTR) of DNMT3b mRNA, miR-148 family regulates this de novo DNMT and might be responsible for the several different splice variants of DNMT3b (Duursma et al., 2008). Interestingly, since miR-148a is also epigenetically regulated and its promoter is frequently hypermethylated in different tumors (Lehmann et al., 2008; Lujambio et al., 2008), a self-amplified epigenetic loop for this epi-miRNA can be hypothesized, although more experimental evidences need to be provided to confirm this model. In cholangiocarcinoma, miR-148a, -152, and -301 can directly target DNMT1, but their expression is silenced by IL-6, which is involved in cholangiocancerogenesis (Braconi et al., 2010).

Epi-miRNAs are also involved in regulating the expression of HDACs and PRC genes. For instance, HDAC4 is a direct target of both miR-1 and miR-140 (Chen et al., 2006; Tuddenham et al., 2006), while miR-449a binds to the 3'-UTR region of HDAC1 (Noonan et al., 2009). HDAC1 is upregulated in several kind of cancers, and miR-449a reexpression in prostate cancer cells induces cell-cycle arrest, apoptosis, and a senescent-like phenotype by reducing the levels of HDAC1 (Noonan et al., 2009). EZH2 is the catalytic subunit of the PRC2, and is responsible for heterochromatin formation by trimethylating histone H3 lysine 27 (H3K27me3), leading to the silencing of several TSGs. Varambally et al. showed that in prostate cancer cell lines and primary tumors, the expression of miR-101 decreases during cancer progression, inversely correlating with an increase in EZH2. These findings are suggestive of a role as epi-miRNA for miR-101, a hypothesis which was tested and confirmed by showing that miR-101 directly targets EZH2 both in prostate and in bladder cancer models (Friedman et al., 2009; Varambally et al., 2008). Moreover, miR-101-mediated suppression of EZH2 inhibits cancer cell proliferation and colony formation, revealing a TSG role for miR-101, mediated by its modulatory effects on cancer epigenome (Friedman et al., 2009).

IV. CONCLUSION

Epigenetics plays a major role in gene expression regulation. Aberrancies in cell epigenome are implicated in human cancerogenesis. Similarly, variations of the miRNome have been documented in cancer cells with respect to the normal cell counterpart. Recently, the identification of an epigenetic regulation of miRNAs and the existence of epi-miRNAs has shown that miRNAs and epigenetics are intertwined biological effectors of gene regulation. Several aspect of this interaction still needs to be clarified. For instance, why only specific miRNAs (even within the same cluster) are responsive to the treatment with epigenetic drugs? Moreover, epi-miRNA directly targeting DNMTs induces an overall global hypomethylation of the cancerous cell, but how does this translate into a selective reexpression of hypermethylated TSGs and not of oncogenes? Further studies will address these and other concerns. Decoding the connection miR-Nome-epigenome will lead to a better understanding of gene regulation, and ultimately, of human cancerogenesis, therefore allowing the translation of this knowledge into new treatments for cancer patients.

Acknowledgments

M. F. is recipient of a 2009 Kimmel Scholar Award. G. A. C. is supported as a Fellow at The University of Texas M. D. Anderson Research Trust, as a Fellow of The University of Texas System Regents Research Scholar and by NIH, DOD and by 2009 Seena Magowitz—Pancreatic Cancer Action Network—AACR Pilot Grant.

References

Agirre, X., Vilas-Zornoza, A., Jimenez-Velasco, A., Ignacio Martin-Subero, J., Cordeu, L., Garate, L., San Jose-Eneriz, E., Abizanda, G., Rodriguez-Otero, P., Fortes, P., *et al.* (2009). Epigenetic silencing of the tumor suppressor microRNA Hsa-miR-124a regulates CDK6 expression and confers a poor prognosis in acute lymphoblastic leukemia. *Cancer Res.* **69**(10), 4443–4453.

Ambros, V., and Lee, R. C. (2004). Identification of microRNAs and other tiny noncoding RNAs by cDNA cloning. *Methods Mol. Biol.* **265**, 131–158.

Ando, T., Yoshida, T., Enomoto, S., Asada, K., Tatematsu, M., Ichinose, M., Sugiyama, T., and Ushijima, T. (2009). DNA methylation of microRNA genes in gastric mucosae of gastric cancer patients: Its possible involvement in the formation of epigenetic field defect. *Int. J. Cancer* **124**, 2367–2374.

Barbarotto, E., Schmittgen, T. D., and Calin, G. A. (2008). MicroRNAs and cancer: Profile, profile, profile. *Int. J. Cancer* **122**, 969–977.

Bartel, D. P. (2004). MicroRNAs: Genomics, biogenesis, mechanism, and function. *Cell* **116**, 281–297.

Benetti, R., Gonzalo, S., Jaco, I., Munoz, P., Gonzalez, S., Schoeftner, S., Murchison, E., Andl, T., Chen, T., Klatt, P., *et al.* (2008). A mammalian microRNA cluster controls DNA methylation and telomere recombination via Rbl2-dependent regulation of DNA methyltransferases. *Nat. Struct. Mol. Biol.* **15**, 998.

Bernstein, B. E., Meissner, A., and Lander, E. S. (2007). The mammalian epigenome. *Cell* **128**, 669–681.

Braconi, C., Huang, N., and Patel, T. (2010). MicroRNA-dependent regulation of DNA methyltransferase-1 and tumor suppressor gene expression by interleukin-6 in human malignant cholangiocytes. *Hepatology (Baltimore, MD)* **51**, 881–890.

Brueckner, B., Stresemann, C., Kuner, R., Mund, C., Musch, T., Meister, M., Sultmann, H., and Lyko, F. (2007). The human let-7a-3 locus contains an epigenetically regulated microRNA gene with oncogenic function. *Cancer Res.* **67**, 1419–1423.

Cai, X., Hagedorn, C. H., and Cullen, B. R. (2004). Human microRNAs are processed from capped, polyadenylated transcripts that can also function as mRNAs. *RNA* **10**, 1957–1966.

Calin, G. A., and Croce, C. M. (2006). MicroRNA signatures in human cancers. *Nat. Rev.* **6**, 857–866.

Calin, G. A., Sevignani, C., Dumitru, C. D., Hyslop, T., Noch, E., Yendamuri, S., Shimizu, M., Rattan, S., Bullrich, F., Negrini, M., *et al.* (2004). Human microRNA genes are frequently located at fragile sites and genomic regions involved in cancers. *Proc. Natl. Acad. Sci. USA* **101**, 2999–3004.

Calin, G. A., Cimmino, A., Fabbri, M., Ferracin, M., Wojcik, S. E., Shimizu, M., Taccioli, C., Zanesi, N., Garzon, R., Aqeilan, R. I., *et al.* (2008). MiR-15a and miR-16-1 cluster functions in human leukemia. *Proc. Natl. Acad. Sci. USA* **105**, 5166–5171.

Carleton, M., Cleary, M. A., and Linsley, P. S. (2007). MicroRNAs and cell cycle regulation. *Cell Cycle* **6**, 2127–2132.

Chen, J. F., Mandel, E. M., Thomson, J. M., Wu, Q., Callis, T. E., Hammond, S. M., Conlon, F. L., and Wang, D. Z. (2006). The role of microRNA-1 and microRNA-133 in skeletal muscle proliferation and differentiation. *Nat. Genet.* **38,** 228–233.

Croce, C. M. (2009). Causes and consequences of microRNA dysregulation in cancer. *Nat. Rev.* **10,** 704–714.

Cullen, B. R. (2004). Transcription and processing of human microRNA precursors. *Mol. Cell* **16,** 861–865.

Datta, J., Kutay, H., Nasser, M. W., Nuovo, G. J., Wang, B., Majumder, S., Liu, C. G., Volinia, S., Croce, C. M., Schmittgen, T. D., *et al.* (2008). Methylation mediated silencing of MicroRNA-1 gene and its role in hepatocellular carcinogenesis. *Cancer Res.* **68,** 5049–5058.

Duursma, A. M., Kedde, M., Schrier, M., le Sage, C., and Agami, R. (2008). mir-148 targets human DNMT3b protein coding region. *RNA (New York, NY)* **14,** 872–877.

Fabbri, M., Garzon, R., Cimmino, A., Liu, Z., Zanesi, N., Callegari, E., Liu, S., Alder, H., Costinean, S., Fernandez-Cymering, C., *et al.* (2007). MicroRNA-29 family reverts aberrant methylation in lung cancer by targeting DNA methyltransferases 3A and 3B. *Proc. Natl. Acad. Sci. USA* **104,** 15805–15810.

Fabbri, M., Croce, C. M., and Calin, G. A. (2009). MicroRNAs in the ontogeny of leukemias and lymphomas. *Leuk. Lymphoma* **50,** 160–170.

Fazi, F., Racanicchi, S., Zardo, G., Starnes, L. M., Mancini, M., Travaglini, L., Diverio, D., Ammatuna, E., Cimino, G., Lo-Coco, F., *et al.* (2007). Epigenetic silencing of the myelopoiesis regulator microRNA-223 by the AML1/ETO oncoprotein. *Cancer Cell* **12,** 457–466.

Friedman, J. M., Liang, G., Liu, C. C., Wolff, E. M., Tsai, Y. C., Ye, W., Zhou, X., and Jones, P. A. (2009). The putative tumor suppressor microRNA-101 modulates the cancer epigenome by repressing the polycomb group protein EZH2. *Cancer Res.* **69,** 2623–2629.

Garzon, R., Fabbri, M., Cimmino, A., Calin, G. A., and Croce, C. M. (2006). MicroRNA expression and function in cancer. *Trends Mol. Med.* **12,** 580–587.

Garzon, R., Liu, S., Fabbri, M., Liu, Z., Heaphy, C. E., Callegari, E., Schwind, S., Pang, J., Yu, J., Muthusamy, N., *et al.* (2009). MicroRNA -29b induces global DNA hypomethylation and tumor suppressor gene re-expression in acute myeloid leukemia by targeting directly DNMT3A and 3B and indirectly DNMT1. *Blood* **113**(25), 6411–6418.

Grady, W. M., Parkin, R. K., Mitchell, P. S., Lee, J. H., Kim, Y. H., Tsuchiya, K. D., Washington, M. K., Paraskeva, C., Willson, J. K., Kaz, A. M., *et al.* (2008). Epigenetic silencing of the intronic microRNA hsa-miR-342 and its host gene EVL in colorectal cancer. *Oncogene* **27,** 3880–3888.

Han, L., Witmer, P. D., Casey, E., Valle, D., and Sukumar, S. (2007). DNA methylation regulates microRNA expression. *Cancer Biol. Ther.* **6,** 1284–1288.

He, L., and Hannon, G. J. (2004). MicroRNAs: Small RNAs with a big role in gene regulation. *Nat. Rev. Genet.* **5,** 522–531.

Herman, J. G., and Baylin, S. B. (2003). Gene silencing in cancer in association with promoter hypermethylation. *N. Engl. J. Med.* **349,** 2042–2054.

Khoshnaw, S. M., Green, A. R., Powe, D. G., and Ellis, I. O. (2009). MicroRNA involvement in the pathogenesis and management of breast cancer. *J. Clin. Pathol.* **62,** 422–428.

Lee, Y., Kim, M., Han, J., Yeom, K. H., Lee, S., Baek, S. H., and Kim, V. N. (2004). MicroRNA genes are transcribed by RNA polymerase II. *EMBO J.* **23,** 4051–4060.

Lee, E. M., Shin, S., Cha, H. J., Yoon, Y., Bae, S., Jung, J. H., Lee, S. M., Lee, S. J., Park, I. C., Jin, Y. W., *et al.* (2009a). Suberoylanilide hydroxamic acid (SAHA) changes microRNA expression profiles in A549 human non-small cell lung cancer cells. *Int. J. Mol. Med.* **24,** 45–50.

Lee, K. H., Lotterman, C., Karikari, C., Omura, N., Feldmann, G., Habbe, N., Goggins, M. G., Mendell, J. T., and Maitra, A. (2009b). Epigenetic silencing of microRNA miR-107 regulates cyclin-dependent kinase 6 expression in pancreatic cancer. *Pancreatology* **9,** 293–301.

Lehmann, U., Hasemeier, B., Christgen, M., Muller, M., Romermann, D., Langer, F., and Kreipe, H. (2008). Epigenetic inactivation of microRNA gene hsa-mir-9-1 in human breast cancer. *J. Pathol.* **214,** 17–24.

Li, E., Bestor, T. H., and Jaenisch, R. (1992). Targeted mutation of the DNA methyltransferase gene results in embryonic lethality. *Cell* **69,** 915–926.

Li, E., Beard, C., and Jaenisch, R. (1993). Role for DNA methylation in genomic imprinting. *Nature* **366,** 362–365.

Lowery, A. J., Miller, N., Devaney, A., McNeill, R. E., Davoren, P. A., Lemetre, C., Benes, V., Schmidt, S., Blake, J., Ball, G., *et al.* (2009). MicroRNA signatures predict oestrogen receptor, Progesterone receptor and HER2/neu receptor status in breast cancer. *Breast Cancer Res.* **11,** R27.

Lu, L., Katsaros, D., de la Longrais, I. A., Sochirca, O., and Yu, H. (2007). Hypermethylation of let-7a-3 in epithelial ovarian cancer is associated with low insulin-like growth factor-II expression and favorable prognosis. *Cancer Res.* **67,** 10117–10122.

Lujambio, A., Ropero, S., Ballestar, E., Fraga, M. F., Cerrato, C., Setien, F., Casado, S., Suarez-Gauthier, A., Sanchez-Cespedes, M., Git, A., *et al.* (2007). Genetic unmasking of an epigenetically silenced microRNA in human cancer cells. *Cancer Res.* **67,** 1424–1429.

Lujambio, A., Calin, G. A., Villanueva, A., Ropero, S., Sanchez-Cespedes, M., Blanco, D., Montuenga, L. M., Rossi, S., Nicoloso, M. S., Faller, W. J., *et al.* (2008). A microRNA DNA methylation signature for human cancer metastasis. *Proc. Natl. Acad. Sci. USA* **105,** 13556–13561.

Meng, F., Wehbe-Janek, H., Henson, R., Smith, H., and Patel, T. (2008). Epigenetic regulation of microRNA-370 by interleukin-6 in malignant human cholangiocytes. *Oncogene* **27,** 378–386.

Noonan, E. J., Place, R. F., Pookot, D., Basak, S., Whitson, J. M., Hirata, H., Giardina, C., and Dahiya, R. (2009). miR-449a targets HDAC-1 and induces growth arrest in prostate cancer. *Oncogene* **28,** 1714–1724.

Okano, M., Bell, D. W., Haber, D. A., and Li, E. (1999). DNA methyltransferases Dnmt3a and Dnmt3b are essential for de novo methylation and mammalian development. *Cell* **99,** 247–257.

Ortholan, C., Puissegur, M. P., Ilie, M., Barbry, P., Mari, B., and Hofman, P. (2009). MicroRNAs and lung cancer: New oncogenes and tumor suppressors, new prognostic factors and potential therapeutic targets. *Curr. Med. Chem.* **16,** 1047–1061.

Pasquinelli, A. E., Hunter, S., and Bracht, J. (2005). MicroRNAs: A developing story. *Curr. Opin. Genet. Dev.* **15,** 200–205.

Pekarsky, Y., Calin, G. A., and Aqeilan, R. (2005). Chronic lymphocytic leukemia: Molecular genetics and animal models. *Curr. Top. Microbiol. Immunol.* **294,** 51–70.

Plasterk, R. H. (2006). Micro RNAs in animal development. *Cell* **124,** 877–881.

Roman-Gomez, J., Agirre, X., Jimenez-Velasco, A., Arqueros, V., Vilas-Zornoza, A., Rodriguez-Otero, P., Martin-Subero, I., Garate, L., Cordeu, L., San Jose-Eneriz, E., *et al.* (2009). Epigenetic regulation of microRNAs in acute lymphoblastic leukemia. *J. Clin. Oncol.* **27,** 1316–1322.

Saito, Y., Liang, G., Egger, G., Friedman, J. M., Chuang, J. C., Coetzee, G. A., and Jones, P. A. (2006). Specific activation of microRNA-127 with downregulation of the proto-oncogene BCL6 by chromatin-modifying drugs in human cancer cells. *Cancer Cell* **9,** 435–443.

Saito, Y., Friedman, J. M., Chihara, Y., Egger, G., Chuang, J. C., and Liang, G. (2009). Epigenetic therapy upregulates the tumor suppressor microRNA-126 and its host gene EGFL7 in human cancer cells. *Biochem. Biophys. Res. Commun.* **379,** 726–731.

Scott, G. K., Mattie, M. D., Berger, C. E., Benz, S. C., and Benz, C. C. (2006). Rapid alteration of microRNA levels by histone deacetylase inhibition. *Cancer Res.* **66,** 1277–1281.

Sinkkonen, L., Hugenschmidt, T., Berninger, P., Gaidatzis, D., Mohn, F., Artus-Revel, C. G., Zavolan, M., Svoboda, P., and Filipowicz, W. (2008). MicroRNAs control de novo DNA methylation through regulation of transcriptional repressors in mouse embryonic stem cells. *Nat. Struct. Mol. Biol.* **15,** 259–267.

Toyota, M., Suzuki, H., Sasaki, Y., Maruyama, R., Imai, K., Shinomura, Y., and Tokino, T. (2008). Epigenetic silencing of microRNA-34b/c and B-cell translocation gene 4 is associated with CpG island methylation in colorectal cancer. *Cancer Res.* **68**, 4123–4132.

Tuddenham, L., Wheeler, G., Ntounia-Fousara, S., Waters, J., Hajihosseini, M. K., Clark, I., and Dalmay, T. (2006). The cartilage specific microRNA-140 targets histone deacetylase 4 in mouse cells. *FEBS Lett.* **580**, 4214–4217.

Varambally, S., Cao, Q., Mani, R. S., Shankar, S., Wang, X., Ateeq, B., Laxman, B., Cao, X., Jing, X., Ramnarayanan, K., *et al.* (2008). Genomic loss of microRNA-101 leads to overexpression of histone methyltransferase EZH2 in cancer. *Science (New York, NY)* **322**, 1695–1699.

Vasudevan, S., Tong, Y., and Steitz, J. A. (2007). Switching from repression to activation: micro-RNAs can up-regulate translation. *Science (New York, NY)* **318**, 1931–1934.

Vatolin, S., Navaratne, K., and Weil, R. J. (2006). A novel method to detect functional microRNA targets. *J. Mol. Biol.* **358**, 983–996.

Vrba, L., Jensen, T. J., Garbe, J. C., Heimark, R. L., Cress, A. E., Dickinson, S., Stampfer, M. R., and Futscher, B. W. (2010). Role for DNA methylation in the regulation of miR-200c and miR-141 expression in normal and cancer cells. *PLoS One* **5**, e8697.

Weber, M., Hellmann, I., Stadler, M. B., Ramos, L., Paabo, S., Rebhan, M., and Schubeler, D. (2007). Distribution, silencing potential and evolutionary impact of promoter DNA methylation in the human genome. *Nat. Genet.* **39**, 457–466.

Yanaihara, N., Caplen, N., Bowman, E., Seike, M., Kumamoto, K., Yi, M., Stephens, R. M., Okamoto, A., Yokota, J., Tanaka, T., *et al.* (2006). Unique microRNA molecular profiles in lung cancer diagnosis and prognosis. *Cancer Cell* **9**, 189–198.

5

Interplay Between Different Epigenetic Modifications and Mechanisms

Rabih Murr

Friedrich Miescher Institute for Biomedical Research, Maulbeerstrasse 66, 4058 Basel, Switzerland

ABSTRACT

Cellular functions including transcription regulation, DNA repair, and DNA replication need to be tightly regulated. DNA sequence can contribute to the regulation of these mechanisms. This is examplified by the consensus sequences that allow the binding of specific transcription factors, thus regulating

Advances in Genetics, Vol. 70
Copyright 2010, Elsevier Inc. All rights reserved.

0065-2660/10 $35.00
DOI: 10.1016/S0065-2660(10)70005-5

transcription rates. Another layer of regulation resides in modifications that do not affect the DNA sequence itself but still results in the modification of chromatin structure and properties, thus affecting the readout of the underlying DNA sequence. These modifications are dubbed as "epigenetic modifications" and include, among others, histone modifications, DNA methylation, and small RNAs. While these events can independently regulate cellular mechanisms, recent studies indicate that joint activities of different epigenetic modifications could result in a common outcome. In this chapter, I will attempt to recapitulate the best known examples of collaborative activities between epigenetic modifications. I will emphasize mostly on the effect of crosstalks between epigenetic modifications on transcription regulation, simply because it is the most exposed and studied aspect of epigenetic interactions. I will also summarize the effect of epigenetic interactions on DNA damage response and DNA repair. The involvement of epigenetic crosstalks in cancer formation, progression, and treatment will be emphasized throughout the manuscript. Due to space restrictions, additional aspects involving histone replacements [Park, Y. J., and Luger, K. (2008). Histone chaperones in nucleosome eviction and histone exchange. *Curr. Opin. Struct. Biol.* **18**, 282–289.], histone variants [Boulard, M., Bouvet, P., Kundu, T. K., and Dimitrov, S. (2007). Histone variant nucleosomes: Structure, function and implication in disease. *Subcell. Biochem.* **41**, 71–89; Talbert, P. B., and Henikoff, S. (2010). Histone variants—Ancient wrap artists of the epigenome. *Nat. Rev. Mol. Cell Biol.* **11**, 264–275.], and histone modification readers [de la Cruz, X., Lois, S., Sanchez-Molina, S., and Martinez-Balbas, M. A. (2005). Do protein motifs read the histone code? *Bioessays* **27**, 164–175; Grewal, S. I., and Jia, S. (2007). Heterochromatin revisited. *Nat. Rev. Genet.* **8**, 35–46.] will not be addressed in depth in this chapter, and the reader is referred to the reviews cited here.

I. INTRODUCTION

In the eukaryotic cell, DNA is never naked but wrapped around an octamer of histone proteins. Together with other accessory factors, this forms a DNA–protein structure termed chromatin. It is thought that this packaging of DNA helps to preserve the genetic information and the integrity of the genome. Moreover, it has the potential to assist any DNA-templated event such as replication, transcription, and DNA repair. The chromatin structure must be dynamic and amendable to remodeling in order to timely and spatially regulate these processes in various cell types of multicellular organisms (Reik, 2007). Cells have evolved mechanisms that alter and modify the structure of chromatin and thus can impact the readout of the underlying DNA sequence. These include, among others, DNA methylation, histone modifications, chromatin remodeling, and small RNAs. While some of these chromatin modifications have indeed been shown to be heritable and stable over several rounds of cell division, most compelling evidence is still lacking. Nevertheless, over the past years, a wide variety of products and events have been lumped into the umbrella term "epigenetics," which by strict definition would only encompass heritable modifications of chromatin or DNA that are inherited through mitosis and thus have the potential to convey information without altering the underlying DNA sequence (Bird, 2002, 2007). Numerous reports raised the possibility that many of these chromatin and DNA modifications are interdependent. For example, it was put forward that histone modifications can form the so-called "histone code," which means that combination of different modifications may be predictive for consistent cellular outcomes in distinct cell types (Jenuwein and Allis, 2001; Strahl and Allis, 2000). In the following sections, the crosstalk and interdependence between epigenetic modifications, as well as their impact on cellular processes and diseases, such as transcription regulation, DNA repair, and cancer will be discussed. Histone modifications will be referred to as follows: acetylation, ac; methylation, me; phosphorylation, ph; monoubiquitination, ub. For example, H3K4me1/me2/ me3 represent mono/di/trimethylated forms of lysine 4 of histone H3.

II. CROSSTALK BETWEEN HISTONE MODIFICATIONS AND TRANSCRIPTION

The classical view of the effect of histone modifications was that a specific modification on a specific residue of a histone results in a defined outcome (e.g., transcription activation). This certainly holds true for some modifications. For example, there is abundant evidence that acetylation is activating, whereas SUMOylation seems to be repressing (Garcia-Dominguez and Reyes, 2009; MacDonald and Howe, 2009; Nathan *et al.*, 2006). However, in most cases the effects of histone modifications on a certain process are more complex and could only be explained by their crosstalk. Crosstalks between histone modifications could be divided into three major classes: crosstalks between modifications occurring on the same residue (*in situ*), on the same histone (*in cis*), or on different histones (*in trans*) (Latham and Dent, 2007).

A. Crosstalk between histone modifications *in situ* (Fig. 5.1)

The first "logical" crosstalk one could think of is the mutual exclusivity of histone marks that occur on the same residue, the so-called "*in situ*" cross-regulation (Fig. 5.1). "*In situ*" is a Latin term that means "in place" and it implies that addition of one modification to a residue can chemically and/or sterically block additional modification of that particular amino acid (Latham and Dent, 2007). For example, H3K9 cannot accommodate both methyl and acetyl groups in the

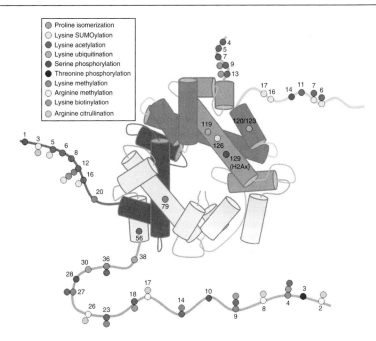

Figure 5.1. Histone modifications and their crosstalks "*in situ.*" Schematic representation of a nucleosome with the structure of the top four polypeptides of core histones. Only one histone of each duplex is shown. Blue, H2A; green, H2B; yellow, H3; magenta, H4; and faint yellow, portion of H3 from the bottom half of the nucleosome. The position of the residues along the histone tails as well as on histone cores is arbitrary. The structure of the nucleosome was adapted from Wolffe and Hayes (1999). Here, are shown all reported histone modifications as well as the possible interactions "*in situ.*"

same time (Fig. 5.1). These mutually exclusive marks have opposing roles in gene regulation: while H3K9ac occurs mostly on promoters and 5′ regions of active genes, H3K9me is related to heterochromatin and gene repression (Nakayama *et al.*, 2001; Rea *et al.*, 2000). In addition, lysine acetylation seems to be in competition with SUMOylation (Nathan *et al.*, 2006) and biotinylation (Camporeale *et al.*, 2007; Hassan and Zempleni, 2006; Fig. 5.1). Although the role of these latter modifications is still poorly understood, their competition with acetylation indicates that they could have an effect on gene regulation. For example, H4K12 is biotinylated in heterochromatin, while its acetylation was related to active transcription (Camporeale *et al.*, 2007). Likewise, several lysines on H2B and H4 cannot be acetylated and SUMOylated in the same time (Fig. 5.1), and SUMOylation was related to repression in *yeast* but this awaits further confirmation in higher eukaryotes (Nathan *et al.*, 2006). Additional individual lysines on histones H2A, H3, and H4 can be acetylated, methylated, or biotinylated (Hassan and Zempleni, 2006; Fig. 5.1). Different modifications of the same residue do not always have opposing roles, but rather sometimes help defining several steps of the same process. This is exemplified by H3K36me and H3K36ac which are both related to active transcription, but while H3K36me is enriched in bodies of active genes, acetylation is enriched on promoters. An overview of crosstalk *in situ* is illustrated in Fig. 5.1.

B. Crosstalk between histone modifications *in cis* (Fig. 5.2)

1. Crosstalk between H3S10ph and other H3 modifications

One of the earliest hints of crosstalk between histone modifications came in 1994 when Barratt and colleagues studied immediate-early (IE) gene response to diverse agents, including growth factors (Barratt *et al.*, 1994). This leads to a rapid transcriptional activation of a group of genes including proto-oncogenes such as *c-fos* and *c-jun* and it further entails an early phosphorylation of histone H3. The data showed that phosphorylation is restricted to a subset of histone H3 that is susceptible to later hyperacetylation, indicating a possible functional relationship between histone acetylation and phosphorylation (Barratt *et al.*, 1994). Later on, the three-dimensional (3D) structure of the histone acetyltransferase (HAT) domain of *Tetrahymena* Gcn5 in ternary complex with the cofactor CoA and a substrate peptide of the H3 tail (tGCN5/CoA/histoneH3) were determined (Rojas *et al.*, 1999). It revealed that, although the H3 substrate was bound to GCN5 through an extensive set of hydrogen bonds and van der Waals interactions, the majority of these interactions were mediated through the backbone region of the peptide. In particular, K9, S10, and T11 in the amino-terminal tail were not visible in the 3D structure. This indicated that modifications on these three residues, in particular H3S10ph, might have an effect on GCN5 binding to H3. Follow-up *in vitro* experiments showed that GCN5 has a 10-fold stronger affinity to phosphorylated H3 in comparison to its unphosphorylated counterpart (Cheung *et al.*, 2000). Moreover, the laboratory of Shelley Berger proved that H3S10ph is required for transcriptional activation linked to histone acetylation (Lo *et al.*, 2000). Although the colocalization of these two posttranslational modifications (PTMs) was confirmed, it was still unknown whether they occur together on the same histone tail. To address this issue, an elegant biochemical study used an antibody specifically recognizing histone H3 tails that are simultaneously modified by phosphorylation and acetylation (Clayton *et al.*, 2000). The study provides unambiguous evidence that phosphorylation and acetylation of H3 occur on the same histone tail. In a later study, the Berger laboratory tested fractionated *yeast* whole-cell extract for kinase activity on H3S10 and identified Snf1 as the kinase specific for H3S10ph. Interestingly, the authors further observed that K14ac and expression of the activator INO1 were dependent on H3S10ph. Why then are both modifications needed for activation of the same gene promoter? One could predict that the two modifications act synergistically to facilitate the binding of a protein that contains both a bromodomain specific to the acetylated lysines and a phosphorserine-binding motif specific for the phosphorylation. However, only the 14-3-3 protein was identified to bind to both modifications at the same time and eventually serve as an integrating function for the combination of these two modifications (Walter et al., 2008). Nevertheless, the fact that H3K14ac does not increase 14-3-3 binding (Macdonald et al., 2005) and that 14-3-3 does not contain a bromodomain indicates that additional proteins may be involved in the phosphoacetylation-based gene activation.

Besides its effect on H3K14ac, H3S10ph also seems to affect H3K9 methylation (Fig. 5.2). Indeed, when H3 peptides were in vitro phosphorylated on S10, this prevented the SUV39H1-dependent methylation of K9 and vice versa, methylation of H3K9 prevented H3S10ph (Rea et al., 2000). When murine Suv39h1 and Suv39h2 were deleted in mouse embryonic fibroblasts (MEFs), there was about threefold increase of the characteristic, heterochromatin-associated H3ph foci. These results were further confirmed by measuring the relative abundance of H3ph in Suv39h double null cells, which showed higher levels when compared to wild-type cells indicating that H3K9me and H3S10ph are mutually exclusive.

2. Crosstalk between H4S1ph and H4ac in cis (Fig. 5.2)

In yeast, casein kinase-2 (CK2) phosphorylates H4S1 and this phosphorylation event antagonizes H4 acetylation by the NuA4 HAT complex in vitro (Utley et al., 2005). In addition, CK2 associates with and probably helps recruitment of Sin3-Rpd3 histone deacetylase (HDAC) in vivo leading to further reduction of acetylation and suggesting that H4S1 phosphorylation and hypoacetylation of H4 may be co-regulated (Utley et al., 2005).

3. Crosstalk between arginine methylation and lysine acetylation in cis

a. On H3
The identification of specific histone methyltransferases (HMTs) helped uncovering many interactions between histone methylation and acetylation. In particular, arginine methylation was thought to be an activation mark as

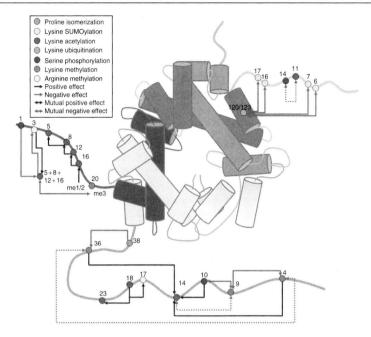

Figure 5.2. Histone modification crosstalks "*in cis*." Schematic representation of a nucleosome with the structure of the top four polypeptides of core histones. Only one histone of each duplex is shown. Blue, H2A; green, H2B; yellow, H3; magenta, H4; and faint yellow, portion of H3 from the bottom half of the nucleosome. The position of the residues along the histone tails as well as on histone cores is arbitrary. The structure of the nucleosome was adapted from Wolffe and Hayes (1999). Dotted lanes indicate interactions that were not physically confirmed or that only occur in *yeast* or *Drosophila*. Me1, monomethylation and me2/3, di-/trimethylation.

suggested by the fact that protein arginine methyltransferases are recruited to promoters by transcription factors (Lee *et al.*, 2005a). This suggested a possible interaction with acetylation which is also an activating mark. The first evidence came from a study of the Stallcup laboratory in which a yeast two-hybrid screen was used to identify proteins that interact with p160 coactivator. The study led to the identification of the arginine methyltransferase, coactivator-associated arginine methyltransferase 1 (CARM1; Chen *et al.*, 1999). It was found that CARM1 could enhance transcriptional activation by nuclear receptors, such as p160. As p160 was shown to activate transcription through the recruitment of HATs such as p300, CBP (CREB binding protein), and p/CAF (P300/CBP-associated factor)

(Chen *et al.*, 1997; Swope *et al.*, 1996), the authors proposed that methylation of histone H3 by CARM1 may cooperate with histone acetylation to activate transcription. In a following study performed in the same laboratory, the functional and mechanistic relationship between CARM1 and p300 was further analyzed (Chen *et al.*, 2000). This revealed that the p160 coactivator interacts with p300 and CARM1 via its AD1 and AD2 domains, respectively. In addition, both bindings are required for activation, are interdependent and synergistic. More recently, the kinetics of the interdependence between CARM1 and CBP were revisited at the pS2 promoter in human breast cancer cell lines stimulated with estrogen (Daujat *et al.*, 2002). The first event was acetylation of K18 by CBP, followed by the acetylation of K23 which then both lead to the recruitment of CARM1 and the methylation of R17 (Fig. 5.2). The requirement of the acetyl-transferase activity of CBP in the recruitment of CARM1 was confirmed by the use of acetyltransferase-deficient CBP which was unable to induce recruitment of CARM1. Finally, H3K18ac and K23ac but not H3K14ac were shown to increase the affinity of CARM1 to its substrates (Daujat *et al.*, 2002).

b. On H4

A crosstalk between H4R3 methylation and H4 acetylation was also reported, nevertheless it appears more complex than the crosstalk between H3R17me and H3ac (Fig. 5.2). Using HeLa cell extracts, H4R3-specific HMT, PRMT1 was identified (Wang *et al.*, 2001b). Unlike for histone H3, the acetylation of H4 seems to inhibit PRMT1-dependent methylation of H4R3 both *in vitro* and *in vivo*. Conversely, H4R3me seems to increase acetylation by p300, in human cells, leading to transcriptional activity (Strahl *et al.*, 2001; Wang *et al.*, 2001b). This mode of action looks like a negative feedback loop: H4R3me induces H4ac and gene activation and H4ac removes R3 methylation which would lead to a decrease in histone acetylation probably preventing constitutive overexpression. However, the effect of H4R3me in humans seems to depend on the identity of the methyltransferase used to deposit the methyl group. For example, methyla-tion of H4R3 by PRMT5 represses the expression of a cyclin E1 promoter-driven reporter gene (Fabbrizio *et al.*, 2002). Similarly, dimethylation of H4R3 in *yeast* is associated with heterochromatin and gene repression (Yu *et al.*, 2006). The fact that in human cells only PRMT5-dependent methylation is associated with gene repression suggests that PRMT5 could have a direct effect on gene expression. Indeed, PRMT5 was shown to be associated with the mSin3/HDAC, NuRD HDAC and Brg1/hBrm complexes and is recruited to genes involved in the control of cell proliferation (e.g., c-Myc target gene: cad and tumor suppressors: ST7 and NM23) in correlation with their repression (Le Guezennec *et al.*, 2006; Pal *et al.*, 2003, 2004).

4. Interdependence between different lysine methylation and their crosstalk with lysine acetylation *in cis*

Using *Tetrahymena* as a model, Strahl *et al.*, 1999 separated nuclei into micronuclei (transcriptionally inactive) and macronuclei (active), and discovered a specific H3K4 methylase activity which is only related to active macronuclei. This activity was confirmed in *yeast* and HeLa cells. Although, in *Tetrahymena*, no interaction between HMTs (histone methyltransferases) and HATs and no colocalization of methyl and acetyl marks were found, the situation was different in *yeast* where H3K4 methylation marks occur preferentially at acetylated nucleosomes. A very detailed study using chromatin immunoprecipitation (ChIP) at thyroid hormone receptor (TR) target genes in Xenopus oocytes revealed that inactive promoters show high levels of H3K9me and low levels of H3K4me, H3R17ph, and H3S10ph/K14ac (Li *et al.*, 2002). Upon activation, H3K9me decreases while H3R17ph, and H3S10ph/K14ac increase. Surprisingly, activation was also accompanied by an extensive decrease in H3K4me, a mark traditionally linked to activation. The mutual exclusivity between H3K9me and H3K14ac was also reported in *Schizosaccharomyces pombe* (Nakayama *et al.*, 2001) and in chicken erythrocytes (Fig. 5.2). However, based on mass spectroscopy analysis, these two marks were shown to be able to coexist on the same histone in human cells and in plants (Johnson *et al.*, 2004; Thomas *et al.*, 2006; Zhang *et al.*, 2004).

The discovery of SET7, as a H3K4-specific HMT, was an opportunity to study the relationship between H3K4 methylation, H3K9 methylation, and H3/H4 acetylation (Wang *et al.*, 2001a). Results showed that H3K4 and H3K9 methylation are mutually exclusive and, while H3K4me can successfully increase the recruitment of p300 and subsequent H3 acetylation, H3K9me is in contrast inhibiting histone acetylation (Wang *et al.*, 2001a). The same H3K4-specific HMT was identified one year later by the group of Danny Reinberg and was termed SET9 instead of SET7 (Nishioka *et al.*, 2002a). This study could show that H3K4me and H3K9me are mutually exclusive; however, the presence of H3K4me mark inhibits the recruitment of SUV39h1 but not G9a, both of which are H3K9-specific HMTs.

5. Involvement of RNAi in the recruitment of histone-modifying activities

Two mechanisms could serve as the initial trigger for H3K9me targeting to chromatin: DNA-binding factors such as transcription factors or RNAi. Evidence for direct targeting of H3K9me by RNAi came first from studies on the core RNAi machinery which includes Dicer (Dcr), Argonaute (Ago), and RNA-dependent RNA polymerase (RdRP) in *S. pombe* (Allshire *et al.*, 1995).

Later, several studies on different organisms demonstrated the role of RNAi in heterochromatin establishment (Cam *et al.*, 2005; Grewal and Elgin, 2007; Grewal and Jia, 2007; Hall *et al.*, 2002; Kato *et al.*, 2005; Matzke and Birchler, 2005; Mochizuki *et al.*, 2002; Pal-Bhadra *et al.*, 2004; Reinhart and Bartel, 2002; Taverna *et al.*, 2002; Volpe *et al.*, 2002), suggesting a novel mode of crosstalk between histone modifications and RNAi. Moreover, the interdependence of H3K9me and RNAi is mutual and forms a positive feedback loop. Indeed, in *S. pombe*, deletion of Clr4 (the H3K9 methyltransferase) abolishes the generation of short interfering RNAs (siRNAs; Buhler *et al.*, 2006). The binding of the RNA-induced initiation of transcriptional gene silencing (RITS) complex to centromeres requires H3K9 methylation and the presence of the chromatin factor Swi6 (the *S. pombe* HP1 homolog; Buhler *et al.*, 2006).

Various other studies suggested the involvement of RNA-dependent pathways in establishing chromatin modification patterns. For example, in the ciliated protozoan *Tetrahymena*, EZL1-driven H3K27 methylation is dependent on the RNAi pathway (Liu *et al.*, 2007). In *Drosophila*, components of the RNAi machinery are involved in polycomb group (PcG)-mediated transcriptional repression at the polycomb repressive elements (PREs) of the Fab-7 regulatory element. In mammals, microRNAs (miRNAs) and RNAi pathway components have been suggested to participate to the repression of target promoters, the recruitment of PcG components, and the methylation of H3K27 (Kim *et al.*, 2008a).

6. Effects of the crosstalk between different lysine methylation *in cis* on transcription

Lysine methylation has been suggested to modulate transcriptional activation, elongation, and repression, depending on the modified histone residue. H3K4me1, me2, and me3 have been found on active promoters and linked to transcription initiation and elongation (Bernstein *et al.*, 2002; Briggs *et al.*, 2002; Krogan *et al.*, 2003; Ng *et al.*, 2003; Santos-Rosa *et al.*, 2002), whereas HK36me2 and me3 have been correlated with transcription elongation (Kizer *et al.*, 2005; Krogan *et al.*, 2003; Li *et al.*, 2003; Strahl and Allis, 2000; Xiao *et al.*, 2003). Methylation of H3K79 was also implicated in transcription activation and elongation (Pokholok *et al.*, 2005; Schubeler *et al.*, 2004; Steger *et al.*, 2008). On the other hand, three lysine methylation sites are connected to transcriptional repression: H3K9, H3K27, and H4K20. To some extent, the known factors that bind to these methylation marks may explain how the same chemical modification can have opposite effects when present at different residues. For example, H3K4me can form a docking site for histone-modifying complexes or chromatin-remodeling factors, such as Taf3 (Vermeulen *et al.*, 2007) which would bind to the methylation mark through their Plant Homeo Domain (PHD), thus activating transcription. The *yeast* Spt-Ada-Gcn5-acetyltransferase (SAGA) complex, which contains the H3 acetyltransferase Gcn5, associates with the chromodomain protein and chromatin remodeler Chd1. Chd1 binds to H3K4me, assisting in the recruitment of SAGA and leading to enhanced acetylation (Pray-Grant *et al.*, 2005; Sims *et al.*, 2005). Methylation of H3K4 is also associated with increased acetylation of H3 by p300 and other HATs (Wang *et al.*, 2001a). Furthermore, the degree of H3K4 methylation reflects on the extent of H3 acetylation: H3K4me1 is less acetylated than H3K4me3 (Nightingale *et al.*, 2007). In particular, H3K14ac dependence on H3K4me was studied in details. In budding *yeast*, NuA3 binds H3K4me3 through the Yng1 subunit. This interaction, as well as methylation of H3K36, is essential for the acetylation of H3K14 (Martin *et al.*, 2006a,b; Taverna *et al.*, 2006). H3K4 and H3K36 methylation may be also required for H4K8 acetylation (Morillon *et al.*, 2005). Interestingly, H3 acetylation can in return promote methylation, as the Gcn5 HAT complex stimulates H3K4 trimethylation in *Saccharomyces cerevisiae* (Govind *et al.*, 2007). The mechanism of H3K4me-dependent activation of gene expression may, in addition, involve a decrease in the recruitment of nucleosome-remodeling and histone deacetylase (NURD) complex. Notably, Mi2, a member of the NURD complex can bind unmodified H3 tails and hence H3K4 methylation leads to the inhibition of this binding resulting in increased acetylation and subsequent activation (Nishioka *et al.*, 2002a). Conversely, HDAC activity is recruited by H3K36me leading to deacetylation of histones H3 and H4 which helps to prevent spurious transcription initiation at cryptic sites in gene bodies. This mechanism is dependent on the activity of Rpd3 and mSin3A–HDAC complexes in *yeast* and humans, respectively (Brown *et al.*, 2006; Joshi and Struhl, 2005; Keogh *et al.*, 2005; Lee and Shilatifard, 2007; Li *et al.*, 2007a). H3K9 methyltransferases also interact with HDACs (Czermin *et al.*, 2001; Vaute *et al.*, 2002); this finding and the fact that HDACs interact with HP1 as well led to the proposal of a positive feedback loop model for heterochromatin formation. In this model, deacetylation of H3K9 facilitates K9 methylation, which in turn promotes HP1 binding and recruitment of additional HDACs and HMTs (Czermin *et al.*, 2001; Vaute *et al.*, 2002; Zhang *et al.*, 2002a).

To obtain a more detailed picture on histone methylation distribution along genes, mapping using ChIP-Chip and/or ChIP-seq (next-generation sequencing) experiments was performed and revealed that H3K4me3 and H3K4me2 peak at 5′- ends and at promoter proximal regions of active genes, while H3K4me1 is enriched in gene bodies of active genes. Conversely, H3K36me2/3 marks are enriched in active gene bodies and mostly toward the 3′- end of active genes (Bannister *et al.*, 2005; Barski *et al.*, 2007; Bell *et al.*, 2007; Kizer *et al.*, 2005; Lee and Shilatifard, 2007; Mikkelsen *et al.*, 2007; Pokholok *et al.*, 2005; Rao *et al.*, 2005). Why would domains of H3K4/K9/K27/K36me not overlap? One possibility is that one mark can recruit an enzyme that removes the other mark. For example, the JMJD2A protein, a lysine demethylase that targets H3K9me and H3K36me, associates with H3K4me3, through its tandem tudor domains (Huang et al., 2006). On the other hand, H3K9 methyltransferase might recruit H3K4 demethylase activity. Indeed, in fission yeast, heterochromatin is enriched with H3K9 methylation catalyzed by Clr4, the homolog of the mammalian histone methyltransferase SUV39H1 and Clr4 binds the JmjC domain protein Lid2, a H3K4-specific demethylase. Li et al. demonstrated that Lid2 promotes heterochromatin formation by removing the activating H3K4 methyl marks and facilitating the H3K9 methylation activity of the complex (Li et al., 2008). Lid2 may also coordinate the Set1 methyltransferase and the lysine-specific demethylase 1 (LSD1) to regulate H3K4 and H3K9 methylation levels in euchromatin, although this function may not depend on Lid2 enzymatic activity (Li et al., 2008). Another example is the mammalian MLL3/4 Set1 complex which contains H3K4 methylation activity, as well as UTX, which removes the transcriptionally repressive trimethyl mark from H3K27 (Cho et al., 2007).

Several studies have shown that some developmental genes which are not expressed in embryonic stem (ES) cells have both repressive (H3K27me3) and active (H3K4me3) marks at their promoters, forming so-called "bivalent domains". During cellular differentiation, bivalently marked genes are partially resolved into monovalent by getting rid of one associated mark and thus get either stably activated or stably repressed. Therefore, these bivalent domains were thought to keep genes repressed at a certain developmental window but poised for activation in another subsequent developmental stage (Bernstein et al., 2002; Pan et al., 2007; Zhao et al., 2007). Although very intuitive, this model currently lacks experimental support.

A recent paper provided evidence that all main H3K9 HMTs, namely G9a, GLP, SetDB1, and Suv39h1 could exist in the same mega complex (Fritsch et al., 2010). Deletion of one HMT of this complex seems to affect its integrity and the stability of the other HMTs. To gain further insights into the functional relevance of the interaction between the four H3K9 HMTs, the authors used the retinoic acid ES cell differentiation system and tested the cooperation effect of these HMTs in silencing the Oct3/4 expression. The results showed that deletion of any HMT in the complex does not have an effect on H3K9me and expression of Oct3/4. However, the deletion of all four HMTs efficiently reduces H3K9me and induces the reexpression of Oct3/4. One could propose a model of sequential methylation in which mono/dimethyltransferases G9a and GLP directly interact with trimethyltransferases Suv39h1 and SetDB1 to assure trimethylation and further repression. But it remains to be determined why is it functionally beneficial to have two redundant mono/dimethyltransferases and two trimethyltransferases in the same complex and whether other histone modifiers belong to this complex.

7. Crosstalk between H4K20me and H4ac

H4K20 methylation was linked to both, gene activation and repression and this depends on the degree of methylation. For example, H4K20me3 is linked to repression by a mechanism that is similar to that of H3K9me-dependent repression. Indeed, H4K20me3 inhibits acetylation of H4 and, vice versa, H4 hyperacetylation antagonizes H4K20me3 (Kourmouli et al., 2004; Nishioka et al., 2002b; Sarg et al., 2004; Fig. 5.2). On the other hand, H4K20me1 by SET8 is associated with active gene expression. H4K20me1 correlates with hyperacetylation of H4 at the genes encoding β-globin and Myc in mouse erythroleukemia cells (Talasz et al., 2005). In contrast, H4 acetylation seems not to affect H4K20me1 (Yin et al., 2005). H4K20me2 is also related to hyperacetylation based on mass spectrometry analysis which indicated that H4K5, K8, K12, and K16 are acetylated only when H4K20 is dimethylated (Zhang et al., 2002b). This H4K20me2-dependent acetylation is hierarchal: H4K12ac is dependent on H4K16ac and allows the acetylation of H4K8 and finally all three modifications are required for acetylation of H4K5 (Zhang et al., 2002b).

8. Crosstalk between H2B modifications

H2B K6, K7, K16, and K17 can be subject to SUMOylation, a modification generally related to gene repression in humans and heterochromatin stability in *S. pombe* (Shiio and Eisenman, 2003; Shin et al., 2005). Moreover, mutation of SUMOylated lysines to alanines increases basal transcription of several genes in *yeast*. In contrast, tethering of SUMO to H2B inhibits the induction of the GAL1 gene (Nathan et al., 2006). On the other hand, H2BK123 can be ubiquitinated and this was related to gene activation. Therefore, one could expect that SUMOylation and ubiquitination of H2B are mutually exclusive. Indeed, SUMOylation of K6 and K7 opposes monoubiquitination of H2BK123 (Nathan et al., 2006). In the same way, H2BK123 monoubiquitination opposes SUMOylation of H2B.

One of the best examples of the effect of crosstalk between histone modifications on cell fate decisions comes from a study showing the role of different H2B modifications in apoptosis (Ahn et al., 2006). H2B can be phosphorylated at S14 in humans (S10 in *yeast*; Ahn et al., 2005). Interestingly, studies on apoptosis induction in *yeast* showed that H2BS10ph plays a direct role in mediating apoptosis by promoting chromatin compaction. Mutation experiments confirmed the causal effect of H2BS10ph on apoptosis. Indeed, H2B S10E phosphosite mutant mimics a constitutive phosphorylation and promotes chromatin condensation and cell death. In contrast, H2B S10A mutants, which cannot be phosphorylated, are resistant to cell death (Ahn et al., 2005). H2B was also shown to be acetylated at K11 in logarithmically growing *yeast* (Suka et al., 2001). A very obvious question was if this acetylation would prevent or reduce apoptosis by counteracting the chromatin condensation induced by H3S10ph. Recent study showed that H2BK11ac inhibits the phosphorylation of the adjacent site S10, pointing to a mutually exclusive existence of the K11ac and S10ph marks on the tail of H2B. Moreover, apoptotic *yeast* lacks K11 acetylation, but displays high levels of H2BS10ph. It was additionally found that H2BK11 is deacetylated by the Hos3 HDAC, which is essential to induce phosphorylation of H2BS10 and subsequent apoptosis. Mutational studies showed that *yeast* having H2B K11Q, an acetyl-mimic mutant in H2B, failed to activate the apoptotic pathway, while H2B K11R, a mutant mimicking the deacetylation state, could phosphorylate H2BS10 and induce apoptosis upon treatment with H2O2 (Ahn *et al.*, 2006). Further studies are necessary in order to uncover the mechanism underlying the involvement of H2B acetylation and phosphorylation in inducing apoptosis. Moreover, it has yet to be determined whether this crosstalk is conserved in higher eukaryotes, although there are striking similarities to the mechanism in *yeast*. For example, H2BS14ph is important for the execution of apoptotic pathways (Cheung *et al.*, 2003). Additionally, H2BK15 is acetylated in logarithmically growing HeLa cells, similar to the acetylation of H2BK11 in logarithmically growing *yeast* (Thorne *et al.*, 1990). Together, these findings suggest that crosstalk between acetylation and phosphorylation in H2B may be conserved in higher eukaryotes.

9. Crosstalk between H3 methylation and isomerization

Proline isomerization was only recently identified as a potential new chromatin modification. It is catalyzed by the proline isomerase Frp4 in *S. cerevisiae*, resulting in a conformational shift of the adjacent residues and inhibition of H3K36me3 by Set2, *in vitro*. Thus, it might be that isomerization of H3 proline 38 (H3P38) can affect H3K36 methylation (Nelson *et al.*, 2006). Indeed, *in vivo* loss of Frp4 results in increased H3K36 methylation and decreased transcription of specific *yeast* genes (Nelson *et al.*, 2006).

C. Crosstalk between histone modifications *in trans* (Fig. 5.3)

1. Crosstalk between H2BK123ub, H3K4me, and H3K79me

In addition to its role in controlling its own SUMOylation, H2B ubiquitination modulates posttranslational events on other histones. In *yeast*, H2B is mono-ubiquitinated at K123 by the ubiquitin-conjugating enzyme Rad6. Early studies showed that Rad6 and H2BK123ub seem to mediate H3K4 methylation by the Set1 methyltransferase subunit of the COMPASS complex (complex proteins associated with Set1) (Dover *et al.*, 2002; Sun and Allis, 2002). H2Bub was also shown to be required for methylation of H3K79 by the Dot1 methyltransferase

(Briggs *et al.*, 2001; Ng *et al.*, 2002). These interactions seem to be unidirectional as the H3 K4R mutation does not have an effect on H2Bub (Sun and Allis, 2002) and as H3K4me still occurs in the Dot1-deleted and H3K79 mutant strains (Briggs *et al.*, 2001; Ng *et al.*, 2002). As both H2Bub-dependent methylation marks, namely H3K4me and H3K79me, are involved in transcription repression at telomeric regions for example, it was proposed that this crosstalk could be necessary to induce transcriptional silencing. However, we now know that H3K4me is mostly linked to transcriptional activation, and subsequently the role of H2Bub could be also linked to transcriptional activation, thus controlling the balance between activation and silencing by the induction of different methylation events. Indeed, substitution of the H2BK123 led to reduction of H3K4me3, an increase of H3K36me2, and lower transcription of certain genes (Henry *et al.*, 2003). It has been proposed that ubiquitination of H2B occurs mostly in euchromatin leading to H3K4 and H3K79 methylation which would prevent Sir proteins from association with active euchromatic regions, thereby restricting Sir proteins to heterochromatic regions to mediate silencing (van Leeuwen *et al.*, 2002). At the same time, the ubiquitination would activate the transcription by methylating H3K4 and facilitating transcriptional elongation (Krogan *et al.*, 2003; Ng *et al.*, 2003). Further details on the role of H2BK123 ubiquitination in controlling methylation events on H3 came from studies in *yeast* using ChIP and mass spectrometry which showed that H2Bub does not have an effect on monomethylation of H3K4 and H3K79 by Set1 and Dot1, respectively, but rather on the processivity of these enzymes for catalyzing subsequent di- and trimethylation (Dehe *et al.*, 2005; Schneider *et al.*, 2005; Shahbazian *et al.*, 2005; Fig. 5.3). Furthermore, mutations in the BUR(bypass UAS requirement) complex, which reduce H2B ubiquitination, only affect trimethylation, and not dimethylation, of H3K4 (Laribee *et al.*, 2005). Recent studies provided insight into the mechanism underlying this crosstalk (Lee *et al.*, 2007a). The COMPASS complex is only active when it contains the Cps35 subunit. Cps35 binds to H2Bub and then associates with and activates the COMPASS complex. However, H2Bub exclusively affects the processive activity of the COMPASS complex since only H3K4me2 and me3 are affected in the absence of H2Bub. Although H3K79me3 is also dependent on Cps35 in budding *yeast*, suggesting a mechanism similar to that observed with COMPASS (Lee *et al.*, 2007a), recent *in vitro* experiments showed an additional mechanism for direct H2Bub–H3K79me interaction in human cells via activation of hDot1L by the ubiquitination mark (McGinty *et al.*, 2008).

The relationship between H2Bub and H3K4/K79me3 was further shown to be dependent on proteins involved in RNA and protein stability. Two of the proteasomal adenosine triphosphatases (ATPases) of 19S regulatory complex of the proteasome, Rpt4 and Rpt6, are recruited through H2Bub to the promoters of genes after activation (Ezhkova and Tansey, 2004). Mutations affecting these ATPase subunits reduce global K4 and K79me2/me3. Recent studies have also

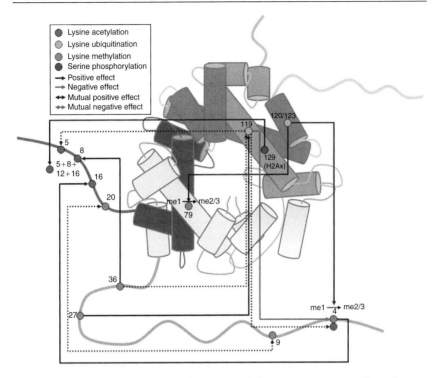

Figure 5.3. Histone modification crosstalks "*in trans.*" Schematic representation of a nucleosome with the structure of the top four polypeptides of core histones. Only one histone of each duplex is shown. Blue, H2A; green, H2B; yellow, H3; magenta, H4; and faint yellow, portion of H3 from the bottom half of the nucleosome. The position of the residues along the histone tails as well as on histone cores is arbitrary. The structure of the nucleosome was adapted from Wolffe and Hayes (1999). Dotted lanes indicate interactions that were not physically confirmed or that only occur in *yeast* or *Drosophila*. Me1, monomethylation and me2/3, di-/trimethylation.

implicated the Ccr4–Not mRNA production and processing complex in this crosstalk (Laribee *et al.*, 2007b; Mulder *et al.*, 2007). The Ccr4–Not complex is involved in gene repression through mRNA degradation and transcription regulation. Components of this complex interact with the 19S regulatory complex of the proteasome and the BUR complex and mutations affecting subunits of Ccr4–Not, reduce H3K4me3 (Laribee *et al.*, 2007a; Mulder *et al.*, 2007).

Although a direct interaction between H2Bub and H3K36me was not found, studies on H2B deubiquitination by Ubp8 ubiquitin protease suggested the presence of a negative interaction between these two modifications. In budding *yeast*, deubiquitination of H2B by Ubp8 allows the localization of Ctk1 kinase to the 5'- end of genes leading to phosphorylation of RNA-polII. This event allows the association of Set2 methyltransferase to phosphorylated RNA-polII and subsequent methylation of H3K36 in the ORFs of active genes (Wyce *et al.*, 2007). Ubp8 also links H2B (de)ubiquitination to histone acetylation. Indeed, Ubp8 is a component of the SAGA–HAT complex (Daniel *et al.*, 2004; Henry *et al.*, 2003) and SAGA integrity is required for Ubp8 deubiquitination activity, as mutations disrupting the complex increase the level of H2Bub (Daniel *et al.*, 2004; Henry *et al.*, 2003; Ingvarsdottir *et al.*, 2005; Lee *et al.*, 2005b; Powell *et al.*, 2004). Recently, Ubp8 orthologs were identified in *Drosophila* (Nonstop) and humans (USP22; Weake *et al.*, 2008; Zhang *et al.*, 2008; Zhao *et al.*, 2008) and they are also components of the SAGA and TFTC(TBP-free TAF complex)/STAGA (SPT3/ TAF9/GCN5 acetyltransferase) complexes, respectively. In addition, Sus1, a component of SAGA, is required for H2B deubiquitination (Kohler *et al.*, 2006; Kurshakova *et al.*, 2007; Zhao *et al.*, 2008). Thus, it seems that histone acetylation is required for H2B deubiquitination but it has yet to be determined whether H2B deubiquitination is required for acetylation.

From the data presented above, one can propose a model that resumes the crosstalk between H2Bub and H3K4me (the same mechanism possibly applies for H3K79me; Lee *et al.*, 2007a): COMPASS is recruited to phosphorylated RNA-polII. In the absence of H2B ubiquitination, COMPASS can only monomethylate H3K4. The BUR complex would then activate Rad6 which associates with RNA-polII and COMPASS leading to H2BK123 monoubiquitination. This allows the recruitment of Cps35 to the COMPASS complex and the transformation of H3K4me1 into me2 and me3. Components of the 19S regulatory proteasome complex and the Ccr4–Not mRNA processing complex probably facilitate this activity. Then, SAGA HAT complex is recruited, acetylates H3, and deubiquitinates H2Bub via the ubiquitin-specific protease Ubp8.

H2B deubiquitination seems to be connected to H3K27me and H3K9me as well. In *Drosophila*, USP7, which catalyzes deubiquitination of H2Bub *in vitro*, has been implicated in polycomb-mediated silencing (van der Knaap *et al.*, 2005). Similarly, loss of the H2B deubiquitinase SUP32/UBP26 in plants is associated with a reduction of the levels of H3K9me2 and a reduction in promoter DNA methylation (Sridhar *et al.*, 2007).

2. Crosstalk between H2Aub and H3/H4 histone modifications (Fig. 5.3)

PcG protein members represent a well-suited example of close interaction between histone-modifying enzymes (Schuettengruber *et al.*, 2007; Schwartz and Pirrotta, 2007). PcG complexes consist of three separate protein complexes (polycomb repressive complex 1 (PRC1), polycomb repressive complex 2 (PRC2), and

Pho-repressive complex (PhoRC)) that assemble on chromatin and coordinate H2AK119 and H2AK120 ubiquitination in *Drosophila* and vertebrates, respectively, and H3K27 methylation in both (Schwartz and Pirrotta, 2007). *Drosophila* PRC1 contains several subunits, including dRING (RING1A and RING1B in humans) that functions as an E3 ubiquitin ligase and monoubiquitinates H2AK119. PRC2, or the E(Z)/ESC (extra sexcombs) complex, contains the E (Z) H3K27 methyltransferase (Schwartz and Pirrotta, 2007) and SU(Z)12. The PhoRC complex, identified in *Drosophila*, contains the pleiohomeotic/pleiohomeotic-like (PHO/PHOL) proteins. Studies on *Drosophila* (Martin and Zhang, 2005) demonstrated that the first complex to bind to DNA is the PhoRC, through the DNA-binding activity of PHO and PHOL. This leads to the recruitment of the PRC2 complex which trimethylates H3K27. H3K27me3 marks then form docking sites for PRC1 complex which mediates ubiquitination of H2AK119. Knockdown of SU(Z)12 in PRC2 reduces H3K27me3 and leads to a concomitant decrease in H2Aub localization at silenced promoters (Cao et al., 2005), further confirming the role of H3K27me3 in the recruitment of PRC2 complex. A fourth polycomb complex, similar to PRC1, was recently discovered in *Drosophila*. It was called dRING-associated factors (dRAF) and it additionally contains the dKDM2 protein which is a K4 and K36 demethylase. The H3K36 demethylase activity of dKDM2 was shown to stimulate H2A ubiquitination (Lagarou et al., 2008). Thus, the dRAF complex represents yet another example of coordination between the removal of an active mark, H3K36me2, and the addition of a repressive mark, H2AK119ub, although these two activities appears not to depend on each other. A similar complex containing Ring1A and Ring1B was identified in mammals using proteomics approach (Gearhart et al., 2006; Sanchez et al., 2007). The FBXL10–BcoR complex contains additional proteins, including FBXL10/Jhdm1b (KDM2B). This complex was able to ubiquitinate H2A *in vivo* and to demethylate H3K4 through KDM2B but so far the interdependence between H3K4me (demethylation) and H2Aub is still not clear (Frescas et al., 2007).

H2A ubiquitination by the PcGs results in repression of transcription. Several studies tried to unravel the mechanism governing this repression. H2Aub appears to have no effect on other modifications involved in silencing such as H3K9me or H3K27me (Nakagawa et al., 2008; Wang et al., 2004) but rather regulates activating marks such as H3K36me and H3K4me, as discussed earlier. Indeed, H2Aub seems to specifically inhibit MLL3-mediated H3K4 methylation, thus repressing transcription initiation, *in vitro* (Nakagawa et al., 2008). However, during meiosis and mitosis, *Drosophila* H2Aub seems to regulate H3 and H4 acetylation *in trans* and the mutation of H2AT119 causes concomitant loss of H3K14ac and H4K5ac (Ivanovska et al., 2005; Fig. 5.3).

Recently, three major deubiquitinases specific for H2Aub were identified: Ubp-M, 2A-DUB, and USP21 (Joo et al., 2007; Nakagawa et al., 2008; Zhu et al., 2007) and this allowed unraveling novel links between H2A (de)ubiquitination and other histone marks. For example, Ubp-M is essential for H3S10ph mediated by the Aurora B kinase and is required for chromosome segregation during mitosis (Joo et al., 2007). This finding confirms earlier observation that H3S10ph and H2Aub levels inversely correlate during the cell cycle and that H2Aub is absent from isolated metaphase chromosomes (Joo et al., 2007; Matsui et al., 1979; Mueller et al., 1985; Wu and Bonner, 1981). The second H2A deubiquitinase, 2A-DUB interacts with the HAT p300/CBP-associated factor (PCAF/KAT2B) and preferentially deubiquitinates hyperacetylated nucleosomes in vitro (Zhu et al., 2007).

3. Crosstalk between H3me, H4me, and H4ac

In addition to its effect on H3K14ac, H3K4me may also promote acetylation of H4 in trans (Fig. 5.3). A stable complex containing the HMT responsible for the methylation of H3K4 (MLL1) and the HAT enzyme specific for the acetylation of H4K16 (MOF (males absent on the first)) has been purified in human cells. The two histone-modifying activities of the complex were shown to be needed for the activation of transcription both in vitro and in vivo, indicating a possible crosstalk between these two modifications during transcription (Dou et al., 2005). Surprisingly, although H3K4me3 was linked to active transcription, it can also recruit the activity of human ING2, a member of the mSin3a–HDAC1 complex which could result in deacetylation and subsequent repression (Shi et al., 2006). One explanation that reconciles the coexistence of the two opposed events of activation (H3K4me3) and repression (HDAC) came from a study by Hazzalin and Mahadevan (2005). They found that while H3K4me is stable on promoters of active genes, histone acetylation is in continuous dynamic turnover. When this turnover was inhibited, the expression of certain genes was not anymore possible, despite the increase in histone acetylation levels at their promoters. This indicates that cycles of H3 acetylation and deacetylation are important for induction of certain genes in mammalian cells. Another indication of these cycles of acetylation came from a study showing that the H3K4me1 and H3K4me2 demethylase, LSD1 is a member of the CoREST repressor complex, which also contains HDAC1 and HDAC2 (Lee et al., 2005c). LSD1-dependent deacetylation probably occurs before demethylation since LSD1 is the most active toward hypoacetylated H3 and its activity is decreased upon treatment with HDAC inhibitors (Lee et al., 2005c, 2006; Shi et al., 2005). Moreover, the removal of H3K4me3 in human cells by the JARID1d family was associated with the repressive activity of the polycomb-like protein Ring6a (Lee et al., 2007b). Based on these data, one can propose a model of activation/repression cycles. Upon activation signal, H3K4 is first mono- and dimethylated leading to the recruitment of activating complexes and the appearance of H3K4me3 which recruits MOF leading to of H4K16ac. At the end of the activating cycle, H3K4me3 recruits mSin3a–HDAC1 and possibly HDAC2 starting a cycle of repression. The H3K4me3 is then removed by JARID1, which further enhances repression by recruiting the Ring6a polycomb protein. HDAC1 and HDAC2 lead to the recruitment of LSD1 and the demethylation of H3K4me1/2 converting active gene expression to a repressed state.

In a recent study, H3K9me and H4K20me were shown to colocalize revealing a yet undescribed possible crosstalk between these two modifications (Sims et al., 2006; Fig. 5.2). In this study, Sims and colleagues have shown using immunofluorescence assays that H4K20me and H3K9me overlap in MEFs. In addition, a higher resolution mapping of chromosomes showed that H4K20me1 and H3K9me1 are located in facultative heterochromatin, whereas H4K20me3 and H3K9me3 are found together in constitutive pericentric heterochromatin. Finally, they showed that monomethylation of H3K9 and H4K20 can coexist on the same nucleosome (Sims et al., 2006).

III. CROSSTALK BETWEEN DIFFERENT HISTONE MODIFICATIONS IN RESPONSE TO DNA DAMAGE

Besides their role in transcription, histone modifications are also involved in DNA damage response and DNA repair. Rapid phosphorylation of H2AX, at serine 129 (γH2AX) by the PI3K kinases at double strand break (DSB) sites, is one of the first and most easily detectable DNA damage signaling posttranslational events. This event leads to a concert of interconnected epigenetic modifications. First, phosphorylation of serine 129 of H2AX anticorrelates with the phosphorylation of tyrosine (Y) 142 of H2AX, a modification that inhibits the recruitment of phospho-ATM (ataxia telangiectasia mutated) and mediator of DNA damage checkpoint protein 11 (MDC11) to sites of DNA damage (Cook et al., 2009; Xiao et al., 2009). The murine Trrap/Tip60 HAT complex, like its yeast homolog Tra1/NuA4, binds to γH2AX after induction of DSBs (Downs et al., 2004; Downs and Cote, 2005; Loizou et al., 2006; Murr et al., 2006, 2007) leading to hyperacetylation of H4 in the chromatin surrounding the break site. Gcn5 and Esa1 HATs, which can acetylate H3, are also recruited to chromatin around a DSB in yeast (Tamburini and Tyler, 2005). The recruitment of NuA4 HAT complex to γH2AX is mediated by Arp4 which is, in addition, a subunit of ATP-dependent chromatin-remodeling complex INO80/SWR1. Hence, this complex is also recruited to DNA breaks in a γH2AX-dependent fashion and its remodeling activity seems to be required for the repair of DNA DSBs (Downs et al., 2004; Morrison et al., 2004; van Attikum et al., 2004; van Attikum and Gasser, 2009). The hyperacetylation of histones, along with the activity of several chromatin-remodeling complexes, leads to the relaxation of the chromatin structure, thus facilitating the recruitment of repair proteins. Hence, it would appear that cells can utilize the activities of both histone-modifying and remodeling complexes in order to facilitate DNA repair. However, the precise role of γH2AX in direct recruitment of repair proteins is still under debate. Currently, its role seems to be dispensable for the original recruitment of DNA repair factors but indispensable for the accumulation/retention of these factors at DNA break sites (Celeste *et al.*, 2003; Kobayashi *et al.*, 2002; Rogakou *et al.*, 1999). Two mechanisms for the reversal of H2AX phosphorylation were proposed. First, γH2AX can be dephosphorylated through the activity of two phosphatases: Pph3 in yeast and protein phosphatase 2A in mammalian cells (Chowdhury *et al.*, 2005; Keogh *et al.*, 2006). Second, work by the Workman's laboratory showed that in *Drosophila melanogaster*, exchange of phosphorylated H2Av, the homolog of mammalian H2AX, requires prior acetylation of phospho-H2Av by the dTip60 complex (Kusch *et al.*, 2004). Ikura *et al.* (2007) reported a similar mechanism for the removal of H2AX in HeLa cells. Upon DNA damage, H2AX gets acetylated by TIP60, a modification which is necessary for its ubiquitination that, in turn, may lead to its removal from the chromatin. This removal was proposed to help chromatin dynamics during DNA repair. It remains to be established if this mechanism applies for the removal of γH2AX after DNA repair in mammalians.

While H3K4me is linked to HAT recruitment during gene activation, the situation is different upon DNA breaks. In response to DNA damage, the Sin3–HDAC1 deacetylation complex binds to H3K4me3 through the HD domain of the Ing2 (Pena *et al.*, 2006; Shi *et al.*, 2006). This leads to active repression of genes involved in proliferation allowing for DNA repair to take place.

IV. CROSSTALKS INVOLVING DNA METHYLATION AND TRANSCRIPTION

A. Crosstalk between histone modifications and DNA methylation

DNA methylation patterns in human cancer cells are considerably distorted. Typically, cancer cells exhibit a general hypomethylation, notably in intergenic regions and at oncogenes as well as hypermethylation of tumor suppressor genes. This misregulation of DNA methylation may result from aberrant targeting of DNA methyltransferases (DNMTs). Several studies have shown the importance of histone modifications in targeting DNA methylation (McCabe *et al.*, 2009). An important link between H3K9me and DNA methylation in mammals was identified when Suv39h1 HMT-directed H3K9me3 was shown to be required for recruiting Dnmt3b-dependent DNA methylation to pericentromeric repeats (Lehnertz *et al.*, 2003). This crosstalk is dependent on a nuclear ubiquitin-like protein called UHRF1 (ubiquitin-like protein containing PHD and RING domain 1) or Np95 (nuclear protein 95) in mouse and ICBP90 (inverted CCAAT box binding protein 90) in humans. This protein is tightly bound to chromatin and was first defined to be an important determinant in cell-cycle progression, to regulate the expression of topoisomerase IIα and to play a role in DNA damage response (Bonapace et al., 2002; Fujimori et al., 1998; Hopfner et al., 2000; Muto et al., 2002). Through its RING domain, Np95 can ubiquitinate all core histones in vitro. In vivo assays, however, showed a preference toward H3 (Citterio et al., 2004). A first link between this protein and DNA methylation came from studies of human cells in which ICBP90 was found to preferentially bind to methylated CpG through its Set- and RING finger-associated (SRA) domain (Unoki et al., 2004). The SRA domain also interacts with HDAC1 complex, which indicates that ICBP90 may have a repressive effect on gene expression (Unoki et al., 2004). Indeed, it was shown that ICBP90 interacts with the retinoblastoma protein (pRB) and negatively regulates its expression (Jeanblanc et al., 2005). Np95–HDAC1 interaction occurs on methylated promoter regions of various tumor suppressor genes, including p16INK4A and p14ARF, in cancer cells (Unoki et al., 2004). Moreover, it is overexpressed and accumulates in breast-cancer cells, where it might suppress the expression of tumor suppressor genes through deacetylation of histones via HDAC1 (Mousli et al., 2003). These data suggest that ICBP90 could act as an oncogene. During the replication of pericentric heterochromatin, acetylation of histones should be kept at minimum levels to avoid disruption of the heterochromatic structure. The fact that Np95 binds methylated DNA and recruits HDACs indicated that it might have a role in the underacetylation of H4 during heterochromatin replication. Indeed, its deletion results in DNA demethylation of interspersed repeats and altered DNA methylation at imprinted loci and tandem repeats in ES cells. In addition, it causes a hyperacetylation of heterochromatin H4K8, 12, and 16 and an increase of pericentromeric major satellite transcription, in NIH-3T3 and 293T cells (Lehnertz et al., 2003; Papait et al., 2007). Later on, Np95 was shown to preferentially bind to hemimethylated DNA via its SRA domain, similar to Dnmt1's preference for hemimethylated substrates. Moreover, Np95 directly interacts with Dnmt1 indicating that it might actually help recruiting it to newly synthesized DNA during replication. Confirming the role of Np95 in heterochromatin maintenance, recent studies have shown that it is recruited to the heterochromatin through its interaction with H3K9me marks, and it is highly bound to hemimethylated DNA and recruits Dnmt1 allowing maintenance of DNA methylation and heterochromatin during replication (Bostick et al., 2007; Gopalakrishnan et al., 2008; Karagianni et al., 2008; Papait et al., 2008; Rottach et al., 2009; Sharif et al., 2007). In addition to its interaction with maintenance Dnmt1, Np95 was recently shown to interact with de novo methyltransferases Dnmt3a and 3b in mouse cells (Meilinger et al., 2009).

To investigate possible functions of this interaction, an epigenetic silencing assay using fluorescent reporters was developed in ES cells. This showed a similar kinetics to the Np95–Dnmt1 interaction. Indeed, G9a-dependent H3K9 methylation and Np95 recruitment precede the recruitment of Dnmt3a and 3b and are sufficient for silencing. Recent study further confirmed the mechanistic importance of the cooperation between Dnmt3 and G9a in regulating gene expression in *zebrafish* (Rai *et al.*, 2010). Similar to Dnmt1-Suv39h1 cooperativity, Dnmt3 and G9a seem to function together for tissue-specific development. *De novo* DNA methylation is additionally affected by the activity of lymphoid-specific helicase (Lsh) protein, a member of the SNF2-helicase family of chromatin-remodeling protein (De La Fuente *et al.*, 2006). Deletion of *Lsh* perturbs DNA methylation patterns in mice causing reduced *de novo* methylation without affecting maintenance methylation (Zhu *et al.*, 2006). Similar to Np95, Lsh seems to be able to recruit Dnmt1, Dnmt3b, and HDACs and establish a transcriptionally repressive chromatin structure (Myant and Stancheva, 2008). Recent studies indicated that Lsh might even bring together DNA methylation and PcG-dependent histone modifications. Indeed, Lsh was shown to be able to associate with *Hox* genes and regulate Dnmt3b binding, DNA methylation, and silencing of *Hox* genes during development. Moreover, *Lsh* inactivation results in decreased DNMT3B and PcG binding leading to the loss of PRC1- and PRC2-associated chromatin marks (H2Bub and H3K27me3), thus suggesting that Lsh and DNA methylation are important for complete assembly and activity of the polycomb complexes (Xi *et al.*, 2007). Altogether, these data present an example on how different histone modifications (H3K9me, H2Bub, and H3K27me) and histone-modifying proteins (Np95) can crosstalk with DNA methylation and Dnmts as well as chromatin-modifying complexes (Lsh) in order to guarantee a common function (transcription repression and formation of heterochromatin). However, it is important to confirm these data in a physiological *in vivo* setup. Moreover, it is still not clear whether the Np95-dependent H3ub would contribute to this crosstalk, and whether Np95 directly interacts with Lsh and PcGs.

Crosstalk between H3K9me and DNA methylation may be crucial for the silencing of genes in cancer. For example, 5-aza-2'-deoxycytidine treatment leads to decrease in both H3K9me2 and DNA methylation on tumor suppressor gene promoters and their subsequent reactivation (McGarvey *et al.*, 2006; Nguyen *et al.*, 2002). Furthermore, G9a is enriched at the promoters of aberrantly methylated genes in cancer cells, and co-recruitment of G9a, DNMT1, and HP1 to the promoter of the *survivin* gene stimulates H3K9me2 and DNA hypermethylation (Smallwood *et al.*, 2007). Recent evidence showed that inhibition of G9a alone led to concurrent inhibition of DNMT1 and reactivation of silenced metastasis suppressor genes in cancer cells (Wozniak *et al.*, 2007).

Another relationship between lysine methylation and DNA methylation is the mutual exclusivity of H3K4me2/3 and DNA methylation (Meissner *et al.*, 2008; Mohn *et al.*, 2008). This is possibly due to the fact that H3K4me inhibits the binding of DNMT regulatory factor DNMT3L that shows high affinity for unmethylated H3K4. DNMT3L was shown to be a stimulator of *de novo* DNA methylation by DNMT3A/B (Ooi *et al.*, 2007). Although this mechanism was reported to function during maternal imprinting, it may also contribute to cancer-associated hypermethylation. In cancer cells, CpG islands of aberrantly methylated genes are deprived of H3K4me and ac and enriched with repressive marks (Kapoor-Vazirani *et al.*, 2008; Lin *et al.*, 2007).

B. Crosstalk between DNA methylation and polycomb group (PcG) complexes

The data shown above about interaction between DNA methylation and PcG complexes constitute only the "tip of the iceberg." Indeed, a large number of studies reported crosstalks between these two epigenetic regulators (McCabe *et al.*, 2009). For example, EZH2, an H3K27 methyltransferase of PRC2 and PRC3 complexes, was suggested to interact with all three Dnmts and to be required for DNA methylation at some EZH2 target genes in cancer cell lines (Vire *et al.*, 2006). However, the requirement of EZH2 for DNA methylation seems to be transient. EZH2 was proposed to only trigger initial DNA methylation and once promoters acquire dense DNA hypermethylation, EZH2 is no longer required (McGarvey *et al.*, 2007). The interaction between EZH2 and DNMT1 was shown to be necessary for the recruitment of PRC1 complex through its BMI subunit, and this crosstalk is enforced by the interaction between BMI1 and the Dnmt1-associated protein 1 (DMAP1; Hernandez-Munoz *et al.*, 2005; Negishi *et al.*, 2007). On the other hand, another subunit of PRC1, Cbx4 interacts with Dnmt3a and 3b (Kim *et al.*, 2008b; Li *et al.*, 2007b) leading to their SUMOylation. Cbx4 could therefore function as a linker between PRC1-mediated histone ubiquitination and DNA methylation. Cbx4 also recognizes H3K9me3 through its N-terminal chromodomain (Bernstein *et al.*, 2006), which might help the targeting of Dnmt3a/b to regions enriched for H3K9me3. However, this hypothesis awaits further confirmation especially since SUMOylation of dnmt3a seems to reduce its inhibitory activity on transcription (Ling *et al.*, 2004).

Additional studies have revealed connections between DNA methylation and polycomb-mediated H3K27me3 in ES cells that further underscored a potential link between pluripotency and tumorigenesis (Ohm and Baylin, 2007; Schlesinger *et al.*, 2007; Widschwendter *et al.*, 2007). Why would studies on ES cells provide information on tumor initiation and development? The answer resides in the popular cancer-stem-cell model which states that cancers may arise from adult stem cells that undergo a series of mutations or events in response to various environmental or endogenous insults which ultimately leads to their oncogenic transformation. Genome-wide studies comparing cancer tissue and cell lines to ES cells found a relationship between DNA methylation and H3K27me3. Notably, genes commonly targeted for promoter DNA hypermethylation in cancer were not DNA methylated in ES cells but were enriched for the H3K27me3 mediated by PRC2 (Ehrich et al., 2008; Ohm et al., 2007; Schlesinger et al., 2007; Widschwendter et al., 2007). Based on this correlation, it was suggested that H3K27me3 could aberrantly target DNA methylation to tumor suppressor genes in cancer (Ehrich et al., 2008; Ohm et al., 2007; Schlesinger et al., 2007; Widschwendter et al., 2007). The connection between PRCs and DNA methylation may not be limited to an effect of H3K27me3. Recent work has indicated that PRC2 recruits the H3K4me2/3 demethylase Rbp2 (Jarid1a) to its target genes, which may promote DNA methylation through demethylation of H3K4 (Pasini et al., 2008).

These data suggest that one of the events that occur in the transformation of adult stem cells into cancer stem cells is the conversion of repressed but poised state of polycomb targets into a permanently repressed state. These genes would become therefore unresponsive to normal developmental cues and likely provide the cell with a growth advantage. Taken together, these results suggest a potential "targeting" mechanism for aberrant DNA methylation of selected genes in cancer, namely that genes normally repressed by polycomb proteins are preferentially susceptible to acquiring aberrant DNA methylation in cancer. Remarkably, such a crosstalk between H3K27me3 and DNA methylation also seems to play a role during "normal" cellular differentiation. A recent study reported an increased de novo DNA methylation frequency during differentiation of ES cells (Mohn et al., 2008). Similar to cancer development, promoters marked by H3K27me3 in ES cells frequently become DNA methylated in differentiated neuronal cells. These data indicate that also during normal development H3K27me3 and DNA methylation might act together; however, in cancer this crosstalk might be misregulated and contributes to the increased frequency of polycomb targets which become aberrantly DNA methylated and thus stably silenced.

Based on the data provided here, it seems clear that methylation driving of tumorigenesis is not sufficient and that a complex interaction between DNMTs and chromatin-associated factors such as polycomb proteins is necessary for initiation and progression of tumors.

C. Crosstalk between DNA methylation and small RNAs

The recruitment of DNMT activities to DNA is not only dependent on histone modifications. Another area of recent interest is the contribution of small RNA species, such as siRNAs and miRNAs to recruitment of DNA methylation. While most solid evidence comes from plants, there is an increasing amount of reports in mammals suggesting direct and indirect crosstalks between small

RNAs and DNA methylation. si/miRNAs are processed and matured by the DICER RNase III family nuclease (Benetti et al., 2008; Sinkkonen et al., 2008; Ting et al., 2008). The involvement of Dicer in the maintenance of DNA methylation was first suggested by the fact that $Dicer^{-/-}$ ES cells show a significant loss of DNA methylation on heterochromatin marked by H3K9me (Benetti et al., 2008; McCabe et al., 2009). On the other hand, miRNAs might indirectly activate the expression of DNMTs. Indeed, miR-290 family targets and represses Rbl2 which otherwise represses E2F-mediated transcription of DNMTs (Benetti et al., 2008). The implication of miRNA-dependent DNA methylation was also found in cancer. For example, miRNAs were essential for the hypermethylation of several CpG rich promoters in a human colon cancer cell line (Ting et al., 2008). However, a recent study has shown that different families of miRNAs may have different effects on DNA methylation. For example, miR-29 represses the DNMT3A and DNMT3B transcripts directly, and miR-29 is hypermethylated in lung cancers cells. Reexpression of miR-29 in these cells induced DNA demethylation and reduced cell proliferation and tumorigenicity (Fabbri et al., 2007). Recent technologies, such as next-generation sequencing, will allow the identification of novel miRNAs and their corresponding targets, helping to further understand the crosstalk between DNA methylation and miRNAs.

V. CONCLUSIONS AND PERSPECTIVES

Classical interpretation of epigenetic modification effects on cellular processes, which implies that a single epigenetic event is independent on other modifications and has a predictable and stable outcome, has now evolved. The new model is that epigenetic events are interdependent and dynamic and their resulting effect is highly malleable (Gopalakrishnan et al., 2008; Latham and Dent, 2007; Loizou et al., 2006; McCabe et al., 2009; Sawan et al., 2008; Suganuma and Workman, 2008; Vaissiere et al., 2008). In this chapter, several examples of the crosstalks between different epigenetic modifications and their roles in the regulation of cellular processes and tumorigenesis were given. Although most of the interactions were discussed here, there are certainly much more examples that could not be included because of space limitations. Indeed, our knowledge of cross-regulation and interdependence between epigenetic events was extremely enlarged in the last few years. From the very first studies in the early 1990s reporting possible interdependence between histone modifications, there are now many examples of networks that involve several different epigenetic players. Additionally, we acquired more knowledge about the mechanisms and the enzymes governing these crosstalks.

To date, detection of epigenetic crosstalk between specific histone modifications has been accomplished using site-specific antibodies that recognize the modifications or mass spectroscopy. These approaches require, by default, a prior educated prediction of possible interactions and are not efficient for the detection of crosstalks *in trans*. The emergence of microarrays and next-generation sequencing will definitely exponentially increase the number of crosstalks identified in the context of the epigenome. Additionally, no prior presumption will be needed and new interactions which were not necessarily expected will be identified. Likewise, advances in the sensitivity of mass spectroscopic approaches will allow for the definition of new histone modifications and of concomitant modifications on the same histone molecule. Importantly, one has to point out that many currently known crosstalks are based on *in vitro* experiments or works in transformed cancer cell lines. Thus, many if not most of these putative interactions need to be further validated and confirmed *in vivo* to estimate their biological relevance.

Deregulated epigenetic mechanisms are strongly implicated in cancer development and it has become apparent that cancer is as much a disease of abnormal epigenetics as it is a disease of genetic mutations. The above discussion underscores the complex relationship that exists between the histone code and susceptibility to DNA methylation and suggests that misregulation of this relationship may be crucial to the development and targeting of DNA hypermethylation in cancer cells. Unlike genetic alterations, epigenetic changes are reversible. This represents a promising therapeutical potential. A multifaceted strategy targeting multiple components of the epigenetic machinery may be more effective than targeting individual modifications. A few enzymes, drugs and small molecules that influence epigenetic modifications, such as HMT and HDAC inhibitors as well as DNA methylating and demethylating agents, are already in use or in clinical trials, but many more might be potential candidates for cancer treatment.

Acknowledgments

I apologize to colleagues whose relevant publications were not cited because of space limitation. I thank Dr. F. Mohn for critical comments and suggestions. R. M. is supported by an EMBO long-term fellowship.

References

Ahn, S. H., Cheung, W. L., Hsu, J. Y., Diaz, R. L., Smith, M. M., and Allis, C. D. (2005). Sterile 20 kinase phosphorylates histone H2B at serine 10 during hydrogen peroxide-induced apoptosis in *S. cerevisiae. Cell* **120,** 25–36.

Ahn, S. H., Diaz, R. L., Grunstein, M., and Allis, C. D. (2006). Histone H2B deacetylation at lysine 11 is required for yeast apoptosis induced by phosphorylation of H2B at serine 10. *Mol. Cell* **24,** 211–220.

Allshire, R. C., Nimmo, E. R., Ekwall, K., Javerzat, J. P., and Cranston, G. (1995). Mutations derepressing silent centromeric domains in fission yeast disrupt chromosome segregation. *Genes Dev.* **9,** 218–233.

Bannister, A. J., Schneider, R., Myers, F. A., Thorne, A. W., Crane-Robinson, C., and Kouzarides, T. (2005). Spatial distribution of di- and tri-methyl lysine 36 of histone H3 at active genes. *J. Biol. Chem.* **280,** 17732–17736.

Barratt, M. J., Hazzalin, C. A., Cano, E., and Mahadevan, L. C. (1994). Mitogen-stimulated phosphorylation of histone H3 is targeted to a small hyperacetylation-sensitive fraction. *Proc. Natl. Acad. Sci. USA* **91,** 4781–4785.

Barski, A., Cuddapah, S., Cui, K., Roh, T. Y., Schones, D. E., Wang, Z., Wei, G., Chepelev, I., and Zhao, K. (2007). High-resolution profiling of histone methylations in the human genome. *Cell* **129,** 823–837.

Bell, O., Wirbelauer, C., Hild, M., Scharf, A. N., Schwaiger, M., MacAlpine, D. M., Zilbermann, F., van Leeuwen, F., Bell, S. P., Imhof, A., *et al.* (2007). Localized H3K36 methylation states define histone H4K16 acetylation during transcriptional elongation in *Drosophila. EMBO J.* **26,** 4974–4984.

Benetti, R., Gonzalo, S., Jaco, I., Munoz, P., Gonzalez, S., Schoeftner, S., Murchison, E., Andl, T., Chen, T., Klatt, P., *et al.* (2008). A mammalian microRNA cluster controls DNA methylation and telomere recombination via Rbl2-dependent regulation of DNA methyltransferases. *Nat. Struct. Mol. Biol.* **15,** 998.

Bernstein, B. E., Humphrey, E. L., Erlich, R. L., Schneider, R., Bouman, P., Liu, J. S., Kouzarides, T., and Schreiber, S. L. (2002). Methylation of histone H3 Lys 4 in coding regions of active genes. *Proc. Natl. Acad. Sci. USA* **99,** 8695–8700.

Bernstein, E., Duncan, E. M., Masui, O., Gil, J., Heard, E., and Allis, C. D. (2006). Mouse polycomb proteins bind differentially to methylated histone H3 and RNA and are enriched in facultative heterochromatin. *Mol. Cell. Biol.* **26,** 2560–2569.

Bird, A. (2002). DNA methylation patterns and epigenetic memory. *Genes Dev.* **16,** 6–21.

Bird, A. (2007). Perceptions of epigenetics. *Nature* **447,** 396–398.

Bonapace, I. M., Latella, L., Papait, R., Nicassio, F., Sacco, A., Muto, M., Crescenzi, M., and Di Fiore, P. P. (2002). Np95 is regulated by E1A during mitotic reactivation of terminally differentiated cells and is essential for S phase entry. *J. Cell Biol.* **157,** 909–914.

Bostick, M., Kim, J. K., Esteve, P. O., Clark, A., Pradhan, S., and Jacobsen, S. E. (2007). UHRF1 plays a role in maintaining DNA methylation in mammalian cells. *Science* **317,** 1760–1764.

Boulard, M., Bouvet, P., Kundu, T. K., and Dimitrov, S. (2007). Histone variant nucleosomes: Structure, function and implication in disease. *Subcell. Biochem.* **41,** 71–89.

Briggs, S. D., Bryk, M., Strahl, B. D., Cheung, W. L., Davie, J. K., Dent, S. Y., Winston, F., and Allis, C. D. (2001). Histone H3 lysine 4 methylation is mediated by Set1 and required for cell growth and rDNA silencing in *Saccharomyces cerevisiae. Genes Dev.* **15,** 3286–3295.

Briggs, S. D., Xiao, T., Sun, Z. W., Caldwell, J. A., Shabanowitz, J., Hunt, D. F., Allis, C. D., and Strahl, B. D. (2002). Gene silencing: Trans-histone regulatory pathway in chromatin. *Nature* **418,** 498.

Brown, M. A., Sims, R. J., 3rd, Gottlieb, P. D., and Tucker, P. W. (2006). Identification and characterization of Smyd2: A split SET/MYND domain-containing histone H3 lysine 36-specific methyltransferase that interacts with the Sin3 histone deacetylase complex. *Mol. Cancer* **5,** 26.

Buhler, M., Verdel, A., and Moazed, D. (2006). Tethering RITS to a nascent transcript initiates RNAi- and heterochromatin-dependent gene silencing. *Cell* **125,** 873–886.

Cam, H. P., Sugiyama, T., Chen, E. S., Chen, X., FitzGerald, P. C., and Grewal, S. I. (2005). Comprehensive analysis of heterochromatin- and RNAi-mediated epigenetic control of the fission yeast genome. *Nat. Genet.* **37,** 809–819.

Camporeale, G., Oommen, A. M., Griffin, J. B., Sarath, G., and Zempleni, J. (2007). K12-biotinylated histone H4 marks heterochromatin in human lymphoblastoma cells. *J. Nutr. Biochem* **18,** 760–768.

Cao, R., Tsukada, Y., and Zhang, Y. (2005). Role of Bmi-1 and Ring1A in H2A ubiquitylation and Hox gene silencing. *Mol. Cell* **20,** 845–854.

Celeste, A., Fernandez-Capetillo, O., Kruhlak, M. J., Pilch, D. R., Staudt, D. W., Lee, A., Bonner, R. F., Bonner, W. M., and Nussenzweig, A. (2003). Histone H2AX phosphorylation is dispensable for the initial recognition of DNA breaks. *Nat. Cell Biol.* **5,** 675–679.

Chen, H., Lin, R. J., Schiltz, R. L., Chakravarti, D., Nash, A., Nagy, L., Privalsky, M. L., Nakatani, Y., and Evans, R. M. (1997). Nuclear receptor coactivator ACTR is a novel histone acetyltransferase and forms a multimeric activation complex with P/CAF and CBP/p300. *Cell* **90,** 569–580.

Chen, D., Ma, H., Hong, H., Koh, S. S., Huang, S. M., Schurter, B. T., Aswad, D. W., and Stallcup, M. R. (1999). Regulation of transcription by a protein methyltransferase. *Science* **284,** 2174–2177.

Chen, D., Huang, S. M., and Stallcup, M. R. (2000). Synergistic, p160 coactivator-dependent enhancement of estrogen receptor function by CARM1 and p300. *J. Biol. Chem.* **275,** 40810–40816.

Cheung, P., Tanner, K. G., Cheung, W. L., Sassone-Corsi, P., Denu, J. M., and Allis, C. D. (2000). Synergistic coupling of histone H3 phosphorylation and acetylation in response to epidermal growth factor stimulation. *Mol. Cell* **5,** 905–915.

Cheung, W. L., Ajiro, K., Samejima, K., Kloc, M., Cheung, P., Mizzen, C. A., Beeser, A., Etkin, L. D., Chernoff, J., Earnshaw, W. C., et al. (2003). Apoptotic phosphorylation of histone H2B is mediated by mammalian sterile twenty kinase. *Cell* **113,** 507–517.

Cho, Y. W., Hong, T., Hong, S., Guo, H., Yu, H., Kim, D., Guszczynski, T., Dressler, G. R., Copeland, T. D., Kalkum, M., et al. (2007). PTIP associates with MLL3- and MLL4-containing histone H3 lysine 4 methyltransferase complex. *J. Biol. Chem.* **282,** 20395–20406.

Chowdhury, D., Keogh, M. C., Ishii, H., Peterson, C. L., Buratowski, S., and Lieberman, J. (2005). gamma-H2AX dephosphorylation by protein phosphatase 2A facilitates DNA double-strand break repair. *Mol. Cell* **20,** 801–809.

Citterio, E., Papait, R., Nicassio, F., Vecchi, M., Gomiero, P., Mantovani, R., Di Fiore, P. P., and Bonapace, I. M. (2004). Np95 is a histone-binding protein endowed with ubiquitin ligase activity. *Mol. Cell. Biol.* **24,** 2526–2535.

Clayton, A. L., Rose, S., Barratt, M. J., and Mahadevan, L. C. (2000). Phosphoacetylation of histone H3 on c-fos- and c-jun-associated nucleosomes upon gene activation. *EMBO J.* **19,** 3714–3726.

Cook, P. J., Ju, B. G., Telese, F., Wang, X., Glass, C. K., and Rosenfeld, M. G. (2009). Tyrosine dephosphorylation of H2AX modulates apoptosis and survival decisions. *Nature* **458,** 591–596.

Czermin, B., Schotta, G., Hulsmann, B. B., Brehm, A., Becker, P. B., Reuter, G., and Imhof, A. (2001). Physical and functional association of SU(VAR)3-9 and HDAC1 in *Drosophila. EMBO Rep.* **2,** 915–919.

Daniel, J. A., Torok, M. S., Sun, Z. W., Schieltz, D., Allis, C. D., Yates, J. R., 3rd, and Grant, P. A. (2004). Deubiquitination of histone H2B by a yeast acetyltransferase complex regulates transcription. *J. Biol. Chem.* **279,** 1867–1871.

Daujat, S., Bauer, U. M., Shah, V., Turner, B., Berger, S., and Kouzarides, T. (2002). Crosstalk between CARM1 methylation and CBP acetylation on histone H3. *Curr. Biol.* **12,** 2090–2097.

de la Cruz, X., Lois, S., Sanchez-Molina, S., and Martinez-Balbas, M. A. (2005). Do protein motifs read the histone code? *Bioessays* **27,** 164–175.

De La Fuente, R., Baumann, C., Fan, T., Schmidtmann, A., Dobrinski, I., and Muegge, K. (2006). Lsh is required for meiotic chromosome synapsis and retrotransposon silencing in female germ cells. *Nat. Cell Biol.* **8,** 1448–1454.

Dehe, P. M., Pamblanco, M., Luciano, P., Lebrun, R., Moinier, D., Sendra, R., Verreault, A., Tordera, V., and Geli, V. (2005). Histone H3 lysine 4 mono-methylation does not require ubiquitination of histone H2B. *J. Mol. Biol.* **353**, 477–484.

Dou, Y., Milne, T. A., Tackett, A. J., Smith, E. R., Fukuda, A., Wysocka, J., Allis, C. D., Chait, B. T., Hess, J. L., and Roeder, R. G. (2005). Physical association and coordinate function of the H3 K4 methyltransferase MLL1 and the H4 K16 acetyltransferase MOF. *Cell* **121**, 873–885.

Dover, J., Schneider, J., Tawiah-Boateng, M. A., Wood, A., Dean, K., Johnston, M., and Shilatifard, A. (2002). Methylation of histone H3 by COMPASS requires ubiquitination of histone H2B by Rad6. *J. Biol. Chem.* **277**, 28368–28371.

Downs, J. A., and Cote, J. (2005). Dynamics of chromatin during the repair of DNA double-strand breaks. *Cell Cycle* **4**, 1373–1376.

Downs, J. A., Allard, S., Jobin-Robitaille, O., Javaheri, A., Auger, A., Bouchard, N., Kron, S. J., Jackson, S. P., and Cote, J. (2004). Binding of chromatin-modifying activities to phosphorylated histone H2A at DNA damage sites. *Mol. Cell* **16**, 979–990.

Ehrich, M., Turner, J., Gibbs, P., Lipton, L., Giovanneti, M., Cantor, C., and van den Boom, D. (2008). Cytosine methylation profiling of cancer cell lines. *Proc. Natl. Acad. Sci. USA* **105**, 4844–4849.

Ezhkova, E., and Tansey, W. P. (2004). Proteasomal ATPases link ubiquitylation of histone H2B to methylation of histone H3. *Mol. Cell* **13**, 435–442.

Fabbri, M., Garzon, R., Cimmino, A., Liu, Z., Zanesi, N., Callegari, E., Liu, S., Alder, H., Costinean, S., Fernandez-Cymering, C., et al. (2007). MicroRNA-29 family reverts aberrant methylation in lung cancer by targeting DNA methyltransferases 3A and 3B. *Proc. Natl. Acad. Sci. USA* **104**, 15805–15810.

Fabbrizio, E., El Messaoudi, S., Polanowska, J., Paul, C., Cook, J. R., Lee, J. H., Negre, V., Rousset, M., Pestka, S., Le Cam, A., et al. (2002). Negative regulation of transcription by the type II arginine methyltransferase PRMT5. *EMBO Rep.* **3**, 641–645.

Frescas, D., Guardavaccaro, D., Bassermann, F., Koyama-Nasu, R., and Pagano, M. (2007). JHDM1B/ FBXL10 is a nucleolar protein that represses transcription of ribosomal RNA genes. *Nature* **450**, 309–313.

Fritsch, L., Robin, P., Mathieu, J. R., Souidi, M., Hinaux, H., Rougeulle, C., Harel-Bellan, A., Ameyar-Zazoua, M., and Ait-Si-Ali, S. (2010). A subset of the histone H3 lysine 9 methyltransferases Suv39h1, G9a, GLP, and SETDB1 participate in a multimeric complex. *Mol. Cell* **37**, 46–56.

Fujimori, A., Matsuda, Y., Takemoto, Y., Hashimoto, Y., Kubo, E., Araki, R., Fukumura, R., Mita, K., Tatsumi, K., and Muto, M. (1998). Cloning and mapping of Np95 gene which encodes a novel nuclear protein associated with cell proliferation. *Mamm. Genome* **9**, 1032–1035.

Garcia-Dominguez, M., and Reyes, J. C. (2009). SUMO association with repressor complexes, emerging routes for transcriptional control. *Biochim. Biophys. Acta* **1789**, 451–459.

Gearhart, M. D., Corcoran, C. M., Wamstad, J. A., and Bardwell, V. J. (2006). Polycomb group and SCF ubiquitin ligases are found in a novel BCOR complex that is recruited to BCL6 targets. *Mol. Cell. Biol.* **26**, 6880–6889.

Gopalakrishnan, S., Van Emburgh, B. O., and Robertson, K. D. (2008). DNA methylation in development and human disease. *Mutat. Res.* **647**, 30–38.

Govind, C. K., Zhang, F., Qiu, H., Hofmeyer, K., and Hinnebusch, A. G. (2007). Gcn5 promotes acetylation, eviction, and methylation of nucleosomes in transcribed coding regions. *Mol. Cell* **25**, 31–42.

Grewal, S. I., and Elgin, S. C. (2007). Transcription and RNA interference in the formation of heterochromatin. *Nature* **447**, 399–406.

Grewal, S. I., and Jia, S. (2007). Heterochromatin revisited. *Nat. Rev. Genet.* **8**, 35–46.

Hall, I. M., Shankaranarayana, G. D., Noma, K., Ayoub, N., Cohen, A., and Grewal, S. I. (2002). Establishment and maintenance of a heterochromatin domain. *Science* **297**, 2232–2237.

Hassan, Y. I., and Zempleni, J. (2006). Epigenetic regulation of chromatin structure and gene function by biotin. *J. Nutr.* **136**, 1763–1765.

Hazzalin, C. A., and Mahadevan, L. C. (2005). Dynamic acetylation of all lysine 4-methylated histone H3 in the mouse nucleus: Analysis at c-fos and c-jun. *PLoS Biol.* **3**, e393.

Henry, K. W., Wyce, A., Lo, W. S., Duggan, L. J., Emre, N. C., Kao, C. F., Pillus, L., Shilatifard, A., Osley, M. A., and Berger, S. L. (2003). Transcriptional activation via sequential histone H2B ubiquitylation and deubiquitylation, mediated by SAGA-associated Ubp8. *Genes Dev.* **17**, 2648–2663.

Hernandez-Munoz, I., Taghavi, P., Kuijl, C., Neefjes, J., and van Lohuizen, M. (2005). Association of BMI1 with polycomb bodies is dynamic and requires PRC2/EZH2 and the maintenance DNA methyltransferase DNMT1. *Mol. Cell. Biol.* **25**, 11047–11058.

Hopfner, R., Mousli, M., Jeltsch, J. M., Voulgaris, A., Lutz, Y., Marin, C., Bellocq, J. P., Oudet, P., and Bronner, C. (2000). ICBP90, a novel human CCAAT binding protein, involved in the regulation of topoisomerase IIalpha expression. *Cancer Res.* **60**, 121–128.

Huang, Y., Fang, J., Bedford, M. T., Zhang, Y., and Xu, R. M. (2006). Recognition of histone H3 lysine-4 methylation by the double tudor domain of JMJD2A. *Science* **312**, 748–751.

Ikura, T., Tashiro, S., Kakino, A., Shima, H., Jacob, N., Amunugama, R., Yoder, K., Izumi, S., Kuraoka, I., Tanaka, K., *et al.* (2007). DNA damage-dependent acetylation and ubiquitination of H2AX enhances chromatin dynamics. *Mol. Cell. Biol.* **27**, 7028–7040.

Ingvarsdottir, K., Krogan, N. J., Emre, N. C., Wyce, A., Thompson, N. J., Emili, A., Hughes, T. R., Greenblatt, J. F., and Berger, S. L. (2005). H2B ubiquitin protease Ubp8 and Sgf11 constitute a discrete functional module within the *Saccharomyces cerevisiae* SAGA complex. *Mol. Cell. Biol.* **25**, 1162–1172.

Ivanovska, I., Khandan, T., Ito, T., and Orr-Weaver, T. L. (2005). A histone code in meiosis: The histone kinase, NHK-1, is required for proper chromosomal architecture in Drosophila oocytes. *Genes Dev.* **19**, 2571–2582.

Jeanblanc, M., Mousli, M., Hopfner, R., Bathami, K., Martinet, N., Abbady, A. Q., Siffert, J. C., Mathieu, E., Muller, C. D., and Bronner, C. (2005). The retinoblastoma gene and its product are targeted by ICBP90: A key mechanism in the G1/S transition during the cell cycle. *Oncogene* **24**, 7337–7345.

Jenuwein, T., and Allis, C. D. (2001). Translating the histone code. *Science* **293**, 1074–1080.

Johnson, L., Mollah, S., Garcia, B. A., Muratore, T. L., Shabanowitz, J., Hunt, D. F., and Jacobsen, S. E. (2004). Mass spectrometry analysis of Arabidopsis histone H3 reveals distinct combinations of post-translational modifications. *Nucleic Acids Res.* **32**, 6511–6518.

Joo, H. Y., Zhai, L., Yang, C., Nie, S., Erdjument-Bromage, H., Tempst, P., Chang, C., and Wang, H. (2007). Regulation of cell cycle progression and gene expression by H2A deubiquitination. *Nature* **449**, 1068–1072.

Joshi, A. A., and Struhl, K. (2005). Eaf3 chromodomain interaction with methylated H3-K36 links histone deacetylation to Pol II elongation. *Mol. Cell* **20**, 971–978.

Kapoor-Vazirani, P., Kagey, J. D., Powell, D. R., and Vertino, P. M. (2008). Role of hMOF-dependent histone H4 lysine 16 acetylation in the maintenance of TMS1/ASC gene activity. *Cancer Res.* **68**, 6810–6821.

Karagianni, P., Amazit, L., Qin, J., and Wong, J. (2008). ICBP90, a novel methyl K9 H3 binding protein linking protein ubiquitination with heterochromatin formation. *Mol. Cell. Biol.* **28**, 705–717.

Kato, H., Goto, D. B., Martienssen, R. A., Urano, T., Furukawa, K., and Murakami, Y. (2005). RNA polymerase II is required for RNAi-dependent heterochromatin assembly. *Science* **309**, 467–469.

Keogh, M. C., Kurdistani, S. K., Morris, S. A., Ahn, S. H., Podolny, V., Collins, S. R., Schuldiner, M., Chin, K., Punna, T., Thompson, N. J., et al. (2005). Cotranscriptional set2 methylation of histone H3 lysine 36 recruits a repressive Rpd3 complex. Cell 123, 593–605.

Keogh, M. C., Kim, J. A., Downey, M., Fillingham, J., Chowdhury, D., Harrison, J. C., Onishi, M., Datta, N., Galicia, S., Emili, A., et al. (2006). A phosphatase complex that dephosphorylates gammaH2AX regulates DNA damage checkpoint recovery. Nature 439, 497–501.

Kim, D. H., Saetrom, P., Snove, O., Jr., and Rossi, J. J. (2008a). MicroRNA-directed transcriptional gene silencing in mammalian cells. Proc. Natl. Acad. Sci. USA 105, 16230–16235.

Kim, S. H., Park, J., Choi, M. C., Park, J. H., Kim, H. P., Lee, J. H., Oh, D. Y., Im, S. A., Bang, Y. J., and Kim, T. Y. (2008b). DNA methyltransferase 3B acts as a co-repressor of the human polycomb protein hPc2 to repress fibroblast growth factor receptor 3 transcription. Int. J. Biochem. Cell Biol. 40, 2462–2471.

Kizer, K. O., Phatnani, H. P., Shibata, Y., Hall, H., Greenleaf, A. L., and Strahl, B. D. (2005). A novel domain in Set2 mediates RNA polymerase II interaction and couples histone H3 K36 methylation with transcript elongation. Mol. Cell. Biol. 25, 3305–3316.

Kobayashi, J., Tauchi, H., Sakamoto, S., Nakamura, A., Morishima, K., Matsuura, S., Kobayashi, T., Tamai, K., Tanimoto, K., and Komatsu, K. (2002). NBS1 localizes to gamma-H2AX foci through interaction with the FHA/BRCT domain. Curr. Biol. 12, 1846–1851.

Kohler, A., Pascual-Garcia, P., Llopis, A., Zapater, M., Posas, F., Hurt, E., and Rodriguez-Navarro, S. (2006). The mRNA export factor Sus1 is involved in Spt/Ada/Gcn5 acetyltransferase-mediated H2B deubiquitinylation through its interaction with Ubp8 and Sgf11. Mol. Biol. Cell 17, 4228–4236.

Kourmouli, N., Jeppesen, P., Mahadevhaiah, S., Burgoyne, P., Wu, R., Gilbert, D. M., Bongiorni, S., Prantera, G., Fanti, L., Pimpinelli, S., et al. (2004). Heterochromatin and tri-methylated lysine 20 of histone H4 in animals. J. Cell Sci. 117, 2491–2501.

Krogan, N. J., Kim, M., Tong, A., Golshani, A., Cagney, G., Canadien, V., Richards, D. P., Beattie, B. K., Emili, A., Boone, C., et al. (2003). Methylation of histone H3 by Set2 in Saccharomyces cerevisiae is linked to transcriptional elongation by RNA polymerase II. Mol. Cell. Biol. 23, 4207–4218.

Kurshakova, M., Maksimenko, O., Golovnin, A., Pulina, M., Georgieva, S., Georgiev, P., and Krasnov, A. (2007). Evolutionarily conserved E(y)2/Sus1 protein is essential for the barrier activity of Su(Hw)-dependent insulators in Drosophila. Mol. Cell 27, 332–338.

Kusch, T., Florens, L., Macdonald, W. H., Swanson, S. K., Glaser, R. L., Yates, J. R., 3rd, Abmayr, S. M., Washburn, M. P., and Workman, J. L. (2004). Acetylation by Tip60 is required for selective histone variant exchange at DNA lesions. Science 306, 2084–2087.

Lagarou, A., Mohd-Sarip, A., Moshkin, Y. M., Chalkley, G. E., Bezstarosti, K., Demmers, J. A., and Verrijzer, C. P. (2008). dKDM2 couples histone H2A ubiquitylation to histone H3 demethylation during Polycomb group silencing. Genes Dev. 22, 2799–2810.

Laribee, R. N., Krogan, N. J., Xiao, T., Shibata, Y., Hughes, T. R., Greenblatt, J. F., and Strahl, B. D. (2005). BUR kinase selectively regulates H3 K4 trimethylation and H2B ubiquitylation through recruitment of the PAF elongation complex. Curr. Biol. 15, 1487–1493.

Laribee, R. N., Fuchs, S. M., and Strahl, B. D. (2007a). H2B ubiquitylation in transcriptional control: A FACT-finding mission. Genes Dev. 21, 737–743.

Laribee, R. N., Shibata, Y., Mersman, D. P., Collins, S. R., Kemmeren, P., Roguev, A., Weissman, J. S., Briggs, S. D., Krogan, N. J., and Strahl, B. D. (2007b). CCR4/NOT complex associates with the proteasome and regulates histone methylation. Proc. Natl. Acad. Sci. USA 104, 5836–5841.

Latham, J. A., and Dent, S. Y. (2007). Cross-regulation of histone modifications. Nat. Struct. Mol. Biol. 14, 1017–1024.

Le Guezennec, X., Vermeulen, M., Brinkman, A. B., Hoeijmakers, W. A., Cohen, A., Lasonder, E., and Stunnenberg, H. G. (2006). MBD2/NuRD and MBD3/NuRD, two distinct complexes with different biochemical and functional properties. *Mol. Cell. Biol.* **26,** 843–851.

Lee, J. S., and Shilatifard, A. (2007). A site to remember: H3K36 methylation a mark for histone deacetylation. *Mutat. Res.* **618,** 130–134.

Lee, D. Y., Teyssier, C., Strahl, B. D., and Stallcup, M. R. (2005a). Role of protein methylation in regulation of transcription. *Endocr. Rev.* **26,** 147–170.

Lee, K. K., Florens, L., Swanson, S. K., Washburn, M. P., and Workman, J. L. (2005b). The deubiquitylation activity of Ubp8 is dependent upon Sgf11 and its association with the SAGA complex. *Mol. Cell. Biol.* **25,** 1173–1182.

Lee, M. G., Wynder, C., Cooch, N., and Shiekhattar, R. (2005c). An essential role for CoREST in nucleosomal histone 3 lysine 4 demethylation. *Nature* **437,** 432–435.

Lee, M. G., Wynder, C., Bochar, D. A., Hakimi, M. A., Cooch, N., and Shiekhattar, R. (2006). Functional interplay between histone demethylase and deacetylase enzymes. *Mol. Cell. Biol.* **26,** 6395–6402.

Lee, J. S., Shukla, A., Schneider, J., Swanson, S. K., Washburn, M. P., Florens, L., Bhaumik, S. R., and Shilatifard, A. (2007a). Histone crosstalk between H2B monoubiquitination and H3 methylation mediated by COMPASS. *Cell* **131,** 1084–1096.

Lee, M. G., Norman, J., Shilatifard, A., and Shiekhattar, R. (2007b). Physical and functional association of a trimethyl H3K4 demethylase and Ring6a/MBLR, a polycomb-like protein. *Cell* **128,** 877–887.

Lehnertz, B., Ueda, Y., Derijck, A. A., Braunschweig, U., Perez-Burgos, L., Kubicek, S., Chen, T., Li, E., Jenuwein, T., and Peters, A. H. (2003). Suv39h-mediated histone H3 lysine 9 methylation directs DNA methylation to major satellite repeats at pericentric heterochromatin. *Curr. Biol.* **13,** 1192–1200.

Li, J., Lin, Q., Yoon, H. G., Huang, Z. Q., Strahl, B. D., Allis, C. D., and Wong, J. (2002). Involvement of histone methylation and phosphorylation in regulation of transcription by thyroid hormone receptor. *Mol. Cell. Biol.* **22,** 5688–5697.

Li, B., Howe, L., Anderson, S., Yates, J. R., 3rd, and Workman, J. L. (2003). The Set2 histone methyltransferase functions through the phosphorylated carboxyl-terminal domain of RNA polymerase II. *J. Biol. Chem.* **278,** 8897–8903.

Li, B., Gogol, M., Carey, M., Lee, D., Seidel, C., and Workman, J. L. (2007a). Combined action of PHD and chromo domains directs the Rpd3S HDAC to transcribed chromatin. *Science* **316,** 1050–1054.

Li, B., Zhou, J., Liu, P., Hu, J., Jin, H., Shimono, Y., Takahashi, M., and Xu, G. (2007b). Polycomb protein Cbx4 promotes SUMO modification of de novo DNA methyltransferase Dnmt3a. *Biochem. J.* **405,** 369–378.

Li, F., Huarte, M., Zaratiegui, M., Vaughn, M. W., Shi, Y., Martienssen, R., and Cande, W. Z. (2008). Lid2 is required for coordinating H3K4 and H3K9 methylation of heterochromatin and euchromatin. *Cell* **135,** 272–283.

Lin, J. C., Jeong, S., Liang, G., Takai, D., Fatemi, M., Tsai, Y. C., Egger, G., Gal-Yam, E. N., and Jones, P. A. (2007). Role of nucleosomal occupancy in the epigenetic silencing of the MLH1 CpG island. *Cancer Cell* **12,** 432–444.

Ling, Y., Sankpal, U. T., Robertson, A. K., McNally, J. G., Karpova, T., and Robertson, K. D. (2004). Modification of de novo DNA methyltransferase 3a (Dnmt3a) by SUMO-1 modulates its interaction with histone deacetylases (HDACs) and its capacity to repress transcription. *Nucleic Acids Res.* **32,** 598–610.

Liu, Y., Taverna, S. D., Muratore, T. L., Shabanowitz, J., Hunt, D. F., and Allis, C. D. (2007). RNAi-dependent H3K27 methylation is required for heterochromatin formation and DNA elimination in Tetrahymena. *Genes Dev.* **21,** 1530–1545.

Lo, W. S., Trievel, R. C., Rojas, J. R., Duggan, L., Hsu, J. Y., Allis, C. D., Marmorstein, R., and Berger, S. L. (2000). Phosphorylation of serine 10 in histone H3 is functionally linked in vitro and in vivo to Gcn5-mediated acetylation at lysine 14. *Mol. Cell* **5,** 917–926.

Loizou, J. I., Murr, R., Finkbeiner, M. G., Sawan, C., Wang, Z. Q., and Herceg, Z. (2006). Epigenetic information in chromatin: The code of entry for DNA repair. *Cell Cycle* **5,** 696–701.

MacDonald, V. E., and Howe, L. J. (2009). Histone acetylation: Where to go and how to get there. *Epigenetics* **4,** 139–143.

Macdonald, N., Welburn, J. P., Noble, M. E., Nguyen, A., Yaffe, M. B., Clynes, D., Moggs, J. G., Orphanides, G., Thomson, S., Edmunds, J. W., *et al.* (2005). Molecular basis for the recognition of phosphorylated and phosphoacetylated histone h3 by 14-3-3. *Mol. Cell* **20,** 199–211.

Martin, C., and Zhang, Y. (2005). The diverse functions of histone lysine methylation. *Nat. Rev. Mol. Cell. Biol.* **6,** 838–849.

Martin, D. G., Grimes, D. E., Baetz, K., and Howe, L. (2006a). Methylation of histone H3 mediates the association of the NuA3 histone acetyltransferase with chromatin. *Mol. Cell. Biol.* **26,** 3018–3028.

Martin, D. G., Baetz, K., Shi, X., Walter, K. L., MacDonald, V. E., Wlodarski, M. J., Gozani, O., Hieter, P., and Howe, L. (2006b). The Yng1p plant homeodomain finger is a methyl-histone binding module that recognizes lysine 4-methylated histone H3. *Mol. Cell. Biol.* **26,** 7871–7879.

Matsui, S. I., Seon, B. K., and Sandberg, A. A. (1979). Disappearance of a structural chromatin protein A24 in mitosis: Implications for molecular basis of chromatin condensation. *Proc. Natl. Acad. Sci. USA* **76,** 6386–6390.

Matzke, M. A., and Birchler, J. A. (2005). RNAi-mediated pathways in the nucleus. *Nat. Rev. Genet.* **6,** 24–35.

McCabe, M. T., Brandes, J. C., and Vertino, P. M. (2009). Cancer DNA methylation: Molecular mechanisms and clinical implications. *Clin. Cancer Res.* **15,** 3927–3937.

McGarvey, K. M., Fahrner, J. A., Greene, E., Martens, J., Jenuwein, T., and Baylin, S. B. (2006). Silenced tumor suppressor genes reactivated by DNA demethylation do not return to a fully euchromatic chromatin state. *Cancer Res.* **66,** 3541–3549.

McGarvey, K. M., Greene, E., Fahrner, J. A., Jenuwein, T., and Baylin, S. B. (2007). DNA methylation and complete transcriptional silencing of cancer genes persist after depletion of EZH2. *Cancer Res.* **67,** 5097–5102.

McGinty, R. K., Kim, J., Chatterjee, C., Roeder, R. G., and Muir, T. W. (2008). Chemically ubiquitylated histone H2B stimulates hDot1L-mediated intranucleosomal methylation. *Nature* **453,** 812–816.

Meilinger, D., Fellinger, K., Bultmann, S., Rothbauer, U., Bonapace, I. M., Klinkert, W. E., Spada, F., and Leonhardt, H. (2009). Np95 interacts with de novo DNA methyltransferases, Dnmt3a and Dnmt3b, and mediates epigenetic silencing of the viral CMV promoter in embryonic stem cells. *EMBO Rep.* **10,** 1259–1264.

Meissner, A., Mikkelsen, T. S., Gu, H., Wernig, M., Hanna, J., Sivachenko, A., Zhang, X., Bernstein, B. E., Nusbaum, C., Jaffe, D. B., *et al.* (2008). Genome-scale DNA methylation maps of pluripotent and differentiated cells. *Nature* **454,** 766–770.

Mikkelsen, T. S., Ku, M., Jaffe, D. B., Issac, B., Lieberman, E., Giannoukos, G., Alvarez, P., Brockman, W., Kim, T. K., Koche, R. P., *et al.* (2007). Genome-wide maps of chromatin state in pluripotent and lineage-committed cells. *Nature* **448,** 553–560.

Mochizuki, K., Fine, N. A., Fujisawa, T., and Gorovsky, M. A. (2002). Analysis of a piwi-related gene implicates small RNAs in genome rearrangement in tetrahymena. *Cell* **110,** 689–699.

Mohn, F., Weber, M., Rebhan, M., Roloff, T. C., Richter, J., Stadler, M. B., Bibel, M., and Schubeler, D. (2008). Lineage-specific polycomb targets and de novo DNA methylation define restriction and potential of neuronal progenitors. *Mol. Cell* **30,** 755–766.

Morillon, A., Karabetsou, N., Nair, A., and Mellor, J. (2005). Dynamic lysine methylation on histone H3 defines the regulatory phase of gene transcription. *Mol. Cell* **18**, 723–734.

Morrison, A. J., Highland, J., Krogan, N. J., Arbel-Eden, A., Greenblatt, J. F., Haber, J. E., and Shen, X. (2004). INO80 and gamma-H2AX interaction links ATP-dependent chromatin remodeling to DNA damage repair. *Cell* **119**, 767–775.

Mousli, M., Hopfner, R., Abbady, A. Q., Monte, D., Jeanblanc, M., Oudet, P., Louis, B., and Bronner, C. (2003). ICBP90 belongs to a new family of proteins with an expression that is deregulated in cancer cells. *Br. J. Cancer* **89**, 120–127.

Mueller, R. D., Yasuda, H., and Bradbury, E. M. (1985). Phosphorylation of histone H1 through the cell cycle of *Physarum polycephalum*. 24 sites of phosphorylation at metaphase. *J. Biol. Chem.* **260**, 5081–5086.

Mulder, K. W., Brenkman, A. B., Inagaki, A., van den Broek, N. J., and Timmers, H. T. (2007). Regulation of histone H3K4 tri-methylation and PAF complex recruitment by the Ccr4-Not complex. *Nucleic Acids Res.* **35**, 2428–2439.

Murr, R., Loizou, J. I., Yang, Y. G., Cuenin, C., Li, H., Wang, Z. Q., and Herceg, Z. (2006). Histone acetylation by Trrap-Tip60 modulates loading of repair proteins and repair of DNA double-strand breaks. *Nat. Cell Biol.* **8**, 91–99.

Murr, R., Vaissiere, T., Sawan, C., Shukla, V., and Herceg, Z. (2007). Orchestration of chromatin-based processes: Mind the TRRAP. *Oncogene* **26**, 5358–5372.

Muto, M., Kanari, Y., Kubo, E., Takabe, T., Kurihara, T., Fujimori, A., and Tatsumi, K. (2002). Targeted disruption of Np95 gene renders murine embryonic stem cells hypersensitive to DNA damaging agents and DNA replication blocks. *J. Biol. Chem.* **277**, 34549–34555.

Myant, K., and Stancheva, I. (2008). LSH cooperates with DNA methyltransferases to repress transcription. *Mol. Cell. Biol.* **28**, 215–226.

Nakagawa, T., Kajitani, T., Togo, S., Masuko, N., Ohdan, H., Hishikawa, Y., Koji, T., Matsuyama, T., Ikura, T., Muramatsu, M., *et al.* (2008). Deubiquitylation of histone H2A activates transcriptional initiation via trans-histone cross-talk with H3K4 di- and trimethylation. *Genes Dev.* **22**, 37–49.

Nakayama, J., Rice, J. C., Strahl, B. D., Allis, C. D., and Grewal, S. I. (2001). Role of histone H3 lysine 9 methylation in epigenetic control of heterochromatin assembly. *Science* **292**, 110–113.

Nathan, D., Ingvarsdottir, K., Sterner, D. E., Bylebyl, G. R., Dokmanovic, M., Dorsey, J. A., Whelan, K. A., Krsmanovic, M., Lane, W. S., Meluh, P. B., *et al.* (2006). Histone sumoylation is a negative regulator in *Saccharomyces cerevisiae* and shows dynamic interplay with positive-acting histone modifications. *Genes Dev.* **20**, 966–976.

Negishi, M., Saraya, A., Miyagi, S., Nagao, K., Inagaki, Y., Nishikawa, M., Tajima, S., Koseki, H., Tsuda, H., Takasaki, Y., *et al.* (2007). Bmi1 cooperates with Dnmt1-associated protein 1 in gene silencing. *Biochem. Biophys. Res. Commun.* **353**, 992–998.

Nelson, C. J., Santos-Rosa, H., and Kouzarides, T. (2006). Proline isomerization of histone H3 regulates lysine methylation and gene expression. *Cell* **126**, 905–916.

Ng, H. H., Xu, R. M., Zhang, Y., and Struhl, K. (2002). Ubiquitination of histone H2B by Rad6 is required for efficient Dot1-mediated methylation of histone H3 lysine 79. *J. Biol. Chem.* **277**, 34655–34657.

Ng, H. H., Robert, F., Young, R. A., and Struhl, K. (2003). Targeted recruitment of Set1 histone methylase by elongating Pol II provides a localized mark and memory of recent transcriptional activity. *Mol. Cell* **11**, 709–719.

Nguyen, C. T., Weisenberger, D. J., Velicescu, M., Gonzales, F. A., Lin, J. C., Liang, G., and Jones, P. A. (2002). Histone H3-lysine 9 methylation is associated with aberrant gene silencing in cancer cells and is rapidly reversed by 5-aza-2′-deoxycytidine. *Cancer Res.* **62**, 6456–6461.

Nightingale, K. P., Gendreizig, S., White, D. A., Bradbury, C., Hollfelder, F., and Turner, B. M. (2007). Cross-talk between histone modifications in response to histone deacetylase inhibitors: MLL4 links histone H3 acetylation and histone H3K4 methylation. *J. Biol. Chem.* **282,** 4408–4416.

Nishioka, K., Chuikov, S., Sarma, K., Erdjument-Bromage, H., Allis, C. D., Tempst, P., and Reinberg, D. (2002a). Set9, a novel histone H3 methyltransferase that facilitates transcription by precluding histone tail modifications required for heterochromatin formation. *Genes Dev.* **16,** 479–489.

Nishioka, K., Rice, J. C., Sarma, K., Erdjument-Bromage, H., Werner, J., Wang, Y., Chuikov, S., Valenzuela, P., Tempst, P., Steward, R., et al. (2002b). PR-Set7 is a nucleosome-specific methyltransferase that modifies lysine 20 of histone H4 and is associated with silent chromatin. *Mol. Cell* **9,** 1201–1213.

Ohm, J. E., and Baylin, S. B. (2007). Stem cell chromatin patterns: An instructive mechanism for DNA hypermethylation? *Cell Cycle* **6,** 1040–1043.

Ohm, J. E., McGarvey, K. M., Yu, X., Cheng, L., Schuebel, K. E., Cope, L., Mohammad, H. P., Chen, W., Daniel, V. C., Yu, W., et al. (2007). A stem cell-like chromatin pattern may predispose tumor suppressor genes to DNA hypermethylation and heritable silencing. *Nat. Genet.* **39,** 237–242.

Ooi, S. K., Qiu, C., Bernstein, E., Li, K., Jia, D., Yang, Z., Erdjument-Bromage, H., Tempst, P., Lin, S. P., Allis, C. D., et al. (2007). DNMT3L connects unmethylated lysine 4 of histone H3 to de novo methylation of DNA. *Nature* **448,** 714–717.

Pal, S., Yun, R., Datta, A., Lacomis, L., Erdjument-Bromage, H., Kumar, J., Tempst, P., and Sif, S. (2003). mSin3A/histone deacetylase 2- and PRMT5-containing Brg1 complex is involved in transcriptional repression of the Myc target gene cad. *Mol. Cell. Biol.* **23,** 7475–7487.

Pal, S., Vishwanath, S. N., Erdjument-Bromage, H., Tempst, P., and Sif, S. (2004). Human SWI/SNF-associated PRMT5 methylates histone H3 arginine 8 and negatively regulates expression of ST7 and NM23 tumor suppressor genes. *Mol. Cell. Biol.* **24,** 9630–9645.

Pal-Bhadra, M., Leibovitch, B. A., Gandhi, S. G., Rao, M., Bhadra, U., Birchler, J. A., and Elgin, S. C. (2004). Heterochromatic silencing and HP1 localization in *Drosophila* are dependent on the RNAi machinery. *Science* **303,** 669–672.

Pan, G., Tian, S., Nie, J., Yang, C., Ruotti, V., Wei, H., Jonsdottir, G. A., Stewart, R., and Thomson, J. A. (2007). Whole-genome analysis of histone H3 lysine 4 and lysine 27 methylation in human embryonic stem cells. *Cell Stem Cell* **1,** 299–312.

Papait, R., Pistore, C., Negri, D., Pecoraro, D., Cantarini, L., and Bonapace, I. M. (2007). Np95 is implicated in pericentromeric heterochromatin replication and in major satellite silencing. *Mol. Biol. Cell* **18,** 1098–1106.

Papait, R., Pistore, C., Grazini, U., Babbio, F., Cogliati, S., Pecoraro, D., Brino, L., Morand, A. L., Dechampesme, A. M., Spada, F., et al. (2008). The PHD domain of Np95 (mUHRF1) is involved in large-scale reorganization of pericentromeric heterochromatin. *Mol. Biol. Cell* **19,** 3554–3563.

Park, Y. J., and Luger, K. (2008). Histone chaperones in nucleosome eviction and histone exchange. *Curr. Opin. Struct. Biol.* **18,** 282–289.

Pasini, D., Hansen, K. H., Christensen, J., Agger, K., Cloos, P. A., and Helin, K. (2008). Coordinated regulation of transcriptional repression by the RBP2 H3K4 demethylase and Polycomb-Repressive Complex 2. *Genes Dev.* **22,** 1345–1355.

Pena, P. V., Davrazou, F., Shi, X., Walter, K. L., Verkhusha, V. V., Gozani, O., Zhao, R., and Kutateladze, T. G. (2006). Molecular mechanism of histone H3K4me3 recognition by plant homeodomain of ING2. *Nature* **442,** 100–103.

Pokholok, D. K., Harbison, C. T., Levine, S., Cole, M., Hannett, N. M., Lee, T. I., Bell, G. W., Walker, K., Rolfe, P. A., Herbolsheimer, E., et al. (2005). Genome-wide map of nucleosome acetylation and methylation in yeast. *Cell* **122,** 517–527.

Powell, D. W., Weaver, C. M., Jennings, J. L., McAfee, K. J., He, Y., Weil, P. A., and Link, A. J. (2004). Cluster analysis of mass spectrometry data reveals a novel component of SAGA. *Mol. Cell. Biol.* **24**, 7249–7259.

Pray-Grant, M. G., Daniel, J. A., Schieltz, D., Yates, J. R., 3rd, and Grant, P. A. (2005). Chd1 chromodomain links histone H3 methylation with SAGA- and SLIK-dependent acetylation. *Nature* **433**, 434–438.

Rai, K., Jafri, I. F., Chidester, S., James, S. R., Karpf, A. R., Cairns, B. R., and Jones, D. A. (2010). Dnmt3 and G9a cooperate for tissue-specific development in zebrafish. *J. Biol. Chem.* **285**, 4110–4121.

Rao, B., Shibata, Y., Strahl, B. D., and Lieb, J. D. (2005). Dimethylation of histone H3 at lysine 36 demarcates regulatory and nonregulatory chromatin genome-wide. *Mol. Cell. Biol.* **25**, 9447–9459.

Rea, S., Eisenhaber, F., O'Carroll, D., Strahl, B. D., Sun, Z. W., Schmid, M., Opravil, S., Mechtler, K., Ponting, C. P., Allis, C. D., *et al.* (2000). Regulation of chromatin structure by site-specific histone H3 methyltransferases. *Nature* **406**, 593–599.

Reik, W. (2007). Stability and flexibility of epigenetic gene regulation in mammalian development. *Nature* **447**, 425–432.

Reinhart, B. J., and Bartel, D. P. (2002). Small RNAs correspond to centromere heterochromatic repeats. *Science* **297**, 1831.

Rogakou, E. P., Boon, C., Redon, C., and Bonner, W. M. (1999). Megabase chromatin domains involved in DNA double-strand breaks in vivo. *J. Cell Biol.* **146**, 905–916.

Rojas, J. R., Trievel, R. C., Zhou, J., Mo, Y., Li, X., Berger, S. L., Allis, C. D., and Marmorstein, R. (1999). Structure of Tetrahymena GCN5 bound to coenzyme A and a histone H3 peptide. *Nature* **401**, 93–98.

Rottach, A., Frauer, C., Pichler, G., Bonapace, I.M., Spada, F., and Leonhardt, H. (2009). The multi-domain protein Np95 connects DNA methylation and histone modification. *Nucleic Acids Res* **38**, 1796–1804.

Sanchez, C., Sanchez, I., Demmers, J. A., Rodriguez, P., Stroboulis, J., and Vidal, M. (2007). Proteomics analysis of Ring1B/Rnf2 interactors identifies a novel complex with the Fbxl10/Jhdm1B histone demethylase and the Bcl6 interacting corepressor. *Mol. Cell. Proteomics* **6**, 820–834.

Santos-Rosa, H., Schneider, R., Bannister, A. J., Sherriff, J., Bernstein, B. E., Emre, N. C., Schreiber, S. L., Mellor, J., and Kouzarides, T. (2002). Active genes are tri-methylated at K4 of histone H3. *Nature* **419**, 407–411.

Sarg, B., Helliger, W., Talasz, H., Koutzamani, E., and Lindner, H. H. (2004). Histone H4 hyper-acetylation precludes histone H4 lysine 20 trimethylation. *J. Biol. Chem.* **279**, 53458–53464.

Sawan, C., Vaissiere, T., Murr, R., and Herceg, Z. (2008). Epigenetic drivers and genetic passengers on the road to cancer. *Mutat. Res.* **642**, 1–13.

Schlesinger, Y., Straussman, R., Keshet, I., Farkash, S., Hecht, M., Zimmerman, J., Eden, E., Yakhini, Z., Ben-Shushan, E., Reubinoff, B. E., *et al.* (2007). Polycomb-mediated methylation on Lys27 of histone H3 pre-marks genes for de novo methylation in cancer. *Nat. Genet.* **39**, 232–236.

Schneider, J., Wood, A., Lee, J. S., Schuster, R., Dueker, J., Maguire, C., Swanson, S. K., Florens, L., Washburn, M. P., and Shilatifard, A. (2005). Molecular regulation of histone H3 trimethylation by COMPASS and the regulation of gene expression. *Mol. Cell* **19**, 849–856.

Schubeler, D., MacAlpine, D. M., Scalzo, D., Wirbelauer, C., Kooperberg, C., van Leeuwen, F., Gottschling, D. E., O'Neill, L. P., Turner, B. M., Delrow, J., *et al.* (2004). The histone modification pattern of active genes revealed through genome-wide chromatin analysis of a higher eukaryote. *Genes Dev.* **18**, 1263–1271.

Schuettengruber, B., Chourrout, D., Vervoort, M., Leblanc, B., and Cavalli, G. (2007). Genome regulation by polycomb and trithorax proteins. *Cell* **128**, 735–745.

Schwartz, Y. B., and Pirrotta, V. (2007). Polycomb silencing mechanisms and the management of genomic programmes. *Nat. Rev. Genet.* **8**, 9–22.

Shahbazian, M. D., Zhang, K., and Grunstein, M. (2005). Histone H2B ubiquitylation controls processive methylation but not monomethylation by Dot1 and Set1. *Mol. Cell* **19**, 271–277.

Sharif, J., Muto, M., Takebayashi, S., Suetake, I., Iwamatsu, A., Endo, T. A., Shinga, J., Mizutani-Koseki, Y., Toyoda, T., Okamura, K., et al. (2007). The SRA protein Np95 mediates epigenetic inheritance by recruiting Dnmt1 to methylated DNA. *Nature* **450**, 908–912.

Shi, Y. J., Matson, C., Lan, F., Iwase, S., Baba, T., and Shi, Y. (2005). Regulation of LSD1 histone demethylase activity by its associated factors. *Mol. Cell* **19**, 857–864.

Shi, X., Hong, T., Walter, K. L., Ewalt, M., Michishita, E., Hung, T., Carney, D., Pena, P., Lan, F., Kaadige, M. R., et al. (2006). ING2 PHD domain links histone H3 lysine 4 methylation to active gene repression. *Nature* **442**, 96–99.

Shiio, Y., and Eisenman, R. N. (2003). Histone sumoylation is associated with transcriptional repression. *Proc. Natl. Acad. Sci. USA* **100**, 13225–13230.

Shin, J. A., Choi, E. S., Kim, H. S., Ho, J. C., Watts, F. Z., Park, S. D., and Jang, Y. K. (2005). SUMO modification is involved in the maintenance of heterochromatin stability in fission yeast. *Mol. Cell* **19**, 817–828.

Sims, R. J., 3rd, Chen, C. F., Santos-Rosa, H., Kouzarides, T., Patel, S. S., and Reinberg, D. (2005). Human but not yeast CHD1 binds directly and selectively to histone H3 methylated at lysine 4 via its tandem chromodomains. *J. Biol. Chem.* **280**, 41789–41792.

Sims, J. K., Houston, S. I., Magazinnik, T., and Rice, J. C. (2006). A trans-tail histone code defined by monomethylated H4 Lys-20 and H3 Lys-9 demarcates distinct regions of silent chromatin. *J. Biol. Chem.* **281**, 12760–12766.

Sinkkonen, L., Hugenschmidt, T., Berninger, P., Gaidatzis, D., Mohn, F., Artus-Revel, C. G., Zavolan, M., Svoboda, P., and Filipowicz, W. (2008). MicroRNAs control de novo DNA methylation through regulation of transcriptional repressors in mouse embryonic stem cells. *Nat. Struct. Mol. Biol.* **15**, 259–267.

Smallwood, A., Esteve, P. O., Pradhan, S., and Carey, M. (2007). Functional cooperation between HP1 and DNMT1 mediates gene silencing. *Genes Dev.* **21**, 1169–1178.

Sridhar, V. V., Kapoor, A., Zhang, K., Zhu, J., Zhou, T., Hasegawa, P. M., Bressan, R. A., and Zhu, J. K. (2007). Control of DNA methylation and heterochromatic silencing by histone H2B deubiquitination. *Nature* **447**, 735–738.

Steger, D. J., Lefterova, M. I., Ying, L., Stonestrom, A. J., Schupp, M., Zhuo, D., Vakoc, A. L., Kim, J. E., Chen, J., Lazar, M. A., et al. (2008). DOT1L/KMT4 recruitment and H3K79 methylation are ubiquitously coupled with gene transcription in mammalian cells. *Mol. Cell. Biol.* **28**, 2825–2839.

Strahl, B. D., and Allis, C. D. (2000). The language of covalent histone modifications. *Nature* **403**, 41–45.

Strahl, B. D., Ohba, R., Cook, R. G., and Allis, C. D. (1999). Methylation of histone H3 at lysine 4 is highly conserved and correlates with transcriptionally active nuclei in Tetrahymena. *Proc. Natl. Acad. Sci. USA* **96**, 14967–14972.

Strahl, B. D., Briggs, S. D., Brame, C. J., Caldwell, J. A., Koh, S. S., Ma, H., Cook, R. G., Shabanowitz, J., Hunt, D. F., Stallcup, M. R., et al. (2001). Methylation of histone H4 at arginine 3 occurs in vivo and is mediated by the nuclear receptor coactivator PRMT1. *Curr. Biol.* **11**, 996–1000.

Suganuma, T., and Workman, J. L. (2008). Crosstalk among histone modifications. *Cell* **135**, 604–607.

Suka, N., Suka, Y., Carmen, A. A., Wu, J., and Grunstein, M. (2001). Highly specific antibodies determine histone acetylation site usage in yeast heterochromatin and euchromatin. *Mol. Cell* **8**, 473–479.

Sun, Z. W., and Allis, C. D. (2002). Ubiquitination of histone H2B regulates H3 methylation and gene silencing in yeast. *Nature* **418**, 104–108.

Swope, D. L., Mueller, C. L., and Chrivia, J. C. (1996). CREB-binding protein activates transcription through multiple domains. *J. Biol. Chem.* **271**, 28138–28145.

Talasz, H., Lindner, H. H., Sarg, B., and Helliger, W. (2005). Histone H4-lysine 20 monomethylation is increased in promoter and coding regions of active genes and correlates with hyperacetylation. *J. Biol. Chem.* **280**, 38814–38822.

Talbert, P. B., and Henikoff, S. (2010). Histone variants—Ancient wrap artists of the epigenome. *Nat. Rev. Mol. Cell. Biol.* **11**, 264–275.

Tamburini, B. A., and Tyler, J. K. (2005). Localized histone acetylation and deacetylation triggered by the homologous recombination pathway of double-strand DNA repair. *Mol. Cell. Biol.* **25**, 4903–4913.

Taverna, S. D., Coyne, R. S., and Allis, C. D. (2002). Methylation of histone h3 at lysine 9 targets programmed DNA elimination in tetrahymena. *Cell* **110**, 701–711.

Taverna, S. D., Ilin, S., Rogers, R. S., Tanny, J. C., Lavender, H., Li, H., Baker, L., Boyle, J., Blair, L. P., Chait, B. T., *et al.* (2006). Yng1 PHD finger binding to H3 trimethylated at K4 promotes NuA3 HAT activity at K14 of H3 and transcription at a subset of targeted ORFs. *Mol. Cell* **24**, 785–796.

Thomas, C. E., Kelleher, N. L., and Mizzen, C. A. (2006). Mass spectrometric characterization of human histone H3: A bird's eye view. *J. Proteome Res.* **5**, 240–247.

Thorne, A. W., Kmiciek, D., Mitchelson, K., Sautiere, P., and Crane-Robinson, C. (1990). Patterns of histone acetylation. *Eur. J. Biochem.* **193**, 701–713.

Ting, A. H., Suzuki, H., Cope, L., Schuebel, K. E., Lee, B. H., Toyota, M., Imai, K., Shinomura, Y., Tokino, T., and Baylin, S. B. (2008). A requirement for DICER to maintain full promoter CpG island hypermethylation in human cancer cells. *Cancer Res.* **68**, 2570–2575.

Unoki, M., Nishidate, T., and Nakamura, Y. (2004). ICBP90, an E2F-1 target, recruits HDAC1 and binds to methyl-CpG through its SRA domain. *Oncogene* **23**, 7601–7610.

Utley, R. T., Lacoste, N., Jobin-Robitaille, O., Allard, S., and Cote, J. (2005). Regulation of NuA4 histone acetyltransferase activity in transcription and DNA repair by phosphorylation of histone H4. *Mol. Cell. Biol.* **25**, 8179–8190.

Vaissiere, T., Sawan, C., and Herceg, Z. (2008). Epigenetic interplay between histone modifications and DNA methylation in gene silencing. *Mutat. Res.* **659**, 40–48.

van Attikum, H., and Gasser, S. M. (2009). Crosstalk between histone modifications during the DNA damage response. *Trends Cell Biol.* **19**, 207–217.

van Attikum, H., Fritsch, O., Hohn, B., and Gasser, S. M. (2004). Recruitment of the INO80 complex by H2A phosphorylation links ATP-dependent chromatin remodeling with DNA double-strand break repair. *Cell* **119**, 777–788.

van der Knaap, J. A., Kumar, B. R., Moshkin, Y. M., Langenberg, K., Krijgsveld, J., Heck, A. J., Karch, F., and Verrijzer, C. P. (2005). GMP synthetase stimulates histone H2B deubiquitylation by the epigenetic silencer USP7. *Mol. Cell* **17**, 695–707.

van Leeuwen, F., Gafken, P. R., and Gottschling, D. E. (2002). Dot1p modulates silencing in yeast by methylation of the nucleosome core. *Cell* **109**, 745–756.

Vaute, O., Nicolas, E., Vandel, L., and Trouche, D. (2002). Functional and physical interaction between the histone methyl transferase Suv39H1 and histone deacetylases. *Nucleic Acids Res.* **30**, 475–481.

Vermeulen, M., Mulder, K. W., Denissov, S., Pijnappel, W. W., van Schaik, F. M., Varier, R. A., Baltissen, M. P., Stunnenberg, H. G., Mann, M., and Timmers, H. T. (2007). Selective anchoring of TFIID to nucleosomes by trimethylation of histone H3 lysine 4. *Cell* **131,** 58–69.

Vire, E., Brenner, C., Deplus, R., Blanchon, L., Fraga, M., Didelot, C., Morey, L., Van Eynde, A., Bernard, D., Vanderwinden, J. M., *et al.* (2006). The polycomb group protein EZH2 directly controls DNA methylation. *Nature* **439,** 871–874.

Volpe, T. A., Kidner, C., Hall, I. M., Teng, G., Grewal, S. I., and Martienssen, R. A. (2002). Regulation of heterochromatic silencing and histone H3 lysine-9 methylation by RNAi. *Science* **297,** 1833–1837.

Walter, W., Clynes, D., Tang, Y., Marmorstein, R., Mellor, J., and Berger, S. L. (2008). 14-3-3 interaction with histone H3 involves a dual modification pattern of phosphoacetylation. *Mol. Cell. Biol.* **28,** 2840–2849.

Wang, H., Cao, R., Xia, L., Erdjument-Bromage, H., Borchers, C., Tempst, P., and Zhang, Y. (2001a). Purification and functional characterization of a histone H3-lysine 4-specific methyltransferase. *Mol. Cell* **8,** 1207–1217.

Wang, H., Huang, Z. Q., Xia, L., Feng, Q., Erdjument-Bromage, H., Strahl, B. D., Briggs, S. D., Allis, C. D., Wong, J., Tempst, P., *et al.* (2001b). Methylation of histone H4 at arginine 3 facilitating transcriptional activation by nuclear hormone receptor. *Science* **293,** 853–857.

Wang, H., Wang, L., Erdjument-Bromage, H., Vidal, M., Tempst, P., Jones, R. S., and Zhang, Y. (2004). Role of histone H2A ubiquitination in Polycomb silencing. *Nature* **431,** 873–878.

Weake, V. M., Lee, K. K., Guelman, S., Lin, C. H., Seidel, C., Abmayr, S. M., and Workman, J. L. (2008). SAGA-mediated H2B deubiquitination controls the development of neuronal connectivity in the Drosophila visual system. *EMBO J.* **27,** 394–405.

Widschwendter, M., Fiegl, H., Egle, D., Mueller-Holzner, E., Spizzo, G., Marth, C., Weisenberger, D. J., Campan, M., Young, J., Jacobs, I., *et al.* (2007). Epigenetic stem cell signature in cancer. *Nat. Genet.* **39,** 157–158.

Wolffe, A. P., and Hayes, J. J. (1999). Chromatin disruption and modification. *Nucleic Acids Res.* **27,** 711–720.

Wozniak, R. J., Klimecki, W. T., Lau, S. S., Feinstein, Y., and Futscher, B. W. (2007). 5-Aza-2'-deoxycytidine-mediated reductions in G9A histone methyltransferase and histone H3 K9 di-methylation levels are linked to tumor suppressor gene reactivation. *Oncogene* **26,** 77–90.

Wu, R. S., and Bonner, W. M. (1981). Separation of basal histone synthesis from S-phase histone synthesis in dividing cells. *Cell* **27,** 321–330.

Wyce, A., Xiao, T., Whelan, K. A., Kosman, C., Walter, W., Eick, D., Hughes, T. R., Krogan, N. J., Strahl, B. D., and Berger, S. L. (2007). H2B ubiquitylation acts as a barrier to Ctk1 nucleosomal recruitment prior to removal by Ubp8 within a SAGA-related complex. *Mol. Cell* **27,** 275–288.

Xi, S., Zhu, H., Xu, H., Schmidtmann, A., Geiman, T. M., and Muegge, K. (2007). Lsh controls Hox gene silencing during development. *Proc. Natl. Acad. Sci. USA* **104,** 14366–14371.

Xiao, B., Wilson, J. R., and Gamblin, S. J. (2003). SET domains and histone methylation. *Curr. Opin. Struct. Biol.* **13,** 699–705.

Xiao, A., Li, H., Shechter, D., Ahn, S. H., Fabrizio, L. A., Erdjument-Bromage, H., Ishibe-Murakami, S., Wang, B., Tempst, P., Hofmann, K., *et al.* (2009). WSTF regulates the H2A.X DNA damage response via a novel tyrosine kinase activity. *Nature* **457,** 57–62.

Yin, Y., Liu, C., Tsai, S. N., Zhou, B., Ngai, S. M., and Zhu, G. (2005). SET8 recognizes the sequence RHRK20VLRDN within the N terminus of histone H4 and mono-methylates lysine 20. *J. Biol. Chem.* **280,** 30025–30031.

Yu, M. C., Lamming, D. W., Eskin, J. A., Sinclair, D. A., and Silver, P. A. (2006). The role of protein arginine methylation in the formation of silent chromatin. *Genes Dev.* **20,** 3249–3254.

Zhang, K., Williams, K. E., Huang, L., Yau, P., Siino, J. S., Bradbury, E. M., Jones, P. R., Minch, M. J., and Burlingame, A. L. (2002a). Histone acetylation and deacetylation: Identification of acetylation and methylation sites of HeLa histone H4 by mass spectrometry. *Mol. Cell. Proteomics* **1**, 500–508.

Zhang, C. L., McKinsey, T. A., and Olson, E. N. (2002b). Association of class II histone deacetylases with heterochromatin protein 1: Potential role for histone methylation in control of muscle differentiation. *Mol. Cell. Biol.* **22**, 7302–7312.

Zhang, K., Yau, P. M., Chandrasekhar, B., New, R., Kondrat, R., Imai, B. S., and Bradbury, M. E. (2004). Differentiation between peptides containing acetylated or tri-methylated lysines by mass spectrometry: An application for determining lysine 9 acetylation and methylation of histone H3. *Proteomics* **4**, 1–10.

Zhang, X. Y., Pfeiffer, H. K., Thorne, A. W., and McMahon, S. B. (2008). USP22, an hSAGA subunit and potential cancer stem cell marker, reverses the polycomb-catalyzed ubiquitylation of histone H2A. *Cell Cycle* **7**, 1522–1524.

Zhao, X. D., Han, X., Chew, J. L., Liu, J., Chiu, K. P., Choo, A., Orlov, Y. L., Sung, W. K., Shahab, A., Kuznetsov, V. A., *et al.* (2007). Whole-genome mapping of histone H3 Lys4 and 27 trimethylations reveals distinct genomic compartments in human embryonic stem cells. *Cell Stem Cell* **1**, 286–298.

Zhao, Y., Lang, G., Ito, S., Bonnet, J., Metzger, E., Sawatsubashi, S., Suzuki, E., Le Guezennec, X., Stunnenberg, H. G., Krasnov, A., *et al.* (2008). A TFTC/STAGA module mediates histone H2A and H2B deubiquitination, coactivates nuclear receptors, and counteracts heterochromatin silencing. *Mol. Cell* **29**, 92–101.

Zhu, H., Geiman, T. M., Xi, S., Jiang, Q., Schmidtmann, A., Chen, T., Li, E., and Muegge, K. (2006). Lsh is involved in de novo methylation of DNA. *EMBO J* **25**, 335–345.

Zhu, P., Zhou, W., Wang, J., Puc, J., Ohgi, K. A., Erdjument-Bromage, H., Tempst, P., Glass, C. K., and Rosenfeld, M. G. (2007). A histone H2A deubiquitinase complex coordinating histone acetylation and H1 dissociation in transcriptional regulation. *Mol. Cell* **27**, 609–621.

Epigenetic Events Underlying Biological Phenomena

6

Genomic Imprinting
Syndromes and Cancer

Derek Hock Kiat Lim*,† and Eamonn Richard Maher*,†
*Department of Medical & Molecular Genetics, School of Clinical and
Experimental Medicine, University of Birmingham College of Medical and
Dental Sciences, Edgbaston, Birmingham, United Kingdom
†West Midlands Regional Genetics Service, Birmingham Women's Hospital,
Edgbaston, Birmingham, United Kingdom

Advances in Genetics, Vol. 70   0065-2660/10 $35.00
Copyright 2010, Elsevier Inc. All rights reserved.   DOI: 10.1016/S0065-2660(10)70006-7

ABSTRACT

Genomic imprinting represents a form of epigenetic control of gene expression in
which one allele of a gene is preferentially expressed according to the parent-of-
origin of the allele. Genomic imprinting plays an important role in normal
growth and development. Disruption of imprinting can result in a number of
human imprinting syndromes and predispose to cancer. In this chapter, we
describe a number of human imprinting syndromes to illustrate the concepts
of genomic imprinting and how loss of imprinting of imprinted genes their
relationship to human neoplasia. © 2010, Elsevier Inc.

I. INTRODUCTION

Genomic imprinting represents a form of epigenetic control of gene expression in
which one allele of a gene is preferentially expressed according to the parent-of-
origin of the allele. Hence, some imprinted genes are expressed from the paternally
inherited allele and others from the maternally inherited allele—though in some
cases imprinting patterns are more complex and some imprinted genes demonstrate
tissue-specific or developmental stage-specific imprinting. Alterations in imprinting
can be associated with tumorigenesis. In this chapter, we review specific examples of
imprinting disorders to illustrate the biology of genomic imprinting and how
disorders of imprinting can result in certain phenotypes ranging from disorders of
growth, developmental defects, and increased risk of cancer.

II. THE IMPORTANCE OF BIPARENTAL CONTRIBUTION TO THE MAMMALIAN GENOME

In 1984, experiments in nuclear transfer in mouse embryos revealed that diploid
gynogenetic and androgenetic embryos cannot develop normally, demonstrating
that the maternal and paternal contributions to the embryonic genome in mammals are not equivalent and that complete embryogenesis requires the

contribution of both male and female parental genomes (McGrath and Solter, 1984; Surani *et al.*, 1984). In humans, the importance of biparental contribution to the genome is evident by the occurrence of abnormal pregnancies from conceptions lacking either paternal or maternal contributions. Hydatidiform molar pregnancies occur as a result of an androgenetic conceptus, therefore lacking maternal contribution to the genome (Kajii and Ohama, 1977; Ohama *et al.*, 1981). By contrast, ovarian teratomas can result from a spontaneous activation of an ovarian oocyte, resulting in a duplication of the maternal genome (gynogenetic conceptus; Ohama *et al.*, 1985). Subsequently, experiments with mice carrying balanced translocations enabled the generation of mouse embryos containing small genomic regions that were derived only from the mother or father (uniparental disomies). This demonstrated that the requirement for biparental inheritance was restricted to specific regions of the genome (Cattanach and Kirk, 1985). These regions were subsequently found to harbor clusters of imprinted genes, and in 1991, the first imprinted genes were reported (Barlow *et al.*, 1991). Genomic imprinting in mammals has also been shown to play an important role in preventing parthenogenetic development in mice. Parthenogenesis is a form of asexual reproduction where growth and development of the embryo occur without fertilization by a male. The successful survival to adulthood of parthenogenetic mice from two maternal genomes (parthenogenetic mice embryos usually die by 10 days of gestation) following the deletion of an imprinted cluster in mice oocyte which resulted in expression of paternally expressed genes (Kono *et al.*, 2004). This provides further proof of the importance of biparental contribution in normal development.

Approximately 100 imprinted genes have been identified to date in the human and/or mouse genomes (www.mgh.har.mrc.ac.uk/imprinting.impstables.html). Most imprinted genes reside in clusters which allows them to be coordinated by imprinting control elements or imprinting centers (IC) (Reik and Walter, 2001). These clusters are rich in CpG (cytosine–phosphoguanine dinucleotide sequences) islands (Neumann *et al.*, 1995).

III. MECHANISMS INVOLVED IN GENOMIC IMPRINTING

Various mechanisms have been implicated in the establishment and maintenance of genomic imprinting, including DNA methylation, insulation, chromatin modification (see previous chapters), and expression of large noncoding RNAs.

DNA methylation plays a crucial role in both the establishment and maintenance of genomic imprinting (Jaenisch and Bird, 2003). Methylation of CpG dinucleotides can affect gene expression by (a) preventing the binding of transcription factors to the methylated gene promoter, (b) preventing the binding of insulator proteins (e.g., CCCTC-binding factor [CTCF]) to differentially methylated regions (DMRs), and (c) inducing changes in chromatin structure by initially attracting methyl-CpG-binding proteins that bind to methylated CpGs and then recruit other proteins such as histone deacetylases and chromatin modeling proteins, which modify histones to produce a more compact, hetero-chromatic, structure that prevents transcription.

Within imprinted gene domains, DMRs occur at which the methylation status of the CpG islands differs according to the parent-of-origin of the allele (one parental allele is methylated and the opposite parental allele is unmethylated). A key feature of parent-specific imprinting marks in DMRs is that they are erased and then reset (established) in the primordial germ cells during gametogenesis to reflect the sex of the parent for the next generation (Lee *et al.*, 2002; Szabo *et al.*, 2002). DMRs mark imprinting control centers which can act as insulator elements regulating the expression of imprinted genes. One example of this is at the 11p15.5 imprinted region involving the *IGF2/H19* locus. The *H19* DMR maps between *H19* and *IGF2* and contains binding sites for a zinc-finger protein, CCCTC-binding factor (CTCF). On the maternal allele, the CTCF protein binds to the unmethylated *H19* DMR and insulates the *IGF2* promoter region from an enhancer downstream of *H19*. This causes the enhancer to preferentially interact with the *H19* promoter and the untranslated *H19* RNA is expressed from the maternal allele, whereas *IGF2* expression is silenced. In contrast, the *H19* DMR is methylated on the paternal allele and CTCF is unable to bind. This allows the downstream enhancer to preferentially interact with the *IGF2* promoter and *IGF2*, but not *H19*, is expressed from the paternal allele (Bell and Felsenfeld, 2000).

Imprinted noncoding RNAs (e.g., *H19*, *KCNQ1OT1*, and *Airn*) are also a frequent feature of imprinted gene clusters and are often in close proximity to oppositely imprinted protein-coding genes, for example, *H19* and *IGF2*, *KCNQ1OT1* and *CDKN1C*, and *Igf2r* and *Airn*. The mechanism of how these noncoding RNAs are implicated in imprinting establishment and maintenance is currently an area of active research. Another area of intense interest is the role of chromatin modifications in the establishment and maintenance of imprinting (Kacem and Feil, 2009).

IV. MECHANISMS LEADING TO DISORDERS OF IMPRINTING

Mutations or epimutations (pathogenic alterations in DNA methylation or chromatin structure without a change in DNA sequence) affecting the function of the expressed imprinted gene allele (but not the silenced allele) can result in imprinting disorders. In addition, loss of imprinting (LOI) (i.e., both alleles are expressed) can also cause disease. Common mechanisms that give rise to an imprinting disorder include (a) uniparental disomy (UPD), (b) intragenic muta-tions or copy number alterations that directly alter the function of an imprinted gene, and (c) mutations or epimutations in imprinting control centers that result in altered imprinting/expression of imprinted gene(s). In imprinting disorders such as Beckwith–Wiedemann syndrome (BWS) and Silver–Russell syndrome (SRS) (Abu-Amero *et al.*, 2008; Cooper *et al.*, 2005) altered gene expression from aberrant loss or gain of methylation at an imprinting center DMR is the most frequent cause of disease. These epimutations result from a primary alteration in a DMR (in *cis*), but in rare cases may be secondary to alterations in remote (in *trans*) factors (Reik and Walter, 2001).

As an inherited mutation in an imprinted gene will only be of functional significance if it is in the expressed allele, imprinting disorders are notable for unusual inheritance patterns with parent-of-origin effects on penetrance and expression and, in some cases, members of the same family may have divergent phenotypes that result from a similar genomic abnormality (e.g., Prader–Willi syndrome (PWS) and Angelman syndrome (AS) may both result from deletions of chromosome 15q11-q13 and the phenotype depends on which parent the deletion is inherited from). Imprinting disorders are often associated with abnormalities of growth and development and there may be an increased risk of malignancy, most notably in BWS a disease that is associated with the 11p15.5 imprinted gene cluster.

V. THE 11P15.5 IMPRINTED REGION

The 11p15.5 imprinted region contains a cluster of imprinted genes including *IGF2*, *H19*, *CDKN1C*, *KCNQ1*, and *KCNQ1OT1* (Fig. 6.1A). The cluster can be subdivided into two domains containing imprinting centers which regulate the expression of these genes. Imprinting center 1 (IC1) controls the imprinting/expression of *IGF2* and *H19* and is located more telomerically. Imprinting center 2 (IC2) is located centromeric to IC1 and regulates imprinting of a number of genes including *CDKN1C*, *KCNQ1*, and *KCNQ1OT1*.

A. Normal imprinting at 11p15.5: Regulated in *cis* by two ICs

IC1 is marked by the *H19* DMR which is located between *H19* and *IGF2* and is normally methylated on the paternal allele. As described previously, the insulator protein CTCF binds to the unmethylated *H19* DMR on the maternal allele resulting in silencing of *IGF2* and preferential expression of *H19* from the maternal allele, whereas on the paternal allele, the *H19* DMR is methylated, CTCF is unable to bind and there is expression of *IGF2* and silencing of *H19* from the paternal allele (Bell and Felsenfeld, 2000; Hark *et al.*, 2000; Maher and Reik, 2000; Wan and Bartolomei, 2008).

Among the genes regulated by IC2 are the candidate tumor suppressor gene *CDKN1C*, *KCNQ1* (a gene that demonstrates tissue-specific imprinting), and an antisense transcript *KCNQ1OT1* (a noncoding RNA previously known

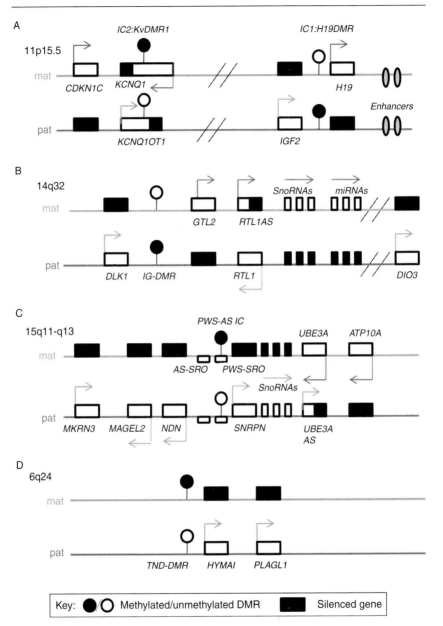

Figure 6.1. The imprinting clusters at (A) 11p15.5, (B) 14q32, (C) 15q11-q13, (D) 6q24. Arrows indicate gene transcription. Diagram not drawn to scale.

as *LIT1*). The DMR involved at IC2 is the KvDMR1 which is located within intron 10 of *KCNQ1*, coincident with the transcriptional start site for *KCNQ1OT1* (Weksberg *et al.*, 2005). KvDMR1 is normally methylated on the maternal allele which results in preferential expression of both *KCNQ1* and *CDKN1C* from the maternal allele. On the paternal allele, KvDMR1 is unmethylated which allows preferential expression of *KCNQ1OT1* and silencing of the maternally expressed imprinted genes including *KCNQ1* and *CDKN1C* from the paternal allele (Diaz-Meyer *et al.*, 2003; Lam *et al.*, 1999; Smilinich *et al.*, 1999). It is unclear whether this noncoding RNA causes silencing directly or as a secondary phenomenon from its transcription.

IGF2 is a fetal growth promoter and is normally expressed on the paternal allele. *CDKN1C* is a growth suppressor which is normally expressed from the maternal allele. Due to normal imprinting, there is "balance" between the expressions of these two genes resulting in normal growth. This is in keeping with Hague's "parental conflict theory of imprinting" whereby it is the interest of the father to derive more maternal resources to his offspring leading to greater survival fitness which would promote the propagation of his genes to successive generations. By contrast, the mother's interest is to produce the maximum number of viable embryos as well as ensure her own nutritive needs for survival and future reproduction. Therefore, embryo growth will be influenced by paternally derived growth promoters and maternally derived growth suppressors (Moore and Haig, 1991).

B. Disordered imprinting at 11p15.5 leading to two opposite clinical phenotypes

Disruption of imprinting at the 11p15.5 imprinted region can be caused by a number of mechanisms and are associated with abnormal growth, developmental defects, and tumor susceptibility but neurological development is mostly unaffected. The two human imprinting syndromes, BWS (an overgrowth syndrome) and SRS (undergrowth syndrome), are associated with abnormal imprinting at 11p15.5.

C. Beckwith–Wiedemann syndrome

1. Features

BWS is a congenital overgrowth disorder characterized by clinical features of pre or postnatal onset of overgrowth, macroglossia (enlarged tongue), and anterior abdominal wall defects. The abdominal wall defects range from mild weakening of the rectus muscles (divarification rectii) to more severe exomphalos requiring corrective surgery. In addition, other clinical features which are often variable include neonatal hypoglycemia, hemihypertrophy, organomegally, ear lobe creases , genitourinaray anomalies, and most significantly embryonal tumors (Cooper et al., 2005; Elliott et al., 1994; Reik and Maher, 1997). The incidence of BWS is estimated to be 1/13,700 (Weksberg et al., 2010).

A number of complications can arise due to the various features of BWS. The pregnancy may be complicated by polyhydramnios or a difficult delivery due to the overgrowth of the baby. Neonatal hypoglycemia may be severe and require treatment with intravenous glucose as untreated severe hypoglycemia may present with seizures and if prolonged may affect neurodevelopment. Feeding difficulties during early life and speech delay may occur as a result of the enlarged tongue which in some cases may require surgical reduction. Learning difficulties may occur as a result of untreated hypoglycemia or if associated with a chromosomal duplication. The overgrowth tends to become less prominent with increasing age and the prognosis is good if exomphalos is treated successfully and embryonal tumors do not occur (Elliott et al., 1994).

2. Risk of tumors in BWS

In, all 5–10% of patients develop tumors which include Wilms tumor (WT; most common), hepatoblastomas, rhabdomysarcomas, adrenocortical carcinoma, and neuroblastomas. Most of the tumors occur in the first 8 years of life with only a few cases being reported after this. Due to the increased risk of embryonal tumors, children with BWS are usually offered surveillance with serial abdominal ultrasound scans and monitoring of serum α-fetoprotein (Beckwith, 1998; Everman et al., 2000). Clinical features in BWS that are associated with a higher risk of tumor development include hemihypertrophy, nephromegaly, and nephrogenic rests (Coppes et al., 1999; DeBaun et al., 1998). In addition, recent studies suggest that the results of molecular genetic testing may be used to determine which children are likely to benefit from screening for WT (Scott et al., 2006).

3. Molecular mechanisms affecting imprinting in BWS

BWS is a heterogeneous disorder and a wide variety of molecular mechanisms can cause disordered imprinting/function of IGF2 and/or CDKN1C and result in BWS. Most cases of BWS are sporadic cases, however familial cases account for around 15% (Cooper et al., 2005).

Paternal uniparental disomy (pUPD, mosaic isodisomy resulting from mitotic recombination) of 11p15.5 accounts for around 20% of cases. pUPD results in two paternal alleles producing biallelic expression of IGF2 and reduced expression of CDKN1C. IC2 epimutations with loss of methylation (LOM) on the maternal allele at KvDMR1 cause biallelic KCNQ1OT1 expression and loss

of *CDKN1C* expression from the maternal allele (Fig. 6.2). This is the common-
est molecular mechanism which accounts for 50% of sporadic cases of BWS
(Cooper *et al.*, 2005). IC1 epimutations with gain of maternal allele methylation
at the *H19* DMR accounts for 5% of BWS cases. This epimutation results in
biallelic expression of *IGF2* and silencing of *H19* expression (Fig.6.3).

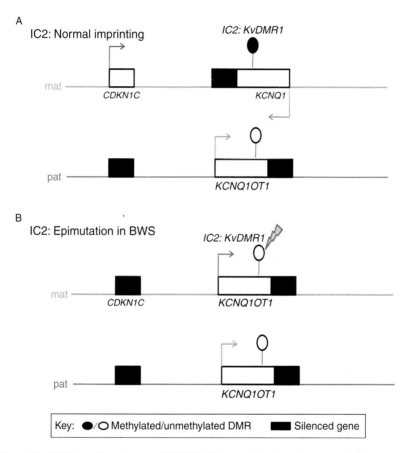

Figure 6.2. (A) Normal imprinting at IC2. KvDMR1 is methylated on the maternal allele resulting
in expression of *CDKN1C* and *KCNQ1* from the maternal allele. On the paternal allele,
KvDMR1 is unmethylated resulting in preferential expression of *KCNQ1OT1* (the
antisense transcript RNA of *KCNQ1*) resulting in silencing of *CDKN1C* and
KCNQ1. (B) IC2 epimutation in BWS. There is loss of methlation at KvDMR1 on
the maternal allele resulting in biallelic expression of *KCNQ1OT1* and silencing of
CDKN1C and *KCNQ1OT1* on both alleles.

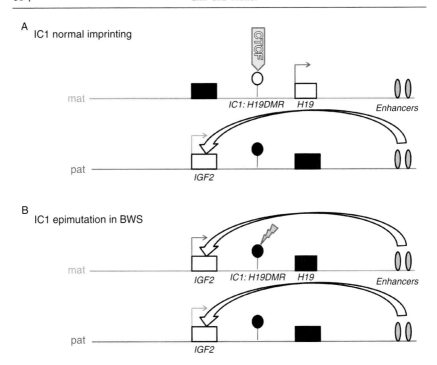

Figure 6.3. (A) Normal imprinting at IC1 at 11p15.5. On the maternal allele, the H19DMR is unmethylated allowing CTCF to bind to the CTCF binding sites at the H19DMR. This acts as a boundary element preventing downstream enhancers access to the IGF2 promoter resulting in preferential expression of H19 and silencing of IGF2 on the paternal allele. On the maternal allele, the H19DMR is methlylated preventing CTCF from binding to the DMR allowing the downstream enhancers access to the IGF2 promoter. This results in preferential expression of IGF2 from the paternal allele. (B) IC1 Epimutation in BWS. There is gain of methylation at the H19DMR on the maternal allele. This now results in biallelic expression of IGF2.

Although IC1 and IC2 epimutations usually occur sporadically, IC1 and IC2 deletions can also occur and lead to familial disease. IC1 deletions on the maternal allele affecting the CTCF binding sites are associated with apparent H19 DMR hypermethylation and biallelic expression of IGF2. An IC2 deletion is associated with silencing of CDKN1C. IC1 deletions are more common than IC2 deletions (Niemitz et al., 2004; Sparago et al., 2004, 2007). Molecular testing with methylation-specific multiplex ligation-dependant probe amplification (MS-MLPA) is widely used to determine both IC1 and IC2 methylation status and copy number so as to inform recurrence risk (Scott et al., 2008a,b).

In addition, cytogenetic abnormalities (rearrangements/translocations) leading to copy number changes at 11p15.5 account for under 2% of BWS cases.

Germ line mutations in *CDKN1C* can be found in approximately 5% of all cases but account for around 50% of familial cases (Hatada *et al.*, 1996; Lam *et al.*, 1999). Familial cases of BWS with *CDKN1C* mutations show autosomal dominant inheritance with parent-of-origin effects such that BWS is only seen with maternally transmitted mutations.

The mechanisms above all affect imprinting in *cis*. Recently, a homozygous germ line mutation in *NLRP2* was found in a mother with two children with BWS due to IC2 epimutations (Meyer *et al.*, 2009). This represents the identification of a *trans* mechanism causing loss of imprinting and BWS (see later).

4. Association of BWS with assisted reproductive technologies

Assisted reproductive technologies (ART) such as *in vitro* fertilization (IVF) and intra-cytoplasmic sperm injection (ICSI) are associated with a ninefold increased risk of BWS (although the absolute risk is low approximately less than 0.1%) (DeBaun *et al.*, 2003; Gicquel *et al.*, 2003; Halliday *et al.*, 2004; Maher *et al.*, 2003). The increased risk of BWS following ART is closely linked to the occurrence of IC2 epimutations with nearly all BWS cases conceived by ART having this molecular defect (Lim *et al.*, 2009). In a subset of BWS patients due to IC2 epimutations, the LOM was not just confined to 11p15 but involved LOM at other non-11p15 maternally methylated DMRs (Bliek *et al.*, 2009; Rossignol *et al.*, 2006). In the authors' experience, this was seen more commonly in ART conceived BWS cases (Lim *et al.*, 2009). Interestingly, the ART conceived cases showed a slightly different phenotype with a lower incidence of exomphalos but slightly higher incidence of non-WT neoplasia compared to naturally conceived BWS IC2 epimutation cases. It may be that these additional epimutations may explain the mild but significant differences in phenotype between ART and naturally conceived BWS children with IC2 epimutations although further study is warranted into this. A number of possible causes for the association have been suggested including ART-associated procedures such as ovarian hyperstimulation or *in vitro* embryo culture and infertility *per se*.

5. An in *trans* cause of BWS

Homozygous mutations in *NLRP2* were found in the mother of two children with BWS and IC2 epimutations (Meyer *et al.*, 2009). *NLRP7* (see later) and *NLRP2* are paralogous genes that map to 19q13.4 and encode members of the CATERPILLAR group of proteins that have been implicated in the innate immune response and are part of the inflammasome (Drenth and van der Meer, 2006; Kinoshita *et al.*, 2005). Both *NLRP7* and *NLRP2* are expressed during oogenesis and women with homozygous inactivating mutations of *NLRP7* or *NLRP2* are at risk of pregnancies with an imprinting disorder (see later NLRP7 and hydatidiform mole) suggesting that the proteins have an in *trans* role in the establishment of imprinting marks in the maternal genome.

6. Genotype (& epigenotype)–phenotype correlations in BWS

Analysis of genotype–phenotype correlations in BWS has identified key associations that provide insights into the pathogenesis of specific developmental abnormalities which can inform clinical management (Cooper et al., 2005). Thus inactivation of *CDKN1C* by germ line mutations or an IC2 epimutation is associated with an increased incidence of exomphalos but has not been associated with WT. In contrast, BWS children with UPD and IC1 deletions or epimutations have a high (>25%) risk of WT and so should be offered regular screening. Increased expression of *IGF2* is shown to be linked to an increased WT's risk in both BWS and sporadic nonsyndromic WT cases. Therefore in BWS, WT is seen more commonly in paternal UPD and IC1 epimutation cases which result in LOI of *IGF2* with resulting biallelic expression of *IGF2* and silencing of *H19*. IC1 epimutations also occur as a somatic event in many sporadic WTs. In addition to an increased WT risk, hemihypertrophy is seen more commonly in paternal UPD cases.

D. Isolated hemihypertrophy

Hemihypertrophy (or hemihyperplasia) refers to asymmetric growth of single or multiple organs or regions of the body. It can be a feature of certain genetic syndromes (such as BWS, Proteus syndrome, and neurofibromatosis type 1) but also may occur in isolation without an underlying syndrome. Patients with isolated hemihypertrophy (IH) have been reported to have an increased risk (5.9%) of developing embryonal tumors (Hoyme et al., 1998). These tumors include WT (most common), hepatoblastoma, neuroblastoma, adrenal cell carcinoma, and leiomyosarcoma of the small bowel. In a subset of patients with IH, somatic mosaicism for epigenetic abnormalities at 11p15.5 (PUPD or epimutations at IC1 or IC2) is detected (Grundy et al., 1991; Shuman et al., 2002; West et al., 2003). In one series, abnormalities in 11p15.5 were detected in 22% of IH cases and PUPD cases were at a higher risk of WT or hepatoblastoma (Shuman et al., 2006).

E. Sporadic nonsyndromic WT

WT can occur in association with genetic overgrowth syndromes such as BWS or occur sporadically. Somatic 11p15.5 defects have been discovered in WT tissue and in some cases in the surrounding normal renal tissue (Ogawa et al., 1993;

Okamoto *et al.*, 1997). Subsequently, Scott and colleagues studied the genomic DNA from lymphocytes of 437 cases of nonsyndromic WT patients and detected constitutional 11p15.5 defects in 3% of cases (and 12% of bilateral cases) (Scott *et al.*, 2008a,b). The defects include *H19* DMR hypermethylation, PUPD, and IC1 mutations. In keeping with the findings of genotype–phenotype correlations in BWS, isolated IC2 defects have not been identified in sporadic WT cases.

F. Silver–Russell syndrome

1. Clinical features

SRS is characterized by pre- and postnatal growth retardation (opposite growth phenotype to BWS). A low birthweight, often two standard deviations below the mean, is typical for SRS babies. The growth retardation relatively spares head growth, and therefore this gives an appearance of relatively large head (macrocephaly) or pseudohydrocephalus when compared to the weight and length of the child. Other clinical features include fifth finger clinodactyly, limb length asymmetry due to hemihypoplasia and genitourinary anomalies including hypospadias. SRS children are described as having a characteristic face which is triangular shaped with a broad forehead, retrognathia (small chin) with downturned corners of the mouth (Abu-Amero *et al.*, 2008; Price *et al.*, 1999; Russell 1954; Silver *et al.*, 1953). The incidence of SRS is reported as variable between 1/3000 and 1/10,000 (Abu-Amero *et al.*, 2008). The limb asymmetry in SRS is due to hemihypoplasia or hemihypotrophy (in contrast to hemihypertrophy which occurs in BWS).

2. Molecular mechanisms in SRS

Most SRS cases are sporadic but rare familial cases have been described (Duncan *et al.*, 1990). Molecular genetic studies in SRS initially focused on chromosome 7 and about 7–10% of SRS cases have maternal UPD for chromosome 7 and a small number of cases have chromosome 7 duplications. *GRB10* at 7p11.2-p12 and *PEG1* at 7q32 are two imprinted genes that were widely investigated as candidate genes for SRS. However, to date, a specific chromosome 7 SRS gene has not been identified. Following reports of maternal duplications of 11p15.5 discovered in patients with a SRS-like phenotype, further investigation discovered that the majority of SRS cases have an 11p15.5 IC1 epimutation opposite to BWS with hypomethylation of the *H19* DMR on the paternal allele (Gicquel *et al.*, 2005). This results in preferential expression of *H19* from both alleles and silencing of *IGF2* (Fig.6.4). This accounts for 60% of SRS cases (Netchine *et al.*, 2007). There has been one report of a child with SRS-like phenotype who inherited a maternal duplication of the centromeric IC2 of 11p15 and a report

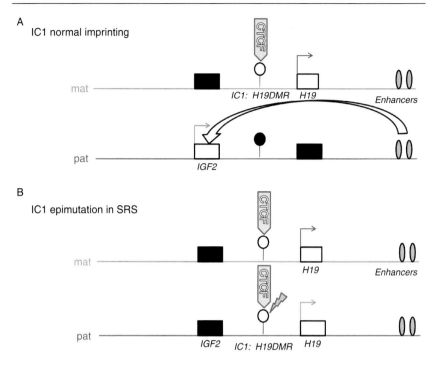

Figure 6.4. (A) Normal imprinting at IC1 at 11p15.5 (B) IC1 epimutation in SRS. Hypomethylation at the *H19*DMR on the paternal allele results in binding of CTCF to the CTCF binding sites at the *H19*DMR and acts as a boundary element preventing downstream enhances access to the *IGF2* promoter resulting in silencing of *IGF2*. This results in biallelic expression of *H19*.

of a case with mosaic maternal UPD11 (Bullman *et al.*, 2008; Schonherr *et al.*, 2007). However, in contrast to the frequency of IC2 epimutations in BWS, epimutations at IC2 in SRS have not been reported.

SRS patients with 11p15 imprinting abnormalities typically present with the "classical" phenotype of SRS with severe fetal growth restriction or failure to thrive, prominent forehead with relative macrocephaly and body asymmetry. SRS cases with mat UPD7 appear to have milder or absent SRS cranial dysmorphology (Abu-Amero *et al.*, 2008; Hannula *et al.*, 2001). Although cases of SRS patients with WT, hepatocellular carcinoma, craniopharyngioma, testicular cancer, and juvenile pilocytic astrocytoma have been reported (Bruckheimer and Abrahamov, 1993; Chitayat *et al.*, 1988; Draznin *et al.*, 1980; Fenton *et al.*, 2008; Weiss and Garnick, 1981). However, a number of studies into SRS have not reported increased risk of tumors (Bartholdi *et al.*, 2009; Bruce *et al.*, 2009; Netchine *et al.*, 2007).

G. The 11p15.5 imprinted region, somatic epimutations, and cancer

Disordered imprinting can occur not only in inherited imprinting syndromes, but also as a somatic event in human cancers. Indeed, LOI of *IGF2* (similar to that seen in a subset of BWS patients) is a frequent observation in some human cancers and can occur at an early stage of tumor development (e.g., WT, colorectal cancer, and ovarian cancer) (Kamikihara *et al.*, 2005; Kaneda and Feinberg, 2005; Yuan *et al.*, 2005).

1. *IGF2* LOI in human cancers

Somatic epimutations causing LOI of *IGF2* are a feature of sporadic embryonal tumor types similar to those seen in BWS (e.g., WT), but also of sporadic adult onset neoplasias such as colorectal cancer, chronic myeloid leukaemias, ovarian tumor, WT, Barret's esophagus and esopheageal cancer, renal cell carcinomas, lung adenocarcinomas, meningiomas, hepatoblastomas, osteosarcomas, rhabdo-myosarcomas, and a number of other cancers (Jelinic and Shaw, 2007). Biallelic expression of *IGF2* functions as an autocrine and paracrine growth factor in many cancers, and can lead to cell hyperplasia and cell proliferation which can be seen in the development of nephrogenic rests in kidneys as well as abnormal crypt development in the colon which are precancerous changes (Ohlsson *et al.*, 1999; Sakatani *et al.*, 2005). IGF2 activates signaling through the IGF1 receptor (IGF1R) leading to activation of signaling cascades including the IRS-1/PI3K/AKT and GRB2/Ras/ERK pathways (Foulstone *et al.*, 2005).

2. Constitutional *IGF2* LOI and colon cancer

A number of observations have implicated LOI of *IGF2* as an important factor in the tumorigenesis of colon cancers. First, a predisposition to more advanced disease, right-sided tumors of the colon and poorly differentiated or mucinous carcinoma, was seen with LOI of *IGF2*. In addition, in some colorectal cancer patients, LOI of *IGF2* is detectable in the surrounding normal colonic tissues and peripheral blood lymphocytes (Cui *et al.*, 2003).

In some cases, partial LOI of *IGF2* can also be detected in normal individuals that can be inherited down the generations (Heijmans *et al.*, 2007). The LOI of *IGF2* appears to be a stable epigenetic phenomenon and not associated with age (Cruz-Correa *et al.*, 2009). It has been suggested that in such individuals, detection of LOI of *IGF2* in the blood might be used as a biomarker for an increased risk of colorectal cancer (Cui *et al.*, 2003).

3. H19 tumor suppressor function

H19 has been proposed as a putative tumor suppressor gene *per se*. The *H19* transcript is a noncoding RNA which recently has been found to be the precursor of the microRNA miR-675 (Cai and Cullen, 2007). Aberrant hypomethylation of the *H19* DMR has been reported in human bladder cancers (Takai *et al.*, 2001). In addition, LOI of *H19* with biallelic expression has also been reported in lung cancer (Kondo *et al.*, 1995).

VI. THE 14q32 IMPRINTING CLUSTER

The imprinted cluster on 14q32.2 contains the maternally expressed genes *GTL2*, *RTL1as*, and the paternally expressed genes *DLK1*, *RTL1*, and *DIO3* (Fig. 6.1B). Imprinting within the cluster is regulated by a DMR (IG-DMR) that is located upstream of *GTL2* and methylated on the paternal allele (Lin *et al.*, 2003). The region has similarities to the *H19-IGF2* domain at 11p15.5 with a paternally methylated DMR and paternally expressed protein-coding gene *DLK1* (cf *IGF2*) and a maternally expressed noncoding RNA *GTL2* (cf *H19*) (Geuns *et al.*, 2007). However, the IG-DMR has no CTCF binding sites and fails to act as a boundary element unlike the *H19* DMR and a paternally inherited deletion of the IG-DMR does not produce a clinical phenotype (Edwards and Ferguson-Smith, 2007). Two distinct phenotypes are seen in 14q32 imprinting disorders: the maternal UPD14 (matUPD14) and paternal UPD14 (patUPD14) phenotypes.

The maternal uniparental disomy for chromosome 14 phenotype presents with pre- and postnatal growth retardation, neonatal hypotonia, precocious puberty, facial dysmorphism. Overlapping features with PWS (neonatal hypotonia and later truncal obesity) mean that often PWS is suspected before molecular testing is performed (Cox *et al.*, 2004). The matUPD14 phenotype can also be caused by 14q32.2 deletions involving the paternal allele (Kagami *et al.*, 2008) and an epimutation (loss of paternal allele methylation) at the IG-DMR (Temple *et al.*, 1991, 2007) and is thought to arise from loss of paternally expressed genes especially *DLK1* and *RTL1*. Mice experimental models looking at a paternally derived *Dlk1* mutation and *Rtl1* deletion produced mice with similar phenotypes to human mat UPD14 (Kagami *et al.*, 2008).

patUPD14 is associated with a more severe phenotype than matUPD14 with severe growth retardation and skeletal abnormalities including thoracic dystrophy ("coat hanger ribs"), characteristic facial anomalies and abdominal muscular defects. In addition to paternal UPD deletions involving the maternal

14q32.3 may also present with the patUPD14 phenotype and experimental mouse models suggest that *RTL1* overexpression has a major role in the development of the patUPD14 phenotype.

A. The 14q32 imprinted cluster and cancer

Although both the matUPD14 and patUPD14 phenotypes do not have an increased risk of cancer, at the 14q32 imprinted genes has been reported in a number of human cancers. Loss of *GTL2* (*MEG3*) expression due to hypermethylation of the GTL2 promoter or hypermethylation of the IG-DMR or both has been detected in some neuroblastomas, phaechromocytomas, WT and, nonfunctioning pituitary adenomas (Astuti *et al.*, 2005; Gejman *et al.*, 2008; Zhao *et al.*, 2005). Although *DLK1* expression was increased in neuroblastoma cell lines with both GTL2 promoter hypermethylation and IG-DMR hypermethylation, this was not due to LOI suggesting a different mechanism of regulation compared to the *IGF2/H19* imprinted domain. Increased *DLK1* expression without LOI was also detected in hepatocellular carcinomas but methylation status of the IG-DMR was not examined (Huang *et al.*, 2007). Increased expression of *DLK1* is also found in small cell lung carcinomas, neuroendocrine tumors, glioblastomas, and lymphomas (Helman *et al.*, 1987; Laborda *et al.*, 1993; Yin *et al.*, 2004, 2006). However, in contrast to these reports, reduced expression of *DLK1* was reported in renal cell carcinomas (Kawakami *et al.*, 2006).

VII. 15q11-q13 IMPRINTED REGION

The chromosome 15q11-q13 imprinted cluster contains paternally expressed genes (including *SNRPN*, *MAGEL2*, *MRKN3*, and *NDN*), maternally expressed genes (including *UBE3A* and *ATP10A*), small nucleolar RNA (snoRNA) clusters, and a complex DMR (the PWS/AS-IC) that is normally methylated on the maternal allele (Fig. 6.1C; Landers *et al.*, 2004; Runte *et al.*, 2001). The two human imprinting disorders associated with abnormal expression or function of 15q11-q13 imprinted cluster genes are AS and PWS.

AS is characterized by severe learning disability, microcephaly, speech difficulties, seizures, ataxia, and typical neurobehavioral features (Williams *et al.*, 2006). The unique behavior seen in AS is often described as a "happy puppet" with laughing, excitability, hand flapping, and frequent smiling.

PWS is also a primarily neurodevelopmental disorder with hypotonia and poor feeding in early life followed by excessive eating (hyperphagia) and obesity later in childhood, hypogonadotrophic hypogonadism and typical physical characteristics (Butler *et al.*, 2007; Cassidy and Driscoll, 2009). The neurodevelopmental delay is less severe than in AS with mild to moderate learning difficulties (Whittington *et al.*, 2004). Short stature may manifest in childhood and is always present by the second decade of life in the absence of replacement growth hormone (Burman *et al.*, 2001). Many imprinting syndromes are characterized by neurodevelopmental and growth abnormalities. However, in contrast to BWS in which overgrowth is accompanied by tumor susceptibility (but usually not neurodevelopmental delay), individuals with AS or PWS do not appear to be at high risk of neoplasia. Both AS and PWS result from abnormalities of the 15q11-13 imprinted gene region. The majority of Angelman syndrome individuals have large (5–7 mb) deletions on the maternal allele and paternal UPD of 15q11 accounts for around 7% of cases. These findings demonstrated that AS resulted from inactivation of maternally expressed gene(s) and the finding of germ line mutations in *UBE3A* demonstrated that loss of *UBE3A* expression or function was the cause of the disorder (Kishino *et al.*, 1997; Matsuura *et al.*, 1997). In contrast, PWS results from loss of expression of paternally expressed 15q11 imprinted genes. Thus majority of PWS cases have a large 5–7 MB deletion on the paternal allele and maternal UPD accounts for around 25–30% of cases. In contrast to AS, attempts to identify germ line mutations in PWS candidate genes have been unsuccessful but recent studies implicate snoRNAs in the pathogenesis of the PWS phenotype (de Smith *et al.*, 2009; Sahoo *et al.*, 2008). Imprinting of the 15q11 imprinted gene cluster is regulated by a bipartite imprinting center and microdeletions or epimutations involving the center can produce AS or PWS (depending on the parent-of-origin of the affected allele and the precise location of the microdeletion) (Horsthemke and Wagstaff, 2008).

A. The 15q11-q13 imprinted cluster and cancer

Currently, there are no reports of tumors occurring in AS in the literature. Isolated case reports of PWS patients with testicular tumor, hepatoblasoma, WT, intratubular germ cell tumor, and a pituitary adenoma have been reported (Bettio *et al.*, 1996; Coppes *et al.*, 1993; Hashizume *et al.*, 1991; Jaffray *et al.*, 1999; Robinson and Jones, 1990). A survey of PWS patients reported an increased incidence of myeloid leukemias (Davies *et al.*, 2003). However, loss of imprinting at the 15q11-q13 region has not been implicated in chronic or acute myeloid leukemias. Therefore, in contrast to the 11p15.5 IC1 element (*H19* DMR), somatic epimutations or deletions in 15q11 have not been reported in human cancers.

VIII. THE 6q24 IMPRINTED REGION

The 6q24 imprinted gene cluster contains two paternally expressed imprinted genes (Fig. 6.1D). *PLAGL1* (also known as *ZAC/LOT1*) encodes a zinc-finger transcription factor that localizes to the nuclear compartment. *HYMAI* is an untranslated RNA of unknown function. The DMR involved at 6q24 is the TND DMR located in the promoter region of *PLAGL1* and is normally methylated on the maternal allele (Arima *et al.*, 2001). Germ line disruption of imprinting at the 6q24 imprinted cluster is associated with the condition transient neonatal diabetes mellitus (TND).

A. Transient neonatal diabetes mellitus

TND is defined as diabetes that occurs within the first 6 weeks of life in a term infant and resolves by 18 months of age. The hallmark feature of TND is the occurrence of transient hyperglycemia with low insulin levels in the neonatal period. Along with the diabetes, affected babies often have intrauterine growth retardation. Other variables that have been reported in TND children include macroglossia and umbilical hernia. The incidence of TND in the UK is estimated at 1/400,000 live births (Temple and Shield, 2002; Temple *et al.*, 2000). The neonatal diabetes is transient but 40% of affected individuals subsequently develop type 2 diabetes. During the remission phase after the transient neonatal diabetes, episodes of hyperglycemia can occur during periods of intercurrent illnesses or may manifest as gestational diabetes in affected pregnant females.

TND is caused by biallelic expression of the paternally expressed genes at 6q24. Seventy percentage of cases of TND are due to paternal UPD6 or duplications involving the paternal 6q24 region. A further 20% of cases demonstrate loss of maternal allele methylation (epimutation) at the TND DMR resulting in biallelic expression of *PLAGL1* and *HYMAI*. Interestingly, about 50% of TND patients with a TND DMR epimutation were found to have LOM at other imprinted loci. The LOM at the other imprinted loci include *PEG1* (7q32), *PEG3* (19q13.4), *KvDMR1* (11p15.5) *GRB10* (7p12-p11.2), and *NESPAS* (20q13) DMRS (Mackay *et al.*, 2006a,b, 2008). This group of patients has been designated as having "hypomethylation of multiple imprinted loci" (HIL). Review of TND patients with HIL and loss of methylation at 11p15.5 IC2 revealed some overlap with the clinical features of BWS including anterior abdominal wall defects (Mackay *et al.*, 2006a,b). Subsequently, six families with TND and HIL were found to have homozygous mutations in the *ZFP57* gene at 6p22. *ZFP57* has six exons encoding a protein with a predicted KRAB domain and seven zinc fingers (Mackay *et al.*, 2008). Individuals with homozygous *ZFP57* mutations all show hypomethylation at the *GRB10* and *PEG3* DMRS. Hypomethylation at those DMRs is seen less commonly in HIL individuals without *ZFP57* mutations. Hypomethylation at *KvDMR1* and *PEG1* DMRs can occur in HIL individuals with or without *ZFP57* mutations. The findings that the hypomethylation at the multiple DMRs are mosaic suggest that *ZFP57* plays a role in the maintenance of methylation during the earliest multicellular stages of development (Mackay *et al.*, 2008), however studies of *Zfp57* knockout mice suggest a possible role both in the establishment of germ line methylation imprints and in the postfertilization maintenance of methylation imprints (Li *et al.*, 2008).

B. The 6q24 imprinted cluster and cancer

To date, there have been no reports in the literature of TND patients with cancer. However, loss of expression of *PLAGL1* due to hypermethylation of its promoter has been demonstrated in breast cancer, ovarian cancer, pituitary adenomas, and basal cell carcinomas (Basyuk *et al.*, 2005; Bilanges *et al.*, 1999; Kamikihara *et al.*, 2005; Pagotto *et al.*, 2000).

IX. GLOBAL GENOMIC IMPRINTING DISORDERS: *NLRP7* AND BIPARENTAL COMPLETE HYDATIDIFORM MOLES

Hydatidiform molar pregnancies are abnormal conceptions characterized by cystic degeneration of the trophoblastic villi without normal embryonic development. Moles occur when there is no maternal contribution to the genome. This can be due to either an androgentic conceptus (where both sets of chromosomes are paternal in origin) or the absence of maternal methylation imprint marks in biparental complete hydatidiform moles. Biparental complete hydatidiform moles have both paternal and maternal chromosome genomes but, as mentioned, there is LOM across maternally methylated imprinted loci resulting in a paternal epigenotype across the genome (El-Maarri *et al.*, 2003; Fisher *et al.*, 2002; Judson *et al.*, 2002).

A. Cancer risk in hydatidiform moles

Hydatidiform molar pregnancies must be managed carefully as it has the potential to cancerous change with local invasion or metastasis. Management of hydatidiform moles includes evacuation of the mole, X-rays and monitoring of hCG levels. There is a 5–15% risk of the moles becoming invasive where it is called persistent trophoblastic disease. The moles may invade through the uterine wall and cause further complications such as hemorrhage and in some cases develop into a choriocarcinoma. Choriocarcinomas are malignant, rapidly growing form of cancer with the ability to metastasize. However, despite those features, prognosis is usually good with chemotherapy treatment (Berkowitz and Goldstein, 2009; Sebire et al., 2002).

B. Familial cases of biparental complete hydatidiform moles

Familial cases of biparental complete hydatidiform moles can result from *NLRP7* homozygous mutations in the mother (Murdoch et al., 2006). Homozygote women with *NLRP7* mutations have normal somatic cell methylation patterns, but imprinting in their germ cells is disrupted. As a result, molar pregnancies and reproductive wastages occur due to the failure to establish or maintain normal maternal epigenotype patterns. Homozygote males on the other hand have normal fertility and normal methylation levels. In contrast to *ZFP57*, homozygous *NLRP7* mutation individuals have normal methylation patterns but the women are at risk of conceptions with methylation defects. The methylation defects in *NLRP7-* associated familial hydatidiform moles are specific for maternally imprinted loci. DNA methylation at nonimprinted genes and genes subjected to X-inactivation are unaffected (Djuric et al., 2006).

Females with *NLRP7* homozygous mutation have also been reported to develop persistent trophoblastic disease with a similar frequency as seen in nonfamilial or nonrecurrent complete hydatidiform molar pregnancies (Wang et al., 2009). Therefore, the management of *NLRP7* women is similar due to the risk of persistent trophoblastic diseases including choriocarcinomas.

Currently, there is no evidence for *NLRP7* inactivation in cancer. However, *NLRP7* has been shown to be significantly activated in testicular seminomas and endometrial cancers (Ohno et al., 2008; Okada et al., 2004).

X. SUMMARY/CONCLUSION

In this chapter, we described a number of human imprinting syndromes to illustrate the concepts of genomic imprinting. Their reported mechanisms which result in disordered imprinting and their relative frequencies and associated cancer risk are summarized in Table 6.1. LOI of imprinted genes is also commonly found in a number of human cancers and appears to be early events in tumorigenesis. We anticipate that with further application of high- throughput genomic techniques to elucidate the cancer genome and epigenome the role of imprinted genes in human cancer will continue to grow.

Table 6.1. Examples of Imprinting Disorders with Reported Mechanisms and Cancer Risk

Imprinted domain	Disease	Cytogenetic abnormality	Uniparental disomy	Gene mutation	Imprinting center epimutation	Deletions/duplications involving imprinting center	Cancer risk
11p15.5	**Beckwith–Wiedemann syndrome (BWS)**	Chromosomal rearrangement (1–2%)	Paternal UPD (20%)	*CDKN1C* on maternal allele (5% sporadic, 50% in familial cases)	IC1 epimutation (*H19DMR* hypermethylation (5%)) IC2 epimutation (KvDMR1 hypomethylation (50%))	IC1 deletion on maternal allele (<5%) IC2 deletion on maternal allele (<2%)	Embryonal tumors: Wilms tumor, Hepatoblastoma, Rhabdomyosarcoma
11p15.5	**Silver–Russell syndrome (SRS)**	Maternal duplication of 11p15 (<2%)	Maternal UPD (1 case)	–	IC1 epimutation (*H19DMR* hypomethylation (up to 70%))	IC2 duplication on maternal allele (1 case)	–
7p12.2, 7q32.2	**Silver–Russell syndrome (SRS)**	Chromosomal rearrangement (<2%) Maternal duplication of 7p	Maternal UPD (5%)	–	–	–	–
6q24	**Transient neonatal diabetes (TND)**	Visible paternal duplication of 6q24 (2%)	Paternal UPD (35–40%)	–	Hypomethylation of the TND (*PLAGL1*) DMR (20%)	Submicroscopic duplication of the 6q24 region on the paternal allele (35–40%)	–
14q32.2	**MatUPD14**	Chromosomal rearrangement	Mat UPD14 (majority of cases)	–	Hypomethylation of the IG-DMR on the paternal allele (7 reported cases)	Deletion involving the 14q32 imprinted region on the paternal allele	–

Locus	Syndrome	Chromosomal rearrangement/deletion	UPD	Mutation	Methylation defect	Imprinting center deletion	Associated condition
14q32.2	patUPD14	Chromosomal rearrangement	pUPD (majority of cases)	—	Hypermethylation of the IG DMR on the maternal allele (three reported cases)	Deletion involving the 14q32 imprinted region on the maternal allele	—
15q11-q13	Angelman syndrome (AS)	5–7 mb visible deletion on maternal allele at 15q11-q13 Chromosomal rearrangement (<1%)	Paternal UPD (7%)	UBE3A on maternal allele (11%)	Hypomethylation of the SNRPN DMR on the maternal allele (80–90% of imprinting defect cases) *2–5% of AS have imprinting defects	Deletion involving the AS-SRO at the PWS/AS-IC on the maternal allele (10–20% of imprinting defect cases) *2–5% of AS have imprinting defects	—
15q11-q13	Prader–Willi Syndrome (PWS)	5–7 mb visible deletion on paternal allele at 15q11-q13 Chromosomal rearrangement	Maternal UPD (25–30%)	—	Hypermethylation of the SNRPN DMR on the paternal allele (75–80% of imprinting defect cases) *1% of PWS are due to imprinting defects	Deletion involving the PWS-SRO at the PWS/AS-IC on the paternal allele (15% of imprinting defect cases) *1% of PWS are due to imprinting defects	Myeloid leukemias
Multiple loci	TND + hypomethylation at multiple imprinted Loci (HIL)	—	—	Homozygous mutations in ZFP57	—	—	—
Mutiple loci in germ cells of females	Familial biparental complete hydatidiform moles	—	—	Homozygous mutations in NLRP7	—	—	Persistent trophoblastic disease, choriocarcinoma

References

Abu-Amero, S., Monk, D., Frost, J., Preece, M., Stanier, P., and Moore, G. E. (2008). The genetic aetiology of Silver-Russell syndrome. *J. Med. Genet.* **45**(4), 193–199.

Arima, T., Drewell, R. A., Arney, K. L., Inoue, J., Makita, Y., Hata, A., Oshimura, M., Wake, N., and Surani, M. A. (2001). A conserved imprinting control region at the HYMAI/ZAC domain is implicated in transient neonatal diabetes mellitus. *Hum. Mol. Genet.* **10**(14), 1475–1483.

Astuti, D., Latif, F., Wagner, K., Gentle, D., Cooper, W. N., Catchpoole, D., Grundy, R., Ferguson-Smith, A. C., and Maher, E. R. (2005). Epigenetic alteration at the DLK1-GTL2 imprinted domain in human neoplasia: Analysis of neuroblastoma, phaeochromocytoma and Wilms' tumour. *Br. J. Cancer* **92**(8), 1574–1580.

Barlow, D. P., Stoger, R., Herrmann, B. G., Saito, K., and Schweifer, N. (1991). The mouse insulin-like growth factor type-2 receptor is imprinted and closely linked to the Tme locus. *Nature* **349** (6304), 84–87.

Bartholdi, D., Krajewska-Walasek, M., Ounap, K., Gaspar, H., Chrzanowska, K. H., Ilyana, H., Kayserili, H., Lurie, I. W., Schinzel, A., and Baumer, A. (2009). Epigenetic mutations of the imprinted IGF2-H19 domain in Silver-Russell syndrome (SRS): Results from a large cohort of patients with SRS and SRS-like phenotypes. *J. Med. Genet.* **46**(3), 192–197.

Basyuk, E., Coulon, V., Le Digarcher, A., Coisy-Quivy, M., Moles, J.-P., Gandarillas, A., and Journot, L. (2005). The candidate tumor suppressor gene zac is involved in keratinocyte differentiation and its expression is lost in basal cell carcinomas. *Mol. Cancer Res.* **3**(9), 483–492.

Beckwith, J. B. (1998). Nephrogenic rests and the pathogenesis of Wilms tumor: Developmental and clinical considerations. *Am. J. Med. Genet.* **79**(4), 268–273.

Bell, A. C., and Felsenfeld, G. (2000). Methylation of a CTCF-dependent boundary controls imprinted expression of the Igf2 gene. *Nature* **405**(6785), 482–485.

Berkowitz, R. S., and Goldstein, D. P. (2009). Molar pregnancy. *N. Engl. J. Med.* **360**(16), 1639–1645.

Bettio, D., Giardino, D., Rizzi, N., Riva, P., Volpi, L., Barantani, E., Tagliaferri, A., and Larizza, L. (1996). Isochromosome 15q of maternal origin in a Prader-Willi patient with pituitary adenoma. *Acta. Genet. Med. Gemellol. (Roma)* **45**(1–2), 213–216.

Bilanges, B., Varrault, A., Basyuk, E., Rodriguez, C., Mazumdar, A., Pantaloni, C., Bockaert, J., Theillet, C., Spengler, D., and Journot, L. (1999). Loss of expression of the candidate tumor suppressor gene ZAC in breast cancer cell lines and primary tumors. *Oncogene* **18**(27), 3979–3988.

Bliek, J., Verde, G., Callaway, J., Maas, S. M., De Crescenzo, A., Sparago, A., Cerrato, F., Russo, S., Ferraiuolo, S., Rinaldi, M. M., *et al.* (2009). Hypomethylation at multiple maternally methylated imprinted regions including PLAGL1 and GNAS loci in Beckwith-Wiedemann syndrome. *Eur. J. Hum. Genet.* **17**(5), 611–619.

Bruce, S., Hannula-Jouppi, K., Peltonen, J., Kere, J., and Lipsanen-Nyman, M. (2009). Clinically distinct epigenetic subgroups in Silver-Russell syndrome: The degree of H19 hypomethylation associates with phenotype severity and genital and skeletal anomalies. *J. Clin. Endocrinol. Metab.* **94**(2), 579–587.

Bruckheimer, E., and Abrahamov, A. (1993). Russell-Silver syndrome and Wilms tumor. *J. Pediatr.* **122**(1), 165–166.

Bullman, H., Lever, M., Robinson, D. O., Mackay, D. J., Holder, S. E., and Wakeling, E. L. (2008). Mosaic maternal uniparental disomy of chromosome 11 in a patient with Silver-Russell syndrome. *J. Med. Genet.* **45**(6), 396–399.

Burman, P., Ritzen, E. M., and Lindgren, A. C. (2001). Endocrine dysfunction in Prader-Willi syndrome: A review with special reference to GH. *Endocr. Rev.* **22**(6), 787–799.

Butler, M. G., Theodoro, M. F., Bittel, D. C., and Donnelly, J. E. (2007). Energy expenditure and physical activity in Prader-Willi syndrome: Comparison with obese subjects. *Am. J. Med. Genet. A* **143**(5), 449–459.

Cai, X., and Cullen, B. R. (2007). The imprinted H19 noncoding RNA is a primary microRNA precursor. *RNA* **13**(3), 313–316.

Cassidy, S. B., and Driscoll, D. J. (2009). Prader-Willi syndrome. *Eur. J. Hum. Genet.* **17**(1), 3–13.

Cattanach, B. M., and Kirk, M. (1985). Differential activity of maternally and paternally derived chromosome regions in mice. *Nature* **315**(6019), 496–498.

Chitayat, D., Friedman, J. M., Anderson, L., and Dimmick, J. E. (1988). Hepatocellular carcinoma in a child with familial Russell-Silver syndrome. *Am. J. Med. Genet.* **31**(4), 909–914.

Cooper, W. N., Luharia, A., Evans, G. A., Raza, H., Haire, A. C., Grundy, R., Bowdin, S. C., Riccio, A., Sebastio, G., Bliek, J., et al. (2005). Molecular subtypes and phenotypic expression of Beckwith-Wiedemann syndrome. *Eur. J. Hum. Genet.* **13**(9), 1025–1032.

Coppes, M. J., Sohl, H., Teshima, I. E., Mutirangura, A., Ledbetter, D. H., and Weksberg, R. (1993). Wilms tumor in a patient with Prader-Willi syndrome. *J. Pediatr.* **122**(5 Pt 1), 730–733.

Coppes, M. J., Arnold, M., Beckwith, J. B., Ritchey, M. L., D'Angio, G. J., Green, D. M., and Breslow, N. E. (1999). Factors affecting the risk of contralateral Wilms tumor development: A report from the National Wilms Tumor Study Group. *Cancer* **85**(7), 1616–1625.

Cox, H., Bullman, H., and Temple, I. K. (2004). Maternal UPD(14) in the patient with a normal karyotype: Clinical report and a systematic search for cases in samples sent for testing for Prader-Willi syndrome. *Am. J. Med. Genet. A* **127A**(1), 21–25.

Cruz-Correa, M., Zhao, R., Oviedo, M., Bernabe, R. D., Lacourt, M., Cardona, A., Lopez-Enriquez, R., Wexner, S., Cuffari, C., Hylind, L., et al. (2009). Temporal stability and age-related prevalence of loss of imprinting of the insulin-like growth factor-2 gene. *Epigenetics* **4**(2), 114–118.

Cui, H., Cruz-Correa, M., Giardiello, F. M., Hutcheon, D. F., Kafonek, D. R., Brandenburg, S., Wu, Y., He, X., Powe, N. R., and Feinberg, A. P. (2003). Loss of IGF2 imprinting: A potential marker of colorectal cancer risk. *Science* **299**(5613), 1753–1755.

Davies, H. D., Leusink, G. L., McConnell, A., Deyell, M., Cassidy, S. B., Fick, G. H., and Coppes, M. J. (2003). Myeloid leukemia in Prader-Willi syndrome. *J. Pediatr.* **142**(2), 174–178.

de Smith, A. J., Purmann, C., Walters, R. G., Ellis, R. J., Holder, S. E., Van Haelst, M. M., Brady, A. F., Fairbrother, U. L., Dattani, M., Keogh, J. M., et al. (2009). A deletion of the HBII-85 class of small nucleolar RNAs (snoRNAs) is associated with hyperphagia, obesity and hypogonadism. *Hum. Mol. Genet.* **18**(17), 3257–3265.

DeBaun, M. R., Siegel, M. J., and Choyke, P. L. (1998). Nephromegaly in infancy and early childhood: A risk factor for Wilms tumor in Beckwith-Wiedemann syndrome. *J. Pediatr.* **132** (3 Pt 1), 401–404.

DeBaun, M. R., Niemitz, E. L., and Feinberg, A. P. (2003). Association of in vitro fertilization with Beckwith-Wiedemann syndrome and epigenetic alterations of LIT1 and H19. *Am. J. Hum. Genet.* **72**(1), 156–160.

Diaz-Meyer, N., Day, C. D., Khatod, K., Maher, E. R., Cooper, W., Reik, W., Junien, C., Graham, G., Algar, E., Der Kaloustian, V. M., et al. (2003). Silencing of CDKN1C (p57KIP2) is associated with hypomethylation at KvDMR1 in Beckwith-Wiedemann syndrome. *J. Med. Genet.* **40**(11), 797–801.

Djuric, U., El-Maarri, O., Lamb, B., Kuick, R., Seoud, M., Coullin, P., Oldenburg, J., Hanash, S., and Slim, R. (2006). Familial molar tissues due to mutations in the inflammatory gene, NALP7, have normal postzygotic DNA methylation. *Hum. Genet.* **120**(3), 390–395.

Draznin, M. B., Stelling, M. W., and Johanson, A. J. (1980). Silver-Russell syndrome and cranio-pharyngioma. *J. Pediatr.* **96**(5), 887–889.

Drenth, J. P., and van der Meer, J. W. (2006). The inflammasome—A linebacker of innate defense. *N. Engl. J. Med.* **355**(7), 730–732.

Duncan, P. A., Hall, J. G., Shapiro, L. R., and Vibert, B. K. (1990). Three-generation dominant transmission of the Silver-Russell syndrome. *Am. J. Med. Genet.* **35**(2), 245–250.

Edwards, C. A., and Ferguson-Smith, A. C. (2007). Mechanisms regulating imprinted genes in clusters. *Curr. Opin. Cell Biol.* **19**(3), 281–289.

Elliott, M., Bayly, R., Cole, T., Temple, I. K., and Maher, E. R. (1994). Clinical features and natural history of Beckwith-Wiedemann syndrome: Presentation of 74 new cases. *Clin. Genet.* **46**(2), 168–174.

El-Maarri, O., Seoud, M., Coullin, P., Herbiniaux, U., Oldenburg, J., Rouleau, G., and Slim, R. (2003). Maternal alleles acquiring paternal methylation patterns in biparental complete hydatidiform moles. *Hum. Mol. Genet.* **12**(12), 1405–1413.

Everman, D. B., Shuman, C., Dzolganovski, B., O'Riordan, M. A., Weksberg, R., and Robin, N. H. (2000). Serum alpha-fetoprotein levels in Beckwith-Wiedemann syndrome. *J. Pediatr.* **137**(1), 123–127.

Fenton, E., Refai, D., See, W., and Rawluk, D. J. (2008). Supratentorial juvenile pilocytic astrocytoma in a young adult with Silver-Russell syndrome. *Br. J. Neurosurg.* **22**(6), 776–777.

Fisher, R. A., Hodges, M. D., Rees, H. C., Sebire, N. J., Seckl, M. J., Newlands, E. S., Genest, D. R., and Castrillon, D. H. (2002). The maternally transcribed gene p57(KIP2) (CDNK1C) is abnormally expressed in both androgenetic and biparental complete hydatidiform moles. *Hum. Mol. Genet.* **11**(26), 3267–3272.

Foulstone, E., Prince, S., Zaccheo, O., Burns, J. L., Harper, J., Jacobs, C., Church, D., and Hassan, A. B. (2005). Insulin-like growth factor ligands, receptors, and binding proteins in cancer. *J. Pathol.* **205**(2), 145–153.

Gejman, R., Batista, D. L., Zhong, Y., Zhou, Y., Zhang, X., Swearingen, B., Stratakis, C. A., Hedley-Whyte, E. T., and Klibanski, A. (2008). Selective loss of MEG3 expression and intergenic differentially methylated region hypermethylation in the MEG3/DLK1 locus in human clinically nonfunctioning pituitary adenomas. *J. Clin. Endocrinol. Metab.* **93**(10), 4119–4125.

Geuns, E., De Temmerman, N., Hilven, P., Van Steirteghem, A., Liebaers, I., and De Rycke, M. (2007). Methylation analysis of the intergenic differentially methylated region of DLK1-GTL2 in human. *Eur. J. Hum. Genet.* **15**(3), 352–361.

Gicquel, C., Gaston, V., Mandelbaum, J., Siffroi, J. P., Flahault, A., and Le Bouc, Y. (2003). In vitro fertilization may increase the risk of Beckwith-Wiedemann syndrome related to the abnormal imprinting of the KCN1OT gene. *Am. J. Hum. Genet.* **72**(5), 1338–1341.

Gicquel, C., Rossignol, S., Cabrol, S., Houang, M., Steunou, V., Barbu, V., Danton, F., Thibaud, N., Le Merrer, M., Burglen, L., *et al.* (2005). Epimutation of the telomeric imprinting center region on chromosome 11p15 in Silver-Russell syndrome. *Nat. Genet.* **37**(9), 1003–1007.

Grundy, P., Telzerow, P., Paterson, M. C., Haber, D., Berman, B., Li, F., and Garber, J. (1991). Chromosome 11 uniparental isodisomy predisposing to embryonal neoplasms. *Lancet* **338**(8774), 1079–1080.

Halliday, J., Oke, K., Breheny, S., Algar, E., and Amor, D. J. (2004). Beckwith-Wiedemann syndrome and IVF: A case-control study. *Am. J. Hum. Genet.* **75**(3), 526–528.

Hannula, K., Kere, J., Pirinen, S., Holmberg, C., and Lipsanen-Nyman, M. (2001). Do patients with maternal uniparental disomy for chromosome 7 have a distinct mild Silver-Russell phenotype? *J. Med. Genet.* **38**(4), 273–278.

Hark, A. T., Schoenherr, C. J., Katz, D. J., Ingram, R. S., Levorse, J. M., and Tilghman, S. M. (2000). CTCF mediates methylation-sensitive enhancer-blocking activity at the H19/Igf2 locus. *Nature* **405**(6785), 486–489.

Hashizume, K., Nakajo, T., Kawarasaki, H., Iwanaka, T., Kanamori, Y., Tanaka, K., Utuki, T., Mishina, J., and Watanabe, T. (1991). Prader-Willi syndrome with del(15)(q11, q13) associated with hepatoblastoma. *Acta Paediatr. Jpn.* **33**(6), 718–722.

Hatada, I., Ohashi, H., Fukushima, Y., Kaneko, Y., Inoue, M., Komoto, Y., Okada, A., Ohishi, S., Nabetani, A., Morisaki, H., *et al.* (1996). An imprinted gene p57KIP2 is mutated in Beckwith-Wiedemann syndrome. *Nat. Genet.* **14**(2), 171–173.

Heijmans, B. T., Kremer, D., Tobi, E. W., Boomsma, D. I., and Slagboom, P. E. (2007). Heritable rather than age-related environmental and stochastic factors dominate variation in DNA methylation of the human IGF2/H19 locus. *Hum. Mol. Genet.* **16**(5), 547–554.

Helman, L. J., Thiele, C. J., Linehan, W. M., Nelkin, B. D., Baylin, S. B., and Israel, M. A. (1987). Molecular markers of neuroendocrine development and evidence of environmental regulation. *Proc. Natl. Acad. Sci. USA* **84**(8), 2336–2339.

Horsthemke, B., and Wagstaff, J. (2008). Mechanisms of imprinting of the Prader-Willi/Angelman region. *Am. J. Med. Genet. A* **146A**(16), 2041–2052.

Hoyme, H. E., Seaver, L. H., Jones, K. L., Procopio, F., Crooks, W., and Feingold, M. (1998). Isolated hemihyperplasia (hemihypertrophy): Report of a prospective multicenter study of the incidence of neoplasia and review. *Am. J. Med. Genet.* **79**(4), 274–278.

Huang, J., Zhang, X., Zhang, M., Zhu, J.-D., Zhang, Y.-L., Lin, Y., Wang, K.-S., Qi, X.-F., Zhang, Q., Liu, G.-Z., *et al.* (2007). Up-regulation of DLK1 as an imprinted gene could contribute to human hepatocellular carcinoma. *Carcinogenesis* **28**(5), 1094–1103.

Jaenisch, R., and Bird, A. (2003). Epigenetic regulation of gene expression: How the genome integrates intrinsic and environmental signals. *Nat. Genet.* **33**(Suppl.), 245–254.

Jaffray, B., Moore, L., and Dickson, A. P. (1999). Prader-Willi syndrome and intratubular germ cell neoplasia. *Med. Pediatr. Oncol.* **32**(1), 73–74.

Jelinic, P., and Shaw, P. (2007). Loss of imprinting and cancer. *J. Pathol.* **211**(3), 261–268.

Judson, H., Hayward, B. E., Sheridan, E., and Bonthron, D. T. (2002). A global disorder of imprinting in the human female germ line. *Nature* **416**(6880), 539–542.

Kacem, S., and Feil, R. (2009). Chromatin mechanisms in genomic imprinting. *Mamm. Genome* **20** (9–10), 544–556.

Kagami, M., Sekita, Y., Nishimura, G., Irie, M., Kato, F., Okada, M., Yamamori, S., Kishimoto, H., Nakayama, M., Tanaka, Y., *et al.* (2008). Deletions and epimutations affecting the human 14q32.2 imprinted region in individuals with paternal and maternal upd(14)-like phenotypes. *Nat. Genet.* **40**(2), 237–242.

Kajii, T., and Ohama, K. (1977). Androgenetic origin of hydatidiform mole. *Nature* **268**(5621), 633–634.

Kamikihara, T., Arima, T., Kato, K., Matsuda, T., Kato, H., Douchi, T., Nagata, Y., Nakao, M., and Wake, N. (2005). Epigenetic silencing of the imprinted gene ZAC by DNA methylation is an early event in the progression of human ovarian cancer. *Int. J. Cancer* **115**(5), 690–700.

Kaneda, A., and Feinberg, A. P. (2005). Loss of imprinting of IGF2: A common epigenetic modifier of intestinal tumor risk. *Cancer Res.* **65**(24), 11236–11240.

Kawakami, T., Chano, T., Minami, K., Okabe, H., Okada, Y., and Okamoto, K. (2006). Imprinted DLK1 is a putative tumor suppressor gene and inactivated by epimutation at the region upstream of GTL2 in human renal cell carcinoma. *Hum. Mol. Genet.* **15**(6), 821–830.

Kinoshita, T., Wang, Y., Hasegawa, M., Imamura, R., and Suda, T. (2005). PYPAF3, a PYRIN-containing APAF-1-like protein, is a feedback regulator of caspase-1-dependent interleukin-1beta secretion. *J. Biol. Chem.* **280**(23), 21720–21725.

Kishino, T., Lalande, M., and Wagstaff, J. (1997). UBE3A/E6-AP mutations cause Angelman syndrome. *Nat. Genet.* **15**(1), 70–73.

Kondo, M., Suzuki, H., Ueda, R., Osada, H., Takagi, K., and Takahashi, T. (1995). Frequent loss of imrpiniting of the H19 gene is often associated with its overexpression in human lung cancers. *Oncogene* **10**(6), 1193–1198.

Kono, T., Obata, Y., Wu, Q., Niwa, K., Ono, Y., Yamamoto, Y., Park, E. S., Seo, J. S., and Ogawa, H. (2004). Birth of parthenogenetic mice that can develop to adulthood. *Nature* **428**(6985), 860–864.

Laborda, J., Sausville, E. A., Hoffman, T., and Notario, V. (1993). dlk, a putative mammalian homeotic gene differentially expressed in small cell lung carcinoma and neuroendocrine tumor cell line. *J. Biol. Chem.* **268**(6), 3817–3820.

Lam, W. W., Hatada, I., Ohishi, S., Mukai, T., Joyce, J. A., Cole, T. R., Donnai, D., Reik, W., Schofield, P. N., and Maher, E. R. (1999). Analysis of germline CDKN1C (p57KIP2) mutations in familial and sporadic Beckwith-Wiedemann syndrome (BWS) provides a novel genotype-phenotype correlation. *J. Med. Genet.* **36**(7), 518–523.

Landers, M., Bancescu, D. L., Le Meur, E., Rougeulle, C., Glatt-Deeley, H., Brannan, C., Muscatelli, F., and Lalande, M. (2004). Regulation of the large (approximately 1000 kb) imprinted murine Ube3a antisense transcript by alternative exons upstream of Snurf/Snrpn. *Nucleic Acids Res.* **32**(11), 3480–3492.

Lee, J., Inoue, K., Ono, R., Ogonuki, N., Kohda, T., Kaneko-Ishino, T., Ogura, A., and Ishino, F. (2002). Erasing genomic imprinting memory in mouse clone embryos produced from day 11.5 primordial germ cells. *Development* **129**(8), 1807–1817.

Li, X., Ito, M., Zhou, F., Youngson, N., Zuo, X., Leder, P., and Ferguson-Smith, A. C. (2008). A maternal-zygotic effect gene, Zfp57, maintains both maternal and paternal imprints. *Dev. Cell* **15**(4), 547–557.

Lim, D., Bowdin, S. C., Tee, L., Kirby, G. A., Blair, E., Fryer, A., Lam, W., Oley, C., Cole, T., Brueton, L. A., *et al.* (2009). Clinical and molecular genetic features of Beckwith-Wiedemann syndrome associated with assisted reproductive technologies. *Hum. Reprod.* **24**(3), 741–747.

Lin, S. P., Youngson, N., Takada, S., Seitz, H., Reik, W., Paulsen, M., Cavaille, J., and Ferguson-Smith, A. C. (2003). Asymmetric regulation of imprinting on the maternal and paternal chromosomes at the Dlk1-Gtl2 imprinted cluster on mouse chromosome 12. *Nat. Genet.* **35**(1), 97–102.

Mackay, D. J., Boonen, S. E., Clayton-Smith, J., Goodship, J., Hahnemann, J. M., Kant, S. G., Njolstad, P. R., Robin, N. H., Robinson, D. O., Siebert, R., *et al.* (2006a). A maternal hypomethylation syndrome presenting as transient neonatal diabetes mellitus. *Hum. Genet.* **120**(2), 262–269.

Mackay, D. J., Hahnemann, J. M., Boonen, S. E., Poerksen, S., Bunyan, D. J., White, H. E., Durston, V. J., Thomas, N. S., Robinson, D. O., Shield, J. P., *et al.* (2006b). Epimutation of the TNDM locus and the Beckwith-Wiedemann syndrome centromeric locus in individuals with transient neonatal diabetes mellitus. *Hum. Genet.* **119**(1–2), 179–184.

Mackay, D. J., Callaway, J. L., Marks, S. M., White, H. E., Acerini, C. L., Boonen, S. E., Dayanikli, P., Firth, H. V., Goodship, J. A., Haemers, A. P., *et al.* (2008). Hypomethylation of multiple imprinted loci in individuals with transient neonatal diabetes is associated with mutations in ZFP57. *Nat. Genet.* **40**(8), 949–951.

Maher, E. R., and Reik, W. (2000). Beckwith-Wiedemann syndrome: Imprinting in clusters revisited. *J. Clin. Invest.* **105**(3), 247–252.

Maher, E. R., Brueton, L. A., Bowdin, S. C., Luharia, A., Cooper, W., Cole, T. R., Macdonald, F., Sampson, J. R., Barratt, C. L., Reik, W., *et al.* (2003). Beckwith-Wiedemann syndrome and assisted reproduction technology (ART). *J. Med. Genet.* **40**(1), 62–64.

Matsuura, T., Sutcliffe, J. S., Fang, P., Galjaard, R. J., Jiang, Y. H., Benton, C. S., Rommens, J. M., and Beaudet, A. L. (1997). De novo truncating mutations in E6-AP ubiquitin-protein ligase gene (UBE3A) in Angelman syndrome. *Nat. Genet.* **15**(1), 74–77.

McGrath, J., and Solter, D. (1984). Completion of mouse embryogenesis requires both the maternal and paternal genomes. *Cell* **37**(1), 179–183.

Meyer, E., Lim, D., Pasha, S., Tee, L. J., Rahman, F., Yates, J. R., Woods, C. G., Reik, W., and Maher, E. R. (2009). Germline mutation in NLRP2 (NALP2) in a familial imprinting disorder (Beckwith-Wiedemann Syndrome). *PLoS Genet.* **5**(3), e1000423.

Moore, T., and Haig, D. (1991). Genomic imprinting in mammalian development: A parental tug-of-war. *Trends Genet.* **7**(2), 45–49.

Murdoch, S., Djuric, U., Mazhar, B., Seoud, M., Khan, R., Kuick, R., Bagga, R., Kircheisen, R., Ao, A., Ratti, B., *et al.* (2006). Mutations in NALP7 cause recurrent hydatidiform moles and reproductive wastage in humans. *Nat. Genet.* **38**(3), 300–302.

Netchine, I., Rossignol, S., Dufourg, M. N., Azzi, S., Rousseau, A., Perin, L., Houang, M., Steunou, V., Esteva, B., Thibaud, N., *et al.* (2007). 11p15 imprinting center region 1 loss of methylation is a common and specific cause of typical Russell-Silver syndrome: Clinical scoring system and epigenetic-phenotypic correlations. *J. Clin. Endocrinol. Metab.* **92**(8), 3148–3154.

Neumann, B., Kubicka, P., and Barlow, D. P. (1995). Characteristics of imprinted genes. *Nat. Genet.* **9**(1), 12–13.

Niemitz, E. L., DeBaun, M. R., Fallon, J., Murakami, K., Kugoh, H., Oshimura, M., and Feinberg, A. P. (2004). Microdeletion of LIT1 in familial Beckwith-Wiedemann syndrome. *Am. J. Hum. Genet.* **75**(5), 844–849.

Ogawa, O., Eccles, M. R., Szeto, J., McNoe, L. A., Yun, K., Maw, M. A., Smith, P. J., and Reeve, A. E. (1993). Relaxation of insulin-like growth factor II gene imprinting implicated in Wilms' tumour. *Nature* **362**(6422), 749–751.

Ohama, K., Kajii, T., Okamoto, E., Fukuda, Y., Imaizumi, K., Tsukahara, M., Kobayashi, K., and Hagiwara, K. (1981). Dispermic origin of XY hydatidiform moles. *Nature* **292**(5823), 551–552.

Ohama, K., Nomura, K., Okamoto, E., Fukuda, Y., Ihara, T., and Fujiwara, A. (1985). Origin of immature teratoma of the ovary. *Am. J. Obstet. Gynecol.* **152**(7 Pt 1), 896–900.

Ohlsson, R., Cui, H., He, L., Pfeifer, S., Malmikumpu, H., Jiang, S., Feinberg, A. P., and Hedborg, F. (1999). Mosaic allelic insulin-like growth factor 2 expression patterns reveal a link between Wilms' tumorigenesis and epigenetic heterogeneity. *Cancer Res.* **59**(16), 3889–3892.

Ohno, S., Kinoshita, T., Ohno, Y., Minamoto, T., Suzuki, N., Inoue, M., and Suda, T. (2008). Expression of NLRP7 (PYPAF3, NALP7) protein in endometrial cancer tissues. *Anticancer Res.* **28**(4C), 2493–2497.

Okada, K., Hirota, E., Mizutani, Y., Fujioka, T., Shuin, T., Miki, T., Nakamura, Y., and Katagiri, T. (2004). Oncogenic role of NALP7 in testicular seminomas. *Cancer Sci.* **95**(12), 949–954.

Okamoto, K., Morison, I. M., Taniguchi, T., and Reeve, A. E. (1997). Epigenetic changes at the insulin-like growth factor II/H19 locus in developing kidney is an early event in Wilms tumorigenesis. *Proc. Natl. Acad. Sci. USA* **94**(10), 5367–5371.

Pagotto, U., Arzberger, T., Theodoropoulou, M., Grubler, Y., Pantaloni, C., Saeger, W., Losa, M., Journot, L., Stalla, G. K., and Spengler, D. (2000). The expression of the antiproliferative gene ZAC is lost or highly reduced in nonfunctioning pituitary adenomas. *Cancer Res.* **60**(24), 6794–6799.

Price, S. M., Stanhope, R., Garrett, C., Preece, M. A., and Trembath, R. C. (1999). The spectrum of Silver-Russell syndrome: A clinical and molecular genetic study and new diagnostic criteria. *J. Med. Genet.* **36**(11), 837–842.

Reik, W., and Maher, E. R. (1997). Imprinting in clusters: Lessons from Beckwith-Wiedemann syndrome. *Trends Genet.* **13**(8), 330–334.

Reik, W., and Walter, J. (2001). Genomic imprinting: Parental influence on the genome. *Nat. Rev. Genet.* **2**(1), 21–32.

Robinson, A. C., and Jones, W. G. (1990). Prader Willi syndrome and testicular tumour. *Clin. Oncol. (R. Coll. Radiol.)* **2**(2), 117.

Rossignol, S., Steunou, V., Chalas, C., Kerjean, A., Rigolet, M., Viegas-Pequignot, E., Jouannet, P., Le Bouc, Y., and Gicquel, C. (2006). The epigenetic imprinting defect of patients with Beckwith-Wiedemann syndrome born after assisted reproductive technology is not restricted to the 11p15 region. *J. Med. Genet.* **43**(12), 902–907.

Runte, M., Huttenhofer, A., Gross, S., Kiefmann, M., Horsthemke, B., and Buiting, K. (2001). The IC-SNURF-SNRPN transcript serves as a host for multiple small nucleolar RNA species and as an antisense RNA for UBE3A. *Hum. Mol. Genet.* **10**(23), 2687–2700.

Russell, A. (1954). A syndrome of intra-uterine dwarfism recognizable at birth with cranio-facial dysostosis, disproportionately short arms, and other anomalies (5 examples). *Proc. R. Soc. Med.* **47** (12), 1040–1044.

Sahoo, T., del Gaudio, D., German, J. R., Shinawi, M., Peters, S. U., Person, R. E., Garnica, A., Cheung, S. W., and Beaudet, A. L. (2008). Prader-Willi phenotype caused by paternal deficiency for the HBII-85 C/D box small nucleolar RNA cluster. *Nat. Genet.* **40**(6), 719–721.

Sakatani, T., Kaneda, A., Iacobuzio-Donahue, C. A., Carter, M. G., de Boom Witzel, S., Okano, H., Ko, M. S., Ohlsson, R., Longo, D. L., and Feinberg, A. P. (2005). Loss of imprinting of Igf2 alters intestinal maturation and tumorigenesis in mice. *Science* **307**(5717), 1976–1978.

Schonherr, N., Meyer, E., Roos, A., Schmidt, A., Wollmann, H. A., and Eggermann, T. (2007). The centromeric 11p15 imprinting centre is also involved in Silver-Russell syndrome. *J. Med. Genet.* **44**(1), 59–63.

Scott, R. H., Walker, L., Olsen, O. E., Levitt, G., Kenney, I., Maher, E., Owens, C. M., Pritchard-Jones, K., Craft, A., and Rahman, N. (2006). Surveillance for Wilms tumour in at-risk children: Pragmatic recommendations for best practice. *Arch. Dis. Child.* **91**(12), 995–999.

Scott, R. H., Douglas, J., Baskcomb, L., Huxter, N., Barker, K., Hanks, S., Craft, A., Gerrard, M., Kohler, J. A., Levitt, G. A., *et al.* (2008a). Constitutional 11p15 abnormalities, including heritable imprinting center mutations, cause nonsyndromic Wilms tumor. *Nat. Genet.* **40**(11), 1329–1334.

Scott, R. H., Douglas, J., Baskcomb, L., Nygren, A. O., Birch, J. M., Cole, T. R., Cormier-Daire, V., Eastwood, D. M., Garcia-Minaur, S., Lupunzina, P., *et al.* (2008b). Methylation-specific multiplex ligation-dependent probe amplification (MS-MLPA) robustly detects and distinguishes 11p15 abnormalities associated with overgrowth and growth retardation. *J. Med. Genet.* **45**(2), 106–113.

Sebire, N. J., Foskett, M., Fisher, R. A., Rees, H., Seckl, M., and Newlands, E. (2002). Risk of partial and complete hydatidiform molar pregnancy in relation to maternal age. *BJOG* **109**(1), 99–102.

Shuman, C., Steele, L., Fei, Y. L., Ray, P. N., Zackai, E., Parisi, M., Squire, J., and Weksberg, R. (2002). Paternal uniparental disomy of 11p15 is associated with isolated hemihyperplasia and expands the Wiedemann-Beckwith syndrome spectrum. *Am. J. Hum. Genet.* **71**(Suppl), 477.

Shuman, C., Smith, A. C., Steele, L., Ray, P. N., Clericuzio, C., Zackai, E., Parisi, M. A., Meadows, A. T., Kelly, T., Tichauer, D., *et al.* (2006). Constitutional UPD for chromosome 11p15 in individuals with isolated hemihyperplasia is associated with high tumor risk and occurs following assisted reproductive technologies. *Am. J. Med. Genet. A* **140A**(14), 1497–1503.

Silver, H. K., Kiyasu, W., George, J., and Deamer, W. C. (1953). Syndrome of congenital hemi-hypertrophy, shortness of stature, and elevated urinary gonadotropins. *Pediatrics* **12**(4), 368–376.

Smilinich, N. J., Day, C. D., Fitzpatrick, G. V., Caldwell, G. M., Lossie, A. C., Cooper, P. R., Smallwood, A. C., Joyce, J. A., Schofield, P. N., Reik, W., *et al.* (1999). A maternally methylated CpG island in KvLQT1 is associated with an antisense paternal transcript and loss of imprinting in Beckwith-Wiedemann syndrome. *Proc. Natl. Acad. Sci. USA* **96**(14), 8064–8069.

Sparago, A., Cerrato, F., Vernucci, M., Ferrero, G. B., Silengo, M. C., and Riccio, A. (2004). Microdeletions in the human H19 DMR result in loss of IGF2 imprinting and Beckwith-Wiedemann syndrome. *Nat. Genet.* **36**(9), 958–960.

Sparago, A., Russo, S., Cerrato, F., Ferraiuolo, S., Castorina, P., Selicorni, A., Schwienbacher, C., Negrini, M., Ferrero, G. B., Silengo, M. C., *et al.* (2007). Mechanisms causing imprinting defects in familial Beckwith-Wiedemann syndrome with Wilms' tumour. *Hum. Mol. Genet.* **16**(3), 254–264.

Surani, M. A., Barton, S. C., and Norris, M. L. (1984). Development of reconstituted mouse eggs suggests imprinting of the genome during gametogenesis. *Nature* **308**(5959), 548–550.

Szabo, P. E., Hubner, K., Scholer, H., and Mann, J. R. (2002). Allele-specific expression of imprinted genes in mouse migratory primordial germ cells. *Mech. Dev.* **115**(1–2), 157–160.

Takai, D., Gonzales, F. A., Tsai, Y. C., Thayer, M. J., and Jones, P. A. (2001). Large scale mapping of methylcytosines in CTCF-binding sites in the human H19 promoter and aberrant hypomethylation in human bladder cancer. *Hum. Mol. Genet.* **10**(23), 2619–2626.

Temple, I. K., and Shield, J. P. (2002). Transient neonatal diabetes, a disorder of imprinting. *J. Med. Genet.* **39**(12), 872–875.

Temple, I. K., Cockwell, A., Hassold, T., Pettay, D., and Jacobs, P. (1991). Maternal uniparental disomy for chromosome 14. *J. Med. Genet.* **28**(8), 511–514.

Temple, I. K., Gardner, R. J., Mackay, D. J., Barber, J. C., Robinson, D. O., and Shield, J. P. (2000). Transient neonatal diabetes: Widening the understanding of the etiopathogenesis of diabetes. *Diabetes* **49**(8), 1359–1366.

Temple, I. K., Shrubb, V., Lever, M., Bullman, H., and Mackay, D. J. (2007). Isolated imprinting mutation of the DLK1/GTL2 locus associated with a clinical presentation of maternal uniparental disomy of chromosome 14. *J. Med. Genet.* **44**(10), 637–640.

Wan, L. B., and Bartolomei, M. S. (2008). Regulation of imprinting in clusters: Noncoding RNAs versus insulators. *Adv. Genet.* **61**, 207–223.

Wang, C. M., Dixon, P. H., Decordova, S., Hodges, M. D., Sebire, N. J., Ozalp, S., Fallahian, M., Sensi, A., Ashrafi, F., Repiska, V., *et al.* (2009). Identification of 13 novel NLRP7 mutations in 20 families with recurrent hydatidiform mole; missense mutations cluster in the leucine-rich region. *J. Med. Genet.* **46**(8), 569–575.

Weiss, G. R., and Garnick, M. B. (1981). Testicular cancer in a Russell-Silver dwarf. *J. Urol.* **126**(6), 836–837.

Weksberg, R., Shuman, C., and Smith, A. C. (2005). Beckwith-Wiedemann syndrome. *Am. J. Med. Genet. C Semin. Med. Genet.* **137C**(1), 12–23.

Weksberg, R., Shuman, C., and Beckwith, J. B. (2010). Beckwith–Wiedemann syndrome. *Eur. J. Hum. Genet.* **18**(1), 8–14.

West, P. M. H., Love, D. R., Stapleton, P. M., and Winship, I. M. (2003). Paternal uniparental disomy in monozygotic twins discordant for hemihypertrophy. *J. Med. Genet.* **40**(3), 223–226.

Whittington, J., Holland, A., Webb, T., Butler, J., Clarke, D., and Boer, H. (2004). Academic underachievement by people with Prader-Willi syndrome. *J. Intellect. Disabil. Res.* **48**(Pt 2), 188–200.

Williams, C. A., Beaudet, A. L., Clayton-Smith, J., Knoll, J. H., Kyllerman, M., Laan, L. A., Magenis, R. E., Moncla, A., Schinzel, A. A., Summers, J. A., *et al.* (2006). Angelman syndrome 2005: Updated consensus for diagnostic criteria. *Am. J. Med. Genet. A* **140**(5), 413–418.

Yin, D., Xie, D., De Vos, S., Liu, G., Miller, C. W., Black, K. L., and Koeffler, H. P. (2004). Imprinting status of DLK1 gene in brain tumors and lymphomas. *Int. J. Oncol.* **24**(4), 1011–1015.

Yin, D., Xie, D., Sakajiri, S., Miller, C. W., Zhu, H., Popoviciu, M. L., Said, J. W., Black, K. L., and Koeffler, H. P. (2006). DLK1: Increased expression in gliomas and associated with oncogenic activities. *Oncogene* **25**(13), 1852–1861.

Yuan, E., Li, C. M., Yamashiro, D. J., Kandel, J., Thaker, H., Murty, V. V., and Tycko, B. (2005). Genomic profiling maps loss of heterozygosity and defines the timing and stage dependence of epigenetic and genetic events in Wilms' tumors. *Mol. Cancer Res.* **3**(9), 493–502.

Zhao, J., Dahle, D., Zhou, Y., Zhang, X., and Klibanski, A. (2005). Hypermethylation of the promoter region is associated with the loss of MEG3 gene expression in human pituitary tumors. *J. Clin. Endocrinol. Metab.* **90**(4), 2179–2186.

7

Epigenetic Codes in Stem Cells and Cancer Stem Cells

Yasuhiro Yamada and Akira Watanabe
Center for iPS Cell Research and Application (CiRA), Kyoto University, Shogoin, Sakyo-ku, Kyoto, Japan

ABSTRACT

Definition of stemness states that a stem cell population should be maintained over long periods of time, while generating all differentiated cell types of the corresponding tissues. Epigenetic regulation plays an important role in such process because the context of genome sequences is generally unchanged by

Advances in Genetics, Vol. 70
0065-2660/10 $35.00
DOI: 10.1016/S0065-2660(10)70007-9

differentiation process. Recent evidence indicates that an abnormal control of cellular differentiation is involved in the process of carcinogenesis [Hochedlinger, K., Yamada, Y., Beard, C., and Jaenisch, R. (2005). Ectopic expression of Oct-4 blocks progenitor-cell differentiation and causes dysplasia in epithelial tissues. Cell 121, 465-477]. Therefore, understanding how cellular differentiation is controlled would be useful for obtaining a better understanding of the mechanisms underlying carcinogenesis. In this chapter, we will describe recent advances in understanding the epigenetic codes that govern differentiation of stem cells, especially focusing on embryonic stem cells. We will also discuss the concept of cancer stem cells, in which the epigenetic regulations control differentiation of tumor cells and such regulations play a central role in the determination of whether a tumor cell is capable of tumor initiation or not. © 2010, Elsevier Inc.

I. EPIGENETIC CODES IN STEM CELLS

Embryonic stem (ES) cells, which are derived from the inner cell mass of blastocyst-stage embryos, have pluripotency, the potential to differentiate into any fetal and adult cell types both *in vitro* and *in vivo* (Jaenisch and Young, 2008; Keller, 2005). Accumulating evidence suggests that the pluripotency of ES cells is maintained through both direct gene regulation by transcription factors and epigenetic regulation such as DNA methylation and histone modifications (Niwa, 2007). Although the core regulatory circuitry of transcription factors involving Oct3/4, Nanog, and Sox2 plays a central role in the maintenance of pluripotency (Niwa, 2007), the contribution of epigenetic modifications in ES cells to the preservation of cell identity while maintaining the genome in a flexible state to allow for differentiation into multiple lineages has been only partially known. The past few years have seen considerable progress in the ability to characterize epigenetic modifications, such as DNA methylation and histone modifications on a global scale, and several interesting findings have begun to emerge with regard to the epigenetic features of ES cells (Fig. 7.1). These studies also revealed unique chromatin states in the lineage-committed cells. Changes of epigenetic modification alter chromatin density and the accessibility of cellular machinery to the DNA, thereby modulating the transcriptional potential of the underlying DNA sequence. Therefore, uncovering the mechanisms by which such unique chromatin influences the gene expression patterns in ES cells should provide valuable insight into the process of cell fate specification, thereby helping to elucidate the differentiation control of tissue-specific stem cells and, furthermore, clarifying the pathogenesis of various diseases including cancer.

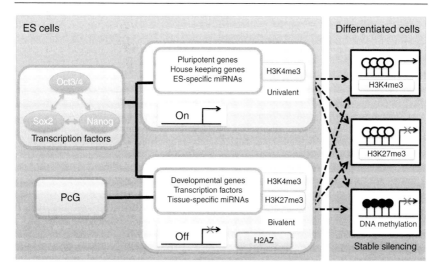

Figure 7.1. A schematic summary of the epigenetic modification and transcriptional regulation in ES cells and differentiated cells. Epigenetic modifications in ES cells are thought to play a role in the preservation of ES cell identity while maintaining the genome in a flexible state to allow for differentiation into multiple lineages. In contrast, a stable silencing mechanism is achieved in differentiated cells by DNA methylation, thus leading to the irreversible repression of the transcriptional state.

A. Epigenetic status in ES cells

1. DNA methylation

Recent comprehensive analyses using new technologies including next-generation sequencing have provided an overview of the epigenetic status including DNA methylation, one of the central epigenetic modifications, which play an essential role in the cellular process including genome regulation, development, and disease (Bestor, 2000; Bird, 2002; Reik and Murrell, 2000). Methylated DNA immunoprecipitation combined with microarrays (MeDIP-chip) or next-generation sequencing (MeDIP-seq) provides global maps of methylated DNA. Reduced representation bisulfite sequencing (RRBS), which covers most CpG islands, a representative sampling of conserved noncoding elements, and transposons generated DNA methylation maps for mouse ES cells at a single-nucleotide resolution (Meissner *et al.*, 2008). The DNA methylation maps show the unique characteristics of DNA methylation patterns in mouse ES cells. The CpG density is generally inversely correlated with the DNA methylation levels. Indeed, CpGs in regions of high CpG density and low density tend to be generally unmethylated and methylated, respectively, in mouse ES cells. However, the methylation maps showed that a subset of CpG shows opposite patterns in mouse ES cells. In contrast, histone methylation patterns (the presence of H3K4 methylation and the absence of H3K9 methylation) are nicely correlated with the DNA methylation patterns, thereby suggesting that the epigenetic modifications themselves, rather than the underlying genome sequence context, may directly determine the DNA methylation pattern.

The genome-wide, single-nucleotide-resolution maps of methylated cytosines in human ES cells also revealed unique DNA methylation patterns (Lister et al., 2009). It was thought that DNA methylation in vertebrates occurs almost exclusively in the context of CpG dinucleotides. However, the whole-genome bisulfite sequencing revealed that nearly one-quarter of all methylation in human ES cells is in a non-CG context (CNG and CNN), whereas a vast majority of methylated cytosine is observed at CpG dinucleotides in differentiated cells. Although the cause of such unique methylation patterns remains unknown, a periodicity of 8–10 bases that correspond to a single turn of the DNA helix (Cokus et al., 2008) is observed in such non-CG methylcytosine, thus suggesting that DNMT3A may be involved in catalyzing the methylation at the non-CG sites (Jia et al., 2007). Since the artificial acquisition of abnormal genomic methylation during in vitro manipulation has been reported (Meissner et al., 2008), it should also be considered that cultured ES cells might differ from the stem cells of the inner cell mass of blastocyst. Although previous studies have revealed an established functional role of non-CpG methylation in plants (Pavlopoulou and Kossida, 2007), it still remains unknown whether such non-CpG methylation plays a functional role in gene regulation in vertebrates. ES cells may use alternative mechanisms of DNA methylation to exhibit a unique type of gene regulation.

2. Histone modifications

The surface of nucleosomes is modified at many sites of histones by methylation, acetylation, and phosphorylation. There are over 60 different residues on histones where modifications have been detected by specific antibodies or by direct detection of modified residues with mass spectrometry (Kouzarides, 2007). Like the global maps of DNA methylation produced by RRBS and MeDIP-chip/seq, immunoprecipitation with antibody against a modified histone combined with either microarray or next-generation sequencing provides global maps of histone modifications (Liu et al., 2005; Pokholok et al., 2005). The general abundance of transcriptionally active chromatin marks, such as trimethylation of Lys 4 of histone H3 (H3K4me3) and acetylation of histone H4 (H4Ac), has been noted. In contrast, the histone marks associated with transcriptionally inactive genes, such as trimethylation of Lys 27 of histone H3 (H3K27me3), have been identified. Although scarce distribution of H3K27me3 is generally observed in

ES cells, enrichment of H3K27me3 is detectable in specific regions of the genome with transcriptionally active chromatin marks (Bernstein *et al.*, 2006; Mikkelsen *et al.*, 2007). These regions with both the repressive and active marks are designated as the "bivalent domains," and frequently overlap the regions in which the binding sites for pluripotency-associated transcription factors are clustered. Genes with bivalent domains are silenced in ES cells, thus suggesting that the repressive mark dominates over the transcription initiation mark and the binding of pluripotency-associated transcription factors. High-CpG-density promoters (HCPs) at housekeeping genes are enriched with the transcription initiation mark H3K4me3 (univalent domains) and tend to be highly expressed in ES cells, whereas those at developmental genes are enriched with both H3K4me3 and the repressive mark H3K27me3 (bivalent domains) and tend to be silent. Therefore, the bivalent domain is generally considered to play a central role in the maintenance of pluripotency in undifferentiated cells by repressing the expression of genes activated immediately after the induction of differentiation.

Evidence from biochemical and genetic screening shows that Polycomb group (PcG) proteins function in two distinct complexes, polycomb-repressing complex 1 (PRC1) and 2 (PRC2), and the core components are conserved from the fruit fly to humans (Ringrose and Paro, 2004). A genome-wide location analysis in mouse ES cells has revealed that PRC1 and PRC2 co-occupy many genes involved in developmental processes, but they silenced these genes in undifferentiated ES cells (Boyer *et al.*, 2006). Consistent with the notion that PRC2 is a key regulator mediating the gene repression marks H3K27me3 (Cao *et al.*, 2002; Czermin *et al.*, 2002; Kuzmichev *et al.*, 2002), such PcG-occupied developmental genes contain nucleosomes enriched with an H3K27me3 modification. Since DNA methylation is thought to be a stable silencing mechanism that is required for irreversibly locking-in the repressed transcriptional state (Jaenisch and Bird, 2003), gene repression that is associated with PcG proteins and the specific histone modifications with bivalent domains is therefore considered to play an important role in the maintenance of the genome in a flexible state to allow for differentiation into multiple lineages.

3. Histone variants

The replacement of conventional histones with specific variants is one of the mechanisms that regulate the chromatin structure (Henikoff and Ahmad, 2005). Histone variants are thought to play important roles in eukaryotes by influencing a wide range of DNA-mediated processes such as genome integrity, X inactivation, DNA repair, and gene regulation (Guillemette and Gaudreau, 2006; Hake and Allis, 2006). Genome-wide maps of Histone variant H2AZ in mouse ES cells revealed that H2AZ is enriched at a large set of silent developmental genes in a manner that is quite similar to that of Suz12, a component of PcG proteins (Creyghton et al., 2008). PcG proteins, which are associated with H3K27me3, are transcriptional repressors that play important roles in regulating developmental gene expression patterns by epigenetic modification of chromatin structure. Importantly, H2AZ depletion in ES cells by RNA interference results in an increased expression of the H2AZ target genes accompanied by the loss of PcG proteins from the promoters, thus suggesting a functional role of H2AZ in gene regulation in conjunction with PcG proteins. It is interesting to note that H2AZ incorporation is suggested to protect genes from DNA methylation in Arabidopsis (Zilberman et al., 2008). H2AZ incorporation associated with the absence of DNA methylation may involve mechanisms allowing developmental genes to remain silent, yet poised for activation in ES cells to maintain the genome in a flexible state. In addition to the involvement of aberrant DNA methylation patterns in most cancers, both H2AZ and PcG proteins have been linked to cancer progression because of their altered expression patterns in many types of cancers (Bracken and Helin, 2009; Dunican et al., 2002). Therefore, it will be of extreme interest to elucidate the mechanisms by which these regulators are recruited to genome sequences, and such clarification may make it possible to uncover the mechanisms to explain how altered patterns of epigenetic modifications are generated in cancer cells.

4. MicroRNAs

Several lines of evidence indicate that miRNAs contribute to the control of normal development through modulating the gene expression of a number of genes in a variety of mammalian cell types (Krek et al., 2005; Lewis et al., 2005; Lim et al., 2005). ES cells deficient in miRNA-processing enzymes show impaired differentiation and proliferation, thus suggesting that miRNAs play a role in the early differentiation of ES cells (Kanellopoulou et al., 2005; Murchison et al., 2005; Wang et al., 2007). A subset of miRNAs is preferentially expressed in ES cells and is downregulated as ES cells start to differentiate (Houbaviy et al., 2003; Suh et al., 2004). A core regulatory circuitry of transcription factors for the maintenance of ES cell pluripotency connect with these promoters for miRNAs that are preferentially expressed in ES cells (Marson et al., 2008). In contrast, a smaller set of miRNAs is repressed in ES cells and is selectively expressed in specific cell type upon early differentiation. Importantly, such silent miRNA genes are occupied by both key transcription factors for ES cell maintenance and PcG proteins. This fact suggests that PcG-mediated repression is involved in the silencing of such miRNA. Therefore, miRNA activation is, at least in part, controlled epigenetically, and the epigenetic control may play a role in the maintenance of a flexible state of the genome ready for rapid activation of the differentiation-related miRNAs upon the early differentiation of ES cells into multiple lineages.

B. Epigenetic changes as ES cells differentiate

The total level of methylated cytosine in mouse ES cells is similar to that in differentiated tissues such as the kidney and the liver *in vivo* (Biniszkiewicz *et al.*, 2002). This is consistent with the observation that the total CpG methylation levels are highly correlated with undifferentiated ES cells and ES cells-derived neural progenitor cells (NPCs) (Meissner *et al.*, 2008). However, the DNA methylation maps revealed that the methylation patterns of CpGs undergo extensive changes during cellular differentiation into NPCs, particularly in regulatory regions outside of core promoters (Meissner *et al.*, 2008). A subset of unmethylated or methylated CpGs become methylated or unmethylated, respectively, when undifferentiated ES cells start to differentiate, and these changes are strongly associated with changes in histone modification patterns. Low-CpG density promoters (LCPs) are generally associated with tissue-specific genes. A small subset of LCPs is enriched with H3K4me3 or H3K4me2 in undifferentiated ES cells, and interestingly none are enriched with the repressive mark H3K27me3 (Mikkelsen *et al.*, 2007). The loss or gain of H3K4 methylation occurs at LCPs when ES cells start to differentiate and such histone methylation is inversely correlated with the CpG methylation levels at these promoters (Meissner *et al.*, 2008). Such inverse correlation between DNA methylation levels and H3K4me2-enriched sites is detectable at highly conserved noncoding elements (HCNE). Significant changes of epigenetic modification are remarkably detectable at HCNEs upon differentiation of ES cells. Loss or gain of H3K4 methylation at HCNEs is accompanied by the gain or loss of CpG methylation levels upon differentiation of ES cells, respectively. Uncovering the functional significance of such altered pattern of genomic methylation at HCNEs may provide insights into the gene regulation underling cell fate specification and the pathogenesis of diseases.

RRBS also revealed that astrocytes obtained from *in vivo* NPCs harbor substantially less genomic methylation at HCPs than those obtained from *in vitro* (ES cells)-derived NPCs (Meissner *et al.*, 2008). Furthermore, in contrast to the observation of only 30 HCPs, which are closely associated with germ line-specific genes, to be methylated in the *in vivo*-derived astrocytes, the *in vitro*-derived astrocytes revealed hypermethylation at the additional 305 HCPs. These findings suggest that the acquisition of DNA methylation may occur depending on the past exposure of different circumstances, and environmental effects can provoke epigenetic responses. Although DNAs from various somatic tissues *in vivo* display few hypermethylated HCPs with the exception of specific loci such as germ line-specific genes, a gradual hypermethylation at a characteristic set of HCPs is obtained in the course of *in vitro* culture. Therefore, these findings suggest that *in vitro* manipulation can cause the artificial acquisition of aberrant genomic methylation patterns. In this context, these findings have raised concerns over the accuracy of cellular models generated *in vitro*. In particular, previous studies demonstrated that cancer cell lines harbor abnormal genomic methylation at a number of HCPs of tumor suppressor genes. Although a number of reviews predict that such abnormal CpG methylation plays a role in silencing of tumor suppressor genes and contribute to the carcinogenesis (Jones and Baylin, 2007), it is necessary to consider that a subset of aberrant methylation patterns in cancer cell lines might be attributable to the secondary effect of *in vitro* manipulations. Given that abnormal DNA methylation is considered to be a therapeutic target, it is important to distinguish secondary acquisition of an aberrant methylation pattern from bona fide abnormal CpG methylation, which thus contributes to carcinogenesis.

C. Epigenetic reprogramming of differentiated cells

Somatic cells can be reprogrammed to a pluripotent state through the ectopic expression of defined factors (Takahashi and Yamanaka, 2006; Takahashi *et al.*, 2007). The generation of such induced pluripotent stem (iPS) cells may provide an attractive source for regenerative medicine. In addition, derivation of patient-specific stem cells is expected to become a useful tool to examine the pathogenesis of diseases (Saha and Jaenisch, 2009). Takahashi and Yamanaka successfully reprogrammed mouse embryonic fibroblasts and adult fibroblasts into pluripotent ES-like cells in 2006 after viral-mediated transduction of the four transcription factors Oct3/4, Sox2, Myc, and Klf4 followed by selection for activation of the Fbx15. These cells were shown to be pluripotent by their ability to form teratomas. However, Fbx15-iPS cells were unable to generate live chimera, suggesting that they are not equivalent to ES cells. Activation of the endogenous Oct3/4 or Nanog genes for selection yields iPS (Oct3/4-iPS or Nanog-iPS) cells that can contribute to postnatal chimeras with the capacity of germ line transmission (Maherali *et al.*, 2007; Okita *et al.*, 2007; Wernig *et al.*, 2007). Therefore, Fbx15-iPS cells are partially reprogrammed cells, whereas Oct3/4-iPS or Nanog-iPS cells are fully reprogrammed cells.

Changes in epigenetic modifications, including DNA methylation and histone modifications, appear to happen progressively during the course of reprogramming, over an extended period of time (Mikkelsen *et al.*, 2008). The endogenous Oct3/4 and Nanog promoters of Fbx15-iPS cells remain to be methylated, which is accompanied by the lower expression levels of these endogenous genes, whereas these promoters in Oct3/4-iPS or Nanog-iPS cells were hypomethylated, suggesting that differences in the epigenetic state may account for the different extent of cellular reprogramming. It is important to understand epigenetic variation among each iPS clone by comparing epigenetic status of ES cells to control the quality of iPS cells because the development of cancer from incorrectly reprogrammed iPS cells is one of major concerns over the feasibility of iPS cells for regenerative medicine.

II. TISSUE-SPECIFIC STEM CELLS AND CANCER STEM CELLS

A minimal definition of stemness states that a stem cell population should be maintained over long periods of time, while generating all differentiated cell types of the corresponding tissues. Previous studies have suggested the presence of tissue-specific stem cells. The underlying mechanisms in which tissue-specific stem cells turn into differentiated cells remain largely unknown. However, epigenetic regulation may play an important role in such process because the context of genome sequences is generally unchanged by differentiation process. Many types of cancer retain the hierarchy of differentiation, which is observed in the originating tissues, implying the heterogeneity of cancer cells. The concept of a cancer stem cell model may provide clue to develop an effective treatment targeting a subset of cancer cells, which give rise to the fatal cancer mass (Dick, 2009). This review will first introduce tissue-specific stem cells identified in various tissues. Then, involvement of tissue-specific stem cells in carcinogenesis will be described and the concept of cancer stem cells will be discussed.

A. Tissue-specific stem cells

1. Hematopoietic stem cells (HSCs)

The first experimental evidence indicating the existence of stem cells was the discovery of a population of clonogenic bone marrow cells capable of generating myelo-erythroid colonies in the spleen of lethally irradiated hosts (Till and Mc, 1961). These colonies occasionally contained clonogenic cells that could be retransplanted into secondary lethally irradiated hosts and reconstitute the immune system. These were proposed to be hematopoietic stem/progenitor cells with the essential characteristics of self-renewal and differentiation potential for all types of blood cells (Passegue et al., 2003). The development of clonal assays for major hematopoietic lineages together with the availability of fluorescence-activated cell sorter (FACS) has enabled the purification of hematopoietic stem cells from mice and the enrichment of human hematopoietic stem cells. Hematopoietic stem/progenitor cells have been extensively characterized using several marker combinations, including Thy1.1, c-Kit, Sca-1, and Flk-2 (Christensen and Weissman, 2001; Morrison et al., 1995). About 1:5000 mouse bone marrow cells have long-term, multilineage, repopulating capability (LT-HSCs), whereas about 1:1000 have a more limited, short-term, multilineage repopulating capability (ST-HSCs) (Passegue et al., 2003). LT-HSCs self-renew for the life of the host, while the derivative ST-HSCs retain self-renewal activity for about 8 weeks and give rise to the self-renewing multipotent progenitors, which differentiate into two oligolineage-restricted progenitors, the common myeloid progenitors and the common lymphoid progenitors. Two such progenitor populations can give rise to all different lineages of the blood.

2. Epithelial stem cells

The location of epithelial tissue stem cells within each organ has been identified, at least in part, by taking advantage of the relative quiescence of stem cells (Blanpain et al., 2007). Label-retaining cells have been considered to be putative stem cells and have been found in discrete locations within these tissues (Bickenbach, 1981). Recent studies using the label-retaining method and stem cell-specific markers have identified the putative epithelial stem cells in various tissues, such as hair follicles, the intestine, and mammary glands (Cotsarelis et al., 1990; Potten et al., 2002; Zeps et al., 1998).

The hair follicle is formed during embryogenesis as an appendage of the epidermis. The stem cells in hair follicles are assumed to reside at a region known as the bulge. Indeed, a label-retaining experiment demonstrated that slow cycling cells localize in the bulge region (Tumbar et al., 2004). The multipotency of stem cells within the bulge was first suggested from transplantation experiments in which a dissected bulge region could be grafted onto immunodeficient mice, resulting in the differentiation of these bulge cells into the complete repertoire of skin epithelial cells. A previous study demonstrated that keratin-15 (K15)-expressing cells at the bulge give rise to the entire hair follicle (Morris et al., 2004). Finally, clonal analyses have demonstrated that single FACS-purified bulge cells could differentiate into multilineages, thus indicating such cells to be multipotent stem cells.

The epithelium of the small intestine is the most rapidly self-renewing cell type. The crypts have long been assumed to contain functional stem cells. Although there is agreement that crypts harbor four to six independent stem cells, two hypotheses exist with regard to their exact identity and location (Clevers, 2009). The first hypothesis is that the stem cells are in a quiescent state and located at position +4 relative to the crypt bottom (Potten, 1977). A previous study showed that Bmi1 expressing cells are located at the position +4 and genetically marked Bmi1+ cells can produce offspring representing the entire crypt epithelium, suggesting that Bmi1+ position +4 cells are stem cells (Sangiorgi and Capecchi, 2008). In contrast, the second hypothesis is based on the identification of the crypt base columnar (CBC) cells, small cycling cells located between the Paneth cells (Cheng and Leblond, 1974a,b). Lgr5 expression is specifically detectable in CBC cells and, like Bmi1+ cells, genetically marked Lgr5+ cells can differentiate into all epithelial cells in the whole crypt (Barker *et al.*, 2007). Interestingly, in contrast to the idea that stem cells are generally in slow-cycling state, Lgr5+ CBC cells proliferate actively. Although a recent study suggested that Bmi1+ cells and Lgr5 expressing CBC cells represent identical populations of stem cells (van der Flier *et al.*, 2009), the hierarchy of the two types of the stem cells still remains controversial.

The existence of multipotent stem cells in the mammary gland was demonstrated by experiments in which fragments of retrovirus-labeled mammary glands were transplanted into the fat pads. The transplanted mammary glands could be serially transplanted and outgrown clonally to create second-generation glands, indicating long-term self-renewal properties (Kordon and Smith, 1998). Mammary stem cells and their transient amplifying progeny are thought to reside in the terminal end buds along the basement membrane and have the capacity to form epithelial precursors that are committed to either a ductal or alveolar fate. A single myoepithelial cell could form a complete mammary gland structure, including the ductal and myoepithelial cell lineages when Lin-CD29hiCD24+ myoepithelial cells in the mammary glands were purified by FACS and individual cells were transplanted into fat pads (Shackleton *et al.*, 2006; Stingl *et al.*, 2006). The putative mammary stem cells appear to be slow-cycling, since they are able to retain nucleotide labeling over an extended chase period. They express a number of genes that are also upregulated in stem cells at the hair follicle.

3. Neural stem/progenitor cells

Previous studies have distinguished four main types of neural stem cells in the developing and adult brain (Conti and Cattaneo, 2010). Neurogenesis in mammals begins with induction of the neuroectoderm, which forms the neural plate and then folds to give rise to the neural tube. These structures are made up by a layer of so-called neuroepithelial progenitors (NEPs) (Gotz and Huttner, 2005). NEPs initially divide symmetrically at the ventricular zone, to increase the pool of progenitor cells but later asymmetrically to generate a progenitor and a daughter cell that migrates outward. NEPs are responsible for the first wave of neurogenesis, after which they give rise to both radial glia and basal progenitors. Radial glia are the main cell type in the developing brain, where they serve both as neural progenitors and as scaffolds for migrating newborn neurons. Radial glia express astroglial markers such as glial fibrillary acidic protein (GFAP) and their potential for differentiation is less broad than that of NEPs. Basal progenitors originate from NEPs at the early stages of development and at later stages form radial glia. Basal progenitors are predominantly present in the subventricular zone (SVZ) in the developing telencephalon, and are considered neurogenic transit-amplifying progenitors during restricted time periods. Adult progenitors are a population of multipotent neural cells mainly present in two specialized niches of the adult mammalian brain, the SVZ of the lateral ventricle wall and the subgranular zone of the dentate gyrus. They are derived directly from radial glia that convert into astrocytic-like neural stem cells in the postnatal brain, which maintain neurogenesis and gliogenesis throughout adult life.

B. Origin of cancer cells

The important question in cancer research is whether cancer arises through mutations in stem cells, or differentiated cells undergo malignant transformation through the process of dedifferentiation. A hallmark of all cancers is the capacity for unlimited self-renewal, which is also a defining characteristic of normal stem cells. Given the shared characteristics between cancer cells and stem cells, it has been proposed that cancers may be initiated by transforming events that take place in tissue-specific stem cells (Hochedlinger et al., 2005). Previous studies of acute myeloid leukemia (AML) showed that, for most AML subtypes, except for the M3 subtype, the only cells capable of transplanting AML in NOD/SCID mice have a CD34+ CD38− phenotype, which is similar to that of normal HSCs (Blair et al., 1997; Bonnet and Dick, 1997). These observations suggest that HSCs rather than committed progenitors are the target for leukemic transformation in most AML subtypes. In contrast, the M3 subtype of AML has been suggested to originate in progenitor populations. The M3-associated fusion gene is present in CD34− CD38+ cell populations but not in CD34+ CD38− cell populations, which are enriched with hematopoietic stem cells (Turhan et al., 1995). The observation contrarily suggests that the transformation process may involve a more differentiated cell type than hematopoietic stem cells. Further evidence indicates that cells lacking self-renewing activity, such as committed progenitors and mature cells, can also be the target cells for cancer development that came from analyses of leukemia-associated genes in the mouse. Transgenic mice expressing BCR-ABL from the promoter specific for committed cells such as neutrophils, monocytes, and their immediate progenitors, common myeloid progenitors and myelomonocytic progenitors, but not for hematopoietic stem cells develop a CML-like disease (Passegue et al., 2003). In contrast, recent study on a mouse model for intestinal tumorigenesis suggests the stem cell origin of cancer (Barker et al., 2009). Apc floxed mice were used to demonstrate that a stem cell-specific deletion of Apc is required for the transformation and subsequent tumor development of the intestinal epithelium. Genome-scale analyses of epigenetic modification also support the stem cell origin of cancer

(Ohm *et al.*, 2007; Schlesinger *et al.*, 2007; Widschwendter *et al.*, 2007). PcG proteins reversibly repress genes encoding transcription factors required for differentiation in stem cells. Previous findings revealed that such PcG target genes in stem cells are also targets of the cancer-specific promoter DNA hypermethylation. Stable silencing of such differentiation-associated genes by DNA methylation may expand stem-like undifferentiated cells, predisposing to them subsequent genetic/epigenetic alterations. The origin of cancer cells remains controversial and further analyses are necessary to determine which types of cancer require a stem cell origin for the carcinogenesis.

C. Heterogeneity of cancer cells and the cancer stem cell hypothesis

Tumors exhibit marked morphological heterogeneity (Dick, 2009; Visvader and Lindeman, 2008). For example, well-differentiated squamous cell carcinomas contain "cancer pearls" consisting of a red-colored keratin mass. Cancer cells are thought to produce this keratin mass through differentiation into keratin-containing cells. It is important that cancer cells near cancer pearls generally do not proliferate, suggesting that cancer cells stop their proliferation according to their differentiation. Such functional heterogeneity of tumor cells has been described in examinations of leukemia blast proliferation kinetics in human AML and acute lymphoblastic leukemia (ALL) patients (Dick, 2009). Importantly, the majority of leukemic blasts are postmitotic and need to be continuously replenished from a relatively small proliferative fraction. Such morphological and functional heterogeneity of cancer cells predicts the hierarchy of the distinct populations within a tumor, and the expected hierarchy is the origin of the concept of cancer stem cells.

The stochastic model for tumor initiation predicts that a tumor is biologically homogeneous and the behavior of the cancer cells is influenced by intrinsic or extrinsic factors, and that every tumor cell has equal tumor initiation potential (Dick, 2009). All tumor cells are equally sensitive to such stochastic intrinsic or extrinsic influences and tumor cells can convert from one state to another, representing the heterogeneity of tumor cells. In contrast, the hierarchy model (cancer stem cell model) predicts that a subset of tumor cells is biologically distinct, and they maintain themselves by their self-renewal activity while they generate progeny, which lack stem cell properties (Dick, 2009) (Fig. 7.2). Accumulating evidence suggests the cancer stem cell model, in which some cell fractions are enriched for tumor-initiating activity while others are completely devoid of the tumor-initiating cells, can be applied for a variety of cancers. It is important to note that the different properties between the tumorigenic stem cells and the nontumorigenic partially differentiated cells are attributable to the different epigenetic regulations and not to the genetic abnormalities. It has long been discussed as how the genetic abnormalities and epigenetic regulations play

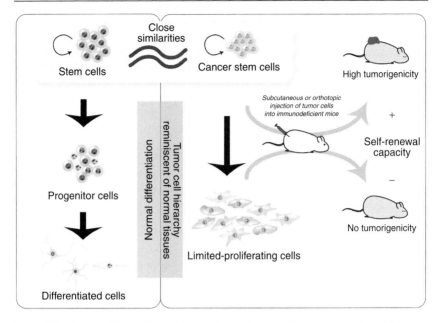

Figure 7.2. Tumor cell hierarchy reminiscent of normal tissue and cancer stem cell model. Tissue-specific stem cells and cancer stem cells may share biological characteristics. Cancer stem cells could be distinguished from other cancer cells by their tumorigenic potential, their ability to generate additional cancer stem cells, thus indicating self-renewal activity, and their ability to generate phenotypically diverse nontumorigenic cancer cells with limited proliferative activity.

a role in each stage of multistep carcinogenesis. The tumor stem cell hypothesis implies that the epigenetic regulation determines whether a tumor cell is capable of tumor initiation or not.

The recent consensus definition of a cancer stem cell is a cell within a tumor that possesses the capacity to self-renew and to generate the heterogeneous lineages of cancer cells that comprise the tumor. Early studies on AML support the existence of tumor cells with stem-cell-like properties (Dick, 2008). Tritium-labeling experiments showed that blood cancers contain a subset of immature cells with different cycling properties from majority of cancer cells. There are two proliferative fractions identified in patients with AML: a larger, fast cycling subset with a 24-h cell cycle time and a smaller, slow cycling fraction with a dormancy estimated to last from weeks to months. The slow cycling fraction is regarded as the leukemic stem cell fraction because they have similar cytokinetic properties to those of normal hematopoietic stem cells. Intriguingly, this slow cycling fraction is generating the fast cycling fraction, which is capable of generating postmitotic progeny, predicted the existence of a leukemic stem cell. Teratocarcinoma cells can spontaneously differentiate into mature quiescent cells (Pierce *et al.*, 1960), thus suggesting that a tumor retains a hierarchy defined by a maturation process in the same as that normal tissue development occurs. Recently, putative cancer stem cells have been identified also in solid tumors. A number of cell surface markers have been useful for the isolation of subsets enriched for tumor-initiating cells in solid tumors. The cell surface marker includes CD133 (also known as PROM1), CD44, CD24, epithelial cell adhesion molecule (epCAM, also known as epithelial specific antigen (ESA) and TACSTD1), THY1, and ATP-binding cassette B5 (ABCB5), as well as Hoechst33342 exclusion by the side population cells (Dick, 2008; Visvader and Lindeman, 2008).

D. Leukemias

The notion of tumorigenic leukemic stem cells has emerged based on several studies showing that only a small subset of leukemic cells was capable of extensive proliferation *in vitro* and *in vivo*. A previous study demonstrated that only 1 in 10,000 to 1 in 100 leukemic cells were able to form colonies *in vitro* in colony-forming assays and only 1–4% of the total number of leukemic cells transplanted *in vivo* could form spleen colonies (Bruce and Van Der Gaag, 1963). Since the read-out of the leukemic cells nicely mirrors the difference in clonogenicity among the normal hematopoietic cells, the clonogenic leukemic cells have been described as leukemic stem cells. Blair *et al.* (1997) and Bonnet and Dick (1997) clearly demonstrated that most of the leukemic cells are unable to proliferate extensively and only a small, defined subset of cells is constantly clonogenic. These studies identified and purified leukemic stem cells for human AML as Thy1-CD34+CD38− cells from various patient samples. Although these cells represent a small variable proportion of the total cells of the AML cells, they are the only cells capable of transferring AML to immunodeficient mice, thus suggesting that they are leukemic stem cells.

E. Breast cancers

The first solid malignant tumor from which tumor-initiating cells were identified and isolated was breast cancer (Al-Hajj *et al.*, 2003). Breast cancer cells contain a subset of CD44+ CD24−/low Lineage-cancer cells. Although CD44+ CD24−/low Lineage-cancer cells and alternative cancer cells exhibit similar cell cycle distribution, as few as 100 CD44+ CD24−/low Lineage-cancer cells are able to form tumors in immunodeficient mice, whereas thousands of alternative cancer cells fail to form tumors. The tumorigenic subpopulation can be serially passaged and can generate new tumors containing additional CD44+ CD24−/low Lineage-cancer cells with the phenotypically diverse mixed populations of nontumorigenic cells found in the initial tumor. Distinct markers combinations for isolating the breast tumor-initiating cells were also reported. A THY1+ CD24+ cancer cell population in MMTV-Wnt1 mammary tumors is highly enriched for tumorigenic activity in comparison to the alternative population (Cho et al., 2008). One in 200 such cells generates tumors, which are phenotypically similar to the original tumor, and these cells can be serially passaged, suggesting the self-renewal activity. A combination of $\beta 1$ integrin, CD24, and CD61 also enriches tumor cells with tumor-initiating capacity (Vaillant et al., 2008). A subpopulation of cells with $\beta 1$ integrin and CD24 expression are also suggested to be enriched for tumor-initiating cells in the Trp53-null mammary tumor model (Zhang et al., 2008).

F. Brain tumors

The CD133+ fraction ranges from 19% to 29% among glioblastomas, and from 6% to 21% among medulloblastomas (Singh et al., 2004). The CD133+ and CD133− cells have the same cytogenetic alterations, including abnormal karyotype and aneuploid pattern, thus suggesting that they are clonally derived (Singh et al., 2003). Importantly, engraftment of tumor cells into the brain of immunodeficient mice revealed that as few as 100 CD133+ cells are sufficient for the formation of human brain tumors in mice. In contrast, the injection of 50,000–100,000 of CD133− cells does not result in the formation of tumors. CD133+ tumor cells can form tumor spheres and differentiate in culture into tumor cells that phenotypically resemble the original tumor. A histological analysis of the CD133+ xenografts recapitulates the histology of the original tumor types. In addition, CD133 immunostaining of the CD133+ xenografts showed that the xenografts consist of a minority of CD133+ cells and a majority of CD133− cells, suggesting that a tumor hierarchy exists in which the CD133+ cells may generate CD133− tumor cells. Serial transplantation experiments from primary CD133+ xenografts revealed that 1000 CD133+ cells from the primary xenografts can constantly give rise to secondary xenografts with identical histology. These results indicate that CD133+ tumor cells have a self-renewal capacity in this population.

G. Colon cancers

Previous studies showed that the expression of CD133 is useful to enrich tumor-initiating cells in human colon cancers similar to brain tumors (O'Brien et al., 2007). The population of CD133+ cells in primary colon cancer tissues ranges from 1.8% to 24.5%. Interestingly, CD133+ colon cancer cells generate tumors when a relatively small number of cells are transplanted into immunodeficient mice, whereas CD133− cells did not possess the tumor-initiating capacity, showing the heterogeneous characteristics of colon cancer cells in terms of the ability of tumor formation. A limiting dilution analysis estimated that one tumor-initiating cell exists in 5.7×10^4 unfractionated tumor cells, whereas 262 CD133+ colon cancer cells contain one tumor-initiating cell, indicating that CD133+ selection can enrich the tumor-initiating cells 200-fold. The histology and degree of differentiation of such xenografts resembles the original tumors from which they were derived. Serial transplantation experiments from primary xenografts demonstrate that only CD133+ and not CD133− cells are capable of initiating tumor formation in serially transplanted secondary and tertiary mice, thus suggesting the self-renewal activity of CD133+ cells. These findings suggest that colon cancers contain a small fraction of cells that are capable of sustaining the tumor mass.

Mouse colonic adenomas arising from Lgr5+ stem cells contained a subset of Lgr5+ adenoma cells, suggesting that a stem/progenitor cell hierarchy is maintained in adenomas (Barker et al., 2009). Although genomic instability is one of the prevailing explanations of heterogeneity of cancer cells, the retaining hierarchy of stem/progenitor cells in such tumors suggests the involvement of epigenetic regulation underlying normal stem/progenitor hierarchy. It is therefore interesting to determine the diversity of epigenetic modifications among cancer cells of a single tumor, which may control a stem/progenitor cell hierarchy and may account for the different drug sensitivities for chemotherapy.

H. Skin tumors

Stem cells in the hair follicle have been demonstrated to reside in the bulge area. Bulge stem cells maintain follicular homeostasis and specifically express CD34. A recent study using DMBA-TPA-induced mouse model for skin carcinogenesis demonstrated that CD34+ cells sorted from cutaneous tumors are 100-fold more potent in initiating tumors than the bulk of tumor (Malanchi et al., 2008). Similar to other organs, a small number of skin tumor cells, namely as few as 1000 cells, induced tumors in syngeneic transplants, whereas 2.5×10^4 CD34− cells could not form such tumors. Importantly, the tumors derived from CD34+ cells exhibit the same hierarchical organization as the parent tumors and the CD34+ subset have self-renewing capacity, thus suggesting that subset of CD34+ tumor cells fulfills the definition of cancer stem cells.

I. Cancer stem cells as targets for therapy

Conventional anticancer approaches are directed predominantly at bulk tumor populations. Such strategies often have limited efficacy because of the intrinsic or acquired drug resistance and resistance to ionizing radiation. From the clinical point of view, the cancer stem cell concept has significant implications (Dean *et al.*, 2005). Increasing evidence reveals that a subset of cancer cells with tumorigenic activity have unique biological properties and represent a subpopulation of cells within cancers that is characterized by increased resistance to chemo- and radiotherapy, thus indicating that conventional anticancer approaches might frequently fail to eradicate the cell subset that initiates and perpetuates tumorigenesis (Visvader and Lindeman, 2008). A study in leukemia stem cells supports the concept that certain cancer stem cells may enter a quiescent state (Ishikawa *et al.*, 2007). The majority of human leukemia stem cells in xenotransplanted mice was found in the G0 phase of the cell cycle and was resistant to chemotherapy. CD133+ tumor-initiating cells in glioblastoma specimens or glioma xenografts irradiated *in vivo* are more resistant to ionizing irradiation than CD133− tumor cells (Bao *et al.*, 2006). Importantly, CD133+ brain tumor cells preferentially activate the DNA damage repair system than the bulk of the tumor, thereby conferring resistance to radiation treatment.

Uncovering the nature of cancer stem cells and controlling differentiation status of such cells will provide a novel efficient strategy to treat cancers. Although a number of studies suggest the involvement of cancer stem/tumor-initiating cells in a wide range of carcinogenesis, whether the cancer stem cell model is universal to all cancers is still controversial. Although the stem/progenitor hierarchy in cancer supports the concept of a cancer stem cell model, it is interesting to investigate the heterogeneity of cancers arising in organs in which the presence of tissue-specific stem cell remains controversial. Understanding of the nature of cancer stem cells might also provide important clue to understand the nature of tissue-specific stem cells.

References

Al-Hajj, M., Wicha, M. S., Benito-Hernandez, A., Morrison, S. J., and Clarke, M. F. (2003). Prospective identification of tumorigenic breast cancer cells. *Proc. Natl. Acad. Sci. USA* **100**, 3983–3988.

Bao, S., Wu, Q., McLendon, R. E., Hao, Y., Shi, Q., Hjelmeland, A. B., Dewhirst, M. W., Bigner, D. D., and Rich, J. N. (2006). Glioma stem cells promote radioresistance by preferential activation of the DNA damage response. *Nature* **444**, 756–760.

Barker, N., van Es, J. H., Kuipers, J., Kujala, P., van den Born, M., Cozijnsen, M., Haegebarth, A., Korving, J., Begthel, H., Peters, P. J., *et al.* (2007). Identification of stem cells in small intestine and colon by marker gene Lgr5. *Nature* **449**, 1003–1007.

Barker, N., Ridgway, R. A., van Es, J. H., van de Wetering, M., Begthel, H., van den Born, M., Danenberg, E., Clarke, A. R., Sansom, O. J., and Clevers, H. (2009). Crypt stem cells as the cells-of-origin of intestinal cancer. *Nature* **457**, 608–611.

Bernstein, B. E., Mikkelsen, T. S., Xie, X., Kamal, M., Huebert, D. J., Cuff, J., Fry, B., Meissner, A., Wernig, M., Plath, K., *et al.* (2006). A bivalent chromatin structure marks key developmental genes in embryonic stem cells. *Cell* **125**, 315–326.

Bestor, T. H. (2000). The DNA methyltransferases of mammals. *Hum. Mol. Genet.* **9**, 2395–2402.

Bickenbach, J. R. (1981). Identification and behavior of label-retaining cells in oral mucosa and skin. *J. Dent. Res.* **60**(Spec. No. C), 1611–1620.

Biniszkiewicz, D., Gribnau, J., Ramsahoye, B., Gaudet, F., Eggan, K., Humpherys, D., Mastrangelo, M. A., Jun, Z., Walter, J., and Jaenisch, R. (2002). Dnmt1 overexpression causes genomic hypermethylation, loss of imprinting, and embryonic lethality. *Mol. Cell. Biol.* **22**, 2124–2135.

Bird, A. (2002). DNA methylation patterns and epigenetic memory. *Genes Dev.* **16**, 6–21.

Blair, A., Hogge, D. E., Ailles, L. E., Lansdorp, P. M., and Sutherland, H. J. (1997). Lack of expression of Thy-1 (CD90) on acute myeloid leukemia cells with long-term proliferative ability in vitro and in vivo. *Blood* **89**, 3104–3112.

Blanpain, C., Horsley, V., and Fuchs, E. (2007). Epithelial stem cells: Turning over new leaves. *Cell* **128**, 445–458.

Bonnet, D., and Dick, J. E. (1997). Human acute myeloid leukemia is organized as a hierarchy that originates from a primitive hematopoietic cell. *Nat. Med.* **3**, 730–737.

Boyer, L. A., Plath, K., Zeitlinger, J., Brambrink, T., Medeiros, L. A., Lee, T. I., Levine, S. S., Wernig, M., Tajonar, A., Ray, M. K., *et al.* (2006). Polycomb complexes repress developmental regulators in murine embryonic stem cells. *Nature* **441**, 349–353.

Bracken, A. P., and Helin, K. (2009). Polycomb group proteins: Navigators of lineage pathways led astray in cancer. *Nat. Rev. Cancer* **9**, 773–784.

Bruce, W. R., and Van Der Gaag, H. (1963). A quantitative assay for the number of murine lymphoma cells capable of proliferation in vivo. *Nature* **199**, 79–80.

Cao, R., Wang, L., Wang, H., Xia, L., Erdjument-Bromage, H., Tempst, P., Jones, R. S., and Zhang, Y. (2002). Role of histone H3 lysine 27 methylation in Polycomb-group silencing. *Science* **298**, 1039–1043.

Cheng, H., and Leblond, C. P. (1974a). Origin, differentiation and renewal of the four main epithelial cell types in the mouse small intestine. I. Columnar cell. *Am. J. Anat.* **141**, 461–479.

Cheng, H., and Leblond, C. P. (1974b). Origin, differentiation and renewal of the four main epithelial cell types in the mouse small intestine. V. Unitarian Theory of the origin of the four epithelial cell types. *Am. J. Anat.* **141**, 537–561.

Cho, R. W., Wang, X., Diehn, M., Shedden, K., Chen, G. Y., Sherlock, G., Gurney, A., Lewicki, J., and Clarke, M. F. (2008). Isolation and molecular characterization of cancer stem cells in MMTV-Wnt-1 murine breast tumors. *Stem Cells* **26**, 364–371.

Christensen, J. L., and Weissman, I. L. (2001). Flk-2 is a marker in hematopoietic stem cell differentiation: A simple method to isolate long-term stem cells. *Proc. Natl. Acad. Sci. USA* **98**, 14541–14546.

Clevers, H. (2009). Searching for adult stem cells in the intestine. *EMBO Mol. Med.* **1**, 255–259.

Cokus, S. J., Feng, S., Zhang, X., Chen, Z., Merriman, B., Haudenschild, C. D., Pradhan, S., Nelson, S. F., Pellegrini, M., and Jacobsen, S. E. (2008). Shotgun bisulphite sequencing of the Arabidopsis genome reveals DNA methylation patterning. *Nature* **452**, 215–219.

Conti, L., and Cattaneo, E. (2010). Neural stem cell systems: Physiological players or in vitro entities? *Nat. Rev. Neurosci.* **11**, 176–187.

Cotsarelis, G., Sun, T. T., and Lavker, R. M. (1990). Label-retaining cells reside in the bulge area of pilosebaceous unit: Implications for follicular stem cells, hair cycle, and skin carcinogenesis. *Cell* **61**, 1329–1337.

Creyghton, M. P., Markoulaki, S., Levine, S. S., Hanna, J., Lodato, M. A., Sha, K., Young, R. A., Jaenisch, R., and Boyer, L. A. (2008). H2AZ is enriched at polycomb complex target genes in ES cells and is necessary for lineage commitment. *Cell* **135**, 649–661.

Czermin, B., Melfi, R., McCabe, D., Seitz, V., Imhof, A., and Pirrotta, V. (2002). Drosophila enhancer of Zeste/ESC complexes have a histone H3 methyltransferase activity that marks chromosomal Polycomb sites. *Cell* **111**, 185–196.

Dean, M., Fojo, T., and Bates, S. (2005). Tumour stem cells and drug resistance. *Nat. Rev. Cancer* **5**, 275–284.

Dick, J. E. (2008). Stem cell concepts renew cancer research. *Blood* **112**, 4793–4807.

Dick, J. E. (2009). Looking ahead in cancer stem cell research. *Nat. Biotechnol.* **27**, 44–46.

Dunican, D. S., McWilliam, P., Tighe, O., Parle-McDermott, A., and Croke, D. T. (2002). Gene expression differences between the microsatellite instability (MIN) and chromosomal instability (CIN) phenotypes in colorectal cancer revealed by high-density cDNA array hybridization. *Oncogene* **21**, 3253–3257.

Gotz, M., and Huttner, W. B. (2005). The cell biology of neurogenesis. *Nat. Rev. Mol. Cell Biol.* **6**, 777–788.

Guillemette, B., and Gaudreau, L. (2006). Reuniting the contrasting functions of H2A.Z. *Biochem. Cell Biol.* **84**, 528–535.

Hake, S. B., and Allis, C. D. (2006). Histone H3 variants and their potential role in indexing mammalian genomes: The "H3 barcode hypothesis". *Proc. Natl. Acad. Sci. USA* **103**, 6428–6435.

Henikoff, S., and Ahmad, K. (2005). Assembly of variant histones into chromatin. *Annu. Rev. Cell Dev. Biol.* **21**, 133–153.

Hochedlinger, K., Yamada, Y., Beard, C., and Jaenisch, R. (2005). Ectopic expression of Oct-4 blocks progenitor-cell differentiation and causes dysplasia in epithelial tissues. *Cell* **121**, 465–477.

Houbaviy, H. B., Murray, M. F., and Sharp, P. A. (2003). Embryonic stem cell-specific MicroRNAs. *Dev. Cell* **5**, 351–358.

Ishikawa, F., Yoshida, S., Saito, Y., Hijikata, A., Kitamura, H., Tanaka, S., Nakamura, R., Tanaka, T., Tomiyama, H., Saito, N., *et al.* (2007). Chemotherapy-resistant human AML stem cells home to and engraft within the bone-marrow endosteal region. *Nat. Biotechnol.* **25**, 1315–1321.

Jaenisch, R., and Bird, A. (2003). Epigenetic regulation of gene expression: How the genome integrates intrinsic and environmental signals. *Nat. Genet.* **33**(Suppl.), 245–254.

Jaenisch, R., and Young, R. (2008). Stem cells, the molecular circuitry of pluripotency and nuclear reprogramming. *Cell* **132**, 567–582.

Jia, D., Jurkowska, R. Z., Zhang, X., Jeltsch, A., and Cheng, X. (2007). Structure of Dnmt3a bound to Dnmt3L suggests a model for de novo DNA methylation. *Nature* **449**, 248–251.

Jones, P. A., and Baylin, S. B. (2007). The epigenomics of cancer. *Cell* **128**, 683–692.

Kanellopoulou, C., Muljo, S. A., Kung, A. L., Ganesan, S., Drapkin, R., Jenuwein, T., Livingston, D. M., and Rajewsky, K. (2005). Dicer-deficient mouse embryonic stem cells are defective in differentiation and centromeric silencing. *Genes Dev.* **19**, 489–501.

Keller, G. (2005). Embryonic stem cell differentiation: Emergence of a new era in biology and medicine. *Genes Dev.* **19**, 1129–1155.

Kordon, E. C., and Smith, G. H. (1998). An entire functional mammary gland may comprise the progeny from a single cell. *Development* **125**, 1921–1930.

Kouzarides, T. (2007). Chromatin modifications and their function. *Cell* **128**, 693–705.

Krek, A., Grun, D., Poy, M. N., Wolf, R., Rosenberg, L., Epstein, E. J., MacMenamin, P., da Piedade, I., Gunsalus, K. C., Stoffel, M., *et al.* (2005). Combinatorial microRNA target predictions. *Nat. Genet.* **37**, 495–500.

Kuzmichev, A., Nishioka, K., Erdjument-Bromage, H., Tempst, P., and Reinberg, D. (2002). Histone methyltransferase activity associated with a human multiprotein complex containing the Enhancer of Zeste protein. *Genes Dev.* **16**, 2893–2905.

Lewis, B. P., Burge, C. B., and Bartel, D. P. (2005). Conserved seed pairing, often flanked by adenosines, indicates that thousands of human genes are microRNA targets. *Cell* **120**, 15–20.

Lim, L. P., Lau, N. C., Garrett-Engele, P., Grimson, A., Schelter, J. M., Castle, J., Bartel, D. P., Linsley, P. S., and Johnson, J. M. (2005). Microarray analysis shows that some microRNAs downregulate large numbers of target mRNAs. *Nature* **433**, 769–773.

Lister, R., Pelizzola, M., Dowen, R. H., Hawkins, R. D., Hon, G., Tonti-Filippini, J., Nery, J. R., Lee, L., Ye, Z., Ngo, Q. M., et al. (2009). Human DNA methylomes at base resolution show widespread epigenomic differences. Nature **462**, 315–322.

Liu, C. L., Kaplan, T., Kim, M., Buratowski, S., Schreiber, S. L., Friedman, N., and Rando, O. J. (2005). Single-nucleosome mapping of histone modifications in S. cerevisiae. PLoS Biol. **3**, e328.

Maherali, N., Sridharan, R., Xie, W., Utikal, J., Eminli, S., Arnold, K., Stadtfeld, M., Yachechko, R., Tchieu, J., Jaenisch, R., et al. (2007). Directly reprogrammed fibroblasts show global epigenetic remodeling and widespread tissue contribution. Cell Stem Cell **1**, 55–70.

Malanchi, I., Peinado, H., Kassen, D., Hussenet, T., Metzger, D., Chambon, P., Huber, M., Hohl, D., Cano, A., Birchmeier, W., et al. (2008). Cutaneous cancer stem cell maintenance is dependent on beta-catenin signalling. Nature **452**, 650–653.

Marson, A., Levine, S. S., Cole, M. F., Frampton, G. M., Brambrink, T., Johnstone, S., Guenther, M. G., Johnston, W. K., Wernig, M., Newman, J., et al. (2008). Connecting microRNA genes to the core transcriptional regulatory circuitry of embryonic stem cells. Cell **134**, 521–533.

Meissner, A., Mikkelsen, T. S., Gu, H., Wernig, M., Hanna, J., Sivachenko, A., Zhang, X., Bernstein, B. E., Nusbaum, C., Jaffe, D. B., et al. (2008). Genome-scale DNA methylation maps of pluripotent and differentiated cells. Nature **454**, 766–770.

Mikkelsen, T. S., Ku, M., Jaffe, D. B., Issac, B., Lieberman, E., Giannoukos, G., Alvarez, P., Brockman, W., Kim, T. K., Koche, R. P., et al. (2007). Genome-wide maps of chromatin state in pluripotent and lineage-committed cells. Nature **448**, 553–560.

Mikkelsen, T. S., Hanna, J., Zhang, X., Ku, M., Wernig, M., Schorderet, P., Bernstein, B. E., Jaenisch, R., Lander, E. S., and Meissner, A. (2008). Dissecting direct reprogramming through integrative genomic analysis. Nature **454**, 49–55.

Morris, R. J., Liu, Y., Marles, L., Yang, Z., Trempus, C., Li, S., Lin, J. S., Sawicki, J. A., and Cotsarelis, G. (2004). Capturing and profiling adult hair follicle stem cells. Nat. Biotechnol. **22**, 411–417.

Morrison, S. J., Uchida, N., and Weissman, I. L. (1995). The biology of hematopoietic stem cells. Annu. Rev. Cell Dev. Biol. **11**, 35–71.

Murchison, E. P., Partridge, J. F., Tam, O. H., Cheloufi, S., and Hannon, G. J. (2005). Characterization of Dicer-deficient murine embryonic stem cells. Proc. Natl. Acad. Sci. USA **102**, 12135–12140.

Niwa, H. (2007). How is pluripotency determined and maintained? Development **134**, 635–646.

O'Brien, C. A., Pollett, A., Gallinger, S., and Dick, J. E. (2007). A human colon cancer cell capable of initiating tumour growth in immunodeficient mice. Nature **445**, 106–110.

Ohm, J. E., McGarvey, K. M., Yu, X., Cheng, L., Schuebel, K. E., Cope, L., Mohammad, H. P., Chen, W., Daniel, V. C., Yu, W., et al. (2007). A stem cell-like chromatin pattern may predispose tumor suppressor genes to DNA hypermethylation and heritable silencing. Nat. Genet. **39**, 237–242.

Okita, K., Ichisaka, T., and Yamanaka, S. (2007). Generation of germline-competent induced pluripotent stem cells. Nature **448**, 313–317.

Passegue, E., Jamieson, C. H., Ailles, L. E., and Weissman, I. L. (2003). Normal and leukemic hematopoiesis: Are leukemias a stem cell disorder or a reacquisition of stem cell characteristics? Proc. Natl. Acad. Sci. USA **100**(Suppl 1), 11842–11849.

Pavlopoulou, A., and Kossida, S. (2007). Plant cytosine-5 DNA methyltransferases: Structure, function, and molecular evolution. Genomics **90**, 530–541.

Pierce, G. B., Jr., Dixon, F. J., Jr., and Verney, E. L. (1960). Teratocarcinogenic and tissue-forming potentials of the cell types comprising neoplastic embryoid bodies. Lab. Invest. **9**, 583–602.

Pokholok, D. K., Harbison, C. T., Levine, S., Cole, M., Hannett, N. M., Lee, T. I., Bell, G. W., Walker, K., Rolfe, P. A., Herbolsheimer, E., et al. (2005). Genome-wide map of nucleosome acetylation and methylation in yeast. Cell **122**, 517–527.

Potten, C. S. (1977). Extreme sensitivity of some intestinal crypt cells to X and gamma irradiation. *Nature* **269,** 518–521.

Potten, C. S., Owen, G., and Booth, D. (2002). Intestinal stem cells protect their genome by selective segregation of template DNA strands. *J. Cell Sci.* **115,** 2381–2388.

Reik, W., and Murrell, A. (2000). Genomic imprinting. Silence across the border. *Nature* **405,** 408–409.

Ringrose, L., and Paro, R. (2004). Epigenetic regulation of cellular memory by the Polycomb and Trithorax group proteins. *Annu. Rev. Genet.* **38,** 413–443.

Saha, K., and Jaenisch, R. (2009). Technical challenges in using human induced pluripotent stem cells to model disease. *Cell Stem Cell* **5,** 584–595.

Sangiorgi, E., and Capecchi, M. R. (2008). Bmi1 is expressed in vivo in intestinal stem cells. *Nat. Genet.* **40,** 915–920.

Schlesinger, Y., Straussman, R., Keshet, I., Farkash, S., Hecht, M., Zimmerman, J., Eden, E., Yakhini, Z., Ben-Shushan, E., Reubinoff, B. E., *et al.* (2007). Polycomb-mediated methylation on Lys27 of histone H3 pre-marks genes for de novo methylation in cancer. *Nat. Genet.* **39,** 232–236.

Shackleton, M., Vaillant, F., Simpson, K. J., Stingl, J., Smyth, G. K., Asselin-Labat, M. L., Wu, L., Lindeman, G. J., and Visvader, J. E. (2006). Generation of a functional mammary gland from a single stem cell. *Nature* **439,** 84–88.

Singh, S. K., Clarke, I. D., Terasaki, M., Bonn, V. E., Hawkins, C., Squire, J., and Dirks, P. B. (2003). Identification of a cancer stem cell in human brain tumors. *Cancer Res.* **63,** 5821–5828.

Singh, S. K., Hawkins, C., Clarke, I. D., Squire, J. A., Bayani, J., Hide, T., Henkelman, R. M., Cusimano, M. D., and Dirks, P. B. (2004). Identification of human brain tumour initiating cells. *Nature* **432,** 396–401.

Stingl, J., Eirew, P., Ricketson, I., Shackleton, M., Vaillant, F., Choi, D., Li, H. I., and Eaves, C. J. (2006). Purification and unique properties of mammary epithelial stem cells. *Nature* **439,** 993–997.

Suh, M. R., Lee, Y., Kim, J. Y., Kim, S. K., Moon, S. H., Lee, J. Y., Cha, K. Y., Chung, H. M., Yoon, H. S., Moon, S. Y., *et al.* (2004). Human embryonic stem cells express a unique set of microRNAs. *Dev. Biol.* **270,** 488–498.

Takahashi, K., and Yamanaka, S. (2006). Induction of pluripotent stem cells from mouse embryonic and adult fibroblast cultures by defined factors. *Cell* **126,** 663–676.

Takahashi, K., Tanabe, K., Ohnuki, M., Narita, M., Ichisaka, T., Tomoda, K., and Yamanaka, S. (2007). Induction of pluripotent stem cells from adult human fibroblasts by defined factors. *Cell* **131,** 861–872.

Till, J. E., and Mc, C. E. (1961). A direct measurement of the radiation sensitivity of normal mouse bone marrow cells. *Radiat. Res.* **14,** 213–222.

Tumbar, T., Guasch, G., Greco, V., Blanpain, C., Lowry, W. E., Rendl, M., and Fuchs, E. (2004). Defining the epithelial stem cell niche in skin. *Science* **303,** 359–363.

Turhan, A. G., Lemoine, F. M., Debert, C., Bonnet, M. L., Baillou, C., Picard, F., Macintyre, E. A., and Varet, B. (1995). Highly purified primitive hematopoietic stem cells are PML-RARA negative and generate nonclonal progenitors in acute promyelocytic leukemia. *Blood* **85,** 2154–2161.

Vaillant, F., Asselin-Labat, M. L., Shackleton, M., Forrest, N. C., Lindeman, G. J., and Visvader, J. E. (2008). The mammary progenitor marker CD61/beta3 integrin identifies cancer stem cells in mouse models of mammary tumorigenesis. *Cancer Res.* **68,** 7711–7717.

van der Flier, L. G., van Gijn, M. E., Hatzis, P., Kujala, P., Haegebarth, A., Stange, D. E., Begthel, H., van den Born, M., Guryev, V., Oving, I., *et al.* (2009). Transcription factor achaete scute-like 2 controls intestinal stem cell fate. *Cell* **136,** 903–912.

Visvader, J. E., and Lindeman, G. J. (2008). Cancer stem cells in solid tumours: Accumulating evidence and unresolved questions. *Nat. Rev. Cancer* **8**, 755–768.

Wang, Y., Medvid, R., Melton, C., Jaenisch, R., and Blelloch, R. (2007). DGCR8 is essential for microRNA biogenesis and silencing of embryonic stem cell self-renewal. *Nat. Genet.* **39**, 380–385.

Wernig, M., Meissner, A., Foreman, R., Brambrink, T., Ku, M., Hochedlinger, K., Bernstein, B. E., and Jaenisch, R. (2007). In vitro reprogramming of fibroblasts into a pluripotent ES-cell-like state. *Nature* **448**, 318–324.

Widschwendter, M., Fiegl, H., Egle, D., Mueller-Holzner, E., Spizzo, G., Marth, C., Weisenberger, D. J., Campan, M., Young, J., Jacobs, I., *et al.* (2007). Epigenetic stem cell signature in cancer. *Nat. Genet.* **39**, 157–158.

Zeps, N., Bentel, J. M., Papadimitriou, J. M., D'Antuono, M. F., and Dawkins, H. J. (1998). Estrogen receptor-negative epithelial cells in mouse mammary gland development and growth. *Differentiation* **62**, 221–226.

Zhang, M., Behbod, F., Atkinson, R. L., Landis, M. D., Kittrell, F., Edwards, D., Medina, D., Tsimelzon, A., Hilsenbeck, S., Green, J. E., *et al.* (2008). Identification of tumor-initiating cells in a p53-null mouse model of breast cancer. *Cancer Res.* **68**, 4674–4682.

Zilberman, D., Coleman-Derr, D., Ballinger, T., and Henikoff, S. (2008). Histone H2A.Z and DNA methylation are mutually antagonistic chromatin marks. *Nature* **456**, 125–129.

8 Inheritance of Epigenetic Aberrations (Constitutional Epimutations) in Cancer Susceptibility

Megan P. Hitchins

Adult Cancer Program, Lowy Cancer Research Centre, Prince of Wales Clinical School, University of New South Wales, Sydney, Australia

Advances in Genetics, Vol. 70   0065-2660/10 $35.00
Copyright 2010, Elsevier Inc. All rights reserved.   DOI: 10.1016/S0065-2660(10)70008-0

ABSTRACT

The pathogenic role for heritable mutations in the DNA sequence of tumor suppressor and DNA repair genes has been well established in familial cancer syndromes. These germ line mutations confer a high risk of developing particular types of cancer, according to the gene affected, at a young age of onset when compared to sporadically arising cancers of a similar type. The widespread role for epigenetic dysregulation in the development and progression of sporadic cancers is also well recognized. However, it has only become apparent in recent years that epigenetic aberrations can also occur constitutionally to confer a similar cancer phenotype as a genetic mutation within the same gene. These epigenetic errors are termed "constitutional epimutations" and are characterized by promoter methylation and transcriptional silencing of a single allele of the gene in normal somatic tissues in the absence of a sequence mutation within the affected locus. This is best exemplified in Lynch syndrome, which is an autosomal dominant cancer susceptibility syndrome characterized by the early development of colo-rectal, uterine, and additional cancers exhibiting microsatellite instability due to impaired mismatch repair. Lynch syndrome is usually caused by heterozygous loss-of-function germ line mutations of the mismatch repair genes, namely *MLH1*, *MSH2*, *MSH6*, and *PMS2*. Tumors develop following an acquired somatic loss of the remaining functional allele. However, a subset of Lynch syndrome cases without genetic mutations instead has a constitutional epimutation of *MLH1* or *MSH2*. These epimutations are associated with distinct patterns of inheritance depending on the nature of the mechanisms underlying them. © 2010, Elsevier Inc.

I. INTRODUCTION

An epimutation describes an epigenetic aberration that alters gene activity to confer a phenotype without any change within the DNA code of the affected gene. The first observation of a naturally occurring epimutation was in 1999 in the toadflax plant (*Linaria vulgaris*; Cubas *et al.*, 1999), which caused a floral pheno-type first described as the "mutant peloric" pattern of radial petal symmetry by the famous botanist Linnaeus in 1749 (Fig. 8.1). In the same year, the first clear example of the inheritance of an altered epigenetic state in mammals was described at the agouti locus in mice, giving rise to a phenotype of yellow coat color, obesity, diabetes, and increased cancer risk (Morgan *et al.*, 1999). The first description of what subsequently became known as a "constitutional epimuta-tion" in humans came 3 years later, with aberrant methylation of the *MLH1* mismatch repair gene detected in peripheral blood, which predisposed to the development of cancer (Fig. 8.2; Gazzoli *et al.*, 2002). By this time, the genetic basis for a number of familial forms of cancer had been well established. A number of autosomal dominant familial cancer syndromes caused by heterozygous germ line mutations have been described, of which Lynch syndrome or hereditary non-polyposis colorectal cancer (HNPCC) is the most common (Lynch, 1999). Mutations in autosomal dominant cancer syndromes give rise to a high risk of cancer development, usually at an early age of onset. While the predisposition to cancer is transmitted in an autosomal dominant pattern, in order for cancer to develop, loss of both copies of the relevant gene is typically required in the

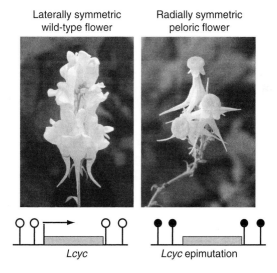

Figure 8.1. Epimutation of the *Lcyc* gene in toadflax causes a change in floral shape. Left, wild-type flowers of toadflax showing lateral symmetry and dorso-ventral asymmetry, and beneath, the *Lcyc* gene is unmethylated and normally expressed. Right, the peloric floral pattern showing radial symmetry of five spike-shaped petals, and beneath, causative methylation of *Lcyc* resulting in gene inactivation. Photograph of flowers kindly provided by Professor Enrico Coen.

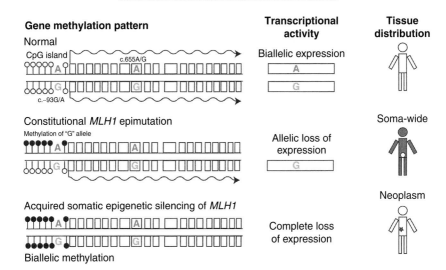

Figure 8.2. Comparison between a constitutional *MLH1* epimutation and acquired somatic methylation in neoplasia. Schematic of the molecular features and tissue distribution of a constitutional *MLH1* epimutation as compared with the normal epigenetic state and biallelically hypermethylated state observed in sporadic microsatellite unstable colorectal cancers. Left, illustration of the *MLH1* gene. The CpG island is shown as lollipops, with methylated CpG sites in black and unmethylated CpGs in white. Exons are depicted as boxes. Transcriptional activity is denoted by the waved arrow. G/A SNP sites that distinguish the two genetic alleles of *MLH1* are shown according to their positions. Center, allelic transcription of *MLH1*, with mRNA transcripts depicted as a rectangle, with expressed alleles differentiated by the exonic G/A SNP. A single allele of *MLH1* is expressed in the somatic tissues in cases with a constitutional epimutation. Right, somatic distribution of the *MLH1* methylation is shown in red. In cases with a constitutional epimutation, methylation is present in all somatic tissues, but erasure of methylation has been demonstrated in the male germ line. Acquired somatic methylation of both *MLH1* alleles is confined to the tumor in colorectal cancer cases. (For interpretation of the references to color in this figure legend, the reader is referred to the Web version of this chapter.)

somatic cells, as originally proposed by Knudson in his "two-hit" hypothesis. In this model, the germ line mutation serves as the "first hit," with an acquired loss of the remaining functional allele in somatic cells through deletion, point mutation, or epigenetic inactivation representing the "second hit" (Knudson, 1996).

However, apparent pathogenic germ line mutations in the relevant disease-associated gene fail to account for every case presenting with the clinical features of the given cancer syndrome. While the hunt for additional genes or cryptic mutations within known genes to identify the causes for disease in these outstanding cases continues, recent attention has been given to investigating the role for epigenetic-based disruption of gene activity in disease causation. Although it has also been well recognized that epigenetic silencing of crucial tumor suppressor and DNA repair genes is a common feature in the development and/or progression of sporadic cancers, a small number of cases with an early-onset cancer syndrome have now been identified, which are attributable to epigenetic silencing of the relevant disease-associated gene in their constitutional DNA. It is interesting to note that cases with constitutional epigenetic silencing of one allele in their normal cells occur in the absence of an identifiable sequence mutation within the genetic code of the gene—the affected locus itself remains intact. The term that has been broadly adopted to describe this novel type of epigenetic-based mechanism is "epimutation." The epimutations identified in early-onset cancer syndromes to date are characterized by promoter methylation and transcriptional silencing of a single allele of the gene in normal somatic tissues in the absence of a sequence alteration within the affected gene locus (Fig. 8.2; Hitchins *et al.*, 2005). Epimutations can thus represent an alternative mechanism for disease causation, or "phenocopy," of a genetically based disease. While genetic mutations and epimutations may produce the same end result—a given disease—there are key distinctions. The first is molecular. In cases with an epimutation, the gene will be modified by an altered pattern of epigenetic moieties, but will not contain a conventional mutation of the DNA sequence. However, some epimutations may arise as a secondary consequence of a linked genetic alteration in the vicinity of the affected gene. Secondly, epimutations may give rise to an altered pattern of inheritance compared to that associated with sequence mutations of the same gene. Epimutations have been shown to give rise to both classic Mendelian or non-Mendelian patterns of inheritance, according to their underlying mechanism (Chan *et al.*, 2006; Hitchins *et al.*, 2007). Where the epigenetic disturbance is dependent upon a preexisting *cis*-acting genetic change located near the affected gene, the pattern of inheritance is likely to be Mendelian. However, the epigenetic modifications themselves are labile, and so where no such fundamental *cis*-genetic basis exists, epimutations may arise spontaneously but then be reversed in the germ line and thus may not be vertically transmissible to the next generation, or may demonstrate non-Mendelian inheritance patterns. This chapter will follow the journey of discovery of epimutations and describe the types of epimutation that have been reported, with a focus on human disease phenotypes, particularly cancer. Furthermore, this chapter will address the underlying origins and mechanisms (or potential mechanisms) of onset of the various types of epimutation, and describe how these correlate with the observed patterns of inheritance.

II. DEFINITIONS

Epimutation   an epigenetic aberration that causes transcrip-
                              tional silencing of a gene that is normally active,
                              or activation of a gene that is normally silent, in
                              the absence of an alteration to the DNA sequence
                              within the affected gene (Holliday, 1987; van
                              Overveld et al., 2003).

Germ line epimutation   an epimutation that is present (with its epigenetic
                              modifications intact) within the gamete to affect
                              a single parental allele (Suter et al., 2004).

Constitutional epimutation   an epimutation that is present in the normal cells
                              of an individual to underlie a disease phenotype,
                              but is not necessarily present in the germ line
                              (Hitchins and Ward, 2009).

Mosaic epimutation   an epimutation that shows variable degrees of
                              allelic methylation and transcriptional silencing
                              and/or tissue distribution. Mosaicism is a common
                              facet of epimutations.

III. EPIMUTATION CONFERRING FLORAL SYMMETRY IN TOADFLAX

The first discovery of a naturally occurring epimutation was in the toadflax plant
(Fig. 8.1; Cubas et al., 1999). As originally described by Linnaeus, the flowers of
the toadflax plant have five petals, but have quite distinctive patterns (Linnaeus,
1749). The wild-type flowers have variously shaped petals that show lateral
symmetry but dorsoventral asymmetry. In contrast, the peloric flowers show
radial symmetry of the petals, with five spike-shaped petals spaced at even
intervals in a circular shape. The cyc gene had previously been shown to control
dorsolateral asymmetry of the flower in snapdragon (Antirrhinum), and inactivating mutations of cyc had been shown to contribute to a similar phenotype of
radial symmetry in snapdragon (Luo et al., 1999). Cubas and colleagues cloned its
homologue in toadflax, Lcyc, and used restriction fragment length polymorphism
(RFLP) and sequence analysis to compare the gene sequence between wild-type
and peloric toadflax plants. Interestingly, while the phenotypically distinct
plants did display fragment length polymorphisms on Southern blots containing
DNA digested by particular restriction endonucleases, direct sequencing failed to
identify any sequence-based differences in the Lcyc gene between the two flower-
ing types that might account for the RFLPs observed. However, some of the

restriction enzymes employed were sensitive to methylation status, suggesting differences in methylation patterns between the wild-type and peloric plants might account for the altered fragment lengths instead. Further RFLP analyses were performed using pairs of isoschizomers, which recognize the same sequence target site, but only one is capable of cleavage if the recognition site is methylated. It was thus demonstrated that the plants with wild-type flowers were unmethylated, whereas peloric plants were methylated and consequently demonstrated loss of *Lcyc* gene activity (Fig. 8.1). Interestingly, some peloric plants produced branches with flowers that appeared wild-type, and methylation analysis by RFLP showed this was due to reversion of the *Lcyc* gene from the methylated to the unmethylated state. Furthermore, some plants displayed mosaicism, with some branches bearing flowers with intermediate patterns, and their degree of morphological change was found to correspond to the level of *Lcyc* methylation. These findings showed that the lack of *Lcyc* gene expression in peloric plants was independent of any sequence change, but was due to an epigenetic change, and moreover, the methylation states were shown to be transmitted from the parental plants to their progeny. However, the varying degrees of somatic reversion of the epimutation giving rise to flowers with intermediate petal shapes showed this phenomenon to be somatically less stable than a genetic mutation. The authors speculated that epigenetic changes could have significant implications in the evolution of plant species (Cubas *et al.*, 1999).

IV. NOMENCLATURE AND ORIGINS OF EPIMUTATIONS IN HUMANS

The assignment of appropriate nomenclature for epimutations has been vociferously debated in the literature. The context of this controversy has surrounded the potential origins of epimutations; whether they arise in the germ line or in the somatic cells of the developing embryo or adult. An "epimutation" was originally defined as an epigenetic aberration that causes transcriptional silencing of a gene that is normally active, or conversely, reactivation of a gene that is normally silent, in the absence of a DNA sequence alteration within the affected gene (Holliday, 1987; van Overveld *et al.*, 2003). The somatic epigenetic aberrations frequently observed in neoplasia, and sometimes detected in foci of preneoplastic cells, could be viewed as "acquired somatic epimutations." Epigenetic silencing in this context often affects both alleles of the gene, and thus represents the singular mechanism of gene inactivation that results in loss of protein expression. The somatic biallelic hypermethylation of the mismatch repair gene *MLH1* in sporadic colorectal cancer is a classic example of this phenomenon (Fig. 8.2; Cunningham *et al.*, 1998; Herman *et al.*, 1998).

Some epimutations may arise in the germ line, with their altered epigenetic state already present in the gamete, and subsequently become disseminated throughout the somatic tissues of the individual. These are referred to as "germ line epimutations" (Suter et al., 2004). However, epigenetic modifications are labile, and as such are usually erased during gametogenesis (Morgan et al., 2005). Thus it is difficult to demonstrate definitively that an epimutation originated in the germ line itself, especially in humans. While epigenetic inheritance through the germ line has been demonstrated in particular mouse strains, as discussed in further detail below, the existence of a true germ line epimutation in humans remains to be proven definitively. Nevertheless, it remains possible that some epimutations that are apparent in constitutional tissues may indeed have arisen in the gamete and be transmissible through the germ line with the epigenetic modifications still intact via a mechanism of "gametic epigenetic inheritance" (previously referred to as "transgenerational epigenetic inheritance"; Whitelaw and Whitelaw, 2008). Alternatively, the epigenetic modifications that are associated with an epimutation in the somatic cells of a carrier may be erased in the germ line, but its underlying mechanism may still be transmissible through the germ line, and thus the epimutation may become reestablished in the next generation postfertilization. The focus of this chapter is on "constitutional epimutations," which are present in the normal somatic tissues, wherein they either cause a disease phenotype or confer a predisposition to the development of disease (Fig. 8.2). Some constitutional epimutations may have arisen with their epigenetic modifications intact within the germ line and thus also represent germ line epimutations. Others may have arisen in the somatic cells of the embryo at a sufficiently early stage that they become widely disseminated through the soma. Constitutional epimutations are distinguishable from the acquired epigenetic alterations that are sometimes observed in small foci of non- or preneoplastic cells, referred to as a "field defect" or "field cancerization" (Chai and Brown, 2009). As originally conceived by Slaughter in 1953, "field cancerization" was proposed to predispose to locally recurrent cancer due to the accumulation of genetic alterations in patches of preneoplastic cells (Slaughter et al., 1953). Epigenetic alterations identified in this context tend to be found in distinct clusters of cells, for example within individual colonic crypts, which likely have their origins within adult stem cells. These may result in clonal amplification of an epigenetically abnormal cell cluster to form precursor lesions. While constitutional epimutations often demonstrate mosaic patterns between individual cells or tissue-types, they affect a single allele of the gene, and are either widely distributed throughout the soma, or throughout a particular tissue-type such as epithelium. They are intrinsic, not acquired later in life.

While the controversy surrounding the precise nomenclature of epimutations might appear semantic, the origins and mechanisms that underlie individual types of epimutation are of key importance, because the inheritance patterns associated with them are dependent upon this. Yet, irrespective of the origins of the different types of epimutation that have been described, the dysregulation of gene activity that ultimately ensues in the normal somatic tissues of an individual plays an etiological role in disease.

V. EPIMUTATION OF THE MISMATCH REPAIR GENES IN CANCER SUSCEPTIBILITY

Constitutional epimutations of the mismatch repair genes *MLH1* and *MSH2* in Lynch syndrome, or hereditary non-polyposis colorectal cancer (HNPCC), represent classic but very distinct examples of this phenomenon. While *MLH1* epimutations typically arise spontaneously and demonstrate non-Mendelian inheritance patterns, in contrast, *MSH2* epimutations arise secondarily due to a nearby *cis*-acting genetic defect and thus demonstrate Mendelian inheritance. These two forms of epimutations represent the two most studied forms of epimutation in human disease and are described in detail herein.

A. Lynch syndrome and the mismatch repair genes

Lynch syndrome (OMIM 120435) is the most common form of familial cancer, characterized by the young onset of colorectal and endometrial cancers primarily, with an elevated risk of gastric, small intestine, ovarian, brain, and other extracolonic cancers (Lynch and Lynch, 2004). Lynch syndrome is usually caused by heterozygous germ line mutations within the mismatch repair genes namely *MLH1*, *MSH2*, *MSH6*, and *PMS2* and demonstrates autosomal dominant inheritance (Lynch and Lynch, 2004). Of all the pathogenic mutations identified in Lynch syndrome, *MLH1* and *MSH2* account for 50% and 40%, respectively, with the remaining 10% in *MSH6* and *PMS2* (Peltomaki and Vasen, 2004). These mutations result in loss of protein function. Tumor formation usually conforms to Knudson's "two-hit" model in which the germ line mutation (first-hit) is followed by somatic loss of the normal allele (second hit), usually via deletion and detected as loss-of-heterozygosity (LOH) in the tumor DNA (Hemminki *et al.*, 1994; Tomlinson *et al.*, 2001). Lynch syndrome-associated tumors typically exhibit microsatellite instability (MSI) as a direct consequence of impaired mismatch repair activity, which is detectable as altered lengths of tandem repeats at designated microsatellite loci within the tumor DNA (Boland *et al.*, 1998). The mismatch repair system functions through the formation of heterodimeric protein complexes, of which the MutL homologue (MLH1 and PMS2) and the MutS homologue (MSH2 and MSH6) comprise the key components (Kolodner, 1995). Immunohistochemical analysis of the mismatch repair proteins often provides a good indication of the mismatch repair gene most likely to harbor a germ line defect (Hendriks *et al.*, 2003; Lindor *et al.*, 2002).

Various sets of clinical criteria have been devised to standardize the clinical definition of Lynch syndrome and identify cases warranting germ line screening for a causative mismatch repair defect. The original Amsterdam I criteria (three relatives with CRC present in at least two successive generations, one of whom must be a first-degree relative of the other two; colorectal cancer diagnosed in at least one relative < 50 years of age) represent the "gold standard" for a clinical diagnosis of Lynch syndrome (Vasen *et al.*, 1991). These criteria have been progressively revised to incorporate additional clinical and molecular characteristics of this syndrome as a fuller picture of the phenotypic spectrum associated with mismatch repair mutations emerged. The current revised Bethesda Guidelines are less stringent, taking into account extracolonic cancers and the lack of a family history due to *de novo* events (colorectal cancer < 50 years; synchronous/metachronous colorectal cancer or other HNPCC-related tumor irrespective of age; colorectal cancer demonstrating MSI < 60 years; colorectal cancer with at least one relative with a Lynch syndrome-related tumor < 50 years; colorectal cancer and at least two first- or second-degree relatives with Lynch syndrome-associated tumors at any age; Umar *et al.*, 2004). The latter guidelines recommend genetic screening for a germ line mutation only if the tumor exhibits MSI and the proband meets one or more of the clinical criteria. Standard genetic screening for germ line mutations involves the combinatorial approach of exonic sequence analysis of the relevant gene(s) for point mutations and multiplex ligation probe amplification (MLPA) to detect structural alterations such as exonic deletions and duplications within *MLH1* and *MSH2* (Grabowski *et al.*, 2005). Nevertheless, despite clear clinical guidelines, the identification of the key genes that encode the various components of the mismatch repair system, as well as the combinatorial approach to germ line screening, pathogenic mismatch repair mutations fail to be identified in up to one-third of Lynch syndrome cases. Constitutional epimutations of the *MLH1* and *MSH2* genes have been identified in a proportion of these mutation-negative cases as described below.

B. Identification and molecular characterization of constitutional MLH1 epimutations in cases with Lynch syndrome

The first case to be reported with what we now know as a constitutional *MLH1* epimutation was in a female proband with no family history of cancer, who had presented with a microsatellite unstable colorectal cancer demonstrating loss of MLH1 immunoexpression at the young age of 25 years (Gazzoli *et al.*, 2002). Germ line screening had revealed no causative mismatch repair mutation of

MLH1, MSH2, or MSH6. However, methylation of the MLH1 promoter was detected in the peripheral blood lymphocytes of this individual. Moreover, the proband was heterozygous for the common c.93G>A SNP (rs1800734) within the MLH1 promoter, allowing the two genetic alleles to be distinguished from one another and the methylation status of each allele to be observed separately. The individual strands of the MLH1 promoter were separated by cloning the amplified fragment encompassing the SNP site into plasmids and transforming and propagating them in competent E. coli bacterial cells. By sequencing the plasmid inserts extracted from individual bacterial colonies, each containing the MLH1 promoter insert derived from a single template strand, the methylation pattern at individual CpG dinucleotides of each cloned allele could be observed, as illustrated in Fig. 8.3. In this manner, the CpG methylation was shown to be associated with just one allele of the SNP, indicating monoallelic methylation of the MLH1 promoter. Loss-of-heterozygosity analysis at the same SNP site in DNA extracted from the tumor showed that the opposite, unmethylated allele, had been eliminated, providing the "second hit" in tumorigenesis. The authors postulated that the methylation constituted a novel mode of germ line inactivation of this cancer susceptibility gene (Gazzoli et al., 2002).

In 2004, four cases of Japanese heritage presenting with sporadic early-onset colorectal and additional cancers were identified with monoallelic methylation (affecting a single allele at the c.93G>A SNP) or hemiallelic methylation (affecting approximately half of the alleles, but the subject was uninformative at the promoter SNP) in their peripheral blood lymphocytes and other normal tissues, including normal colonic or gastric epithelium and endometrium (Miyakura et al., 2004). Loss of heterozygosity of the unmethylated allele was also found in the colorectal tumor of one of the cases.

Subsequently in 2004, Suter et al. identified a further two cases, an unrelated male and female, each with multiple primary carcinoma, and showed definitively that the MLH1 methylation was distributed widely and evenly throughout their soma (Suter et al., 2004). They demonstrated methylation of a single allele at the c.93G>A SNP of the MLH1 promoter was present in somatic tissues that were derived from all three embryonic cell lineages, including in buccal mucosa (endoderm), peripheral blood (mesoderm), and hair follicles (ectoderm). This provided strong evidence that the epimutation had originated either in the germ line or during early embryogenesis, prior to gastrulation when the inner cell mass of the embryo separates into these three distinct embryonic somatic cell lineages. The authors also reported the detection of MLH1 methylation in approximately 1% of the spermatozoa from the male proband. They argued that while the epimutation was erased in most spermatozoa cells, the epimutation could also be carried in a proportion of germ cells, which could confer a proportionally low risk of transmission of the epimutation to the next generation. The finding of monoallelic methylation throughout the

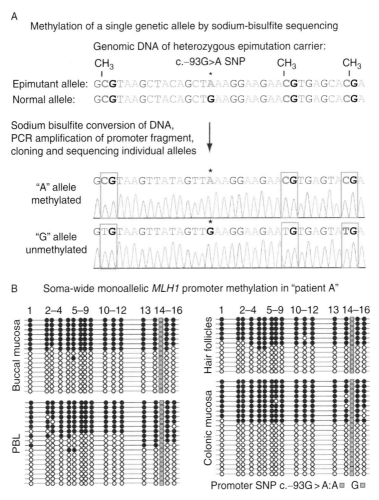

Figure 8.3. Soma-wide methylation of a single *MLH1* allele in a carrier of a constitutional *MLH1* epimutation, as shown by allelic sodium bisulfite sequencing of promoter fragments. (A) Sequencing of individual alleles of the *MLH1* promoter following sodium bisulfite conversion and cloning of the amplified fragments. Treatment with sodium bisulfite converts unmethylated cytosines to thymines via uracil, whereas methylated cytosines at CpG dinucleotides remain unchanged. The common c-93G>A SNP (rs1800734) within the sequenced fragment is indicated by an asterix and the sites of individual CpG dinucleotides within the original DNA are boxed. As shown for epimutation carrier, Patient A, methylation (determined by the presence of CGs in the bisulphite-converted DNA) is monoallelic, and associated with the "A" allele. The "G" allele beneath is unmethylated, as indicated by the conversion to TGs at the CpG sites flanking the SNP site. (B) Soma-wide methylation of the "A" allele in Patient A. Individually sequenced clones, represented by horizontal lines, are aligned according to their genotype and tissue of origin at the G/A SNP, depicted as colored boxes. Individual CpG sites within the sequenced promoter fragment are shown as circles and numbered. Black indicates methylation, white absence of methylation. PBL, peripheral blood lymphocytes.

somatic cells, coupled with their interpretation that methylation was present in a proportion of sperm, suggested the epimutation was likely to have arisen in the germ line. Thus the term "germ line epimutation" was coined to describe this defect (Suter *et al.*, 2004). However, it was subsequently demonstrated in an addendum to this report, that the low levels of *MLH1* methylation initially detected in the spermatozoa cells most likely represented an artifact, attributable to the presence of $\sim 1\%$ contaminating somatic cells in the spermatozoa sample (Hitchins and Ward, 2007). In those subsequent analyses of the same sample of spermatozoa, a control for the detection of somatically derived methylation was incorporated in the form of the *SNRPN* gene, which is subjected to genomic imprinting. The *SNRPN* promoter is methylated specifically on the maternally inherited allele in somatic tissues, whereas it is completely unmethylated on the paternally derived allele in the somatic cells of healthy individuals, as well as in spermatozoa (Geuns *et al.*, 2003). Therefore, any methylation detected at *SNRPN* in the test sample of spermatozoa would be indicative of contamination with somatically derived DNA. Similar levels of $\leq 1\%$ methylation of both *MLH1* and *SNRPN* were indeed detected in the spermatozoa sample of the *MLH1* epimutation carrier. Yet no methylation of *MLH1* or *SNRPN* was detected in a second sample of spermatozoa subsequently donated by the same patient that were subjected to "swim-up" separation prior to methylation testing (Hitchins and Ward, 2007). Since the male carrier of the epimutation had soma-wide monoallelic *MLH1* methylation at near 50% levels, and since it is notoriously difficult to separate cells with a 100% purity, presence of somatically derived DNA among the spermatozoa cells in the original sample would be a more plausible explanation than retention of the epimutation and *SNRPN* methylation in a small proportion of spermatozoa cells. Thus, the new interpretation of these findings was that *MLH1* epimutations are completely reversed in the male germ line.

In 2005, with the identification of a further sporadic case with early-onset cancer, it was shown definitively that the constitutional *MLH1* epimutation resulted in complete allelic loss of transcription (Hitchins *et al.*, 2005). Through the use of an expressible sequence polymorphism within an exon for which the subject was heterozygous as a tracer, it was demonstrated that only one allele was transcribed and hence present among the mRNAs. Furthermore, by genotyping in family members, it was shown that the epimutation arose *de novo* on the maternally derived allele, since the mother was not a carrier, and that the subject's siblings also inherited the same maternal allele, but had no trace of methylation (Hitchins *et al.*, 2005). A similar case of *de novo* occurrence of a constitutional *MLH1* epimutation on the maternally derived allele was subsequently reported in another case with Lynch syndrome, Patient B. Haplotyping of the proband and her family members demonstrated that the proband, her mother, and one of her sons carried the very same *MLH1* allele, but this was only affected by the epimutation in the proband. This indicated that the epimutation arose *de novo* on the maternal allele in the proband, and was then reversed in her own germ line, such that the son inherited the allele unmethylated and transcriptionally reactivated (Fig. 8.4; Hitchins *et al.*, 2007). This family showed definitively that epimutant alleles arose spontaneously and reverted back to the normal functional state in the germ line.

As further cases were identified, it became apparent that in carriers of constitutional *MLH1* epimutations, their phenotype was highly consistent with Lynch syndrome. The cases have presented with early-onset colorectal, endometrial, and other cancers that arise in the Lynch syndrome tumor spectrum, and the tumors have typically demonstrated microsatellite instability and loss of the MLH1 protein. There appears to be no bias in gender or ethnic origin among carriers, with individuals of Northern European, Mediterranean, Japanese, and Chinese heritage reported to date (Gylling *et al.*, 2009; Miyakura *et al.*, 2004; Morak *et al.*, 2008; Valle *et al.*, 2007; Zhou *et al.*, 2008). However, the most notable difference between cases bearing a constitutional epimutation of *MLH1* and those with sequence mutations of the mismatch repair genes is that they tend to arise sporadically, or have no remarkable family history of cancer. By contrast, carriers of sequence mutations within the mismatch repair genes sometimes have an extensive family history due to the faithful transmission of germ line mutation from parent to offspring.

C. Mosaic MLH1 epimutations

Through cloning and sequencing of individual amplified fragments of the *MLH1* promoter in patients informative for the c-93G>A promoter SNP, it was shown that constitutional *MLH1* epimutations generally display dense methylation of a single allele of the promoter. However, there was some variation in allelic methylation patterns (Suter *et al.*, 2004). Some alleles showed that all, or the majority of CpG dinucleotide sites within the fragment were methylated, while others had patchy methylation. A small number of alleles containing the affected SNP genotype were completely unmethylated. In patients with up to 50% of alleles densely or partially methylated, complete transcriptional silencing of the affected *MLH1* allele was observed. In addition, one allele of the *EPM2AIP1* gene, which shares the same bidirectional promoter but lies head-to-head with *MLH1* and is expressed from the antisense strand in the opposite direction, was also inactivated (Hitchins *et al.*, 2007). However, a proportion of cases have been shown to have considerable allelic mosaicism in methylation density, with some cases displaying methylation of as few as 10% of alleles (Morak *et al.*, 2008). Interestingly, in these cases, allelic expression studies at the common expressible

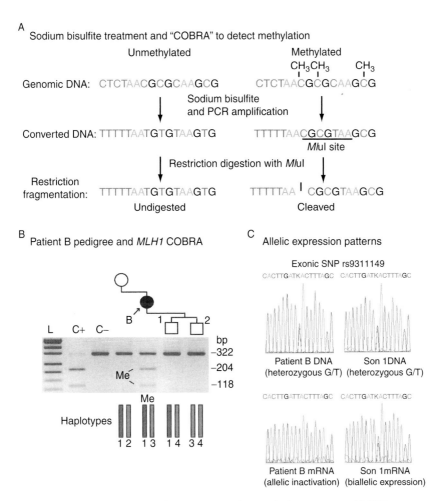

A Sodium bisulfite treatment and "COBRA" to detect methylation

	Unmethylated	Methylated
Genomic DNA:	CTCTAACGCGCAAGCG	CTCTAACGCGCAAGCG

Sodium bisulfite and PCR amplification

Converted DNA: TTTTTAATGTGTAAGTG   TTTTTAACGCGTAAGCG
*Mlu*I site

Restriction digestion with *Mlu*I

Restriction fragmentation: TTTTTAATGTGTAAGTG   TTTTTAA | CGCGTAAGCG
Undigested   Cleaved

B Patient B pedigree and *MLH1* COBRA

L   C+   C−
bp
−322
−204
Me
−118

Me
Haplotypes
1 2   1 3   1 4   3 4

C Allelic expression patterns

Exonic SNP rs9311149
CACTTGATKACTTTAGC   CACTTGATKACTTTAGC

Patient B DNA   Son 1DNA
(heterozygous G/T)   (heterozygous G/T)

CACTTGATTACTTTAGC   CACTTGATKACTTTAGC

Patient B mRNA   Son 1mRNA
(allelic inactivation) (biallelic expression)

Figure 8.4. *De novo* occurrence and intergenerational reversal of a constitutional *MLH1* epimutation, as identified by "COBRA." (A) The process of combined bisulfite and restriction analysis (COBRA) is depicted for a given sequence of genomic DNA, resulting in the digestion of PCR amplified fragments with an enzyme specific to a methylated strand of template DNA. (B) Pedigree of MLH1 epimutation carrier, Patient B. Methylation analysis of family members by COBRA following gel electrophoresis is shown beneath, with lanes corresponding to the subjects directly above. Patient B shows a 50:50 banding pattern of methylated (digested) and unmethylated (undigested) alleles, similar to the methylated control (C+) whereas the unmethylated normal control DNA (C−) remains entirely undigested. The methylated bands (Me) are indicated. L, DNA size ladder. The haplotypes of each family member are depicted and numbered beneath. The red allele (allele 1) is associated with the epimutation, but is only methylated in the proband, indicating it arose spontaneously on the maternal allele, and was demethylated in her son. (C) Sequence traces showing allelic transcription of the *MLH1* gene for Patient B and her eldest son, at an expressible G/T SNP site for which they are both heterozygous in their DNA. Only the "T" allele is expressed in Patient B's mRNA, due to silencing of the "C" allele by epimutation. Both alleles are present in her son's mRNA, due to reversion of the allele to its normal active state in him. (For interpretation of the references to color in this figure legend, the reader is referred to the Web version of this chapter.)

c.655A>G SNP within *MLH1* exon 8 showed an allelic expression imbalance, but with only partial repression of the affected allele, consistent with methylation in just a proportion of cells (Morak et al., 2008).

The reason for some cases showing a partial or mosaic *MLH1* epimutation is unknown. Although epigenetic modifications are mitotically heritable, it is possible that they are partially reversible during somatic cell division, perhaps due to environmental influences. It is well established that the DNA methyltransferase enzymes do not faithfully replicate the methylation pattern at every CpG dinucleotide, especially at pairs of consecutive CpG sites at which one of the pair may escape methylation following DNA replication. It is possible that this mechanism would result in a gradual reversal of methylation. Alternatively, the epimutations in these cases may have arisen postfertilization in the developing embryo and thus be present in a proportion of derivative cells in the adult. Since this phenomenon has been reported primarily in peripheral blood lymphocytes, another explanation would be the expansion of a population of hematopoietic precursor cells that contained unmethylated alleles. Finally, it is also possible that the mechanism underlying the epimutation in these cases is not fully penetrant.

Epigenetic mosaicism in carriers of epimutations is an important phenomenon since it has implications with respect to molecular detection as well as phenotypic severity. Where methylation is present at low levels, it is possible that employment of less sensitive screening methods may fail to positively identify them. Alternatively, if the methylation demonstrates significant variability in levels between different tissue-types, as has been found for patients with constitutional epimutations of *MSH2*, testing for methylation in the peripheral blood of probands may give rise to false-negative results, while methylation levels are high in tissues susceptible to cancer development. In the case of constitutional *MLH1* epimutations, the methylation appears to be widely and evenly distributed throughout the somatic tissues; however, this has not been extensively investigated. It is possible that cases with low levels of *MLH1* methylation, and consequently a more subtle allelic expression imbalance, may have a reduced likelihood of developing cancer, or later age of cancer onset. However, too few cases have been identified to allow phenotypic comparisons to be made at present.

D. Non-Mendelian inheritance of constitutional MLH1 epimutations

MLH1 epimutations were initially considered to pose a low risk of transmission from one generation to the next due to the major epigenetic reprogramming events that occur during the reproductive life-cycle (Jass, 2007), as described in more detail below. Unlike sequence mutations, epigenetic modifications to the DNA are labile and subject to reversal between generations. Furthermore, most carriers of *MLH1* epimutations were sporadic cases of Lynch syndrome due to the *de novo* occurrence of the epimutation (Gazzoli *et al.*, 2002; Hitchins *et al.*, 2005; Miyakura *et al.*, 2004). Although a proportion of cases did have a weak family history of cancer, it was argued that this simply reflected ascertainment bias, since the patients studied had been recruited through Family Cancer Clinics due to the early-onset symptoms suggestive of a familial cancer syndrome (Jass, 2007). It was argued that any positive family history was likely to be coincidental due to the frequency of cancer in the general population and caused by independent mechanisms in the family members. Coupled together, these cogent points argued against the transmission of *MLH1* epimutations from carrier to offspring. This was an important debate, because this defect had presented with a genetic counseling conundrum since its first identification. While carriers of a sequence mutation within a mismatch repair gene and their family members could be counseled that their mutation had a 50% chance of inheritance of the mutation and be advised of their risk of cancer development according to carrier status, the risk of transmission of an *MLH1* epimutation was, and still remains, unknown. However, in 2007, the unprecedented finding of a family in which the *MLH1* epimutation was transmitted from mother to one of her children in a non-Mendelian pattern proved definitively that this defect can be passed from one generation to the next (Hitchins *et al.*, 2007). Subsequently, a second familial case was reported (Morak *et al.*, 2008).

In the first case (Hitchins *et al.*, 2007), "Family A," a female proband who had presented with multiple primary tumors demonstrating MSI and loss of MLH1 expression at a young age, consistent with Lynch syndrome, was found to carry a constitutional *MLH1* epimutation throughout her somatic tissues. The proband had no family history of cancer, and consistent with this, she was negative for sequence mutations of the mismatch repair genes upon germ line screening. The proband was informative for the c-93G>A SNP in the *MLH1* promoter, as well as the benign expressible c.655A>G SNP (rs1799977) within *MLH1* exon 8, allowing comprehensive allelic methylation and expression studies to be conducted. Her methylation was found to be associated with the "A" allele of the promoter SNP and only the "A" allele of the common benign c.655A>G exon 8 SNP was expressed, indicating complete loss of expression of the other allele in her normal somatic tissues, as illustrated in Fig. 8.2. This was thus typical of previously described epimutations of *MLH1*. The proband had four healthy sons, and strikingly, one of the four sons (the second son) was also found to carry an *MLH1* epimutation throughout his somatic tissues. SNP genotyping in the pedigree showed the methylation was present on the same genetic allele as his mother, suggestive of direct transmission of the epimutation (Fig. 8.5A). Allelic expression studies showed that the same *MLH1* allele was also completely inactivated in his somatic tissues. More intriguingly, haplotyping of SNPs and microsatellite repeats around *MLH1* in all family members showed

A "Family A" pedigree and detection of methylation by "COBRA"

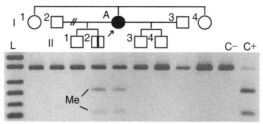

B Genetic haplotypes

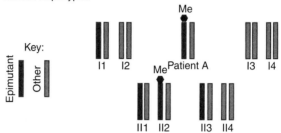

C Summary family members carrying the epimutant allele

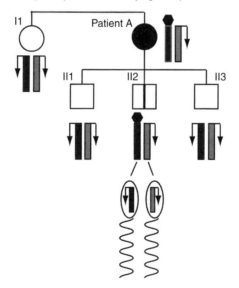

Figure 8.5. Intergenerational transmission of constitutional *MLH1* epimutation in "Family A". Intergenerational transmission of constitutional *MLH1* epimutation in Patient A. Family A pedigree with generations numbered I and II. Beneath is the COBRA gel with lanes according to the family members above. The COBRA shows the presence of methylation (Me) in Patient A and her second son (II2), but absence of methylation in

that three of Patient A's four sons, as well as one of her sisters, also carried the same affected allele (Fig. 8.5B). Although the three sons had inherited the very same maternal *MLH1* allele, methylation was only present in one of them. Therefore the methylation had been passed from the affected mother, Patient A, to just one of her three sons who had inherited the same genetic allele that was affected by the epimutation in her. Sensitive methylation analyses in these two sons (the first and third sons) found no trace of methylation, indicating it had been completely erased in them. Allelic expression studies at informative expressed SNP sites within the *MLH1* and overlapping *EPM2AIP1* locus showed that both alleles were normally expressed again, indicating that with the erasure of methylation, the allele reverted back to its normal functional state. While the mother and her second son both demonstrated soma-wide allelic methylation and transcriptional silencing, no trace of *MLH1* methylation was found in his spermatozoa, and furthermore, the epimutant allele had become reactivated again therein (Fig. 8.5C; Hitchins *et al.*, 2007). This provided clear evidence that while the epimutation could be present soma-wide, it was efficiently erased in the germ cells of this male carrier. This begged the question of how an *MLH1* epimutation could be transmitted from one generation to a proportion of offspring when clear reversal of the epimutation was witnessed in the germ line of the son. While the son who carried the *MLH1* epimutation was asymptomatic at the time of the report, this nevertheless places him at an elevated risk of cancer development.

In the second report of transmission of a constitutional *MLH1* epimutation, a male proband with early-onset colorectal cancer consistent with Lynch syndrome but negative for a germ line mutation was found to have methylation of a single allele of the *MLH1* promoter in his peripheral blood (Morak *et al.*, 2008). However, the levels of methylation detected in his blood were relatively

all other family members. (B) Haplotypes of the family members are depicted, with the genetic allele associated with the epimutation shown in red, with all other alleles in gray. The epimutant allele is present in five family members, including three of Patient A's sons, but is only methylated (black hexagram) in Patient A and one son (II2), indicating erasure of the epimutation in her two other sons (II1 and II3). (C) Analysis of the inherited epimutation in Son II2. Schematic overview of the stochastic transmission and reversal of the *MLH1* epimutation in Family A. Only family members carrying the genetic allele on which the epimutation arose (red) are included. Arrows indicate transcription of the respective alleles. The allele on which the epimutation was borne was transmitted from Patient A to three of her sons. However, the epimutation itself (black hexagon) was transmitted from Patient A to her second son II2 only, causing transcriptional silencing of the allele in them. The allele was transcribed in the two other sons, II1 and II3, and in the spermatozoa of son II2, due to erasure of the epimutation in his germ line. (For interpretation of the references to color in this figure legend, the reader is referred to the Web version of this chapter.)

low at 10%. Although his mother was unaffected by cancer at the age of 64 years on testing, she was also found to have a mosaic pattern of *MLH1* methylation at a level of 8% in her blood (Morak *et al.*, 2008). This was a most surprising finding. That an *MLH1* epimutation present at low levels in the somatic tissues of the parent could be passed to the next generation seemed highly unusual.

It is possible that in other carriers of a constitutional *MLH1* epimutation who had an affected first degree relative also represent unrealized cases of inheritance of this defect, but this could not be established due to the lack of availability of familial samples (Suter *et al.*, 2004).

The mechanisms by which the intergenerational inheritance of the *MLH1* epimutations occurred in these two families remain unknown. Nevertheless, the demonstration of inheritance of an *MLH1* epimutation in two families has confounded genetic counseling. While there is a clear risk of transmission, the non-Mendelian inheritance pattern found in "Family A" renders the precise level of that risk unpredictable at present. The current recommendation is that family members of individuals identified with a constitutional *MLH1* epimutation should be counseled that there is an elevated risk of inheritance, and that each of them should be offered methylation testing to determine their carrier status and further posttest genetic counseling on receipt of their results. The transmission patterns associated with this defect will become more apparent as additional affected and unaffected relatives of probands with an *MLH1* epimutation undergo testing. Although *MLH1* epimutations have now been shown to be erased in the spermatozoa of two male carriers, the potential for paternal transmission of this defect also remains unknown. Only one case has been reported in whom the epimutation arose on the paternally inherited allele, and in this case, the epimutation arose *de novo* (Goel *et al.*, 2010). With this one exception, the *MLH1* epimutations in all the other cases (studied in detail) were shown to have arisen spontaneously on the maternally derived allele or were inherited from the mother. However, the mode of transmission is unclear. If *MLH1* epimutations are reversed in gametes, as we know they are, but then later reestablished postfertilization in a proportion of offspring, it is possible that this defect could be paternally transmitted as well. This is an important question that remains to be addressed. However, until such a time as the precise mechanism that underlies the onset and inheritance of this defect is identified, and its associated inheritance pattern is defined, accurate genetic counseling cannot be proffered.

E. Potential mechanisms underlying constitutional MLH1 epimutations

The potential mechanisms by which constitutional *MLH1* epimutations may arise and be transmitted in a non-Mendelian pattern may either be fundamentally epigenetic- or genetic-based. The arguments for and against these two potential mechanisms are presented below.

If *MLH1* epimutations represent a primary epigenetic-based defect that is entirely independent from genetic sequence, then transmission from parent to offspring would require that the defect is transmitted through the germ line with its epigenetic modifications intact via a mechanism known as "gametic epigenetic inheritance," or through retention of an "epigenetic memory" through the germ line (Chong *et al.*, 2007). Carriage of an epimutation with its CpG methylation and repressive chromatin modifications intact through the gametes seems unlikely given the major epigenetic reprogramming events that occur during mammalian reproduction. The prior demonstration of complete reversal of the *MLH1* epimutation during spermiogenesis in two male carriers of a soma-wide epimutation also argues against this mechanism (Hitchins and Ward, 2007). The epigenetic marks associated with parental allele-specific imprints at genes subject to genomic imprinting in somatic tissues are erased in the primordial germ cells (gamete precursors). A classic example is the H19 gene, which is expressed specifically from the unmethylated maternally derived allele, and monoallelically methylated at the promoter of the silent paternally derived allele, in somatic cells. However, both alleles of imprinted genes including H19 are unmethylated in primordial germ cells, due to this efficient erasure of the repressive epigenetic modifications at these "differentially methylated regions" (often the gene promoter itself). These methylation imprints are then reestablished in the gametes or on the appropriate parental allele during preimplantation embryonic development, according to the sex of the parent from which they originated (Morgan *et al.*, 2005). In spermatozoa cells, many of the histones in the chromatin are replaced by protamines, allowing tight compaction of the genome into the very tip of the sperm head (Sassone-Corsi, 2002). Following fertilization, there is further demethylation specifically of the paternal genome in the male pronucleus (Santos *et al.*, 2002). Since this occurs in the zygote in the absence of DNA replication, this process is referred to as "active" demethylation, and may involve a process of deamination of methylated cytosines followed by base-excision repair (Morgan *et al.*, 2004). During preimplantation embryonic development there is still further demethylation of the genome, referred to as "passive demethylation" as it accompanies cell division. As the blastocyst forms, new somatic patterns of methylation are set in the inner cell mass by a process of "active methylation" (Santos *et al.*, 2002). Interestingly, parent-specific imprints set in the gamete withstand the epigenetic fluctuations that occur postfertilization. And thus it is plausible that an epigenetic aberration of a non imprinted gene such as *MLH1* could arise spontaneously in the gamete and subsequently reach the somatic cells of the offspring without complete reversal during embryogenesis. Interestingly, the vast majority of imprinted genes have their differentially methylated sites on the maternally derived allele, and thus it is conceivable that these might arise as a consequence of "faulty imprinting" in the oocyte. Certainly, this would be consistent with the apparently spontaneous onset of constitutional *MLH1* epimutations and their strong bias in the origin on the maternally derived allele. However, if constitutional *MLH1* epimutations represent a purely epigenetic phenomenon, then one would anticipate their erasure in the primordial germ cells contemporaneously with removal of genomic imprints, and thus any transmission to the subsequent generation would require a failure in epigenetic reprogramming at this stage. Imprinted genes appear to retain an epigenetic "memory" since methylation is first restored on the allele that was previously methylated (Davis *et al.*, 2000; Lucifero *et al.*, 2004). How such a memory is conveyed, whether this can be through chromatin modification other than cytosine methylation, remains unknown. Since histones are replaced by protamines during spermiogenesis (Sassone-Corsi, 2002), this may facilitate the complete erasure of an epigenetic memory in the male germ line, whereas chromatin remodeling may be less efficient during oogenesis, with a resultant bias in matrilineal inheritance of *MLH1* epimutations. If such an epigenetic memory exists, and this fails to be erased in the primordial germ cells, this could be mediated by the altered action of noncoding RNAs such an antisense or microRNAs.

Alternatively, constitutional *MLH1* epimutations may be initiated and transmitted from one generation to the next via an unusual genetically based mechanism, for instance, due to the cosegregation of a predisposing genetic defect in a proportion of offspring. In this scenario, the epimutation need not be transmitted with its epigenetic appendages intact within the gamete, but may be reestablished on the affected allele at a later stage of embryonic development following a genetic cue. In Family A, haplotyping at microsatellite repeat loci for several megabases either side of the *MLH1* gene in the mother and her three sons showed no discernible difference between the affected allele in family members with and without the *MLH1* epimutation itself. Since the three sons thus had the very same maternal allele for some genetic distance around *MLH1*, but only one of them bore the epimutation, this argues against a fully penetrant *cis*-acting mutation in the vicinity of the *MLH1* locus (Hitchins and Ward, 2007). It is possible that a linked genetic alteration exists on this allele, but that it only serves to predispose to the epimutant state. The coinheritance of a *trans*-acting factor in the family members carrying the epimutation is also possible, for example a mutation within a gene encoding an epigenetic modifier, as has been shown to occur in *Drosophila* (Xing *et al.*, 2007). This would allow the intergenerational transmission of an epimutation in a proportion of offspring. If this was the case, an epimutation may be transmissible both maternally and paternally, and the epigenetic state in the gametes themselves would provide no indication of inheritance risk. Since we have now shown that the epimutation is cleared in spermatozoa, an observation of paternal transmission would provide strong evidence for a genetic basis to the onset of *MLH1* epimutations.

Whether *MLH1* epimutations are primarily genetic or epigenetic in origin, there is also the possibility that particular environmental factors also influence the likelihood of transmission of this defect, but such interactions are almost impossible to assess in humans, especially where the defect is also rare in occurrence.

F. Constitutional epimutation of MSH2 in Lynch syndrome

In 2006, Chan and colleagues reported the finding of an *MSH2* epimutation in multiple members of a three-generation Lynch syndrome family of Chinese heritage, which segregated with a particular genetic haplotype in an autosomal dominant pattern (Chan et al., 2006). Three siblings from this family had developed a MSI colorectal or endometrial tumor each demonstrating loss of MSH2 expression below the age of 50 years, and their mother had a number of adenomas removed under colonoscopic surveillance. Germ line testing had failed to identify any sequence mutation or structural alteration of the *MSH2* gene, or other mismatch repair genes, in the proband. However, methylation testing of the *MSH2* promoter in the proband's colorectal tumor and accompanying normal colonic mucosa was positive. The proband was heterozygous for two benign SNPs within the *MSH2* promoter (c.118C>T and c.433G>T) and allelic methylation analyses showed the methylation was associated with a single genetic allele. Methylation testing and haplotyping of SNP and microsatellite markers within MSH2 and the surrounding region in 12 family members from three successive generations revealed that 10 family members spanning all three generations were positive for methylation and this segregated faithfully with a particular genetic haplotype. This strict Mendelian inheritance pattern strongly implicated the existence of an underlying *cis* genetic defect that resulted in the secondary methylation of the *MSH2* promoter, but the *MSH2* gene itself appeared to be genetically intact in this family. Another interesting observation was that the levels of *MSH2* methylation varied considerably between different tissues within the individual members of this family. In the rectal mucosa from the 10 members of the family bearing the epimutation, the methylation levels were at their highest, ranging from approximately 20% to 40%. The buccal mucosa and endometrium also showed elevated *MSH2* methylation levels at 5–14%. However, in peripheral blood lymphocytes, the methylation levels were low, at 2–5%. The finding of an autosomal dominant *MSH2* epimutation in a large family meeting the stringent Amsterdam I criteria for Lynch syndrome was unprecedented (Chan et al., 2006). Clearly the underlying mechanism was faithfully transmitted through the germ line, but the highly mosaic tissue distribution of promoter methylation was quite distinct from the previously reported even distribution of somatic *MLH1* methylation detected in cases with *MLH1* epimutations.

G. *Cis*-genetic deletions of the upstream EPCAM gene underlie MSH2 epimutations

Three years later, in 2009, the genetic mechanism underlying the heritable *MSH2* epimutation in this family and additional families of Dutch heritage was identified as terminal deletions of the *EPCAM* gene located immediately upstream of *MSH2*, which resulted in transcriptional elongation from *EPCAM* directly into *MSH2* (Fig. 8.6; Ligtenberg *et al.*, 2009). This causative genetic defect appears to have initially been identified through serendipity. A Dutch proband with an early-onset MSI cancer with dual loss of the MSH2 and MSH6 proteins, but in whom no germ line mutation of *MSH2*, *MSH6*, or *MLH1* had been identified, was under routine investigation for structural changes of *MSH6* by MLPA. In MLPA, changes in dosage suggestive of deletions or duplications are sought by quantitative hybridization of probes located within each exon of the gene of interest, while additional probes at distinct loci serve as controls for normalization of signal intensity. In the *MSH6* MLPA kit, two such control probes target the *EPCAM* gene (formerly *TACSTD1*) on chromosome 2 located 17 kb upstream of *MSH2*, and a reduced signal was observed for the probe located within the final exon of *EPCAM* in this index patient. Further genetic analyses

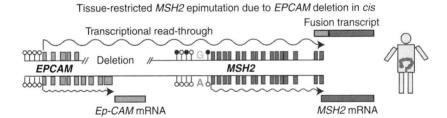

Figure 8.6. Epimutation of *MSH2* caused by transcriptional read-through from the upstream *EPCAM* gene due to *cis*-deletion of the polyadenylation signal. *MSH2* epimutation, as described for "Family HK," caused by a linked interstitial deletion of the terminal exons of the *EPCAM* gene (in blue) located immediately upstream of *MSH2* (in red). Exons are depicted as boxes, CpG sites as lollipops, transcriptional activity a waved arrows, and transcripts as rectangles, with color pertaining to the gene they are derived from. Loss of the polyadenylation of *EPCAM* on the epimutant allele (top) results in transcriptional read-through into *MSH2* and the generation of fused Ep-CAM-MSH2 transcripts. This results in mosaic methylation of the *MSH2* promoter (depicted as black lollipops) on the deleted allele. The c.118C>T SNP (G/A on the complementary strand as analyzed by allelic sodium bisulfite sequencing) within the *MSH2* promoter was used to demonstrate the methylation was monoallelic. Right, distribution of *MSH2* methylation in somatic tissues. The pink background indicates low-level methylation and the red color high levels of methylation, as found in the colonic epithelium. (For interpretation of the references to color in this figure legend, the reader is referred to the Web version of this chapter.)

revealed a heterozygous ~5 kb interstitial deletion that removed the last two exons of EPCAM, while retaining the MSH2 gene entirely intact (Ligtenberg et al., 2009). The same deletion was identified in additional reportedly unrelated Dutch families with a shared haplotype on the affected allele, suggesting a founder deletion. A larger deletion of ~23 kb was identified in two Chinese families, including the originally reported family, which removed the final four exons of EPCAM, with the distal breakpoint upstream of MSH2 (Ligtenberg et al., 2009). Additional Lynch syndrome cases with germ line deletions of the distal portion of EPCAM have since been reported in cases whose tumors demonstrate loss of MSH2 protein expression, including in large autosomal-dominant HNPCC kindreds (Kovacs et al., 2009; Niessen et al., 2009). Each of these germ line deletions encompass the polyadenylation signal of EPCAM but leave the MSH2 gene intact. Loss of the polyadenylation (transcription termination) signal resulted in transcriptional elongation from EPCAM into MSH2, generating abnormal EPCAM-MSH2 fusion transcripts. These transcripts showed splicing from the donor site of the most 3' exon of EPCAM that remained intact directly onto the splice acceptor site at exon 2 of the downstream MSH2. Notably, it was also shown that the tissues in which the highest levels of somatic MSH2 methylation had been detected, which were primarily epithelial based, were those in which the EPCAM gene is most highly expressed. In a proportion of cases with EPCAM deletions, no MSH2 methylation was detectable in the peripheral blood lymphocytes at all, however, methylation was consistently high in colorectal mucosa (Kovacs et al., 2009). The steps in the epigenetic process by which MSH2 methylation is induced have not been defined. Since the degree of methylation reflects the tissue-distribution of EPCAM expression and thus is directly related to the levels of activity of the EPCAM promoter, this may be mediated by an RNA-based mechanism. The resultant fusion transcripts result in premature translation termination and hence will be subject to nonsense-mediated decay. This may in turn induce methylation of the MSH2 promoter. However, the methylation is restricted to the allele on which the germ line deletion is located, and thus must act in cis, arguing against an RNA-based feedback mechanism. It is more plausible in cells expressing high levels of EPCAM, that the MSH2 allele located downstream of the deletion is placed under the control of the activated EPCAM promoter, such that extension of transcription into MSH2 overrides the MSH2 promoter. RNA polymerase activity may displace the transcription factors that would normally bind to the MSH2 promoter, or the simple lack of MSH2 promoter activity, may ultimately result in the accrual of MSH2 methylation as a secondary consequence. In tissues that do not express EPCAM above basal levels, some normal MSH2 expression may be retained on the affected allele, with transcription initiation from the correct site. This possibility has not been tested, but if this is the case, the phenotype associated with EPCAM deletions may be more confined to EPCAM-expressing tissues than the severe, wide-spread tumor distribution frequently observed in cases with germ line mutations within MSH2 itself. Nevertheless, epithelial tissues expressing EPCAM are typically those most susceptible to the development of tumors associated with Lynch syndrome. Thus constitutional loss of MSH2 expression due to linked terminal deletion of EPCAM represents an additional etiological mechanism for Lynch syndrome-associated tumors. Since MSH2 epimutations are secondary to a fundamental genetic defect that is carried faithfully through the germ line from one generation to the next, this gives rise to a 50% risk of transmission to offspring, and has been found to underlie the Lynch syndrome phenotype in large families showing autosomal dominant inheritance. Thus screening for this defect needs to be incorporated into routine molecular diagnosis for Lynch syndrome cases with loss of MSH2 expression in their tumors that cannot be attributed to a germ line mutation of MSH2 itself. Due to the degree of MSH2 methylation mosaicism in EPCAM deletion carriers, some of whom had a complete lack of detectable methylation in their peripheral blood DNA, routine diagnostic screening for this defect cannot rely on detection of this epigenetic modification in genomic DNA isolated from blood, which represents the most common sample source used for germ line screening. Instead screening must be based upon detection of the primary genetic alteration in germ line DNA that gives rise to MSH2 epimutations as a secondary event.

VI. EPIMUTATIONS OF ADDITIONAL CANCER GENES IN CANCER SUSCEPTIBILITY SYNDROMES

In recent years there have been further reports of epimutations in additional cancer-related genes that give rise to other early-onset cancer phenotypes. While these isolated reports may suggest that epimutations may be exceedingly rare, or only affect certain genes, this field of study is nascent and a broader etiological role for epimutations in cancer susceptibility may emerge with further research.

A. Epimutation of the DAPK1 gene in chronic lymphocytic leukemia

Chronic lymphocytic leukemia (CLL) is one of the most common forms of adult leukemia in which the B lymphocytes undergo malignant change. CLL is usually sporadic in occurrence, affecting adults over the age of 60 years, but approximately 10% of cases are familial and associated with a younger age of onset (Yuille et al., 2000). Most familial cases comprise nuclear families with a few affected members, with CLL sometimes cosegregating with other B lymphocytic proliferative conditions as well. Case-control studies have shown the relative risk of CLL in an individual who has a first-degree relative with CLL is elevated up to 8.5-fold, providing evidence of an underlying genetic basis (Brown, 2008). However, the causes of CLL are heterogeneous, which has confounded linkage and association studies in determining the genes involved. Genome-wide linkage analyses in individual families have found little evidence for the involvement of a single locus, although weak evidence for linkage at several distinct loci has been found. However, the success of linkage analysis is based upon the assumption of Mendelian inheritance and the conferral of a substantive disease penetrance by the causative mutation. The general concensus therefore has been that CLL is caused by the coinheritance of low-penetrance genetic alleles that have an additive effect, hence the observation of familial clusters of CLL. Large CLL kindreds with affected members spanning multiple generations are extremely rare, but these cases have been compatible with autosomal dominant inheritance of a single gene defect conferring a high risk of developing disease. In one such family with seven affected members spanning three generations, genome-wide linkage analysis identified a region of chromosome 9 with a high lod score (Raval et al., 2007). This region contained three known genes, of which one, *DAPK1*, seemed a likely candidate given that it encodes the death-associated protein kinase 1, which promotes apoptosis in response to various stimuli and suppresses tumor metastasis *in vivo*. *DAPK1* was found to be frequently downregulated in sporadic CLL samples, as well as leukemic cell lines, due to acquired methylation of the promoter, further implicating it in CLL (Raval et al., 2007). Sequencing of the *DAPK1* gene in the affected members of the apparently autosomal dominant family failed to identify any mutation within the coding region. However, identification of several informative SNPs enabled allelic expression studies to be conducted. These showed significant allelic expression imbalance in affected family members, with suppression of one particular allele, and this was accompanied by allelic methylation of the *DAPK1* promoter (Fig. 8.7; Raval et al., 2007). Shotgun sequencing of the genomic regions either side of the *DAPK1* gene in an affected and unaffected family member, and comparison between the two sequences identified a large number of single nucleotide changes, most of which were ruled out as previously identified benign SNPs, or variants found in the healthy control population. This monumental undertaking finally led to the isolation of a single-nucleotide change, c.-6531A>G, a significant distance upstream of the gene itself (Raval et al., 2007). Functional studies showed that this change resulted in reduced transcriptional output mediated by an increased binding affinity for the HOXB7 transcriptional repressor (Fig. 8.7). Methylation analyses of the peripheral blood mononuclear cells in affected family members showed hypermethylation of the *DAPK1* promoter in a significant proportion of alleles, which in turn likely reduced transcriptional levels further (Raval et al., 2007). While the step-wise mechanism by which this sequence mutation several kilobases upstream of the *DAPK1* gene leads to promoter methylation was not defined, the accrual of

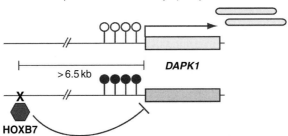

Figure 8.7. *DAPK1* epimutation caused by a single-nucleotide change that increased binding of the HOXB7 repressor in a family with chronic lymphocytic leukemia. The normal *DAPK1* allele is shown in yellow, with horizontal arrow and small rectangles indicating normal transcription of mRNAs. The epimutant allele is shown in gray, with a X indicating the position of the c.6531A>G mutation. The binding affinity of the HOXB7 (blue hexagon) transcriptional repressor is increased by the mutation, which causes transcriptional downregulation and methylation of the mutant allele, as indicated by black lollipops. (For interpretation of the references to color in this figure legend, the reader is referred to the Web version of this chapter.)

methylation as a secondary consequence of reduced transcriptional activity was suggested, ultimately consolidating gene inactivation on hypermethylated alleles. Alternatively, HOXB7 may attract and form complexes with chromatin modifiers. The finding of a heterozygous germ line sequence change in affected members of this family was consistent with the autosomal dominant pattern of disease inheritance observed. However, this case provides an exemplary precedent for the interaction between a cryptic genetic change and the epigenetic machinery in altering gene regulation, with dramatic consequences in disease causation. This begs the question whether epimutations play a more wide-spread role in the etiology of familial clusters of CLL, in view of the small numbers of affected individuals within nuclear families, and the potential for somatic epigenetic mosaicism and intergenerational reversibility of some forms of epimutation.

B. CDH1 epimutation in hereditary diffuse gastric cancer

Hereditary diffuse gastric cancer (HDGC) is an autosomal dominant cancer predisposition syndrome characterized by familial clustering of early-onset diffuse gastric cancer and lobular breast cancer (Pharoah *et al.*, 2001). Heterozygous loss-of-function germ line mutations or large deletions of the E-cadherin (*CDH1*) gene are identified in approximately 40% of HDGC families (Guilford *et al.*, 1998). Cancer development typically occurs following somatic inactivation of the remaining normal *CDH1* allele by LOH, mutation, or epigenetic inactivation, with the resultant tumors displaying either loss of E-cadherin protein expression or altered subcellular localization (Oliveira *et al.*, 2009).

Germ line allelic expression imbalance is often detectable when germ line point mutations or structural alterations within genes result in premature termination of the encoded protein, since they elicit nonsense-mediated decay of the affected mRNA transcript. In a recent study of HDGC probands for whom no causative germ line mutation had been identified, allelic quantification of *CDH1* transcripts was conducted using common exonic SNPs within the *CDH1* gene, for which the probands were heterozygous, to identify any allelic expression imbalance in their peripheral blood mRNA, that would indicate the presence of such frameshift mutations. Several cases demonstrating severe allelic expression imbalance were identified, amongst which a single proband was found to have methylation of the *CDH1* promoter in peripheral blood DNA (Pinheiro *et al.*, 2010). The proband was also heterozygous for a common germ line SNP within the *CDH1* promoter, allowing allelic methylation analyses to be conducted. Methylation was found to be associated with a single allele of this promoter SNP, which was shown to be in phase with the allele of the exonic SNP that was reduced in mRNA, consistent with a consitutional epimutation of the *CDH1* gene in this case. Interestingly, the family did not fulfill the strict criteria for HDGC, but did have a positive family history of gastric cancer, as well as breast and colorectal cancer, which also sometimes occur in the context of HDGC. Unfortunately, aside from the proband in whom a *CDH1* epimutation was identified, the other affected members of the family were deceased, preventing further analyses of additional relatives to determine if this novel defect was the cause of their disease, or its heritability. Further screening of an additional cohort of gastric cancer family porbands failed to identify any further cases with consitutional *CDH1* methylation, suggesting this defect is rare in occurrence (Pinheiro *et al.*, 2010).

C. Constitutional Methylation of BRCA1 in familial breast cancer

Familial early-onset breast cancer accounts for about 10% of all breast cancers and usually demonstrates autosomal dominant inheritance. The major genes associated with familial breast cancer are the *BRCA1* and *BRCA2* genes, and heterozygous germ line inactivating mutations of these two genes together account for approximately half of cases (Peto *et al.*, 1999). *BRCA1* mutations typically give rise to early-onset breast cancer specifically in women, with the tumors often showing characteristic histological features resembling medullary cancer and loss of immunoexpression of the estrogen receptor, the progesterone receptor and HER2, and these characteristics have been used to select cases for *BRCA1* germ line screening (Miolo *et al.*, 2009).

In a recent study, a small number (seven) of probands with familial early-onset breast cancer whose tumors demonstrated the immunohistochemical features associated with *BRCA1* germ line mutations, but for whom standard germ line screening had failed to identify a pathogenic mutation of either of the

BRCA genes, were screened for constitutional methylation of the *BRCA1* promoter using highly sensitive methods. In one proband who had developed two early-onset breast tumors, hypermethylation was detected in 5–14% of alleles in her peripheral blood and buccal mucosa, suggestive of a constitutional epimutation demonstrating somatic mosaicism (Snell *et al.*, 2008). Furthermore, in one of her primary breast tumors, the methylation had reached a level 60%, strongly suggesting the defect played an etiological role in her cancer development. No further study of the affect of this epimutation on gene activity was conducted and family members were not available for testing to determine if the epimutation had been inherited. Intriguingly, low levels of methylation (1%) were identified in two additional cases, and these also showed elevated levels of *BRCA1* methylation in their tumors, whereas the cases without any trace of constituional methylation were negative for *BRCA1* methylation in their tumors (Snell *et al.*, 2008). No healthy controls were included in this study and therefore the levels of constitutional methylation, if any, in the normal population were not assessed. The aetiological role of constitutional epimutations of the BRCA genes in breast cancer thus remains uncertain at present.

VII. CONSTITUTIONAL EPIMUTATIONS IN OTHER HUMAN DISEASES

Additional observations of constitutional epigenetic change resulting in altered gene activity in humans, which might also be regarded as constitutional epimutations, have also been identified in other high-penetrance genes to cause diseases other than cancer. These cases provide important insight into the underlying mechanisms that can give rise to epimutation in disease etiology in general.

A. Familial case of α-thalassemia due to epimutation of the HBA2 gene and the role of antisense expression

In a familial case of α-thalassemia, the disease-associated *HBA2* gene was found to be genetically intact, but the promoter of one allele was hypermethylated and the allele transcriptionally silenced in the peripheral blood lymphocytes of the male proband and his affected mother. Interestingly, no methylation was detected in the proband's spermatozoa, indicating epigenetic reversal in his germ line. The *HBA2* epimutation was found to be linked to an interstitial deletion downstream of *HBA2*, which removed the neighboring globin genes and the distal portion of the *LUC7L* gene that is transcribed from the antisense strand in the opposite direction to *HBA2*, while leaving *HBA2* itself structurally intact (Fig. 8.8A). This placed the tail end of the *HBA2* gene near the oppositely oriented promoter of the *LUC7L* gene (Barbour *et al.*, 2000). Deletion of the final exons of *LUC7L* resulted in the removal of the transcription termination

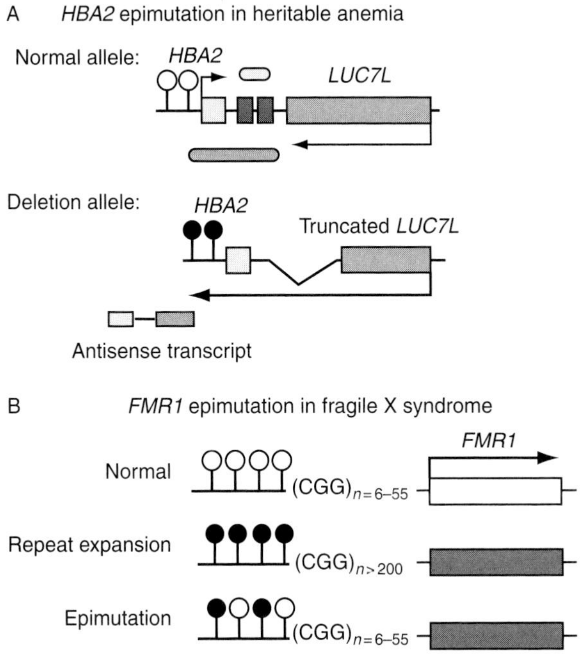

A   *HBA2* epimutation in heritable anemia

B   *FMR1* epimutation in fragile X syndrome

Figure 8.8. Mechanisms of constitutional epigenetic silencing in human disease. (A) Maternally inherited epimutation of *HBA2* in a familial case of α-thalassemia. The *HBA2* gene is shown in yellow, the *LUC7* gene in blue, and intervening globin genes in pink. Transcriptional activity is shown by horizontal arrows and colored rectangles indicate mRNAs from the genes they are transcribed from. Heterozygous deletion of the terminal exons of the *LUC7* gene result in transcriptional elongation up into *HBA2* on the antisense strand, causing *HBA2* promoter methylation (black lollipops) and loss of *HBA2* expression. (B) Epigenetic inactivation of the *FMR1* gene in fragile X syndrome due to methylation induced by repeat expansion (middle), or by mosaic epimutation in the absence of repeat expansion (bottom). The degree of methylation determined the penetrance of the phenotype in different members of a fragile X family. (For interpretation of the references to color in this figure legend, the reader is referred to the Web version of this chapter.)

signal of this gene, and it was subsequently shown that antisense transcripts derived from the *LUC7L* promoter continued through the deletion breakpoint and upstream into *HBA2*. Since the *LUC7L* gene is widely expressed, its promoter activity overrode that of the hematopoietic *HBA2* gene, ultimately resulting in methylation of the *HBA2* promoter. It was interesting to note that the promoter was demethylated in the germ line of the male proband in spite of the presence of the *cis*-genetic deletion underpinning his epimutation. Since the proband had inherited the deletion and consequent epimutation from his mother in classic autosomal dominant pattern, this suggests that the *HBA2* methylation subsequently arose somatically on the allele bearing the deletion. Intricate experiments were conducted in a mouse transgenic model and embryonic stem cells to recapitulate the epigenetic events associated with the deletion. These showed that the *HBA2* promoter became methylated on the epimutant allele during cell differentiation early in embryonic development, contemporaneous with the *de novo* methylation events that establish early somatic epigenetic patterns during implantation. This was the first example of transcriptional silencing of a nonimprinted gene due to antisense transcription (Tufarelli *et al.*, 2003). This mechanism is not dissimilar to that subsequently found for *MSH2* epimutations, in that epimutations of both genes were caused by a failure in transcription termination from a neighboring locus (Ligtenberg *et al.*, 2009).

The role of endogenous antisense transcription in triggering epigenetic silencing may be widespread in carcinogenesis, with over 100 sense–antisense transcription pairs identified among tumor suppressor genes alone (Yu *et al.*, 2008). This mechanism has been definitively proven for the *p15* tumor suppressor gene and its naturally occurring antisense transcript *p15AS*. The transcript pairs appear to be expressed inversely, with *p15AS* normally expressed at low levels in which *p15* is active, but in leukemia cells in which *p15* is silenced by methylation, *p15AS* was found to be highly expressed. Recombinant *p15AS* expression triggered epigenetic silencing of *p15* both in *cis* and in *trans* via heterochromatin formation, but failed to induce *p15* promoter methylation in differentiated cells. Interestingly, however, methylation was induced during differentiation of mouse embryonic stem cells that had been transfected with recombinant *p15AS*. It was postulated that inappropriate elevation of *p15AS* in cancer stem cells could give rise to subsequent methylation of the *p15* sense transcript in neoplastic cells derived from them (Yu *et al.*, 2008). This notion could be extended to the derivation of constitutional epimutations, whereby endogenous antisense expression in gametes or embryonic stem cells could subsequently give rise to epigenetic silencing in the normal somatic tissues in the offspring.

B. Epigenetic dysregulation caused by hypo- and hypermethylation of repeat regions

Various neuromuscular disorders including Huntington's disease, myotonic dystrophy, and fragile X syndrome have been ascribed to the expansion in copy number of triplet repeats within or near the promoter of the disease-associated gene (Dion and Wilson, 2009). Trinucleotide-repeat expansion has been shown to induce cytosine methylation (where the triplet repeat includes a CpG dinucleotide) or the induction of repressive histone modifications where the repeat does not contain a CpG site (Dion and Wilson, 2009). Normally, the tandem

(CGG)n repeat upstream of the *FMR1* gene would have a copy number (n) of 5–65 and is unmethylated. However, in cases with fragile X syndrome, this triplet repeat increases to 200 or more copies in tandem, incurring methylation of the CpG within the repeat unit (Fig. 8.8B). The disease severity correlates with the degree of expansion and choice of X-chromosome inactivation in females (Loesch et al., 2007). Fragile X syndrome is typically maternally transmitted. A familial case of fragile X has been described with normal copy numbers of the triplet repeat unit in the *FMR1* promoter, but in which mosaic methylation patterns were found in the affected mother and her children, with variable phenotypic expression (Fig. 8.8). The level of methylation mosaicism at the otherwise structurally normal *FMR1* locus accounted for the variable penetrance of the phenotype witnessed in the mother and her children (Stoger et al., 1997).

By contrast, contraction of the polymorphic D4Z4 repeat array at the telomeric end of chromosome 4q has been associated with the autosomal dominant myopathy, facioscapulohumeral muscular dystrophy (FSHD1). In healthy individuals, this repeat array comprises 11–150 copies, whereas in FSHD1 individuals this array is reduced to 1–10 units on one allele. This contraction has been shown to result in hypomethylation of the repeats, which in turn is thought to result in transcriptional upregulation of multiple linked loci. However, some FSHD1 cases, including familial cases, have also been described with hypomethylation of the D4Z4 repeat array, even though the copy numbers are normal, suggesting an epigenetic-based phenomenon independent of the genetic context (van Overveld et al., 2003). Maternal inheritance of the phenotype and hypomethylated state was demonstrated (van Overveld et al., 2003). Additional cases that similarly have normal repeat copy numbers and hypomethylation have since been identified (de Greef et al., 2009).

VIII. INHERITANCE OF ALTERED EPIGENETIC STATES IN MOUSE PHENOTYPES

Intergenerational epigenetic inheritance has been observed in particular inbred strains of mice, most notably the A^{vy} yellow-coated mouse and the *Axin-fused* (*Axin^{Fu}*) tail-kink mouse (Fig. 8.9). In both strains, the phenotypes were attributable to the differential epigenetic states at *cis*-acting intra-cisternal A particle (IAP) retroelements (Morgan et al., 1999; Rakyan et al., 2003). When the retroelement is unmethylated and transcriptionally active, it overrides the activity of the endogenous promoter. When the retroelement is methylated, the gene is normally transcribed.

A^{vy} mice have a phenotypic spectrum of yellow fur, obesity, diabetes, and increased risk of cancer, to a dark pseudoagouti coat color with apparently normal phenotype, with intermediates displaying a variegated coat color.

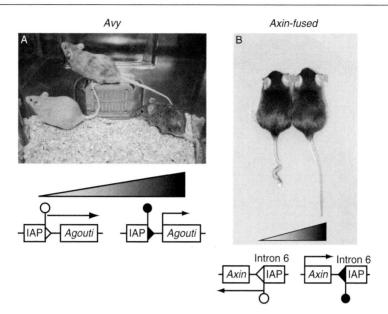

Figure 8.9. Mouse phenotypes induced by the epigenetic state at nearby IAP retrotransposons. (A) A^{vy} mice showing the range of phenotypes from yellow-coated (left) to mottled (middle) and psedoagouti (right) according to the levels of methylation at the differentially methylated IAP retroelement upstream, as indicated by the gradient triangle beneath. A schematic of the IAP insertion site upstream of the *Agouti* gene is shown. When the IAP promoter is unmethylated and active, the *Agouti* gene is constitutively expressed, whereas when the IAP is methylated, *Agouti* is under the control of its endogenous promoter and expressed in appropriate tissues. (B) *Axin-fused* mice showing the tail-kink phenotype (left) and beneath, a schematic of the methylation state of an IAP retro-element inserted within *Axin* intron 6. When the IAP is unmethylated, it generates gain-of-function antisense transcripts that result is axial duplication during embryonic axis formation. When the IAP is methylated and silenced, the *Axin* gene is normally expressed with no observed phenotype (right). Photographs kindly provided by Dr. Nadia Whitelaw and Professor Emma Whitelaw.

A^{vy} mice differ from wild-type *Agouti* (A) mice in that they have an IAP retro-element inserted upstream of the *Agouti* gene placing the *Agouti* locus under the transcriptional control of the retroelement (Fig. 8.9A). However, irrespective of the phenotype observed, A^{vy} mice are genetically identical to one another, and differ only with respect to the epigenetic status of this upstream IAP retro-element. Those with yellow fur and the full disease phenotype are caused by the inappropriate overexpression of the *Agouti* gene, including in tissues in which this gene would not normally be active. This is because the retroelement promoter is unmethylated and consequently overrides the activity of the endog-enous *Agouti* promoter. When the retroelement is fully methylated, the *Agouti* gene is normally expressed and the coat color is dark. Those with mottled coats are epigenetic mosaics, with variable levels of methylation at the retroelement. The methylation levels are proportional to the amount of dark, pseudoagouti coloration. Interestingly, the phenotype among these isogenic A^{vy} mouse strains was shown to be transmissible through the germ line to offspring (Morgan et al., 1999). This is most likely attributable to the generalized resistance of IAP retroelements to epigenetic reprogramming in the germ line, such that the epigenetic state of the retroelement is retained in the gametes (Blewitt et al., 2006; Lane et al., 2003). Interestingly, when the A^{vy} allele was bred in one mouse background strain the intergenerational transmission of the phenotype conferred by the A^{vy} allele demonstrated a significant bias in maternal transmission, which was compounded when also transmitted through the grandmaternal germ line. This maternal effect was separable from any intrauterine environmental effects that may have been invoked, by transferring the fertilized eggs of yellow dams to dark-colored foster mothers, with the same biased outcome in the offspring (Morgan et al., 1999). For this strain, the coat color of the sire appeared to play no contributory role in determining the phenotypes of his offspring, even though it was later demonstrated that the methylation status of the IAP in mature spermatozoa reflected that found in the somatic tissues of the male, confirming the resistance of IAP retroelements to epigenetic reprogramming in the germ line (Rakyan et al., 2003). However, when the A^{vy} allele was bred on another background strain, the phenotypic inheritance of offspring was found to be maternally and paternally contributed, suggesting that genetic background can also play a role in the phenotypic outcome of this epigenetically sensitive allele (Rakyan et al., 2003).

In the $Axin^{Fu}$ mouse strain, the tail-kink phenotype is conferred by the generation of gain-of-action mutant $Axin$ transcripts that originate from an IAP retroelement inserted into intron 6 of the $Axin$ gene, when it is unmethylated (Fig. 8.9B; Rakyan et al., 2003). This results in axial duplication during embryonic axis formation. When the retroelement is methylated and transcriptionally silenced, the phenotype is normal. Both the normal and tail-kink phenotypes of the $Axin^{Fu}$ strain are transmissible through the male and female germ line, once again, due to the retention of the differential epigenetic state of the retroelement in the germ line. Interestingly, the penetrance of the tail-kink phenotype was stronger when it originated from the male parent, presumably due to the overexpression of the mutant IAP-derived transcripts. This may be due to the active demethylation of the male genome within the male pronucleus of the zygote, while the maternal genome remains protected from this epigenetic reprogramming event. With preferential demethylation of the paternally derived IAP retroelement within the $Axin$ gene, upregulation of the mutant transcripts would ensue, fortifying the tail-kink phenotype when paternally transmitted (Rakyan et al., 2003).

The reason the mouse phenotypes associated with the altered epigenetic and transcriptional states of retrotransposable elements are so important is that the human genome is littered with such retroviral elements and thus a similar mechanism of epigenetic dysregulation of genes in humans might also be applicable in clinical conditions. A comparable mechanism involving the variable epigenetic states within retroviral and other types of repetitive sequences could be invoked to explain existing epimutations that are reversed in some offspring but not in others, including constitutional epimutations of *MLH1*.

IX. CONCLUSIONS

Following the completion of the human genome sequence, attention in the new millenium has turned to the epigenome and the role of epigenetics in human disease causation. While the genetic causes of many diseases are now well established, it has become clear that mutations within the coding regions of the relevant disease-associated gene fail to account for all cases with a positive clinical diagnosis. Some of these cases are likely to be attributable to cryptic genetic changes, including mutations within noncoding portions of the known disease genes. However, recent evidence for the role of epimutations as an alternative etiological mechanism for disease has emerged. By definition, these cases bear no sequence mutation within the gene, and the affected locus is intact, but the gene's transcriptional regulation has been altered by epigenetic change.

To date, the identification of constitutional epimutations has been confined to high-penetrance diseases. Gain-of-function epimutations, whereby loss of methylation results in transcriptional activation, appear to act in an autosomal dominant manner, such that only one allele need to be altered for a phenotypic outcome. Inactivating epimutations caused by erroneous constitutional methylation have been identified in a handful of patients with conditions conferred by haploinsufficiency. As discussed in depth, these include hereditary cancer syndromes in which the germ line or constitutional event provides an inherent predisposition to the development of cancer in autosomal dominant fashion, but somatic loss of the remaining functional allele occurs before cancer development ensues. In addition, epimutations of genes subjected to genomic imprinting have also been identified in a handful of cases with a congenital disorder (not covered in this chapter). By definition, only one copy of an imprinted gene would normally be active, and this is dictated by the parental origin of the allele. Inactivation of the single functional allele via deletion, genetic mutation, or altered epigenetic setting (imprintor mutation) in the germ line results in the disease phenotype. Epimutations due to inappropriate loss or gain of methylation of the disease-associated imprinted gene, leading to overexpression or loss of expression from the single active copy, have been described in a small number of cases in the absence of an underlying sequence change (Buiting et al., 1998; Gicquel et al., 2005; Mackay et al., 2006). Once again, these have primarily occurred in sporadic cases, including in monozygotic twins discordant for the disease phenotype, suggesting that they arise as a de novo event during embryogenesis (Bliek et al., 2009; Gicquel et al., 2005; Yamazawa et al., 2008).

Due to the paucity in number of cases identified with constitutional epimutations, it has been widely assumed that this epigenetic mechanism of disease etiology is rare. While this might indeed be the case, this is a nascent field of enquiry. The causative role for MLH1 epimutations in Lynch syndrome is now established, albeit a seemingly minor role in terms of frequency. However, the evidence of an etiological role for epimutations of CDH1 and BRCA1, for example, is scant due to present lack of investigation. To establish the true role and frequency of this type of mechanism would require systematic and comprehensive screening of well-characterized patient cohorts, and the study of additional disease-associated genes for the presence of epimutations in appropriate cases. The level of epigenetic diversity in the healthy population remains to be elucidated. It may be the case that additional genes are affected by epimutation, but these may not confer a disease phenotype, are recessive, or may be associated with a low risk of disease outcome, and therefore remain unrealized. It is plausible that the high-penetrance cases of epimutations identified in recent years represent the tip of the ice-berg, and constitutionally altered epigenetic states may play a more generalized contributory role to late-onset and multifactorial diseases as well. Research into the genetic contribution to complex traits such as diabetes, heart disease, and autoimmunity has focused largely on finding associations with genetic variants that collectively increase the risk of disease. While this may prove to be correct for some of these diseases, the potential role of epigenetic change in elevating disease risk for such traits remains to be investigated.

The heritability of epimutations is variable depending on the underlying mechanism. Where epimutations are genetically mediated by a cis-acting mutation or structural alteration in the vicinity of the gene, even though the sequence of the affected gene is itself normal, the inheritance pattern is consistent with genetic transmission of the defect. While the repressive epigenetic modifications that cause gene inactivation in the soma may not be present in the germ line of the epimutation carrier or their parent, the underlying genetic defect that gives rise to the epimutation secondarily in somatic tissues is nevertheless transmissible within the gametes. Therefore, epimutations associated with an apparent autosomal dominant inheritance pattern are likely to be genetically directed. However, where an epimutation consistently arises spontaneously in sporadic cases of disease or shows stochastic transmission to the next generation, giving rise to non-Mendelian inheritance patterns due to reversion in the germ line, as exemplified by MLH1 epimutations, these may be independent of a linked sequence change. If such a *cis*-genetic change underlies these epimutations, it would be a "leaky" defect, conferring a mere susceptibility of the allele to epimutation. For example, retroviral and repeat sequences, or antisense transcripts, which may themselves be under stochastic epigenetic regulation, could have *cis*-acting effects on genes in the vicinity. Alternatively, the coinheritance of *trans*-acting factors may be responsible for the onset of epimutations and their transmission to a proportion of offspring who have also coinherited the putative *trans* change responsible. In this scenario, it is likely that the epimutations are reversed in the germ line and reestablished postferti-lization in those offspring carrying the *trans*-acting genetic mutation. Until such a time as the mechanisms underlying the onset and inheritance of such defects are defined, these cases present with unprecedented genetic counseling, screen-ing, and ethical issues. This is further compounded by the inherent somatic mosaicism often associated with this type of defect, which has the potential to give rise to false-negative screening results if the defect is undetectable in the source of tissue used for testing. Clearly additional study is required to resolve these confounding factors. Currently, carriers of epimutations need to be advised that there is an elevated risk of vertical transmission, although an accurate numerical value cannot be placed on all cases at present. Where an epimutation is caused by an identifiable linked genetic change, the risk of vertical transmis-sion is 50%, and thus it is important to identify the nature of the change. Current advice is that all first-degree relatives of a confirmed epimutation carrier be offered testing for the epimutation with further posttest genetic counseling. However, much of the work on epimutations is being conducted under research circumstances. This, coupled with the lack of accurate advice that can be given to patients and their families, presents with confounding ethical dilemmas as well. Altogether, understanding the nature of this novel form of disease mecha-nism and its etiological role in human disease is currently in its infancy. Much work remains to be done before the causative role of epimutations becomes fully realized and integrated into mainstream screening and clinical practice.

References

Barbour, V. M., Tufarelli, C., Sharpe, J. A., Smith, Z. E., Ayyub, H., Heinlein, C. A., Sloane-Stanley, J., Indrak, K., Wood, W. G., and Higgs, D. R. (2000). Alpha-thalassemia resulting from a negative chromosomal position effect. *Blood* **96**(3), 800–807.

Blewitt, M. E., Vickaryous, N. K., Paldi, A., Koseki, H., and Whitelaw, E. (2006). Dynamic reprogramming of DNA methylation at an epigenetically sensitive allele in mice. *PLoS Genet.* **2**(4), e49.

Bliek, J., Alders, M., Maas, S. M., Oostra, R. J., Mackay, D. M., Van Der Lip, K., Callaway, J. L., Brooks, A., Van 'T Padje, S., Westerveld, A., *et al.* (2009). Lessons from bws twins: Complex maternal and paternal hypomethylation and a common source of haematopoietic stem cells. *Eur. J. Hum. Genet.* **17**(12), 1625–1634.

Boland, C. R., Thibodeau, S. N., Hamilton, S. R., Sidransky, D., Eshleman, J. R., Burt, R. W., Meltzer, S. J., Rodriguez-Bigas, M. A., Fodde, R., Ranzani, G. N., et al. (1998). A national cancer institute workshop on microsatellite instability for cancer detection and familial predisposition: Development of international criteria for the determination of microsatellite instability in colorectal cancer. Cancer Res. 58(22), 5248–5257.

Brown, J. R. (2008). Inherited predisposition to chronic lymphocytic leukemia. Expert Rev. Hematol. 1(1), 51–61.

Buiting, K., Dittrich, B., Gross, S., Lich, C., Farber, C., Buchholz, T., Smith, E., Reis, A., Burger, J., Nothen, M. M., et al. (1998). Sporadic imprinting defects in prader-willi syndrome and angelman syndrome: Implications for imprint-switch models, genetic counseling, and prenatal diagnosis. Am. J. Hum. Genet. 63(1), 170–180.

Chai, H., and Brown, R. E. (2009). Field effect in cancer-an update. Ann. Clin. Lab. Sci. 39(4), 331–337.

Chan, T. L., Yuen, S. T., Kong, C. K., Chan, Y. W., Chan, A. S., Ng, W. F., Tsui, W. Y., Lo, M. W., Tam, W. Y., Li, V. S., et al. (2006). Heritable germline epimutation of msh2 in a family with hereditary nonpolyposis colorectal cancer. Nat. Genet. 38(10), 1178–1183.

Chong, S., Youngson, N. A., and Whitelaw, E. (2007). Heritable germline epimutation is not the same as transgenerational epigenetic inheritance. Nat. Genet. 39(5), 574–575, author reply 575–576.

Cubas, P., Vincent, C., and Coen, E. (1999). An epigenetic mutation responsible for natural variation in floral symmetry. Nature 401(6749), 157–161.

Cunningham, J. M., Christensen, E. R., Tester, D. J., Kim, C. Y., Roche, P. C., Burgart, L. J., and Thibodeau, S. N. (1998). Hypermethylation of the hmlh1 promoter in colon cancer with microsatellite instability. Cancer Res. 58(15), 3455–3460.

Davis, T. L., Yang, G. J., Mccarrey, J. R., and Bartolomei, M. S. (2000). The h19 methylation imprint is erased and re-established differentially on the parental alleles during male germ cell development. Hum. Mol. Genet. 9(19), 2885–2894.

De Greef, J. C., Lemmers, R. J., Van Engelen, B. G., Sacconi, S., Venance, S. L., Frants, R. R., Tawil, R., and Van Der Maarel, S. M. (2009). Common epigenetic changes of d4z4 in contraction-dependent and contraction-independent fshd. Hum. Mutat. 30(10), 1449–1459.

Dion, V., and Wilson, J. H. (2009). Instability and chromatin structure of expanded trinucleotide repeats. Trends Genet. 25(7), 288–297.

Gazzoli, I., Loda, M., Garber, J., Syngal, S., and Kolodner, R. D. (2002). A hereditary nonpolyposis colorectal carcinoma case associated with hypermethylation of the mlh1 gene in normal tissue and loss of heterozygosity of the unmethylated allele in the resulting microsatellite instability-high tumor. Cancer Res. 62(14), 3925–3928.

Geuns, E., De Rycke, M., Van Steirteghem, A., and Liebaers, I. (2003). Methylation imprints of the imprint control region of the snrpn-gene in human gametes and preimplantation embryos. Hum. Mol. Genet. 12(22), 2873–2879.

Gicquel, C., Rossignol, S., Cabrol, S., Houang, M., Steunou, V., Barbu, V., Danton, F., Thibaud, N., Le Merrer, M., Burglen, L., et al. (2005). Epimutation of the telomeric imprinting center region on chromosome 11p15 in silver-russell syndrome. Nat. Genet. 37(9), 1003–1007.

Goel, A., Nguyen, T. P., Leung, H. C., Nagasaka, T., Rhees, J., Hotchkiss, E., Arnold, M., Banerji, P., Koi, M., Kwok, C. T., et al. (2010). De novo constitutional MLH1 epimutations confer early-onset colorectal cancer in two new sporadic lynch syndrome cases, with derivation of the epimutation on the paternal allele in one. Int. J. Cancer (in press).

Grabowski, M., Mueller-Koch, Y., Grasbon-Frodl, E., Koehler, U., Keller, G., Vogelsang, H., Dietmaier, W., Kopp, R., Siebers, U., Schmitt, W., et al. (2005). Deletions account for 17% of pathogenic germline alterations in mlh1 and msh2 in hereditary nonpolyposis colorectal cancer (hnpcc) families. Genet. Test. 9(2), 138–146.

Guilford, P., Hopkins, J., Harraway, J., Mcleod, M., Mcleod, N., Harawira, P., Taite, H., Scoular, R., Miller, A., and Reeve, A. E. (1998). E-cadherin germline mutations in familial gastric cancer. *Nature* **392**(6674), 402–405.

Gylling, A., Ridanpaa, M., Vierimaa, O., Aittomaki, K., Avela, K., Kaariainen, H., Laivuori, H., Poyhonen, M., Sallinen, S. L., Wallgren-Pettersson, C., *et al.* (2009). Large genomic rearrangements and germline epimutations in lynch syndrome. *Int. J. Cancer* **124**(10), 2333–2340.

Hemminki, A., Peltomaki, P., Mecklin, J. P., Jarvinen, H., Salovaara, R., Nystrom-Lahti, M., De La Chapelle, A., and Aaltonen, L. A. (1994). Loss of the wild type mlh1 gene is a feature of hereditary nonpolyposis colorectal cancer. *Nat. Genet.* **8**(4), 405–410.

Hendriks, Y., Franken, P., Dierssen, J. W., De Leeuw, W., Wijnen, J., Dreef, E., Tops, C., Breuning, M., Brocker-Vriends, A., Vasen, H., *et al.* (2003). Conventional and tissue microarray immunohistochemical expression analysis of mismatch repair in hereditary colorectal tumors. *Am. J. Pathol.* **162**(2), 469–477.

Herman, J. G., Umar, A., Polyak, K., Graff, J. R., Ahuja, N., Issa, J. P., Markowitz, S., Willson, J. K., Hamilton, S. R., Kinzler, K. W., *et al.* (1998). Incidence and functional consequences of hmlh1 promoter hypermethylation in colorectal carcinoma. *Proc. Natl. Acad. Sci. USA* **95**(12), 6870–6875.

Hitchins, M. P., and Ward, R. L. (2007). Erasure of mlh1 methylation in spermatozoa-implications for epigenetic inheritance. *Nat. Genet.* **39**(11), 1289.

Hitchins, M. P., and Ward, R. L. (2009). Constitutional (germline) mlh1 epimutation as an aetiological mechanism for hereditary non-polyposis colorectal cancer. *J. Med. Genet.* **46,** 793–802.

Hitchins, M., Williams, R., Cheong, K., Halani, N., Lin, V. A., Packham, D., Ku, S., Buckle, A., Hawkins, N., Burn, J., *et al.* (2005). Mlh1 germline epimutations as a factor in hereditary nonpolyposis colorectal cancer. *Gastroenterology* **129**(5), 1392–1399.

Hitchins, M. P., Wong, J. J., Suthers, G., Suter, C. M., Martin, D. I., Hawkins, N. J., and Ward, R. L. (2007). Inheritance of a cancer-associated mlh1 germ-line epimutation. *N. Engl. J. Med.* **356**(7), 697–705.

Holliday, R. (1987). The inheritance of epigenetic defects. *Science* **238**(4824), 163–170.

Jass, J. R. (2007). Heredity and DNA methylation in colorectal cancer. *Gut* **56**(1), 154–155.

Knudson, A. G. (1996). Hereditary cancer: Two hits revisited. *J. Cancer Res. Clin. Oncol.* **122**(3), 135–140.

Kolodner, R. D. (1995). Mismatch repair: Mechanisms and relationship to cancer susceptibility. *Trends Biochem. Sci.* **20**(10), 397–401.

Kovacs, M. E., Papp, J., Szentirmay, Z., Otto, S., and Olah, E. (2009). Deletions removing the last exon of tacstd1 constitute a distinct class of mutations predisposing to lynch syndrome. *Hum. Mutat.* **30**(2), 197–203.

Lane, N., Dean, W., Erhardt, S., Hajkova, P., Surani, A., Walter, J., and Reik, W. (2003). Resistance of iaps to methylation reprogramming may provide a mechanism for epigenetic inheritance in the mouse. *Genesis* **35**(2), 88–93.

Ligtenberg, M. J., Kuiper, R. P., Chan, T. L., Goossens, M., Hebeda, K. M., Voorendt, M., Lee, T. Y., Bodmer, D., Hoenselaar, E., Hendriks-Cornelissen, S. J., *et al.* (2009). Heritable somatic methylation and inactivation of msh2 in families with lynch syndrome due to deletion of the 3′ exons of tacstd1. *Nat. Genet.* **41**(1), 112–117.

Lindor, N. M., Burgart, L. J., Leontovich, O., Goldberg, R. M., Cunningham, J. M., Sargent, D. J., Walsh-Vockley, C., Petersen, G. M., Walsh, M. D., Leggett, B. A., *et al.* (2002). Immunohisto-chemistry versus microsatellite instability testing in phenotyping colorectal tumors. *J. Clin. Oncol.* **20**(4), 1043–1048.

Linnaeus, C. (1749). De peloria. Ph. D thesis, Amoenitates Academy, Uppsala, Sweden.

Loesch, D. Z., Bui, Q. M., Huggins, R. M., Mitchell, R. J., Hagerman, R. J., and Tassone, F. (2007). Transcript levels of the intermediate size or grey zone fragile x mental retardation 1 alleles are raised, and correlate with the number of cgg repeats. *J. Med. Genet.* **44**(3), 200–204.

Lucifero, D., Mann, M. R., Bartolomei, M. S., and Trasler, J. M. (2004). Gene-specific timing and epigenetic memory in oocyte imprinting. *Hum. Mol. Genet.* **13**(8), 839–849.

Luo, D., Carpenter, R., Copsey, L., Vincent, C., Clark, J., and Coen, E. (1999). Control of organ asymmetry in flowers of antirrhinum. *Cell* **99**(4), 367–376.

Lynch, H. T. (1999). Hereditary nonpolyposis colorectal cancer (hnpcc). *Cytogenet. Cell Genet.* **86**(2), 130–135.

Lynch, H. T., and Lynch, J. F. (2004). Lynch syndrome: History and current status. *Dis. Markers* **20**(4–5), 181–198.

Mackay, D. J., Hahnemann, J. M., Boonen, S. E., Poerksen, S., Bunyan, D. J., White, H. E., Durston, V. J., Thomas, N. S., Robinson, D. O., Shield, J. P., *et al.* (2006). Epimutation of the tndm locus and the beckwith-wiedemann syndrome centromeric locus in individuals with transient neonatal diabetes mellitus. *Hum. Genet.* **119**(1–2), 179–184.

Miolo, G., Canzonieri, V., De Giacomi, C., Puppa, L. D., Dolcetti, R., Lombardi, D., Perin, T., Scalone, S., Veronesi, A., and Viel, A. (2009). Selecting for brca1 testing using a combination of homogeneous selection criteria and immunohistochemical characteristics of breast cancers. *BMC Cancer* **9**, 360.

Miyakura, Y., Sugano, K., Akasu, T., Yoshida, T., Maekawa, M., Saitoh, S., Sasaki, H., Nomizu, T., Konishi, F., Fujita, S., *et al.* (2004). Extensive but hemiallelic methylation of the hmlh1 promoter region in early-onset sporadic colon cancers with microsatellite instability. *Clin. Gastroenterol. Hepatol.* **2**(2), 147–156.

Morak, M., Schackert, H. K., Rahner, N., Betz, B., Ebert, M., Walldorf, C., Royer-Pokora, B., Schulmann, K., Von Knebel-Doeberitz, M., Dietmaier, W., *et al.* (2008). Further evidence for heritability of an epimutation in one of 12 cases with mlh1 promoter methylation in blood cells clinically displaying hnpcc. *Eur. J. Hum. Genet.* **16**(7), 804–811.

Morgan, H. D., Sutherland, H. G., Martin, D. I., and Whitelaw, E. (1999). Epigenetic inheritance at the agouti locus in the mouse. *Nat. Genet.* **23**(3), 314–318.

Morgan, H. D., Dean, W., Coker, H. A., Reik, W., and Petersen-Mahrt, S. K. (2004). Activation-induced cytidine deaminase deaminates 5-methylcytosine in DNA and is expressed in pluripotent tissues: Implications for epigenetic reprogramming. *J. Biol. Chem.* **279**(50), 52353–52360.

Morgan, H. D., Santos, F., Green, K., Dean, W., and Reik, W. (2005). Epigenetic reprogramming in mammals. *Hum Mol Genet* **14**(Spec. No. 1), R47–R58.

Niessen, R. C., Hofstra, R. M., Westers, H., Ligtenberg, M. J., Kooi, K., Jager, P. O., De Groote, M. L., Dijkhuizen, T., Olderode-Berends, M. J., Hollema, H., *et al.* (2009). Germline hypermethylation of mlh1 and epcam deletions are a frequent cause of lynch syndrome. *Genes Chromosomes Cancer* **48**(8), 737–744.

Oliveira, C., Sousa, S., Pinheiro, H., Karam, R., Bordeira-Carrico, R., Senz, J., Kaurah, P., Carvalho, J., Pereira, R., Gusmao, L., *et al.* (2009). Quantification of epigenetic and genetic 2nd hits in cdh1 during hereditary diffuse gastric cancer syndrome progression. *Gastroenterology* **136** (7), 2137–2148.

Peltomaki, P., and Vasen, H. (2004). Mutations associated with hnpcc predisposition—update of icg-hnpcc/insight mutation database. *Dis. Markers* **20**(4–5), 269–276.

Peto, J., Collins, N., Barfoot, R., Seal, S., Warren, W., Rahman, N., Easton, D. F., Evans, C., Deacon, J., and Stratton, M. R. (1999). Prevalence of brca1 and brca2 gene mutations in patients with early-onset breast cancer. *J. Natl. Cancer Inst.* **91**(11), 943–949.

Pharoah, P. D., Guilford, P., and Caldas, C. (2001). Incidence of gastric cancer and breast cancer in cdh1 (e-cadherin) mutation carriers from hereditary diffuse gastric cancer families. *Gastroenterology* **121**(6), 1348–1353.

Pinheiro, H., Bordeira-Carrico, R., Seixas, S., Carvalho, J., Senz, J., Oliveira, P., Inacio, P., Gusmao, L., Rocha, J., Huntsman, D., et al. (2010). Allele-specific cdh1 downregulation and hereditary diffuse gastric cancer. Hum. Mol. Genet. 19(5), 943–952.

Rakyan, V. K., Chong, S., Champ, M. E., Cuthbert, P. C., Morgan, H. D., Luu, K. V., and Whitelaw, E. (2003). Transgenerational inheritance of epigenetic states at the murine axin(fu) allele occurs after maternal and paternal transmission. Proc. Natl. Acad. Sci. USA 100(5), 2538–2543.

Raval, A., Tanner, S. M., Byrd, J. C., Angerman, E. B., Perko, J. D., Chen, S. S., Hackanson, B., Grever, M. R., Lucas, D. M., Matkovic, J. J., et al. (2007). Downregulation of death-associated protein kinase 1 (dapk1) in chronic lymphocytic leukemia. Cell 129(5), 879–890.

Santos, F., Hendrich, B., Reik, W., and Dean, W. (2002). Dynamic reprogramming of DNA methylation in the early mouse embryo. Dev. Biol. 241(1), 172–182.

Sassone-Corsi, P. (2002). Unique chromatin remodeling and transcriptional regulation in spermatogenesis. Science 296(5576), 2176–2178.

Slaughter, D. P., Southwick, H. W., and Smejkal, W. (1953). Field cancerization in oral stratified squamous epithelium; clinical implications of multicentric origin. Cancer 6(5), 963–968.

Snell, C., Krypuy, M., Wong, E. M., Loughrey, M. B., and Dobrovic, A. (2008). Brca1 promoter methylation in peripheral blood DNA of mutation negative familial breast cancer patients with a brca1 tumour phenotype. Breast Cancer Res. 10(1), R12.

Stoger, R., Kajimura, T. M., Brown, W. T., and Laird, C. D. (1997). Epigenetic variation illustrated by DNA methylation patterns of the fragile-x gene fmr1. Hum. Mol. Genet. 6(11), 1791–1801.

Suter, C. M., Martin, D. I., and Ward, R. L. (2004). Germline epimutation of mlh1 in individuals with multiple cancers. Nat. Genet. 36(5), 497–501.

Tomlinson, I. P., Roylance, R., and Houlston, R. S. (2001). Two hits revisited again. J. Med. Genet. 38(2), 81–85.

Tufarelli, C., Stanley, J. A., Garrick, D., Sharpe, J. A., Ayyub, H., Wood, W. G., and Higgs, D. R. (2003). Transcription of antisense rna leading to gene silencing and methylation as a novel cause of human genetic disease. Nat. Genet. 34(2), 157–165.

Umar, A., Boland, C. R., Terdiman, J. P., Syngal, S., De La Chapelle, A., Ruschoff, J., Fishel, R., Lindor, N. M., Burgart, L. J., Hamelin, R., et al. (2004). Revised bethesda guidelines for hereditary nonpolyposis colorectal cancer (lynch syndrome) and microsatellite instability. J. Natl. Cancer Inst. 96(4), 261–268.

Valle, L., Carbonell, P., Fernandez, V., Dotor, A. M., Sanz, M., Benitez, J., and Urioste, M. (2007). Mlh1 germline epimutations in selected patients with early-onset non-polyposis colorectal cancer. Clin. Genet. 71(3), 232–237.

Van Overveld, P. G., Lemmers, R. J., Sandkuijl, L. A., Enthoven, L., Winokur, S. T., Bakels, F., Padberg, G. W., Van Ommen, G. J., Frants, R. R., and Van Der Maarel, S. M. (2003). Hypomethylation of d4z4 in 4q-linked and non-4q-linked faciosacpulohumeral muscular dystrophy. Nat. Genet. 35(4), 315–317.

Vasen, H. F., Mecklin, J. P., Khan, P. M., and Lynch, H. T. (1991). The international collaborative group on hereditary non-polyposis colorectal cancer (icg-hnpcc). Dis. Colon Rectum 34(5), 424–425.

Whitelaw, N. C., and Whitelaw, E. (2008). Transgenerational epigenetic inheritance in health and disease. Curr. Opin. Genet. Dev. 18(3), 273–279.

Xing, Y., Shi, S., Le, L., Lee, C. A., Silver-Morse, L., and Li, W. X. (2007). Evidence for transgenerational transmission of epigenetic tumor susceptibility in Drosophila. PLoS Genet. 3(9), 1598–1606.

Yamazawa, K., Kagami, M., Fukami, M., Matsubara, K., and Ogata, T. (2008). Monozygotic female twins discordant for silver-russell syndrome and hypomethylation of the h19-dmr. J. Hum. Genet. 53(10), 950–955.

Yu, W., Gius, D., Onyango, P., Muldoon-Jacobs, K., Karp, J., Feinberg, A. P., and Cui, H. (2008). Epigenetic silencing of tumour suppressor gene p15 by its antisense rna. *Nature* **451**(7175), 202–206.

Yuille, M. R., Matutes, E., Marossy, A., Hilditch, B., Catovsky, D., and Houlston, R. S. (2000). Familial chronic lymphocytic leukaemia: A survey and review of published studies. *Br. J. Haematol.* **109**(4), 794–799.

Zhou, H. H., Yan, S. Y., Zhou, X. Y., Du, X., Zhang, T. M., Cai, X., Lu, Y. M., Cai, S. J., and Shi, D. R. (2008). Mlh1 promoter germline-methylation in selected probands of chinese hereditary non-polyposis colorectal cancer families. *World J. Gastroenterol.* **14**(48), 7329–7334.

Cancer Epigenome

9

Cancer Epigenome

Matthias Lechner, Chris Boshoff, and Stephan Beck

UCL Cancer Institute, University College London, London, United Kingdom

I. Introduction
   A. Human epigenome project
   B. Cancer genome project
   C. The cancer genome atlas
   D. NIH roadmap epigenomics program
   E. International cancer genome consortium
   F. International human epigenome consortium
II. Epigenome Analysis Approaches
   A. Array-based DNA methylation analysis
   B. Sequencing-based DNA methylation analysis
   C. Chromatin analysis
III. Cancer Epigenomes: Current State of Affairs
   A. Colon cancer
   B. Testicular cancer
   C. Acute lymphoblastic leukemia
   D. Follicular lymphoma
   E. Melanoma
   F. Hepatocellular carcinoma
   G. Breast cancer
IV. Potential Applications of Cancer Epigenomic Data for Risk
   Prediction, Diagnosis, Prognosis, and Therapy
   A. Potential applications for risk prediction, diagnosis, and
      prognosis
   B. Potential applications for therapy

Advances in Genetics, Vol. 70
0065-2660/10 $35.00
DOI: 10.1016/S0065-2660(10)70009-2

ABSTRACT

Cancer is a heterogeneous disease caused largely by abnormalities of the genome
and the epigenome. Typically, such abnormalities include genetic changes such as
mutations and other genomic rearrangements or epigenetic changes such as
aberrant DNA methylation and histone modifications that are frequently
mediated by exposure to environmental or lifestyle factors. Therefore, comprehensive genetic and epigenetic analysis of cancer genomes is the most effective way to
identify causative changes involved in tumorigenesis, irrespective of whether they
are inherited or acquired. In this chapter, we review recent progress in the field and
discuss some of the pilot studies that have already established epigenomic analysis
as integral part of modern cancer research and present a major step toward
personalized treatment of this disease in the future. © 2010, Elsevier Inc.

I. INTRODUCTION

Compared to the genome, the epigenome is much more dynamic, reflecting many
different functional states in time and space. This dynamics is governed in part by
reversible covalent modifications of the epigenome such as DNA methylation and
histone tail modifications. While additional noncovalent modifications such as
microRNAs (miRNAs) and chromatin-remodeling complexes can also modulate
the epigenome, they do not constitute epigenetic marks *per se* and will not be
discussed further here. For the purpose of this chapter, we define "epigenome" as
the complete set of chemical modifications of chromatin constituents and "methylome" as the complete set of DNA methylation modifications of the genome.

Recent progress in technology now enables epigenetic variation to be
studied at the genome level. However, compared to the analysis of cancer
genomes which is well underway as part of multiple national and international

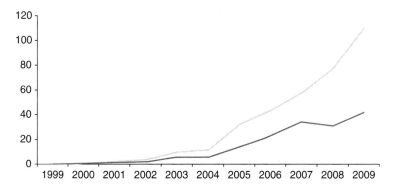

Figure 9.1. Number of publications indexed in PubMed (www.pubmed.gov) for the keywords "epigenome" (total of 368; blue line) and "cancer and epigenome" (total of 166; red line).

efforts, the analysis of cancer epigenomes is still at an early stage. Nevertheless, these advances started to boost the number of publications on epigenomes and cancer epigenomes over the past decade (Fig. 9.1).

DNA methylation in the context of CpG dinucleotides represents the most extensively studied epigenetic modification and was first linked to cancer in 1983 (Feinberg and Vogelstein, 1983). In mammals, it plays a key role in regulating gene expression and other biological processes such as X-chromosome inactivation in female mammals (Chow *et al.*, 2005), parent-of-origin-specific gene expression (imprinting) (Morison *et al.*, 2005), and epigenetic reprograming during mammalian development (Reik, 2007). For instance, a critical step in the genesis of Wilms' tumor is hypomethylation-induced loss of imprinting at *IGF2* (Ogawa *et al.*, 1993), a gene which is also involved in colon cancer (Cui *et al.*, 2003). More recently, CpG and non-CpG methylation has also been shown to be involved in the reprograming of somatic cells into induced pluripotent stem cells (Lister *et al.*, 2009), raising the prospect for cancer cells to become reprogrammable as well in the future (Carette *et al.*, 2010). Please refer to Chapters 6 and 7 for further reading as well as to Chapters 1 and 2, Vol. 71 to learn more about the various extrinsic and intrinsic factors that influence the epigenomic pattern of a cell.

In addition to DNA methylation, a growing number of histone modifications including acetylation, phosphorylation, ubiquitination, and methylation are known to be involved in the same biological processes as well as disease, including cancer (Jones and Baylin, 2007). Working in concert, DNA methylation and histone modifications play key roles in embryonic development, cellular differentiation, and maintenance of cell-type-specific programs in adults, thus defining the phenotype of a cell. For example, gene silencing is frequently associated with DNA hypermethylation at gene promoters and methylation at lysine residues 9 (H3K9me) and 27 (H3K27me) of histone H3. In contrast, activation of gene expression is associated with hypomethylation at gene promoters, local acetylation of histones, and methylation at lysine residues 4 (H3K4me), 36 (H3K36me), and 79 (H3K79me) of histone H3 (Chapter 3). Together, these modifications affect the chromatin structure through a cascade of chemical reactions and interactions with additional proteins and protein complexes. Perturbation of these mechanisms can result in the silencing of tumor suppressor genes as well as the activation of oncogenes, thus initiating carcinogenesis. For example, abnormal growth advantages have been linked to hypomethylation and consequent activation of oncogenes *R-RAS* and *MAP-SIN* in gastric cancer, *S-100* in colon cancer, and *MAGE* in melanoma (Wilson *et al.*, 2007).

The silencing of tumor suppressor genes mediated by site-specific DNA hypermethylation is a recognized mechanism of epigenome-driven tumorigenesis. Tumor suppressor genes encode for proteins that are essential for cell-cycle maintenance, DNA repair, cell adhesion, apoptosis, and angiogenesis. Perturbation of these processes by genetic or epigenetic mechanisms can lead to the development of cancer. Examples of tumor suppressor genes include *RB*, *MLH1*, *BRCA1*, and *P16* (Baylin, 2005). DNA hypermethylation of one allele of a tumor suppressor gene is a key component of the Knudson's second-hit hypothesis or two-hit model, where DNA methylation constitutes the second hit, resulting in the complete loss of gene function (Jones and Laird, 1999). It has to be kept in mind though that not all tumor suppressor genes play such direct roles in carcinogenesis. DNA hypermethylation can also exert indirect effects on a number of effector molecules. By silencing transcription factors or repair proteins, for instance, other downstream targets can be affected and thereby contribute to the malignant transformation of a cell. One example of this is the hypermethylation-induced silencing of the transcription factor *RUNX3* in esophageal cancer (Long *et al.*, 2007).

Although recent studies suggest that the number of epigenetic variations in cancer is similar or even higher than that of genetic variations, it is currently unknown what mechanisms cause a tumor cell to accumulate such widespread epigenetic variations and which of these are causal or consequential. Therefore, it will be important to define experiments and analyses that are capable to distinguish between them, analogous to the respective "driver" and "passenger" mutations, identified as part of cancer genome projects (CGP) (Greenman *et al.*, 2007). It is of course possible that the very early stages of malignant transformation are triggered by "driver" mutations which could include mutations of the epigenome modifier machinery. These driver mutations could then predispose the cells to acquire further epigenetic changes. Conversely, the initial event could also be an epigenetic change resulting in subsequent genetic mutations.

Ability to distinguish "driver" from "passenger" epigenetic changes is crucial for our understanding of neoplastic transformation. Increasing evidence suggests that there are more epigenetic than genetic changes during the transformation from a normal to a malignant cell and that these epigenetic changes occur early in the process and are important driving forces (Jones and Baylin, 2007). In the stem-cell model of carcinogenesis, for instance, cancer stem or progenitor cells are thought to represent the cells of origin of human cancer. An alternative model is that already differentiated cells acquire the respective carcinogenic changes. The stem-cell model of cancer is supported by the following observations: progenitor cells in normal tissue of cancer patients were shown to be genetically and epigenetically altered (Matsubayashi et al., 2003). Moreover, this model also explains why so many heterogeneous cell populations are found within a tumor (Feinberg et al., 2006; Tirino et al., 2008). The acquisition of epigenetic changes plays a key role during these processes. Hypermethylation-induced inactivation of genes that are involved in the self renewal of stem and progenitor cells (e.g., the SFRP family members, APC, CTNNB1, AXIN2, and P16) have been reported in colon and lung cancers (Jones and Baylin, 2007; Kim et al., 2001; Suzuki et al., 2004). These genes are often referred to as "epigenetic gatekeeper genes" as permanent silencing of these variably hypermethylated genes is thought to lock affected cells into a continuous state of self renewal, rendering them incapable of undergoing programmed cell death. While acquiring further genetic and epigenetic mutations, this pool of abnormal stem/precursor cells is believed to undergo positive selection, eventually resulting in complete transformation into cancer cells (Baylin and Ohm, 2006). Please also refer to Chapters 7 and 11.

Irrespective of its origin, it is clear that cancer is a disease caused by widespread genomic and epigenomic alterations involving the dysregulation of hundreds to thousands of genes. Only comprehensive analysis of normal and cancer genomes and epigenomes will identify the full extent of (epi)genomic changes driving carcinogenesis and, thus, promises the development of more effective treatment strategies for the many types of human cancer. In this context, it is important to remember that only epigenomic changes are potentially reversible and consequently hold great promise for therapeutic drug development. In theory, therefore, identification of epigenetic driver mutations could lead to the development of target-specific drugs capable to reverse such epigenetic changes and restore normal gene expression. Over the past decade, several key initiatives (illustrated in Fig. 9.2) have been set up paving the way for comprehensive analysis of cancer (epi)genomes in the near future. For links to each initiative, see Section VII.

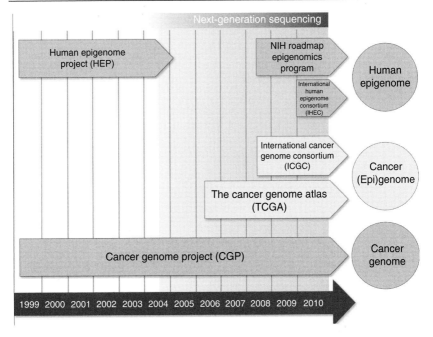

Figure 9.2. Landmark initiatives toward comprehensive analysis of cancer (epi)genomes. The emergence of next-generation sequencing (indicated by blue shading) since 2004 has clearly galvanized the field.

A. Human epigenome project

Established in 1999 (Beck *et al.*, 1999), the Human Epigenome Project (HEP) was the first international epigenome initiative. It was a public/private collabo-ration with the aim to generate human DNA methylation reference profiles based on the identification and cataloging of methylation variable positions (MVPs) in healthy human tissues and cell types. Using high-throughput bisulfite PCR Sanger sequencing, the HEP pioneered high-resolution DNA methylation profiling for three human chromosomes in 12 tissues and cell types, demonstrat-ing among other findings that DNA methylation is tissue-specific, most frequent-ly altered in evolutionary conserved regions and ontogenetically more stable than previously thought (Eckhardt *et al.*, 2006; Rakyan *et al.*, 2004). Building on this success, post-HEP efforts included the generation of genome-wide DNA methylation profiles of 16 tissues/cell types using array-based technology (MeDIP-chip) and the first whole-genome (methylome) profile using a next-generation sequencing-based platform (MeDIP-seq) (Down *et al.*, 2008; Rakyan *et al.*, 2008). At the time of writing, these data collectively constitute the human DNA methylation reference data in the Ensembl genome browser (Flicek *et al.*, 2010).

B. Cancer genome project

Established in the same year as the HEP, the CGP was the first high-throughput CGP. Based at the Wellcome Trust Sanger Institute, the CGP focuses on the identification of somatic mutations and genomic rearrangements in cancer cell lines and, more recently, in primary tumors as well. The CGP has pioneered many experimental and computational approaches for cancer genome analysis and maintains a popular database (Bamford *et al.*, 2004) and census of cancer mutations and genes (Santarius *et al.*, 2010), respectively. Following its first success, the discovery that *BRAF* is mutated in a wide range of human cancers but particularly in melanoma (Davies *et al.*, 2002), the CGP played a major role in the realization that the repertoire of cancer genes and their patterns of somatic mutations are much more complex than previously anticipated (Greenman *et al.*, 2007).

C. The cancer genome atlas

The US-based the Cancer Genome Atlas (TCGA) is the first cancer (epi) genome project in that it has the explicit aim to investigate both genomic and epigenomic (DNA methylation) changes in three types of human cancers: glioblastoma multiforme, lung, and ovarian cancer. For instance, integration of mutation, DNA methylation, and clinical treatment data revealed dysregulation in a mismatch repair pathway in treated glioblastomas, demonstrating the potential power of analyzing (epi)genomes (McLendon *et al.*, 2008). Following the success of the pilot phase, the scope of TCGA was expanded in 2009 to include more than 20 tumor types and thousands of samples over the next 5 years. Thus, TCGA can be expected to make major contributions to our understanding of the human cancer (epi)genome in the coming years.

D. NIH roadmap epigenomics program

This program aims to develop comprehensive reference epigenome maps with focus on human embryonic stem (ES) cells and derivatives as well as the development of new technologies for comprehensive epigenomic analyses. In order to achieve this goal, five specific aims were developed to (1) create an international committee; (2) develop standardized platforms, procedures, and reagents for epigenomics research; (3) conduct demonstration projects to evaluate how epigenomes change; (4) develop new technologies for single cell epigenomic analysis and *in vivo* imaging of epigenetic activity; and (5) create a public data resource to accelerate the application of epigenomics approaches. The first success is a milestone publication (see also Section III), describing the first complete human methylomes at single base-pair resolution of ES cells and differentiated fibroblasts (Lister *et al.*, 2009). The program is complemented by disease-specific projects, including cancer and other common diseases.

E. International cancer genome consortium

The International Cancer Genome Consortium (ICGC) was formed in 2008 with the aim to obtain a comprehensive description of genomic, transcriptomic, and epigenomic changes in 50 different cancer types and/or subtypes (International Cancer Genome Consortium, 2010). Initially, the ICGC will focus on cancers of blood, brain, breast, kidney, liver, pancreas, stomach, oral cavity, and ovary. Based on the anticipated analysis of 500 matched tumor and blood samples per cancer (which translates into the eventual analysis of 50,000 samples), this project makes International Human Epigenome Consortium (IHEC) the definitive initiative for cancer (epi)genomes. While work has started on the genomic analysis for several common cancers, different approaches for epigenomic analysis were still being evaluated at the time of writing.

F. International human epigenome consortium

The IHEC is the most recent initiative, launched in January 2010. Conceptually, it is modeled on ICGC and it aims to serve as an umbrella organization for existing (e.g., NIH Roadmap Epigenomics Program) and new national and international epigenome projects. The aim of this project is to generate 1000 comprehensive human and nonhuman epigenome maps in the context of refe-rence tissues/cell types, stem cells, environment, disease (including cancer), and aging. At the time of writing, there was not yet a dedicated IHEC Web site but preliminary information was hosted on the Web site of the Epigenome Network of Excellence (see Section VII). Although this initiative was not included here, it deserves a special mentioning as it is the main portal and virtual institute for European epigenetics research. It has three main aims: (1) to advance scientific discoveries via a joint research program; (2) to integrate young colleagues via the Network of Excellence Training program; and (3) to establish an open dialog by building interactive Web sites for both the scientific community and general public.

II. EPIGENOME ANALYSIS APPROACHES

The following section provides a brief overview of current and emerging approaches for epigenome analysis and, in particular, for DNA methylation and chromatin profiling. For a more comprehensive treatise, please see recent reviews by Beck and Rakyan (2008) and Laird (2010).

A. Array-based DNA methylation analysis

Until recently, microarrays have been the most widely used tool for comprehensive DNA methylation analysis and new platforms are still being developed for genome-wide or whole-genome screens (e.g., Roche Nimblegen's 2.1M-feature whole-genome tiling array). Advantages of using microarrays include ease of use, high levels of multiplexing and parallel processing to facilitate high-throughput and cost-effective analysis as well as the possibility to capture specific target regions for subsequent array- or sequencing-based analysis (Hodges *et al.*, 2009). Being essentially an analogue technology, disadvantages include lower sensitivity and specificity than achievable by digital technologies such as sequencing. Nevertheless, the first methylomes to be reported were generated using microarray-based approaches (Zhang *et al.*, 2006; Zilberman *et al.*, 2007). However, the introduction of next-generation sequencing is increasingly replacing microarrays in the coming years. A likely exception is so-called beadarrays (e.g., in conjunction with the Illumina Infinium assay) which allow highly sensitive and quantitative methylation analysis at the single base-pair level (Bibikova *et al.*, 2009). While this platform currently only allows to interrogate around 27K CpG sites, Illumina recently announced the release of a new 450K beadarray which is a big step toward its application for more comprehensive methylome analysis as the sites are highly selected and will include many more features than are covered on the 27K beadarray. In the meantime, first publications have shown this platform to be highly informative for the identification of methylation signatures in the context of cancer and aging (Rakyan *et al.*, 2010; Teschendorff *et al.*, 2009, 2010).

B. Sequencing-based DNA methylation analysis

Developed in the early 1990s, bisulfite Sanger sequencing (Frommer *et al.*, 1992) has been the gold-standard for DNA methylation analysis for the past two decades and was used successfully in the early phase of the HET as described in Section I.A. It is based on chemical conversion of unmethylated cytosines to uracils (and thymines following PCR amplification) but leaves methylated cytosines unchanged. However, its gold-standard status has recently been called into question because it cannot distinguish 5-methylcytosine from the recently discovered 5-hydroxymethylcytosine (Huang et al., 2010). Bisulfite Sanger sequencing has been successfully adapted to next-generation sequencing, including the Roche 454 platform (Taylor et al., 2007), the Illumina GA platform (Cokus et al., 2008), and the Applied Biosystems SOLiD platform (Bormann Chung et al., 2010).

These next-generation sequencing techniques have been successfully combined with different capture techniques for methylated DNA, resulting, for example, in AutoMeDIP-Seq (Butcher and Beck, 2010), and MBD-seq (Li et al., 2010). Please also refer to recently published specialist reviews on this subject (Boerno et al., 2010; Reinders and Paszkowski, 2010; Tolstorukov et al., 2010). While these platforms are currently being successfully used as part of the initiatives illustrated in Fig. 9.2, further advances can be expected from future-generation sequencing platforms which will be capable of single-molecule sequencing and direct readout of methylation status without the need of bisulfite conversion or other modifications. Among others, these future platforms include the Helicos tSMS (Harris et al., 2008), the Pacific Biosciences SMRT (Eid et al., 2009), and the Oxford Nanopore Technologies Nanopore Sequencing (Clarke et al., 2009) technologies. There can be little doubt that next- and future-generation sequencing-based platforms will become the approach of choice for future efforts to map and characterize cancer methylomes.

C. Chromatin analysis

Exploiting the tremendous progress in sequencing technology, chromatin immunoprecipitation (ChIP) coupled with next-generation sequencing (ChIP-seq) has already replaced array-based analyses (ChIP-chip) for chromatin profiling, including histone modifications, nucleosome positioning, and DNA-binding proteins (Park, 2009). ChIP-seq offers higher resolution, less noise, and greater coverage than its array-based predecessor and requires less sequencing than methylome profiling. It was developed in 2007 to describe high-resolution profiling of histone modifications (Barski et al., 2007) and chromatin states (Mikkelsen et al., 2007) in the human and mouse genomes, respectively. The studies demonstrated that chromatin profiling can discriminate between genes that are expressed, poised for expression, or stably repressed. In the context of cancer epigenomes, the Barski et al. study further showed that certain histone marks (e.g., H3K4 methylation) are associated with chromosome breakpoints in T cell cancers. Both studies built on previous work using ChIP-chip by Bernstein et al. which resulted in the discovery of bivalent (juxtaposition of active and inactive) histone marks (Bernstein et al., 2006) which are now thought to mark gene promoters to become hypermethylated in cancer (Rodriguez et al., 2008). ChIP-seq can further be used to map protein–DNA interactions across mammalian genomes. Johnson and colleagues, for instance, used this approach to connect transcription factors with their direct targets to form a gene network scaffold (Johnson et al., 2007). More recently, it has also become possible to explore next-generation sequencing for the analysis of any chromosomal interactions, revealing first insights into the three-dimensional nuclear architecture of the (epi)genome in health and disease (Gondor and Ohlsson, 2009).

III. CANCER EPIGENOMES: CURRENT STATE OF AFFAIRS

As evident from the previous sections, the field of cancer epigenomics is still at an early stage and, therefore, it is hardly surprising that no comprehensive analysis has yet been conducted of any cancer epigenome. However, tremendous progress has already been made and several studies are well underway. In this section, therefore, we first briefly discuss recent noncancer epigenome studies before reviewing the current state of affairs for a number of selected cancers.

Because DNA methylation profiling is most widely used and arguably most advanced, methylome analysis will be discussed first. For the purpose of this discussion, a methylation study will qualify as methylome analysis if more than 50% of the approximately 28 million CpG cytosines in the haploid human genome have been analyzed. Using this definition, only a handful of human methylomes have been reported to date. The first (at 100 bp resolution) was reported for mature human spermatozoa in 2008 as proof of feasibility during the development of the MeDIP-seq method (Down et al., 2008). The second (but first at single base-pair resolution) was reported in 2009 (Lister et al., 2009). In their study, Lister et al. analyzed human ES cells and fetal fibroblasts. The main finding—which may have direct relevance to cancer stem cells as well—was that about 25% of the DNA methylation identified in ES cells did not occur in the usual CpG context and may be a special feature of ES and other pluripotent cells, including cancer stem cells. The non-CpG methylation was found to be enriched in gene bodies and depleted in protein-binding sites and enhancers. While DNA methyltransferase DNMT3A is highly expressed in ES cells, it is currently unknown if this enzyme catalyzes the observed non-CpG methylation. In an independent study, Laurent et al. (2010) reported another three methylomes for human ES cells, a fibroblastic differentiated derivative of these ES cells, and neonatal fibroblasts, essentially confirming the results of the Lister et al. study. These and other findings show that comprehensive methylome analysis is highly informative and holds considerable significance for the proposed stem-cell origin of human cancer.

In addition to methylome analysis, profiling of histone modifications constitutes the other main approach in cancer epigenomics. Histone modifications can exert their function independent of or in concert with DNA methylation changes to alter gene expression during tumorigenesis. As previously mentioned in Section II.C, whole-genome histone profiling is already a very well-established approach and it would go beyond the scope of this chapter to review all published studies. Therefore, only a small number exemplar studies will be discussed here that are particularly relevant in the general context of cancer. Complementing a DNA methylation study (Widschwendter et al., 2007), a study by Ohm and colleagues suggested that a stem cell-like chromatin pattern predisposes tumor suppressor genes (particularly polycomb group target genes) to become de novo methylated in tumor cells (Ohm et al., 2007). In early development, polycomb group target genes are frequently marked with H3K27me3 in normal cells and it was shown that the presence of this repressive histone mark is involved in the recruitment of DNA methyltransferases, leading to the described de novo methylation (Schlesinger et al., 2007). The resulting site-specific hypermethylation then leads to permanent silencing of these key regulatory genes, contributing to increased cell proliferation and malignant transformation, a process referred to as "epigenetic switching" (Gal-Yam et al., 2008). In another genome-wide study, Wen et al. (2009) showed that large (up to 4.9 Mb) chromatin blocks marked by H3K9me2 can distinguish differentiated from ES cells. These so-called large organized chromatin K9 modifications (LOCKs) require histone methyltransferase G9a, are inversely related to gene expression within the regions, and are substantially lost in cancer cell lines, hence provide a potential cell type-heritable mechanism for phenotypic plasticity in cancer.

The following is a brief review of the current state of affairs with respect to epigenomic (methylome) analyses for a selected range of cancers.

A. Colon cancer

Colon cancer is the third most common cancer worldwide (Parkin et al., 2001). Epigenetic changes of candidate loci have been studied extensively in this cancer, resulting in the identification of numerous DNA methylation markers (reviewed in Kim et al., 2010), including the CpG island methylator phenotype (CIMP) which allows to classify cancers according to their degree of DNA methylation (Issa, 2004; Toyota et al., 1999) and "Epi-proColon™," the first commercial early detection marker based on detecting aberrantly methylated DNA of the SEPT9 gene in blood plasma. Because of these early successes, it is not surprising that the first cancer methylome analysis was carried out on colon cancer (Irizarry et al., 2009). In this landmark study, Irizarry and colleagues examined the colon methylome using the comprehensive high-throughput array-based relative methylation (CHARM) assay. Analyzing 13 colon cancers and matched normal mucosa from the same individuals, they identified 2707 cancer-specific differentially methylated regions (C-DMRs). Of these, 56% were hypermethylated and 44% were hypomethylated. When analyzed for location, they found that 52% of these C-DMRs overlapped tissue-specific DMRs (T-DMRs) and the majority of T-DMRs (76%) mapped to regions adjacent to (but not within) CpG islands which they termed "CpG island shores." The methylation of these CpG island shores was strongly correlated to gene expression and highly conserved across species. By analyzing these regions in human and mouse tissues, it was possible to discriminate tissue types regardless of their species of origin. Moreover, shore-linked silencing was shown to be reversed by methyltransferase inhibition. In a subsequent study, the same group showed that many of the same differentially methylated CpG island shores were also involved in the reprograming of somatic cells into induced pluripotent stem (iPS) cells (Doi et al., 2009; Pollard et al., 2009). They termed such reprogramed DMRs (R-DMRs). However, the colocalization of hypomethylated R-DMRs and hypermethylated C-DMRs suggested different mechanisms for epigenetic reprograming in iPS cells and cancer. Taken together, these data are consistent with the epigenetic stem/progenitor model of cancer (Feinberg et al., 2006) which proposes that "epigenetic alterations affecting tissue-specific differentiation are the predominant mechanism by which epigenetic changes cause cancer."

B. Testicular cancer

Testicular cancer is the most common malignancy in young men. Over his lifetime, a man's risk of developing testicular cancer is roughly 1 in 250 (0.4%) (Manecksha and Fitzpatrick, 2009). Until now only a limited number of genes has been known to be epigenetically altered in testicular cancer as reviewed by Veltri and Makarov (2006). However, in view of the findings in other cancers, it is not surprising that widespread changes in the DNA methylation are also found in testicular cancer, including in a recent genome-wide study (Cheung et al., 2010). In their study, Cheung and colleagues used methylated DNA immunoprecipitation (MeDIP) and genome-wide tiling arrays to identify epigenetically regulated genes and analyzed ncRNA in testicular germ cell tumors. DMRs were first identified in an in vitro model culture system (Ntera2, Tera-1, and a normal human testis cell line) using this genome-wide approach. In all, 22,452 hypermethylated and 12,756 hypomethylated regions were found in the cancer genome (a total of 35,208 DMRs which represent C-DMRs in accordance with the recent literature). The results were further validated by randomly selecting a number of these C-DMRs and subsequent bisulfite sequencing. As anticipated, the identified C-DMRs were not evenly distributed in the genome with various chromosomal regions being predominantly hypermethylated or hypomethylated.

Only a small number of these regions mapped to promoters. Most hypermethylated (92.9%) and hypomethylated regions (88.2%) were located in genomic regions without any gene annotation (intergenic regions). Only a low percentage of C-DMRs 1.9% (414) of hypermethylated and 2.3% (279) of hypomethylated regions mapped to promoter regions of known genes. As described above, these promoter regions are known to play an important role in gene transcription and methylation of these regions may cause altered transcriptional activity. This was assessed by genome-wide gene-expression analysis which revealed a group of differentially expressed genes (including *APOLD1*, *PCDH10*, and *RGAG1*) that were regulated by DNA methylation. *APOLD1*, for instance, was previously mapped to the susceptibility locus for testicular germ cell tumor (12p13.1), presumably playing an important role in testicular carcinogenesis (Crockford *et al.*, 2006). Cheung *et al.* also observed methylation changes at loci encoding noncoding RNAs. miRNAs represent a class of noncoding RNAs which destabilize or repress translation of messenger RNA at the posttranscriptional level. hsa-mir-199a2, hsa-mir-124a-2, and hsa-mir-184 were found to be linked to hypermethylated DMRs. Furthermore, hsa-mir-199a2 was downregulated both in testicular germ cell tumor samples and in the *in vitro* model culture system. These data clearly demonstrate that widespread DNA methylation changes play a key role in testicular carcinogenesis. Additional studies will be necessary to explain the regulatory function of nonrepetitive intergenic and intronic regions which may have a role in regulating noncoding RNAs/miRNAs.

C. Acute lymphoblastic leukemia

Acute lymphoblastic leukemia (ALL) represents the most common malignancy in childhood with a peak incidence at 2–5 years of age and another peak in old age (Redaelli *et al.*, 2005). Various changes in the DNA methylation of this type of leukemia cells have been reported to play an important role in carcinogenesis, including hypermethylation in the promoter regions of the Fanconi genes, *FANCC* and *FANCL* (Hess *et al.*, 2008). A representative example of an ALL methylation study is the study by Davidsson *et al.* (2009) who examined 20 samples from children with B cell precursor ALL. Their ALL cohort consisted of 10 samples with hyperdiploidy (51–67 chromosomes) and 10 with the *ETV6/RUNX1* fusion. MeDIP was used in combination with 32K bacterial artificial chromosome (BAC) arrays for analysis. Global gene-expression analysis was performed by hybridization of labeled, reversely transcribed RNA to 27K cDNA microarray slides. A total of 8662 different genes with significant 5′ promoter CpG island methylation scores were identified. This accounts for around 30% of all genes covered by this platform. Using these methylation patterns, it was possible to distinguish between the two subgroups of ALL. Furthermore, the promoter-specific CpG island analysis revealed that several B cell- and neoplasia-associated genes were hypermethylated as well as underexpressed (shown by parallel global gene-expression analysis). These genes are known to be mainly involved in apoptosis, cell signaling, transcription regulation, cell proliferation, DNA repair, and cell adhesion, suggesting that the observed DNA methylation changes play an important role in the carcinogenesis of this highly aggressive disease. A total of 58 DMRs with more than four hypermethylated genes per 4 Mb were identified. These hotspots were associated with chromosome bands and predicted to harbor imprinted genes. Moreover, the authors were able to show that tri-/tetrasomic chromosomes in the high-hyperploid ALLs had a lower methylation rate than their disomic counterparts. Although this phenomenon has not been described previously and is an important finding of this study as it may not only have implications for our understanding of the pathogenesis of ALL but also of the pathogenesis of diseases in which constitutional or acquired numerical chromosome anomalies are a feature. This may also have important implications for the treatment of this deadly disease, particularly in view of the emerging epigenetic therapeutic agents (described in Section IV.B).

D. Follicular lymphoma

Follicular lymphoma is a common type of Non-Hodgkin Lymphoma which is a slow growing lymphoma, arising from B cells. Although not really a methylome analysis, the study by Killian et al. (2009) is interesting as it demonstrates with the use of formalin-fixed paraffin-embedded (FFPE) samples. In their study, Killian and colleagues performed DNA methylation analysis and gene-expression profiling using beadarrays on follicular lymphoma samples. A total of 259 differentially methylated targets were identified which were distributed over 183 unique genes. Both FFPE and matched frozen surgical pathology replicates of lymph nodes were screened. The follicular lymphoma DNA methylation profile was compared with the one from reactive lymph nodes as well as with the one from purified populations of peripheral blood B and T cells and transformed lymphoblastoid cell lines to assess them for differences in their DNA methylation patterns. The results showed that the pattern of differentially methylated targets in follicular lymphoma is divergent from normal lymphoid populations. Moreover, they showed complete preservation of the identified methylation signature among different archival tissue specimens. However, due to the limitation of this method, only a small number of all CpG sites was covered by the analysis in comparison to the more comprehensive analyses described above.

E. Melanoma

Although representing a less common form of skin cancer, melanoma is the deadliest of all, accounting for around three quarters of skin cancer-related deaths (Sladden et al., 2009). Epigenetic mechanisms have been described to play a key role in the gene regulation of human melanoma, including the identification of several putative tumor suppressor genes and oncogenes (Howell et al., 2009). Recently, first attempts were reported to comprehensively map genome-wide DNA methylation changes in melanoma cells, including the study by Koga et al. (2009). In their study, Koga and colleagues used MeDIP coupled with a genome-wide promoter array to conduct an integrative analysis of promoter methylation and gene expression in melanoma with the aim to identify potential diagnostic and predictive markers. Eight human melanoma cell strains were compared with melanocytes from newborns and adults. They identified 76 melanoma methylation markers (68 of the markers were hypermethylated and 8 were hypomethylated). They further validated promoter methylation and differential gene expression of five markers, namely COL1A2, NPM2, HSPB6, DDIT4L, and MTIG by bisulfite sequencing and real-time reverse transcriptase PCR. The results also suggest that at least four of these genes may be useful to assess tumor progression. In conclusion, this study represents one of the first genome-wide methylation analyses for melanoma, resulting in the identification of novel genes potentially implicated in this deadly disease. Further studies are now needed to also assess regions beyond promoters which have already been shown to be involved in other cancers.

F. Hepatocellular carcinoma

Hepatocellular carcinoma (HCC) is among the five most common causes of cancer death worldwide and carries a poor prognosis, especially in advanced disease (El-Serag and Mason, 1999). However, prognosis is influenced by stage in that patients whose disease is detected at an early stage and treated by transplantation, resection, or local ablation have 5-year survival rates of 50–70%, but patients with intermediate disease who are treated with transarterial-chemoembolization (TACE) or advanced disease, treated with systemic agents, have median survival of only 18 and 9 months, respectively (Llovet and Bruix, 2003). Therefore, the development of new markers is vital to improve early detection and monitoring of the disease. One of the most comprehensive DNA methylation studies for HCC to date was conducted by Arai et al. (2009). In their study, they used MeDIP in combination with a BAC array (containing 4361 BAC clones) to profile 126 tissue samples (HCCs, noncancerous liver tissue obtained from patients with HCCs, and normal liver tissue). They found that the average number of DMRs (hyper- or hypomethylated) increased from noncancerous liver tissue (of patients with HCC) to HCC. The DNA methylation aberrations in the noncancerous liver tissue were shown to be positively correlated with the future development of HCC. Moreover, a panel of 41 DMRs was shown to be an independent predictor of overall outcome, looking at cancer-free and overall survival rates of patients with HCC. These results provide further evidence that aberrant DNA methylation is widespread in HCC, suggesting that informative DNA methylation markers can be developed. In this respect, patients treated with TACE may be particularly suited for the development of a blood-based DNA methylation marker. TACE causes acute ischemic tumor necrosis and will result in a sudden release of tumor DNA into the blood stream from where this so-called cell-free DNA can be isolated for methylation analysis using a minimal invasive procedure.

G. Breast cancer

Breast cancer is one of the most common cancers worldwide—1 in every 10 women develops breast cancer (Jemal et al., 2005). Breast carcinogenesis involves genetic and epigenetic alterations of which epigenetic changes occur at a very early stage of breast tumorigenesis (Tommasi et al., 2009). These altered DNA methylation profiles have been suggested to potentially serve as biomarkers for risk prediction, early detection, assessment of prognosis, the monitoring of therapy, as well as the prediction of therapy response (Lo and Sukumar, 2008; Martens et al., 2009). In the most recent study, Flanagan and colleagues used MeDIP with a commercial promoter array to investigate genome-wide methylation profiles in 33 familial breast cancers (Flanagan et al., 2010). The aim of this study was to identify patterns of methylation specific to the different mutation groups (BRCA1, BRCA2, and BRCAx) or intrinsic subtypes of breast cancer (basal, luminal A, luminal B, HER2-amplified, and normal-like). The obtained methylation profiles predicted tumor mutation status but did not accurately predict the previously described intrinsic subtypes. In an independent validation set of 47 tumors, these findings were confirmed by pyrosequencing and Epityper analysis. Additionally, the samples were also analyzed for gene expression and copy-number variation and the resulting data were integrated with the methylation data. The meta-analysis revealed that genes found to be frequently hypermethylated also displayed loss of heterozygosity as well as copy-number gains, suggesting a possible mechanism for expression dosage compensation. These results extend our current understanding of possible mechanisms leading to familial breast cancer and provide one of the first integrated data sets of DNA methylation, gene expression, and copy-number variation. The obtained data will help to guide the development of new strategies for diagnosis and treatment of familial breast cancer.

In conclusion, these studies describe many new and interesting findings but are all still restricted by technical limitations allowing only a part of the corresponding cancer methylomes to be interrogated. Following the discovery of additional widespread DNA methylation in non-CpG context (Lister et al., 2009), there is little doubt that future cancer methylome and epigenome analyses will be sequencing- rather than array-based and will also need to include histone modifications, protein–DNA interactions, and miRNA profiles as well as comprehensive gene-expression analysis.

IV. POTENTIAL APPLICATIONS OF CANCER EPIGENOMIC DATA FOR RISK PREDICTION, DIAGNOSIS, PROGNOSIS, AND THERAPY

A. Potential applications for risk prediction, diagnosis, and prognosis

DNA methylation markers that are specific for various cancer types have already been described by a huge number of studies, but so far a few if any of these markers have made the transition into the clinic. However, early diagnosis is of utmost importance for the successful treatment of and survival from many types of human cancer. Too often, cancer is still being detected and diagnosed at an already advanced stage, reducing survival times and treatment options which are then frequently limited to palliative measures only. In this section, we discuss progress toward using DNA methylation markers for risk prediction, diagnosis, and prognosis, bearing in mind that most of the published studies to date have been delimited by the number of markers that could be detected with the technology used at the time.

A recent example for risk prediction is the study by Teschendorff and colleagues, using the Infinium 27K beadarray technology to interrogate 27,578 CpG sites in white blood cells from patients with ovarian cancer and healthy controls (Teschendorff *et al.*, 2009). They identified a DNA methylation signature that predicted the presence of ovarian cancer with an "area under the curve" (AUC) value of 0.8, using blind test sets. These results were confirmed in an independent validation set. What makes this study particularly interesting is the fact that the signature was detected in blood cells (e.g., tumor-unrelated cells). At this point, it is not clear (and it does not really matter in this context) whether the observed DNA methylation changes are the cause or consequence of the cancer. Irrespective of the underlying mechanism(s), however, this study demonstrates that genome-wide DNA methylation profiling, for example, of blood-derived DNA may be useful for risk prediction of ovarian and possibly other cancers.

Of the three categories of markers discussed in this section, the development of markers for early detection and diagnosis is probably the most challenging. In a clinical context, the corresponding test will need to have high sensitivity and specificity and will need to be informative on easily accessible samples, such as blood, urine, stool, or swabs. The current state of such diagnostic markers has recently been reviewed for breast, ovarian, endometrial, and cervical cancer, indicating that up to 82% sensitivity and 100% specificity are currently achievable (Jones *et al.*, 2010). Similar results have been reported and replicated for colon cancer based on a now commercially available blood plasma test called *Epi-proColon*™ (Grutzmann *et al.*, 2008).

An example of genome-wide DNA methylation profiles serving as prognostic indicators is illustrated by a study that is part of the TCGA Project (see also Section I.C) (Noushmehr *et al.*, 2010). Profiling of highly malignant brain tumor glioblastoma multiforme samples allowed identification of a distinct subset of samples showing similar levels of hypermethylation at a large number of loci, termed glioma-CIMP (G-CIMP). The presence of this phenotype was confirmed in a separate validation set of glioblastomas and low-grade gliomas. By determining the clinicopathological features, it was shown that G-CIMP tumors belong to the proneural subgroup and are more prevalent among lower grade gliomas. Moreover, the authors demonstrated the presence of distinct copy-number variations for this phenotype and a significant association of this subgroup with *IDH1* somatic mutations. Patients suffering from glioblastoma of this subgroup are usually younger at the time of diagnosis and, most interestingly, show a significantly improved outcome, underscoring the prognostic value of detecting this specific methylation profile.

B. Potential applications for therapy

The reversible nature of epigenetic aberrations has led to the emergence of the promising field of epigenetic therapy. The aim of this new treatment option is to reverse those epigenetic changes which contribute to carcinogenesis and to restore a "normal epigenome."

Two groups of drugs targeting the epigenome have been reported:

- DNA methylation inhibitors, for example, 5-azacytidine (also known as Vidaza) and 5-aza-2'-deoxycytidine (also known as Decitabine) have already been approved by the Food and Drug Administration (FDA) for treatment of certain hematological malignancies.
- Histone deacetylase (HDAC) inhibitors, for example, suberoylanilide hydroxamic acid (SAHA) also known as Vorinostat has already been approved by the FDA for treatment of various solid tumors and cutaneous T-cell lymphoma (CTCL).

Although not yet target-specific, some of these agents have recently been approved by the FDA for cancer treatment and can be expected to play a major role in epigenetic cancer therapy in the future (Cortez and Jones, 2008). Epigenomic profiling of body fluids (e.g., urine or saliva) or cell-free DNA isolated from blood may further help to monitor the efficacy of these and other therapeutic agents, even if cancer tissue is not readily available for analysis.

In this context, please see also Chapters 12 and 13. In summary, this is clearly a rapidly evolving field and the challenge for the future is to develop target-specific therapeutic epigenetic drugs as well as the tools to effectively monitor them.

V. HEAD AND NECK: EXAMPLE OF AN EMERGING CANCER EPIGENOME PROJECT

A. Head and neck cancer

For the majority of cancers, in particular of rare cancers, there are currently no ongoing epigenome projects. However, this can be expected to change in the near future. An example of such an emerging cancer epigenome project is head and neck cancer (including oral cancer) which is the sixth most common cancer in the United Kingdom and worldwide (Hunter *et al.*, 2005). The most common type of head and neck cancer is squamous-cell carcinoma (HNSCC). As an umbrella term, HNSCC includes cancers found at several locations, for example, in the oral cavity, pharynx, and larynx. Although these cancers have different etiologies and prognoses, they share similar risk factors. Known risk factors include tobacco smoking, alcohol consumption, or human papillomavirus (HPV) infection. Heavy smokers under age 46 have a 20-fold increased risk of oral and pharyngeal cancer, whereas heavy drinkers are reported to have a five-fold increased risk. A combination of heavy smoking and drinking leads to an almost 50-fold increased risk of oral and pharyngeal cancer (Rodriguez *et al.*, 2004). Despite recent advances in the treatment and in the understanding of cancer biology, the 5-year survival rate of 50% for patients with HNSCC has remained largely unchanged for the past 20–30 years with only some advances since the 1990s (Conley, 2006). Thus, it is of utmost importance to develop new strategies in order to increase the currently poor survival rate of head and neck cancer. The UCL head and neck epigenome project (described below) represents one such effort to achieve this goal, mainly focusing on an interesting subgroup: HPV-associated head and cancer.

B. HPV and head and neck cancer

Viruses are associated with 15–20% of human cancers worldwide and have been shown to affect the host epigenome and display changes of their own epigenomes during disease progression (Fernandez and Esteller, 2010). HPV is known to represent a major independent risk factor for HNSCC (D'Souza *et al.*, 2007; Tran *et al.*, 2007). Over 50% of oropharyngeal cancers test positive for HPV-16, with the expression of the E6 and E7 viral oncogenes. As HPV-associated cancers respond better to chemotherapy (82% vs. 55% response rate for HPV-negative cases) and have a better overall survival (95% vs. 62% at 2 years), they represent a major opportunity to stratify the treatment of these patients and to elucidate the mechanisms of drug sensitivity and resistance in all HNSCC (Fakhry et al., 2008).

C. Current state of affairs of head and neck cancer epigenomics

As for other cancer types, numerous DNA methylation studies have been conducted in head and neck cancer, assessing various risk factors such as alcohol, smoking, and HPV infection. In Table 9.1, we have summarized some of the larger scale studies which still only cover a small proportion of the head and neck methylome.

D. Toward a head and neck cancer epigenome project

The aim of this project is to carry out a comprehensive methylome analysis of both HPV-positive and HPV-negative head and HNSCC samples, precancerous cells and nondiseased matched mucosal cells using the latest sequencing technology and combining these data with gene expression and miRNA data. Through integration and comparative analysis of these multidimensional data, we will aim to define the epigenome of HPV-positive and HPV-negative head and neck tumor cells and identify the epigenetic changes at oncogenic loci associated with progression to carcinoma in HPV-positive HNSCC which are currently not well characterized and, consequently, the reason for their better survival is not known. Similar projects are likely to emerge for other types of cancer in the near future.

VI. CONCLUSION AND FUTURE PERSPECTIVES

Carcinogenesis is a highly complex process, involving abundant genomic and epigenomic alterations that eventually lead to the malignant transformation of a normal cell. While the field of cancer genomics is already well established and starts having clinical impact on cancer classification and diagnosis, cancer epigenomics is still at a comparatively early stage. However, key initiatives and programs have already been put in place that can be expected to advance the field rapidly and facilitate system-level integration which has the highest potential for translation into patient benefit and, eventually, for personalized cancer treatment. One very recent effort that is pushing the boundaries in this direction is the TREAT 1000 Project (www.treat1000.org). This project aims to completely sequence the genome, epigenome, and transcriptome of 1000 cancer patients, integrate the data, and use advanced computational modeling to predict the best

Table 9.1. Selected Studies Describing Aberrant DNA Methylation in Head and Neck Cancer

Type of cells/ tissue	Genes reported to be differentially methylated	DNA methylation profiling
Cancer tissue	FANCF (Marsit et al., 2004); SOCS3, SOCS1 (Weber et al., 2005); APC, DAPK1, RARβ, RASSF1, SFRP1, SFRP2, SFRP4, SFRP5, CDH1, p16 (CDKN2A) (Marsit et al., 2006); LHX6 (Estecio et al., 2006); DCC (Carvalho et al., 2006); TCF21 (Smith et al., 2006); C/EBPα (Bennett et al., 2007); TIMP3, ECAD, p16 (CDKN2A), MGMT, DAPK1, RASSF1 with matched oral/ salivary samples (Righini et al., 2007); MLH1, MSH2 (Sengupta et al., 2007); TIMP3, DAPK1 (Nayak et al., 2007); p16 (CDKN2A), MGMT, DAPK1, ECAD (Dikshit et al., 2007); p16 (CDKN2A), DAPK1, MGMT (Martone et al., 2007); p16 (CDKN2A), MGMT, DAPK, RARβ, MLH1, CDH1, RASSF5, MST1 (Steinmann et al., 2009)	Using the Illumina Golden Gate assay (covering 1536 CpG sites), specific DNA methylation profiles were shown to be associated with certain risk factors such as tobacco smoking and alcohol abuse as well as tumor stage (Marsit et al., 2009) Pyrosequencing-based profiling showed LINE hypomethylation to be more pronounced in HPV-negative than in HPV-positive tumors. Genomic instability was greater in HNSCC samples with more pronounced LINE hypomethylation (Richards et al., 2009)
Blood sample (serum or whole blood)	DCC, DAPK1, TIMP3, ESR, CCNA1, CCND2, MINT1, MINT31, CDH1, AIM1, MGMT, p16 (CDKN2A), PGP9.5, RARβ, HIC1 investigated as a panel (Carvalho et al., 2008)	Using combined bisulfite restriction analysis of the LRE1 sequence it was shown that DNA hypomethylation in whole blood is associated with 1.6-fold increased risk of HNSCC (Hsiung et al., 2007)
Oral/salivary rinses	DCC, DAPK1, TIMP3, ESR, CCNA1, CCND2, MINT1, MINT31, CDH1, AIM1, MGMT, p16 (CDKN2A), PGP9.5, RARβ, HIC1 investigated as a panel (Carvalho et al., 2008); CD44 (Franzmann et al., 2007); TIMP3, ECAD, p16 (CDKN2A), MGMT, DAPK1, RASSF1 with matched tumor samples (Righini et al., 2007); LINE-1/COBRALINE-1 (Subbalekha et al., 2009)	No published reports yet

treatment. Integration of multidimensional genomic data has emerged as a powerful tool to dissect complex traits and phenotypes (Mikkelsen et al., 2008). The underlying concept is to treat the entirety of the data as a unique personal biomarker rather than assess common biomarkers in the context of individual tumors and patients. As every tumor and every patient is different, this approach may overcome the long-standing cancer biomarker problem (Sawyers, 2008). If successful, this and similar efforts will pave the way for comprehensive molecular analysis of each and every cancer patient to become an integral part of frontline treatment in personalized medicine.

VII. USEFUL LINKS

Human Epigenome Project (HEP): http://www.epigenome.org
Cancer Genome Project (CGP): http://www.sanger.ac.uk/genetics/CGP/
The Cancer Genome Atlas (TCGA): http://cancergenome.nih.gov/
NIH Roadmap Epigenomics Program: http://www.roadmapepigenomics.org/
International Cancer Genome Consortium (ICGC): http://www.icgc.org/
International Human Epigenome Consortium (IHEC): http://www.epigenome-noe.net/ihec/
Epigenome Network of Excellence: http://www.epigenome-noe.net/
Treat 1000 Project: http://www.treat1000.org/

Acknowledgments

M. L. was supported by the Austrian Science Fund (J2856-B13). C. B. was supported by Cancer Research UK and the Comprehensive Biomedical Research Centre at UCL/UCLH. S. B. was supported by the Wellcome Trust (084071).

References

Arai, E., Ushijima, S., Gotoh, M., Ojima, H., Kosuge, T., Hosoda, F., Shibata, T., Kondo, T., Yokoi, S., Imoto, I., Inazawa, J., Hirohashi, S., et al. (2009). Genome-wide DNA methylation profiles in liver tissue at the precancerous stage and in hepatocellular carcinoma. Int. J. Cancer 125, 2854–2862.

Bamford, S., Dawson, E., Forbes, S., Clements, J., Pettett, R., Dogan, A., Flanagan, A., Teague, J., Futreal, P. A., Stratton, M. R., and Wooster, R. (2004). The COSMIC (Catalogue of Somatic Mutations in Cancer) database and website. Br. J. Cancer 91, 355–358.

Barski, A., Cuddapah, S., Cui, K., Roh, T. Y., Schones, D. E., Wang, Z., Wei, G., Chepelev, I., and Zhao, K. (2007). High-resolution profiling of histone methylations in the human genome. Cell 129, 823–837.

Baylin, S. B. (2005). DNA methylation and gene silencing in cancer. Nat. Clin. Pract. Oncol. 2 (Suppl. 1), S4–S11.

Baylin, S. B., and Ohm, J. E. (2006). Epigenetic gene silencing in cancer—A mechanism for early oncogenic pathway addiction? Nat. Rev. Cancer 6, 107–116.

Beck, S., and Rakyan, V. K. (2008). The methylome: Approaches for global DNA methylation profiling. Trends Genet. 24, 231–237.

Beck, S., Olek, A., and Walter, J. (1999). From genomics to epigenomics: A loftier view of life. *Nat. Biotechnol.* **17,** 1144.

Bennett, K. L., Hackanson, B., Smith, L. T., Morrison, C. D., Lang, J. C., Schuller, D. E., Weber, F., Eng, C., and Plass, C. (2007). Tumor suppressor activity of CCAAT/enhancer binding protein alpha is epigenetically down-regulated in head and neck squamous cell carcinoma. *Cancer Res.* **67,** 4657–4664.

Bernstein, B. E., Mikkelsen, T. S., Xie, X., Kamal, M., Huebert, D. J., Cuff, J., Fry, B., Meissner, A., Wernig, M., Plath, K., Jaenisch, R., Wagschal, A., *et al.* (2006). A bivalent chromatin structure marks key developmental genes in embryonic stem cells. *Cell* **125,** 315–326.

Bibikova, M., Le, J., Barnes, B., Saedinia-Melnyk, S., Shen, R., and Gunderson, K. L. (2009). Genome-wide DNA methylation profiling using Infinium® assay. *Epigenomics* **1,** 177–200.

Boerno, S. T., Grimm, C., Lehrach, H., and Schweiger, M. R. (2010). Next-generation sequencing technologies for DNA methylation analyses in cancer genomics. *Epigenomics* **2,** 199–207.

Bormann Chung, C. A., Boyd, V. L., McKernan, K. J., Fu, Y., Monighetti, C., Peckham, H. E., and Barker, M. (2010). Whole methylome analysis by ultra-deep sequencing using two-base encoding. *PLoS One* **5,** e9320.

Butcher, L. M., and Beck, S. (2010). AutoMeDIP-seq: A high-throughput, whole genome, DNA methylation assay. *Methods* (in press).

Carette, J. E., Pruszak, J., Varadarajan, M., Blomen, V. A., Gokhale, S., Camargo, F. D., Wernig, M., Jaenisch, R., and Brummelkamp, T. R. (2010). Generation of iPSCs from cultured human malignant cells. *Blood* **115**(20), 4039–4042.

Carvalho, A. L., Chuang, A., Jiang, W. W., Lee, J., Begum, S., Poeta, L., Zhao, M., Jeronimo, C., Henrique, R., Nayak, C. S., Park, H. L., Brait, M. R., *et al.* (2006). Deleted in colorectal cancer is a putative conditional tumor-suppressor gene inactivated by promoter hypermethylation in head and neck squamous cell carcinoma. *Cancer Res.* **66,** 9401–9407.

Carvalho, A. L., Jeronimo, C., Kim, M. M., Henrique, R., Zhang, Z., Hoque, M. O., Chang, S., Brait, M., Nayak, C. S., Jiang, W. W., Claybourne, Q., Tokumaru, Y., *et al.* (2008). Evaluation of promoter hypermethylation detection in body fluids as a screening/diagnosis tool for head and neck squamous cell carcinoma. *Clin. Cancer Res.* **14,** 97–107.

Cheung, H. H., Lee, T. L., Davis, A. J., Taft, D. H., Rennert, O. M., and Chan, W. Y. (2010). Genome-wide DNA methylation profiling reveals novel epigenetically regulated genes and non-coding RNAs in human testicular cancer. *Br. J. Cancer* **102,** 419–427.

Chow, J. C., Yen, Z., Ziesche, S. M., and Brown, C. J. (2005). Silencing of the mammalian X chromosome. *Annu. Rev. Genomics Hum. Genet.* **6,** 69–92.

Clarke, J., Wu, H. C., Jayasinghe, L., Patel, A., Reid, S., and Bayley, H. (2009). Continuous base identification for single-molecule nanopore DNA sequencing. *Nat. Nanotechnol.* **4,** 265–270.

Cokus, S. J., Feng, S., Zhang, X., Chen, Z., Merriman, B., Haudenschild, C. D., Pradhan, S., Nelson, S. F., Pellegrini, M., and Jacobsen, S. E. (2008). Shotgun bisulphite sequencing of the Arabidopsis genome reveals DNA methylation patterning. *Nature* **452,** 215–219.

Conley, B. A. (2006). Treatment of advanced head and neck cancer: What lessons have we learned? *J. Clin. Oncol.* **24,** 1023–1025.

Cortez, C. C., and Jones, P. A. (2008). Chromatin, cancer and drug therapies. *Mutat. Res.* **647,** 44–51.

Crockford, G. P., Linger, R., Hockley, S., Dudakia, D., Johnson, L., Huddart, R., Tucker, K., Friedlander, M., Phillips, K. A., Hogg, D., Jewett, M. A., Lohynska, R., *et al.* (2006). Genome-wide linkage screen for testicular germ cell tumour susceptibility loci. *Hum. Mol. Genet.* **15,** 443–451.

Cui, H., Cruz-Correa, M., Giardiello, F. M., Hutcheon, D. F., Kafonek, D. R., Brandenburg, S., Wu, Y., He, X., Powe, N. R., and Feinberg, A. P. (2003). Loss of IGF2 imprinting: A potential marker of colorectal cancer risk. *Science* **299,** 1753–1755.

Davidsson, J., Lilljebjorn, H., Andersson, A., Veerla, S., Heldrup, J., Behrendtz, M., Fioretos, T., and Johansson, B. (2009). The DNA methylome of pediatric acute lymphoblastic leukemia. *Hum. Mol. Genet.* **18,** 4054–4065.

Davies, H., Bignell, G. R., Cox, C., Stephens, P., Edkins, S., Clegg, S., Teague, J., Woffendin, H., Garnett, M. J., Bottomley, W., Davis, N., Dicks, E., *et al.* (2002). Mutations of the BRAF gene in human cancer. *Nature* **417,** 949–954.

Dikshit, R. P., Gillio-Tos, A., Brennan, P., De Marco, L., Fiano, V., Martinez-Penuela, J. M., Boffetta, P., and Merletti, F. (2007). Hypermethylation, risk factors, clinical characteristics, and survival in 235 patients with laryngeal and hypopharyngeal cancers. *Cancer* **110,** 1745–1751.

Doi, A., Park, I. H., Wen, B., Murakami, P., Aryee, M. J., Irizarry, R., Herb, B., Ladd-Acosta, C., Rho, J., Loewer, S., Miller, J., Schlaeger, T., *et al.* (2009). Differential methylation of tissue- and cancer-specific CpG island shores distinguishes human induced pluripotent stem cells, embryonic stem cells and fibroblasts. *Nat. Genet.* **41,** 1350–1353.

Down, T. A., Rakyan, V. K., Turner, D. J., Flicek, P., Li, H., Kulesha, E., Graf, S., Johnson, N., Herrero, J., Tomazou, E. M., Thorne, N. P., Backdahl, L., *et al.* (2008). A Bayesian deconvolution strategy for immunoprecipitation-based DNA methylome analysis. *Nat. Biotechnol.* **26,** 779–885.

D'Souza, G., Kreimer, A. R., Viscidi, R., Pawlita, M., Fakhry, C., Koch, W. M., Westra, W. H., and Gillison, M. L. (2007). Case-control study of human papillomavirus and oropharyngeal cancer. *N. Engl. J. Med.* **356,** 1944–1956.

Eckhardt, F., Lewin, J., Cortese, R., Rakyan, V. K., Attwood, J., Burger, M., Burton, J., Cox, T. V., Davies, R., Down, T. A., Haefliger, C., Horton, R., *et al.* (2006). DNA methylation profiling of human chromosomes 6, 20 and 22. *Nat. Genet.* **38,** 1378–1385.

Eid, J., Fehr, A., Gray, J., Luong, K., Lyle, J., Otto, G., Peluso, P., Rank, D., Baybayan, P., Bettman, B., Bibillo, A., Bjornson, K., *et al.* (2009). Real-time DNA sequencing from single polymerase molecules. *Science* **323,** 133–138.

El-Serag, H. B., and Mason, A. C. (1999). Rising incidence of hepatocellular carcinoma in the United States. *N. Engl. J. Med.* **340,** 745–750.

Estecio, M. R., Youssef, E. M., Rahal, P., Fukuyama, E. E., Gois-Filho, J. F., Maniglia, J. V., Goloni-Bertollo, E. M., Issa, J. P., and Tajara, E. H. (2006). LHX6 is a sensitive methylation marker in head and neck carcinomas. *Oncogene* **25,** 5018–5026.

Fakhry, C., Westra, W. H., Li, S., Cmelak, A., Ridge, J. A., Pinto, H., Forastiere, A., and Gillison, M. L. (2008). Improved survival of patients with human papillomavirus-positive head and neck squamous cell carcinoma in a prospective clinical trial. *J. Natl. Cancer Inst.* **100,** 261–269.

Feinberg, A. P., and Vogelstein, B. (1983). Hypomethylation distinguishes genes of some human cancers from their normal counterparts. *Nature* **301,** 89–92.

Feinberg, A. P., Ohlsson, R., and Henikoff, S. (2006). The epigenetic progenitor origin of human cancer. *Nat. Rev. Genet.* **7,** 21–33.

Fernandez, A. F., and Esteller, M. (2010). Viral epigenomes in human tumorigenesis. *Oncogene* **29,** 1405–1420.

Flanagan, J. M., Cocciardi, S., Waddell, N., Johnstone, C. N., Marsh, A., Henderson, S., Simpson, P., da Silva, L., Khanna, K., Lakhani, S., Boshoff, C., and Chenevix-Trench, G. (2010). DNA methylome of familial breast cancer identifies distinct profiles defined by mutation status. *Am. J. Hum. Genet.* **86,** 420–433.

Flicek, P., Aken, B. L., Ballester, B., Beal, K., Bragin, E., Brent, S., Chen, Y., Clapham, P., Coates, G., Fairley, S., Fitzgerald, S., Fernandez-Banet, J., *et al.* (2010). Ensembl's 10th year. *Nucleic Acids Res.* **38,** D557–D562.

Franzmann, E. J., Reategui, E. P., Pedroso, F., Pernas, F. G., Karakullukcu, B. M., Carraway, K. L., Hamilton, K., Singal, R., and Goodwin, W. J. (2007). Soluble CD44 is a potential marker for the early detection of head and neck cancer. *Cancer Epidemiol. Biomarkers Prev.* **16,** 1348–1355.

Frommer, M., McDonald, L. E., Millar, D. S., Collis, C. M., Watt, F., Grigg, G. W., Molloy, P. L., and Paul, C. L. (1992). A genomic sequencing protocol that yields a positive display of 5-methylcytosine residues in individual DNA strands. *Proc. Natl. Acad. Sci. USA* **89**, 1827–1831.

Gal-Yam, E. N., Egger, G., Iniguez, L., Holster, H., Einarsson, S., Zhang, X., Lin, J. C., Liang, G., Jones, P. A., and Tanay, A. (2008). Frequent switching of Polycomb repressive marks and DNA hypermethylation in the PC3 prostate cancer cell line. *Proc. Natl. Acad. Sci. USA* **105**, 12979–12984.

Gondor, A., and Ohlsson, R. (2009). Chromosome crosstalk in three dimensions. *Nature* **461**, 212–217.

Greenman, C., Stephens, P., Smith, R., Dalgliesh, G. L., Hunter, C., Bignell, G., Davies, H., Teague, J., Butler, A., Stevens, C., Edkins, S., O'Meara, S., *et al.* (2007). Patterns of somatic mutation in human cancer genomes. *Nature* **446**, 153–158.

Grutzmann, R., Molnar, B., Pilarsky, C., Habermann, J. K., Schlag, P. M., Saeger, H. D., Miehlke, S., Stolz, T., Model, F., Roblick, U. J., Bruch, H. P., Koch, R., *et al.* (2008). Sensitive detection of colorectal cancer in peripheral blood by septin 9 DNA methylation assay. *PLoS One* **3**, e3759.

Harris, T. D., Buzby, P. R., Babcock, H., Beer, E., Bowers, J., Braslavsky, I., Causey, M., Colonell, J., Dimeo, J., Efcavitch, J. W., Giladi, E., Gill, J., *et al.* (2008). Single-molecule DNA sequencing of a viral genome. *Science* **320**, 106–109.

Hess, C. J., Ameziane, N., Schuurhuis, G. J., Errami, A., Denkers, F., Kaspers, G. J., Cloos, J., Joenje, H., Reinhardt, D., Ossenkoppele, G. J., Zwaan, C. M., and Waisfisz, Q. (2008). Hypermethylation of the FANCC and FANCL promoter regions in sporadic acute leukaemia. *Cell Oncol.* **30**, 299–306.

Hodges, E., Smith, A. D., Kendall, J., Xuan, Z., Ravi, K., Rooks, M., Zhang, M. Q., Ye, K., Bhattacharjee, A., Brizuela, L., McCombie, W. R., Wigler, M., *et al.* (2009). High definition profiling of mammalian DNA methylation by array capture and single molecule bisulfite sequencing. *Genome Res.* **19**, 1593–1605.

Howell, P. M., Jr., Liu, S., Ren, S., Behlen, C., Fodstad, O., and Riker, A. I. (2009). Epigenetics in human melanoma. *Cancer Control* **16**, 200–218.

Hsiung, D. T., Marsit, C. J., Houseman, E. A., Eddy, K., Furniss, C. S., McClean, M. D., and Kelsey, K. T. (2007). Global DNA methylation level in whole blood as a biomarker in head and neck squamous cell carcinoma. *Cancer Epidemiol. Biomarkers Prev.* **16**, 108–114.

Huang, Y., Pastor, W. A., Shen, Y., Tahiliani, M., Liu, D. R., and Rao, A. (2010). The behaviour of 5-hydroxymethylcytosine in bisulfite sequencing. *PLoS One* **5**, e8888.

Hunter, K. D., Parkinson, E. K., and Harrison, P. R. (2005). Profiling early head and neck cancer. *Nat. Rev. Cancer* **5**, 127–135.

International Cancer Genome Consortium (2010). International network of cancer genome projects. *Nature* **464**, 993–998.

Irizarry, R. A., Ladd-Acosta, C., Wen, B., Wu, Z., Montano, C., Onyango, P., Cui, H., Gabo, K., Rongione, M., Webster, M., Ji, H., Potash, J. B., *et al.* (2009). The human colon cancer methylome shows similar hypo- and hypermethylation at conserved tissue-specific CpG island shores. *Nat. Genet* **41**, 178–186.

Issa, J. P. (2004). CpG island methylator phenotype in cancer. *Nat. Rev. Cancer* **4**, 988–993.

Jemal, A., Murray, T., Ward, E., Samuels, A., Tiwari, R. C., Ghafoor, A., Feuer, E. J., and Thun, M. J. (2005). Cancer statistics, 2005. *CA Cancer J. Clin.* **55**, 10–30.

Johnson, D. S., Mortazavi, A., Myers, R. M., and Wold, B. (2007). Genome-wide mapping of in vivo protein-DNA interactions. *Science* **316**, 1497–1502.

Jones, P. A., and Baylin, S. B. (2007). The epigenomics of cancer. *Cell* **128**, 683–692.

Jones, P. A., and Laird, P. W. (1999). Cancer epigenetics comes of age. *Nat. Genet.* **21**, 163–167.

Jones, A., Lechner, M., Fourkala, E. O., Kristeleit, R., and Widschwendter, M. (2010). Emerging promise of epigenetics and DNA methylation for the diagnosis and management of women's cancers. *Epigenomics* **2**, 9–38.

Killian, J. K., Bilke, S., Davis, S., Walker, R. L., Killian, M. S., Jaeger, E. B., Chen, Y., Hipp, J., Pittaluga, S., Raffeld, M., Cornelison, R., Smith, W. I., Jr., *et al.* (2009). Large-scale profiling of archival lymph nodes reveals pervasive remodeling of the follicular lymphoma methylome. *Cancer Res.* **69**, 758–764.

Kim, D. H., Nelson, H. H., Wiencke, J. K., Zheng, S., Christiani, D. C., Wain, J. C., Mark, E. J., and Kelsey, K. T. (2001). p16(INK4a) and histology-specific methylation of CpG islands by exposure to tobacco smoke in non-small cell lung cancer. *Cancer Res.* **61**, 3419–3424.

Kim, M. S., Lee, J., and Sidransky, D. (2010). DNA methylation markers in colorectal cancer. *Cancer Metastasis Rev.* **29**, 181–206.

Koga, Y., Pelizzola, M., Cheng, E., Krauthammer, M., Sznol, M., Ariyan, S., Narayan, D., Molinaro, A. M., Halaban, R., and Weissman, S. M. (2009). Genome-wide screen of promoter methylation identifies novel markers in melanoma. *Genome Res.* **19**, 1462–1470.

Laird, P. W. (2010). Principles and challenges of genome-wide DNA methylation analysis. *Nat. Rev. Genet.* **11**, 191–203.

Laurent, L., Wong, E., Li, G., Huynh, T., Tsirigos, A., Ong, C. T., Low, H. M., Kin Sung, K. W., Rigoutsos, I., Loring, J., and Wei, C. L. (2010). Dynamic changes in the human methylome during differentiation. *Genome Res.* **20**, 320–331.

Li, N., Ye, M., Li, Y., Yan, Z., Butcher, L. M., Sun, J., Han, X., Chen, Q., Zhang, X., and Wang, J. (2010). Whole genome DNA Methylation analysis based on high throughput sequencing technology. *Methods* (in press).

Lister, R., Pelizzola, M., Dowen, R. H., Hawkins, R. D., Hon, G., Tonti-Filippini, J., Nery, J. R., Lee, L., Ye, Z., Ngo, Q. M., Edsall, L., Antosiewicz-Bourget, J., *et al.* (2009). Human DNA methylomes at base resolution show widespread epigenomic differences. *Nature* **462**, 315–322.

Llovet, J. M., and Bruix, J. (2003). Systematic review of randomized trials for unresectable hepatocellular carcinoma: Chemoembolization improves survival. *Hepatology* **37**, 429–442.

Lo, P. K., and Sukumar, S. (2008). Epigenomics and breast cancer. *Pharmacogenomics* **9**, 1879–1902.

Long, C., Yin, B., Lu, Q., Zhou, X., Hu, J., Yang, Y., Yu, F., and Yuan, Y. (2007). Promoter hypermethylation of the RUNX3 gene in esophageal squamous cell carcinoma. *Cancer Invest.* **25**, 685–690.

Manecksha, R. P., and Fitzpatrick, J. M. (2009). Epidemiology of testicular cancer. *BJU Int.* **104**, 1329–1333.

Marsit, C. J., Liu, M., Nelson, H. H., Posner, M., Suzuki, M., and Kelsey, K. T. (2004). Inactivation of the Fanconi anemia/BRCA pathway in lung and oral cancers: Implications for treatment and survival. *Oncogene* **23**, 1000–1004.

Marsit, C. J., Houseman, E. A., Christensen, B. C., Eddy, K., Bueno, R., Sugarbaker, D. J., Nelson, H. H., Karagas, M. R., and Kelsey, K. T. (2006). Examination of a CpG island methylator phenotype and implications of methylation profiles in solid tumors. *Cancer Res.* **66**, 10621–10629.

Marsit, C. J., Christensen, B. C., Houseman, E. A., Karagas, M. R., Wrensch, M. R., Yeh, R. F., Nelson, H. H., Wiemels, J. L., Zheng, S., Posner, M. R., McClean, M. D., Wiencke, J. K., *et al.* (2009). Epigenetic profiling reveals etiologically distinct patterns of DNA methylation in head and neck squamous cell carcinoma. *Carcinogenesis* **30**, 416–422.

Martens, J. W., Margossian, A. L., Schmitt, M., Foekens, J., and Harbeck, N. (2009). DNA methylation as a biomarker in breast cancer. *Future Oncol.* **5**, 1245–1256.

Martone, T., Gillio-Tos, A., De Marco, L., Fiano, V., Maule, M., Cavalot, A., Garzaro, M., Merletti, F., and Cortesina, G. (2007). Association between hypermethylated tumor and paired surgical margins in head and neck squamous cell carcinomas. *Clin. Cancer Res.* **13**, 5089–5094.

Matsubayashi, H., Sato, N., Fukushima, N., Yeo, C. J., Walter, K. M., Brune, K., Sahin, F., Hruban, R. H., and Goggins, M. (2003). Methylation of cyclin D2 is observed frequently in pancreatic cancer but is also an age-related phenomenon in gastrointestinal tissues. *Clin. Cancer Res.* **9,** 1446–1452.

McLendon, R., Friedman, A., Bigner, D., Van Meir, E. G., Brat, D. J., Mastrogianakis, G. M., Olson, J. J., Mikkelsen, T., Lehman, N., Aldape, K., Alfred Yung, W. K., Bogler, O., et al. (2008). Comprehensive genomic characterization defines human glioblastoma genes and core pathways. *Nature* **455,** 1061–1068.

Mikkelsen, T. S., Ku, M., Jaffe, D. B., Issac, B., Lieberman, E., Giannoukos, G., Alvarez, P., Brockman, W., Kim, T. K., Koche, R. P., Lee, W., Mendenhall, E., et al. (2007). Genome-wide maps of chromatin state in pluripotent and lineage-committed cells. *Nature* **448,** 553–560.

Mikkelsen, T. S., Hanna, J., Zhang, X., Ku, M., Wernig, M., Schorderet, P., Bernstein, B. E., Jaenisch, R., Lander, E. S., and Meissner, A. (2008). Dissecting direct reprogramming through integrative genomic analysis. *Nature* **454,** 49–55.

Morison, I. M., Ramsay, J. P., and Spencer, H. G. (2005). A census of mammalian imprinting. *Trends Genet.* **21,** 457–465.

Nayak, C. S., Carvalho, A. L., Jeronimo, C., Henrique, R., Kim, M. M., Hoque, M. O., Chang, S., Jiang, W. W., Koch, W., Westra, W., Sidransky, D., and Califano, J. (2007). Positive correlation of tissue inhibitor of metalloproteinase-3 and death-associated protein kinase hypermethylation in head and neck squamous cell carcinoma. *Laryngoscope* **117,** 1376–1380.

Noushmehr, H., Weisenberger, D. J., Diefes, K., Phillips, H. S., Pujara, K., Berman, B. P., Pan, F., Pelloski, C. E., Sulman, E. P., Bhat, K. P., Verhaak, R. G., Hoadley, K. A., et al. (2010). Identification of a CpG island methylator phenotype that defines a distinct subgroup of glioma. *Cancer Cell* **17**(5), 510–522.

Ogawa, O., Eccles, M. R., Szeto, J., McNoe, L. A., Yun, K., Maw, M. A., Smith, P. J., and Reeve, A. E. (1993). Relaxation of insulin-like growth factor II gene imprinting implicated in Wilms' tumour. *Nature* **362,** 749–751.

Ohm, J. E., McGarvey, K. M., Yu, X., Cheng, L., Schuebel, K. E., Cope, L., Mohammad, H. P., Chen, W., Daniel, V. C., Yu, W., Berman, D. M., Jenuwein, T., et al. (2007). A stem cell-like chromatin pattern may predispose tumor suppressor genes to DNA hypermethylation and heritable silencing. *Nat. Genet.* **39,** 237–242.

Park, P. J. (2009). ChIP-seq: Advantages and challenges of a maturing technology. *Nat. Rev. Genet.* **10,** 669–680.

Parkin, D. M., Bray, F., Ferlay, J., and Pisani, P. (2001). Estimating the world cancer burden: Globocan 2000. *Int. J. Cancer* **94,** 153–156.

Pollard, S. M., Stricker, S. H., and Beck, S. (2009). A shore sign of reprogramming. *Cell Stem Cell* **5,** 571–572.

Rakyan, V. K., Hildmann, T., Novik, K. L., Lewin, J., Tost, J., Cox, A. V., Andrews, T. D., Howe, K. L., Otto, T., Olek, A., Fischer, J., Gut, I. G., et al. (2004). DNA methylation profiling of the human major histocompatibility complex: A pilot study for the human epigenome project. *PLoS Biol.* **2,** e405.

Rakyan, V. K., Down, T. A., Thorne, N. P., Flicek, P., Kulesha, E., Graf, S., Tomazou, E. M., Backdahl, L., Johnson, N., Herberth, M., Howe, K. L., Jackson, D. K., et al. (2008). An integrated resource for genome-wide identification and analysis of human tissue-specific differentially methylated regions (tDMRs). *Genome Res.* **18,** 1518–1529.

Rakyan, V. K., Down, T. A., Maslau, S., Andrew, T., Yang, T. P., Beyan, H., Whittaker, P., McCann, O. T., Finer, S., Valdes, A. M., Leslie, R. D., Deloukas, P., et al. (2010). Human aging-associated DNA hypermethylation occurs preferentially at bivalent chromatin domains. *Genome Res.* **20,** 434–439.

Redaelli, A., Laskin, B. L., Stephens, J. M., Botteman, M. F., and Pashos, C. L. (2005). A systematic literature review of the clinical and epidemiological burden of acute lymphoblastic leukaemia (ALL). *Eur. J. Cancer Care (Engl.)* **14**, 53–62.

Reik, W. (2007). Stability and flexibility of epigenetic gene regulation in mammalian development. *Nature* **447**, 425–432.

Reinders, J., and Paszkowski, J. (2010). Bisulfite methylation profiling of large genomes. *Epigenomics* **2**, 209–220.

Richards, K. L., Zhang, B., Baggerly, K. A., Colella, S., Lang, J. C., Schuller, D. E., and Krahe, R. (2009). Genome-wide hypomethylation in head and neck cancer is more pronounced in HPV-negative tumors and is associated with genomic instability. *PLoS One* **4**, e4941.

Righini, C. A., de Fraipont, F., Timsit, J. F., Faure, C., Brambilla, E., Reyt, E., and Favrot, M. C. (2007). Tumor-specific methylation in saliva: A promising biomarker for early detection of head and neck cancer recurrence. *Clin. Cancer Res.* **13**, 1179–1185.

Rodriguez, T., Altieri, A., Chatenoud, L., Gallus, S., Bosetti, C., Negri, E., Franceschi, S., Levi, F., Talamini, R., and La Vecchia, C. (2004). Risk factors for oral and pharyngeal cancer in young adults. *Oral Oncol.* **40**, 207–213.

Rodriguez, J., Munoz, M., Vives, L., Frangou, C. G., Groudine, M., and Peinado, M. A. (2008). Bivalent domains enforce transcriptional memory of DNA methylated genes in cancer cells. *Proc. Natl. Acad. Sci. USA* **105**, 19809–19814.

Santarius, T., Shipley, J., Brewer, D., Stratton, M. R., and Cooper, C. S. (2010). A census of amplified and overexpressed human cancer genes. *Nat. Rev. Cancer* **10**, 59–64.

Sawyers, C. L. (2008). The cancer biomarker problem. *Nature* **452**, 548–552.

Schlesinger, Y., Straussman, R., Keshet, I., Farkash, S., Hecht, M., Zimmerman, J., Eden, E., Yakhini, Z., Ben-Shushan, E., Reubinoff, B. E., Bergman, Y., Simon, I., *et al.* (2007). Polycomb-mediated methylation on Lys27 of histone H3 pre-marks genes for de novo methylation in cancer. *Nat. Genet.* **39**, 232–236.

Sengupta, S., Chakrabarti, S., Roy, A., Panda, C. K., and Roychoudhury, S. (2007). Inactivation of human mutL homolog 1 and mutS homolog 2 genes in head and neck squamous cell carcinoma tumors and leukoplakia samples by promoter hypermethylation and its relation with microsatellite instability phenotype. *Cancer* **109**, 703–712.

Sladden, M. J., Balch, C., Barzilai, D. A., Berg, D., Freiman, A., Handiside, T., Hollis, S., Lens, M. B., and Thompson, J. F. (2009). Surgical excision margins for primary cutaneous melanoma. *Cochrane Database Syst. Rev.* (4), CD004835.

Smith, L. T., Lin, M., Brena, R. M., Lang, J. C., Schuller, D. E., Otterson, G. A., Morrison, C. D., Smiraglia, D. J., and Plass, C. (2006). Epigenetic regulation of the tumor suppressor gene TCF21 on 6q23-q24 in lung and head and neck cancer. *Proc. Natl. Acad. Sci. USA* **103**, 982–987.

Steinmann, K., Sandner, A., Schagdarsurengin, U., and Dammann, R. H. (2009). Frequent promoter hypermethylation of tumor-related genes in head and neck squamous cell carcinoma. *Oncol. Rep.* **22**, 1519–1526.

Subbalekha, K., Pimkhaokham, A., Pavasant, P., Chindavijak, S., Phokaew, C., Shuangshoti, S., Matangkasombut, O., and Mutirangura, A. (2009). Detection of LINE-1s hypomethylation in oral rinses of oral squamous cell carcinoma patients. *Oral Oncol.* **45**, 184–191.

Suzuki, H., Watkins, D. N., Jair, K. W., Schuebel, K. E., Markowitz, S. D., Chen, W. D., Pretlow, T. P., Yang, B., Akiyama, Y., Van Engeland, M., Toyota, M., Tokino, T., *et al.* (2004). Epigenetic inactivation of SFRP genes allows constitutive WNT signaling in colorectal cancer. *Nat. Genet.* **36**, 417–422.

Taylor, K. H., Kramer, R. S., Davis, J. W., Guo, J., Duff, D. J., Xu, D., Caldwell, C. W., and Shi, H. (2007). Ultradeep bisulfite sequencing analysis of DNA methylation patterns in multiple gene promoters by 454 sequencing. *Cancer Res.* **67**, 8511–8518.

Teschendorff, A. E., Menon, U., Gentry-Maharaj, A., Ramus, S. J., Gayther, S. A., Apostolidou, S., Jones, A., Lechner, M., Beck, S., Jacobs, I. J., and Widschwendter, M. (2009). An epigenetic signature in peripheral blood predicts active ovarian cancer. *PLoS One* **4,** e8274.

Teschendorff, A. E., Menon, U., Gentry-Maharaj, A., Ramus, S. J., Weisenberger, D. J., Shen, H., Campan, M., Noushmehr, H., Bell, C. G., Maxwell, A. P., Savage, D. A., Mueller-Holzner, E., *et al.* (2010). Age-dependent DNA methylation of genes that are suppressed in stem cells is a hallmark of cancer. *Genome Res.* **20,** 440–446.

Tirino, V., Desiderio, V., d'Aquino, R., De Francesco, F., Pirozzi, G., Graziano, A., Galderisi, U., Cavaliere, C., De Rosa, A., Papaccio, G., and Giordano, A. (2008). Detection and characterization of CD133+ cancer stem cells in human solid tumours. *PLoS One* **3,** e3469.

Tolstorukov, M. Y., Kharchenko, P. V., and Park, P. J. (2010). Analysis of the primary structure of chromatin with next-generation sequencing. *Epigenomics* **2,** 187–197.

Tommasi, S., Karm, D. L., Wu, X., Yen, Y., and Pfeifer, G. P. (2009). Methylation of homeobox genes is a frequent and early epigenetic event in breast cancer. *Breast Cancer Res.* **11,** R14.

Toyota, M., Ahuja, N., Ohe-Toyota, M., Herman, J. G., Baylin, S. B., and Issa, J. P. (1999). CpG island methylator phenotype in colorectal cancer. *Proc. Natl. Acad. Sci. USA* **96,** 8681–8686.

Tran, N., Rose, B. R., and O'Brien, C. J. (2007). Role of human papillomavirus in the etiology of head and neck cancer. *Head Neck* **29,** 64–70.

Veltri, R. W., and Makarov, D. V. (2006). Nucleic acid-based marker approaches to urologic cancers. *Urol. Oncol.* **24,** 510–527.

Weber, A., Hengge, U. R., Bardenheuer, W., Tischoff, I., Sommerer, F., Markwarth, A., Dietz, A., Wittekind, C., and Tannapfel, A. (2005). SOCS-3 is frequently methylated in head and neck squamous cell carcinoma and its precursor lesions and causes growth inhibition. *Oncogene* **24,** 6699–6708.

Wen, B., Wu, H., Shinkai, Y., Irizarry, R. A., and Feinberg, A. P. (2009). Large histone H3 lysine 9 dimethylated chromatin blocks distinguish differentiated from embryonic stem cells. *Nat. Genet.* **41,** 246–250.

Widschwendter, M., Fiegl, H., Egle, D., Mueller-Holzner, E., Spizzo, G., Marth, C., Weisenberger, D. J., Campan, M., Young, J., Jacobs, I., and Laird, P. W. (2007). Epigenetic stem cell signature in cancer. *Nat. Genet.* **39,** 157–158.

Wilson, A. S., Power, B. E., and Molloy, P. L. (2007). DNA hypomethylation and human diseases. *Biochim. Biophys. Acta* **1775,** 138–162.

Zhang, X., Yazaki, J., Sundaresan, A., Cokus, S., Chan, S. W., Chen, H., Henderson, I. R., Shinn, P., Pellegrini, M., Jacobsen, S. E., and Ecker, J. R. (2006). Genome-wide high-resolution mapping and functional analysis of DNA methylation in arabidopsis. *Cell* **126,** 1189–1201.

Zilberman, D., Gehring, M., Tran, R. K., Ballinger, T., and Henikoff, S. (2007). Genome-wide analysis of *Arabidopsis thaliana* DNA methylation uncovers an interdependence between methylation and transcription. *Nat. Genet.* **39,** 61–69.

10

Identification of Driver and Passenger DNA Methylation in Cancer by Epigenomic Analysis

Satish Kalari and Gerd P. Pfeifer

Department of Cancer Biology, Beckman Research Institute of the City of Hope, Duarte, California, USA

ABSTRACT

Human cancer genomes are characterized by widespread aberrations in DNA methylation patterns including DNA hypomethylation of mostly repetitive sequences and hypermethylation of numerous CpG islands. The analysis of DNA methylation patterns in cancer has progressed from single gene studies

Advances in Genetics, Vol. 70
Copyright 2010, Elsevier Inc. All rights reserved.

0065-2660/10 $35.00
DOI: 10.1016/S0065-2660(10)70010-9

examining potentially important candidate genes to a more global analysis where all or almost all promoter and CpG island sequences can be analyzed. We provide a brief overview of these genome-scale methylation-profiling techniques, summarize some of the information that has been obtained with these approaches, and discuss what we have learned about the specificity of methylation aberrations in cancer at a genome-wide level. The challenge is now to identify those methylation changes that are thought to be crucial for the processes of tumor initiation, tumor progression, or metastasis and distinguish these from methylation changes that are merely passenger events that accompany the transformation process but have no effect per se on the process of carcinogenesis. © 2010, Elsevier Inc.

I. INTRODUCTION

DNA methylation is one of the important epigenetic mechanisms that control gene expression, chromatin structure, genome stability, and X chromosome inactivation (Geiman and Robertson, 2002; Jones and Baylin, 2002). Abnormality in DNA methylation can lead to serious imbalance in normal function of cells and can promote pathological conditions. In particular, the genome of cancer cells is known to undergo substantial changes in DNA methylation (Jones and Baylin, 2002). Most notable are genome-wide hypomethylation events that preferentially target repetitive DNA elements, and gene-specific hypermethylation of CpG islands. CpG islands are sequences with greater than normal $G+C$ DNA content (Bird, 1986). Although their exact definition varies, they are usually between 0.2 and 2 kb long and contain a relatively high frequency of CpG dinucleotides. CpG sequences normally are underrepresented in mammalian genomes, owing to mutational pressure and/or lack of efficient DNA repair at methylated CpGs (Pfeifer, 2006). However, in normal tissues and in the germ line, the majority of gene promoter-associated CpG islands remain unmethylated. Accordingly, they are not subject to erosion by mutational events and retain a close to expected frequency of CpG dinucleotides. Methylation of CpG islands becomes aberrant in cancer when many hundreds of CpG islands in individual tumors acquire DNA methylation.

Global DNA hypomethylation and gene-specific hypermethylation are among the prominent hallmarks of cancer genomes (Ehrlich, 2002; Ushijima, 2005). Studies of aberrant methylation emphasize the pervasiveness of these changes in tumorigenesis and tumor progression. The role of DNA hypomethylation is often considered less important due to its global nature, along with limited knowledge of specific genes and genomic regions associated with hypomethylation. However, a cancer-causing role of DNA hypomethylation is clearly suggested by studies in mice carrying hypomorphic alleles of DNA methyltransferase genes, that is, Dnmt1 (Gaudet *et al.*, 2003). These mice develop malignancies, in particular lymphomas and hepatocellular carcinoma, but the effect of Dnmt1 loss can be complicated and may either support or inhibit tumor development (Gaudet *et al.*, 2003; Laird *et al.*, 1995; Yamada *et al.*, 2005). The mechanisms by which DNA hypomethylation is tumor-predisposing are unknown but it is conceivable that reactivation of methylation-silenced repetitive DNA elements and increased genomic instability are involved (Ehrlich, 2002). Most of the literature available on epigenetic factors in initiation and progression of tumorigenesis is dealing with hypermethylation of CpG islands or gene promoters and so is this review.

II. ABERRANT DNA METHYLATION IN CANCER—STARTING FROM SINGLE GENE STUDIES

DNA methylation of promoter CpG islands is strongly associated with gene silencing and is known as a frequent cause of loss of expression of, for example, tumor suppressor genes as well as other genes involved in tumor formation. Much of what is known today about the importance of DNA methylation in cancer was gained earlier through small- and moderate-scale analysis of gene promoters in different tumor types. The very first methodologies employed for the analysis of DNA methylation depended on initial digestion of DNA with methylation-sensitive restriction endonucleases followed by Southern blotting (Bird and Southern, 1978). Later on, sodium bisulfite sequencing and other methods based on that same concept became the methods of choice for single gene analysis (Frommer *et al.*, 1992).

Initial focus on DNA methylation in tumors was centered on the question of methylation-induced silencing of known tumor suppressor genes. During tumorigenesis, both alleles of a tumor suppressor gene need to be inactivated. This can occur by chromosomal deletions or loss-of-function mutations affecting the gene's coding sequence. Alternatively, hypermethylation of CpG islands spanning the promoter regions of tumor suppressor genes (e.g., RB, CDKN2A, VHL, APC, MLH1, RASSF1A, and BRCA1) can lead to gene silencing and thus can be an integral mechanism in tumorigenesis equivalent to gene loss or mutation (Costello *et al.*, 2000; Dammann *et al.*, 2000; Herman and Baylin, 2003; Issa, 1999; Jones and Baylin, 2002; Nephew and Huang, 2003). Since hypermethylation generally leads to permanent inactivation of gene expression, and is thought to be less reversible than altered histone modifications, this epigenetic alteration is considered a key pathway for long-term silencing of genes. To give some examples on one particular type of tumor, we focus on lung cancer. Several specific CpG-island-associated gene methylation events were frequently observed including, for example, CDKN2A, RASSF1A,

RARβ, MGMT, GSTP1, CDH13, APC, DAPK, TIMP3, along with many other genes (Dammann et al., 2005b; Franklin, 2004; Kim et al., 2005; Topaloglu et al., 2004; Toyooka et al., 2003; Yanagawa et al., 2003; Zochbauer-Muller et al., 2001). Genes altered by DNA methylation include those involved in important cellular pathways such as cell cycle regulation (e.g., CDKN2A, and CHFR), proliferation (e.g., CDKN2A and CXCL12), DNA repair (e.g., MGMT), apoptosis (e.g., DAPK, caspase 8, FAS, TRAILR1), RAS signaling (RASSF1A, NORE1A), invasion (e.g., cadherins, ADAMTS1, TIMP3, PTGER2, laminin family), and Wnt signaling (APC, DKK1, SFRP). Some of these pathways affected by epigenetic change are those described as the hallmarks of cancer (Hanahan and Weinberg, 2000). Other studies of non-small cell lung cancer (NSCLC) identified many additional hypermethylated genes (e.g., ARPC1B, DNAH9, FLRT2, G0S2, IRS2, RUNX3, PKP1, SPOCK1, UCHL1, OTX1, BARHL2, MEIS1, and OC2; Bowman et al., 2006; Jin et al., 2009; Rauch et al., 2007, 2008). In the literature, the methylation frequency (i.e., the percentage of tumors analyzed that carry substantially methylated alleles) generally ranges from only a few percent for some genes to more than 80% for other genes. The reported methylation frequencies, even for the same genes, often differ substantially depending on the study population, tumor histology, and/or methodology used to assess CpG island methylation.

The choice of methylation targets analyzed in the numerous single gene studies has often been based on existing knowledge of the presumed function of a particular gene or gene family member, or it was the result of a more or less serendipitous discovery of a particular methylation event. In most cases, CpG islands overlapping the 5′ gene ends or promoters of genes have been analyzed. More recent unbiased genome-wide studies, however, have revealed common tumor-associated methylation of CpG islands outside promoter regions, and it is still unclear whether or not such methylation changes have biological consequences and what exactly these consequences are for tumor formation. Interestingly, there are often cancer-specific methylated CpG islands not associated with any known genes at all. These CpG islands may represent remote regulatory elements or may represent functionally relevant sequences associated with noncoding RNAs.

III. ABERRANT DNA METHYLATION IN CANCER—GENOME-WIDE STUDIES

A much better understanding of the role of DNA methylation in cancer, either as a marker of disease or as an active driver of tumorigenesis, will likely be gained from genome-wide studies of this modification in normal and malignant cells. This goal has become more reachable with the recent introduction of large-scale

Table 10.1. Some Characteristics of Genome-Wide DNA Methylation Detection Techniques

Technique[a]	Sensitivity (μg)	Nature of mCpGs detected	Reference
RLGS	2–5	In NotI sites	Costello *et al.* (2000)
MS-RDA	10	In restriction sites (e.g., HpaII)	Ushijima *et al.* (1997)
DMH	2	In restriction sites (e.g., HpaII, SmaI)	Huang *et al.* (1999)
MCA	5	SmaI restriction sites	Estecio *et al.* (2007)
McrBC	10	Two CpGs separated by 55 bp to 3 kb	Nouzova *et al.* (2004)
MeDIP	2–4	All, CpG density dependent	Weber *et al.* (2005)
MIRA	0.1–0.2	All, CpG-density dependent	Rauch *et al.* (2006)
BS	5	All	Lister *et al.* (2009)

[a]The techniques described are as follows: RLGS, restriction landmark genomic scanning; MS-RDA, methylation-sensitive representational difference analysis; DMH, differential methylation hybridization; MCA, methylated CpG island amplification; McrBC, McrBC nuclease cleavage of methylated DNA; MeDIP, methylated DNA immunoprecipitation; MIRA, methylated-CpG island recovery assay; BS, bisulfite sequencing.

genome analysis methodologies. These techniques have been adopted in various ways to allow for investigation of DNA methylation of many gene loci simultaneously (Table 10.1). In this section, we review several technological advances in genome-wide methylation profiling.

One of the earliest large-scale methylation-profiling techniques developed was methylation-sensitive representational difference analysis (MS-RDA; Smith and Kelsey, 2001; Ushijima and Yamashita, 2009; Ushijima *et al.*, 1997). Genomic DNA is predigested using the methylation-sensitive restriction enzyme *Hpa*II, and a mixture ratio of tester and driver DNAs is optimized to detect differences in methylation status of single copy genes between two tissue samples. Restriction endonuclease digestion-based DNA methylation analysis was modified by Huang and colleagues and developed as differential methylation hybridization (DMH) on array platforms by combining restriction endonucleases and microarrays for high-throughput analysis of the methylation status of CpG islands in human genomes (Huang *et al.*, 1999; Wei *et al.*, 2002; Yan *et al.*, 2001). This method utilizes a restriction enzyme *Mse*I, which recognizes TTAA, a sequence that is rarely present within GC-rich regions, and leaves most CpG islands intact. *Mse*I-generated fragments are ligated to defined synthetic linkers and are further digested, for example, with *Bst*UI, a methylation-sensitive restriction endonuclease. *Bst*UI recognizes and digests the sequence 5′-CGCG within CpG islands when they are unmethylated. CpG islands, which are methylated, resist *Bst*UI restriction digestion, and these methylated fragments can be subsequently amplified by linker-dependent PCR. The resulting PCR products are labeled with fluorescent dyes. To compare genome-level CpG island methylation, equal quantities of BstUI-digested amplicons from two samples (e.g., normal and cancer) are mixed and hybridized onto a microarray. The resulting ratio between the two dyes represents the methylation difference between the two samples.

There are several other methods that are based on restriction endonuclease digestion, such as classical restriction landmark genomic scanning (RLGS; Costello et al., 2000; Hatada et al., 1991), or HpaII tiny fragment enrichment by ligation-mediated PCR (HELP; Figueroa et al., 2009; Khulan et al., 2006). Nouzova et al. (2004) and Lippman et al. (2005) developed a DNA methylation-profiling technique by replacing BstUI or HpaII with McrBC, an unusual restriction enzyme that recognizes and cleaves CpG-methylated DNA. Sites on the DNA recognized by McrBC consist of two half-sites of the form (G/A)mC. These half-sites can be separated by up to 3 kb, but the optimal separation is ∼50–100 base pairs. This method was used to identify a number of hypermethylated regions in an acute promyelocytic leukemia cell line compared to normal peripheral blood mononuclear cells (Nouzova et al., 2004). Irizarry et al. modified the McrBC assay and developed a comprehensive high-throughput array analysis for relative methylation (CHARM; Irizarry et al., 2008). The unmethylated or methylated fractions can be analyzed on microarray or high-throughput sequencing platforms for genome-wide identification of aberrant methylation. Methylated CpG island amplification (MCA) coupled to microarrays also is based on methylation-sensitive restriction enzymes. Target sequences are amplified by PCR using flanking primers followed by sequence analysis or microarray probing. MCA is a powerful approach for simultaneous identification of differentially methylated genomic regions (Estecio et al., 2007; Toyota et al., 1999).

Genome-scale DNA methylation analysis by bisulfite conversion of DNA has now become possible. In bisulfite conversion of DNA, treatment of DNA with sodium bisulfite converts unmethylated cytosines to uracils, whereas methylated cytosines are not affected. The bisulfite-treated samples are then PCR amplified and unmethylated and deaminated cytosines are replaced by thymines during PCR. Then, these samples can be hybridized to microarrays for large-scale analysis of DNA methylation status (Gitan et al., 2002; Hou et al., 2004) or by Illumina sequencing (Gu et al., 2010; Laird, 2010; Lister et al., 2009). The latter approach still is expensive when applied to whole mammalian genomes and requires substantial computational resources. Variations of bisulfite-based approaches include analysis of subareas of the genome (Ball et al., 2009; Meissner et al., 2008), or a highly multiplexed PCR-based approach using the Illumina bead platform (Bibikova et al., 2006).

A third general type of high-throughput approach in methylation analysis is based on affinity purification of methylated DNA. Methylated DNA immunoprecipitation (MeDIP) utilizes mechanical fragmentation of the genomic DNA followed by anti-5mC antibody precipitation to enrich for methylated

DNA fragments (Weber *et al.*, 2005). The immunoprecipitated DNA, enriched in hypermethylated sequences, and total genomic DNA (as input) are labeled with fluorescent dyes Cy5 and Cy3, respectively, and cohybridized onto microarray chips or analyzed by high-throughput sequencing. MeDIP is thus a valuable general fractionation approach, compatible with different analysis platforms to query the level of methylation in genomic sequences at a level of resolution of about 100 bp. One of the crucial factors in this assay is the quality of the anti-5-methylcytosine antibody. Moreover, the MeDIP method is most sensitive for densely methylated sequences, as DNA fragments with many contiguous methylated CpGs are more efficiently precipitated. MeDIP requires effective DNA denaturation before antibody binding.

Affinity purification of methylated DNA by a protein or peptide that can specifically bind to methylated CpGs was initially reported by Cross *et al.* (1994). Among the methods most suitable for genome-wide mapping of DNA methylation, the methylated CpG island recovery assay (MIRA) represents an approach that is based on a methyl-CpG binding protein complex. MIRA depends on the fact that the methyl-CpG-binding protein MBD2B specifically recognizes methylated CpG dinucleotides (Hendrich and Bird, 1998) and that this interaction is strongly enhanced by the MBD3L1 protein (Rauch and Pfeifer, 2005; Rauch *et al.*, 2006, 2007), a heterodimerization partner of MBD2 (Jiang *et al.*, 2004). Among all methyl-CpG-binding proteins known, MBD2B has the highest affinity for methylated DNA and displays the greatest ability to distinguish between methylated and unmethylated DNA. It recognizes a wide range of methylated CpG sequences with little sequence specificity (Fraga *et al.*, 2003). In our lab, lack of a defined sequence specificity of the MBD2B/MBD3L1 complex was confirmed by cloning and random sequencing of MIRA-enriched DNA fragments. Pulldown of methylated fragments is most efficient when a minimum of two methylated CpG sites are present (Rauch *et al.*, 2006). In the MIRA procedure, sonicated genomic DNA is incubated with the MBD2B/MBD3L1 protein complex. Unlike the MeDIP technique, which requires single-stranded DNA for antibody recognition, MIRA works on normal double-stranded DNA; in fact, the complex does not bind to single-stranded DNA. The CpG-methylated DNA is collected from the binding reaction via the GST-tagged MBD2B and glutathione beads, linker ligated, and then PCR amplified. These PCR-amplified MBD-enriched DNA fractions and total genomic DNA (input) are labeled with fluorescent dyes Cy5 and Cy3, respectively, and cohybridized onto microarrays. The ratio of fluorescent intensity (Cy5 to Cy3) indicates the methylation status at each particular sequence analyzed. The MIRA-enrichment method has been proven to be compatible with several types of microarray platforms and high-throughput DNA sequencing platforms and is highly sensitive requiring only 100–200 ng of genomic DNA.

IV. RESULTS FROM DNA METHYLATION PROFILING

The importance and widespread occurrence of CpG island hypermethylation in cancer is becoming increasingly recognized. In initial studies examining a limited number of loci, it has been estimated that between 0.5% and 3% of all genes carrying CpG-rich promoter sequences may be silenced by DNA methylation in several types of cancer (Costello *et al.*, 2000; Shiraishi *et al.*, 2002). Examining all or most CpG islands in the genome, recent reports indicate that generally several hundred to even more than a thousand CpG islands can be methylated in individual tumors (Dudley *et al.*, 2008; Koga *et al.*, 2009; Kuang *et al.*, 2008; Omura *et al.*, 2008; Rauch *et al.*, 2006, 2007, 2008; Tommasi *et al.*, 2009). Table 10.2 summarizes some of the recent studies describing methylation profiling of cancer genomes.

Genome-wide analysis of DNA methylation of lung squamous cell carcinoma (SCC) and matching normal tissue DNA revealed a large number of lung SCC-specific hypermethylated genes. Chromosome tiling array analysis has indicated that all of them were CpG islands or CpG-rich regions, often overlapping or located in close proximity to promoter regions (Rauch *et al.*, 2008). Islands with different CpG densities can become hypermethylated in tumors. It is clear that not all of these hundreds of methylated genes can be tumor suppressor genes. For example, substantial subsets of the methylated genes were represented by a variety of homeobox genes (Rauch *et al.*, 2007). Homeobox gene-associated CpG islands were among the most common stage I disease DNA methylation events identified so far, that is, this methylation event appears in almost every early stage tumor (Rauch *et al.*, 2007, 2008). Genome-wide DNA methylation analysis identified CpG island methylation, for example, in proximity of the OTX1, NR2E1, PAX6, IRX2, OC2, TFAP2A, and EVX2 genes. These genes are tumor-specifically methylated with very little methylation found in normal lung tissue or in blood cell DNA (Rauch *et al.*, 2008).

The frequent methylation of homeobox genes and other developmental genes regulated by the Polycomb complex is a phenomenon observed in many different histological types of human cancer (Ohm *et al.*, 2007; Rauch *et al.*, 2006, 2007; Schlesinger *et al.*, 2007; Widschwendter *et al.*, 2007), as exemplified by several studies, which we will discuss briefly. Genome-wide methylation profiling of ductal carcinoma *in situ*, a premalignant breast lesion with a high potential to progress toward invasive carcinoma identified 108 significant CpG islands that undergo aberrant DNA methylation in ductal carcinoma *in situ* and stage I breast tumors, with methylation frequencies greater than or comparable with those of more advanced invasive carcinoma (50–93%; Tommasi *et al.*, 2009). A substantial fraction of these hypermethylated CpG islands (32% of the annotated CpG islands) was associated with several homeobox genes, such as TLX1, HOXB13, and HNF1B genes. Fifty-three percent of the genes

Table 10.2. Genome-Wide DNA Methylation Studies in Various Cancer Types

Cancer type	Technique used	Notable findings	Reference
Acute myeloid leukemia	Bisulfite Illumina bead array	Genome-wide promoter-associated hyper-methylation associated with improved patient survival	Deneberg et al. (2010)
Brain cancer	Bisulfite Illumina bead array	Genes hypermethylated in glioblastoma were highly enriched for targets of PRC2 in embryonic stem cells	Martinez et al. (2009)
Brain cancer	MIRA	Hypermethylation of neuronal differentia-tion genes in astrocytomas	Wu et al. (2010)
Breast cancer	MIRA	Methylation of homeobox and other developmental genes regulated by the Polycomb complex	Tommasi et al. (2009)
Breast cancer	MeDIP	Agglomerative epigenetic aberrations are frequent events in human breast cancer	Novak et al. (2008)
Breast cancer	MeDIP	Interdependence between DNA methy-lome alterations and morphological changes	Ruike et al. (2010)
Chronic lym-phocytic leukemia	MCA	Methylation of LINE and APP was associated with shorter overall survival	Kuang et al. (2008)
Chronic lym-phocytic leukemia	Bisulfite Illumina bead array	Possible mechanism for pathogenesis	Kanduri et al. (2010)
Colorectal cancer	MeDIP	Three different methylation epigenotypes exist in colorectal cancer	Yagi et al. (2010)
Esophageal squamous cell carcinoma	HELP	Influence of genetic background on DNA methylation	Yang et al. (2010)
Follicular lymphoma	MCA	Extensive hypermethylation in promoters of Polycomb target genes	Bennett et al. (2008)
Hematologic malignancies	MIRA	Methylation status of *TFAP2A* and *EBF2* genes associated with advanced disease in CML	Dunwell et al. (2010)
Head and neck cancer	RLGS	Methylation of genes involved in the transforming growth factor beta signal-ing pathway	Bennett et al. (2008)
Hematologic neoplasms	Bisulfite Illumina bead array	Methylation of *DBC1*, *DIO3*, *FZD9*, *HS3ST2*, *MOS*, and *MYOD1* and their role in development of different hema-tologic neoplasms	Martin-Subero et al. (2009)
Hepatocellular carcinomas	MCA	DNA methylation status was correlated with the cancer-free and overall survival rates of patients	Arai et al. (2009)
Lung cancer	MIRA	Biomarkers for early detection of lung cancer and extensive hypomethylation of repetitive sequences in tumors	Rauch et al. (2008)

(Continues)

Table 10.2. (*Continued*)

Cancer type	Technique used	Notable findings	Reference
Lung cancer	MCA	Differential methylation between meso-thelioma and adenocarcinoma	Goto *et al.* (2009)
Mantle cell lymphoma	HELP	Methylation-based drug targeting	Leshchenko *et al.* (2010)
Ovarian cancer	Bisulfite Illumina bead array	Diagnostic or risk-prediction of ovarian cancer by blood methylation profiling	Teschendorff *et al.* (2009)
Pancreatic cancer	MCA	Identification of aberrantly methylated genes in pancreatic cancers	Omura *et al.* (2008)
Prostate cancer	DMH	Methylation of homeobox or T-box genes	Kron *et al.* (2009)
Skin cancer	MeDIP	Moderate increases of methylation in early and significantly in advanced-stage melanomas	Koga *et al.* (2009)
Testicular germ cell tumors	MeDIP	Function of intergenic and intronic DMRs in the regulation of ncRNAs	Cheung *et al.* (2010)
Urothelial cancer	MCA	DNA methylation as indicator for carcinogenetic risk estimation	Nishiyama *et al.* (2010)

hypermethylated in early-stage breast cancer overlapped with known Polycomb targets and included homeobox genes and other developmental transcription factors (Tommasi *et al.*, 2009). Interestingly, one-third of the CpG islands identified by microarray analysis (26 out of the 81 annotated hits) were associated with members of various homeobox superfamilies (HOX, LHX, NKX, PAX, and so forth) and were preferential targets of *de novo* methylation in early-stage breast cancer (Tommasi *et al.*, 2009). These master regulators control vital functional networks during tissue development and differentiation and are misregulated in a variety of malignancies, including breast cancer (Abate-Shen, 2002; Coletta *et al.*, 2004).

Large-scale DNA methylation analysis of glioblastoma multiforme (GBM) identified 25 hypermethylated genes in more than 20% of the cases studied (Martinez *et al.*, 2009). The most frequently hypermethylated genes were HOXA11, CD81, PRKCDBP, TES, MEST, TNFRSF10A, and FZD9, and these were methylated in more than 50% of the samples. HOXA9, HOXA5, TFAP2C, IGFBP1, and some of the other genes were methylated to a lesser extent (between 25% and 40%) in GBM compared to controls, but were found to be methylated in various other cancers (Martinez *et al.*, 2009). Analyzing biological features of these hypermethylated genes revealed that the group of genes hypermethylated in GBM was highly enriched (41%) for targets of the PRC2 (Polycomb repressive complex 2) in embryonic stem cells. Furthermore, this study identified promoter hypermethylation of the transcription factor gene GATA6 (occurring in 30% of GBM) that was correlated with poor patient survival (Martinez et al., 2009).

We recently completed a study on astrocytoma/glioma patients and analyzed over 28,000 CpG islands in 30 patients (Wu et al., 2010). Several hundred CpG islands undergo specific hypermethylation relative to normal brain with 428 methylation peaks common to more than 25% of the astrocytomas. Genes involved in brain development and neuronal differentiation, such as BMP4, POU4F3, GDNF, OTX2, NEFM, CNTN4, OTP, SIM1, FYN, EN1, CHAT, GSX2, NKX6-1, PAX6, RAX, and DLX2, were strongly enriched among genes frequently methylated in tumors. There was an overrepresentation of homeobox genes, and 31% of the most commonly methylated genes represented targets of the Polycomb complex. We identified several chromosomal loci in which many (sometimes more than 20) consecutive CpG islands were hypermethylated in tumors. Seven of such loci were near homeobox genes, including the HOXC and HOXD clusters, and the BARHL2, DLX1, and PITX2 genes (Wu et al., 2010).

Genome-wide promoter methylation and gene expression analysis of early-passage human melanoma cell lines or tumor specimens compared with melanocytes identified a number of new hypermethylated genes on top of already known promoter-methylated genes in melanoma (e.g., RARB, RASSF1A, and PYCARD; Furuta et al., 2006; Hoon et al., 2004; Spugnardi et al., 2003). A study by Koga et al. (2009) identified the promoter hypermethylated genes COL1A2, NPM2, HSPB6, DDIT4L, MT1G, and SOX3 and also the homeobox genes HOXB13 and HOXA7 in melanoma cells. This study also points out that the frequency of promoter methylation of validated hypermethylated gene promoters (COL1A2, NPM2, HSPB6, DDIT4L, and MT1G) increases moderately in early and significantly in advanced-stage melanomas, using early-passage cell strains and snap-frozen tissues compared with normal melanocytes and nevi (Koga et al., 2009).

Global evaluation of DNA methylation in prostate cancer revealed a large number of hypermethylated genes that were significantly hypermethylated compared to reference samples (Kron et al., 2009). This study found that about 30% of significantly hypermethylated genes of the top 100 methylated genes in prostate cancer were homeobox or T-box genes (e.g., FOXC1, VAX1, SIX6, HOXD3, HHEX, TBX15, HOXD9, GSC, HOXC13, PROX1, TBX4, TBX3, HOXD8, PAX2, IRX6, ALX4, BARX2, BARX, PHOX2A, LBX1, DLX5, DLX6, LHX9, and HOXD8), similar to many other such methylation studies in various cancers (Rauch et al., 2006, 2007; Tommasi et al., 2009).

A genome-wide screen for DNA methylation changes in head and neck squamous cell carcinoma tumors identified five candidate genes, SLC5A8, SEPT9, FUSSEL18, EBF3, and IRX1, as methylated in 27–67% of the HNSCC patient samples tested (Bennett et al., 2008). Genome-wide analysis of promoter-associated CpG island methylation in acute lymphoblastic leukemia identified 404 potential targets of methylation (Kuang et al., 2008). Aberrantly methylated genes identified in this study had methylation frequencies ranging from 23% to 100%. Among the genes validated in primary ALL samples were GIPC2, RSPO1, MAGI1, CAST1, ADCY5, HSPA4L, OCLN, EFNA5, MSX2, GFPT2, GNA14, SALL1, MYO5B, ZNF382, and MN1 (Kuang et al., 2008). A study of DNA hypermethylation in follicular lymphoma also discovered widespread hypermethylation of homeobox genes and previously identified targets of Polycomb repressive complex 2 (PRC2) in cell lines and primary tumors, but not in benign follicular hyperplasia (Bennett et al., 2009).

V. TARGETED METHYLATION OF GENES VERSUS METHYLATION OF TARGET GENES OR "PASSENGER METHYLATION" VERSUS "DRIVER METHYLATION"

Some of the hypermethylated genes in cancer may be bona fide tumor suppressor genes, but it is unlikely that all these numerous methylation changes play a causative role in tumorigenesis. Rather, many promoter CpG islands are probably methylated as a consequence of or in association with carcinogenesis (passenger methylation). It is a challenge to pinpoint those crucial genes that are susceptible to methylation-associated gene silencing and are functionally important in preventing tumorigenesis (driver methylation). The situation is perhaps analogous to the one found for mutational changes in cancer. Genome-wide DNA sequencing of either a large number of coding sequences or entire cancer genomes have revealed a staggering number of mutational changes (Pfeifer and Besaratinia, 2009). Most often, mutations in specific genes occur only a single time among a larger number of tumors sequenced and it is then difficult to determine if that particular mutation is indeed a driver mutation or just a passenger event (Carter et al., 2009). These large-scale sequencing studies have confirmed frequent mutations in known tumor suppressor genes or oncogenes, for example, the p53 or RAS genes, but have occasionally uncovered the existence of novel and likely important driver mutations, for example, in the BRAF gene (Davies et al., 2002) and IDH1 gene (Parsons et al., 2008).

When methylation occurs at the promoter of a known and well-established tumor suppressor gene, for example, CDKN2A, the gene encoding the cyclin-dependent kinase inhibitor protein p16, then it is of course easy to predict that the methylation event at least has the potential to be tumor driving.

However, much of the altered methylation landscape in cancers may be a phenomenon linked to "targeted methylation" whereby a particular gene or chromatin environment predisposes that gene to methylation in cancer and reflects a passenger event.

The mechanisms for aberrant CpG island methylation in cancer are mostly unknown. Generally, CpG island hypermethylation is closely linked to modification of local chromatin architecture serving as one already existing mechanism for silencing transcription. It has been proposed that gene inactivity imposed by changes in chromatin structure or histone modification predisposes to DNA methylation (Bachman *et al.*, 2003; Song *et al.*, 2002). Specific DNA sequences within CpG islands may be associated with the methylation process (Feltus *et al.*, 2003; Keshet *et al.*, 2006). Whether or not these sequences are associated with DNA binding proteins *in vivo* that somehow attract methylation is not clear. Feltus *et al.* developed a method known as Pattern-based Methylation Analysis (PatMAn) based on seven short DNA sequence patterns (TCCCCCNC, TTTCCTNC, TCCNCCNCCC, GGAGNAAG, GAGA-NAAG, GCCACCCC, and GAGGAGGNNG) that discriminated methylation-prone (MP) and methylation-resistant (MR) CpG islands. This classifier predicts CpG islands that are at higher risk for hypermethylation in cancer (Feltus *et al.*, 2003). PatMAn predicted methylation-prone CpG islands associated with embryonic targets of Polycomb-repressive complex 2 (PRC2). McCabe *et al.* further improved PatMAn and developed a second classifier (SUPER-PatMAn) that combines PatMAn DNA patterns with SUZ12-enriched regions as a marker of PRC2 occupancy (McCabe *et al.*, 2009). These studies indicated that both local sequence context and a specific chromatin environment are involved in a large subset of genes undergoing hypermethylation in cancers (McCabe *et al.*, 2009).

Among the genes targeted by Polycomb complexes are many developmental transcription factor genes including homeobox genes but also some known tumor suppressor genes, such as CDKN2A. Aberrant expression of Polycomb group (PcG) and Trithorax group (TrxG) proteins is a common event in many cancers (Esteller, 2007; Pasini *et al.*, 2004; Raaphorst, 2005b; Valk-Lingbeek *et al.*, 2004). Components of Polycomb repressive complex 1 (PRC1; such as BMI1 Valk-Lingbeek *et al.*, 2004) and components of PRC2 (such as EZH2; Bracken *et al.*, 2003) are amplified and/or overexpressed in a broad spectrum of cancers. Aberrant expression of PRC components affects PcG protein complexes (Kuzmichev *et al.*, 2005) thus potentially influencing target gene affinities (Squazzo *et al.*, 2006). Pharmacological disruption or forced expression of PRC2 genes induces apoptosis in cancer cells and provides a proliferative advantage to primary cells, respectively (Sellers and Loda, 2002; Tan *et al.*, 2007). Any change in function of PcG and TrxG proteins, which occur during aging (Pardal *et al.*, 2005; Sharpless and DePinho, 2005) or inflammation (Coussens and Werb, 2002; Lu et al., 2006), may contribute to the development of cancer. However, the mechanism by which Polycomb target genes undergo hypermethylation in cancer is still unknown.

It is difficult to deduce why a large number of homeobox genes become preferential targets of aberrant CpG methylation during tumorigenesis and whether this extensive methylation event can shift their finely tuned homeostasis, thus triggering tumorigenesis, or whether this process is merely associated with the neoplastic event. The widespread and recurrent nature of this phenomenon, however, seems to suggest that a common mechanistic pathway may exist in cancer cells, which promotes de novo methylation of these targets at the onset of tumor development. Paradoxically, however, several homeobox genes are upregulated rather than downregulated in breast cancer and other tumor types, suggesting that several tiers of regulation, in addition to promoter DNA methylation, may concur in determining homeobox misregulation.

Recent data have unraveled the role of Polycomb repressor complexes in targeting and modulating homeobox genes. At least six independent genome-wide studies have identified several common Polycomb targets in vertebrates and flies, most of which are represented by homeobox genes and other developmental transcription factors (Ringrose, 2007). Commonly, most of the homeobox gene-associated methylated CpG islands are embedded in regions other than promoters, consistent with the finding that the PRC2 subunit SUZ12 is distributed across large domains of developmental genes spanning from the promoter up to 2–35 kb into the gene (Lee et al., 2006). SUZ12 is required for the histone H3K27 methyltransferase activity and silencing function of the EED–EZH2 complex and is upregulated in different tumors (Kirmizis et al., 2003). EZH2, the PRC2 catalytic subunit exhibiting histone H3 K27 methyltransferase activity, undergoes gene amplification in several tumor types (Bracken et al., 2003) and is overexpressed in prostate cancer and breast cancer (Ding and Kleer, 2006; Raaphorst, 2005a; Varambally et al., 2002). EZH2 physically interacts with all three DNA methyltransferases in mammalian cells and has been suggested to play a crucial role in regulating de novo DNA methylation and its maintenance at target sequences (Vire et al., 2006) although the maintenance methylation aspect is in question (McGarvey et al., 2007).

Further support for a mechanistic connection between Polycomb silencing and tumor-associated DNA methylation comes from recent studies linking Polycomb occupancy of genes in noncancerous cells and tissues (including embryonic stem cells) with cancer-associated hypermethylation events (Eden et al., 2007; Hahn et al., 2008; Ohm et al., 2007; Rauch et al., 2006, 2007; Schlesinger et al., 2007; Vire et al., 2006; Widschwendter et al., 2007). Both inflammation and aging are associated with methylation of Polycomb target genes indicating that these cancer predisposing scenarios might have a specific epigenetic basis (Hahn et al., 2008; Maegawa et al., 2010).

PcG target genes may be composed of "bivalent" chromatin containing both the active histone mark H3K4me3 and the silencing mark H3K27me3 (Bernstein *et al.*, 2006). Bracken and colleagues have suggested that, in undifferentiated cells, PcG complexes have the potential to target genes poised for silencing as well as target genes predisposed to activation (Bracken *et al.*, 2006). The transition between alternative modes of PcG regulation may require additional signals upon differentiation (and likewise during tumorigenesis), which may include recruitment of additional transcriptional activators and/or competition with PcG antagonists, the TrxG proteins. These signals may have a counteracting effect on PcG-mediated gene repression (Bracken *et al.*, 2006). In addition, recent studies indicate that gene silencing in cancer can occur by histone H3 lysine 27 trimethylation independent of promoter DNA methylation (Kondo *et al.*, 2008).

Specific histone configurations or modifications may either protect from methylation or promote DNA methylation at CpG islands (Fig. 10.1). One possibility is that CpG islands that do not undergo methylation in cancer carry protective factors and that methylation-prone islands lack these factors (Gebhard *et al.*, 2010). Furthermore, it has been shown that genes with high levels of binding of RNA polymerase II, regardless of transcription levels, are resistant to induction of aberrant methylation (Takeshima *et al.*, 2009). Trimethylation of histone H3 lysine 4 (H3K4me3) is associated with active or potentially active genes and unmethylated CpG islands (Barrera *et al.*, 2008). This modification interferes with binding of the *de novo* DNA methyltransferase DNMT3L/DNMT3A complex (Jia *et al.*, 2007; Ooi *et al.*, 2007) and is expected to prevent methylation.

VI. INACTIVATION OF IMPORTANT BIOLOGICAL PATHWAYS BY CANCER-ASSOCIATED METHYLATION

One key question is whether the widespread methylation of Polycomb target genes seen in many types of cancer is functionally important for tumor development. Expression of these genes is often required for certain stages of embryonic development and differentiation. However, we know very little about the importance of these developmental genes in adult somatic stem cells, the cell types from which tumors are most likely derived. One possibility is that the PcG target genes in these stem cells, for example, homeobox genes, are already transcriptionally silent and the methylation event would have no functional consequence (passenger methylation). If the gene of interest is expressed and does have a functional role in stem cells, for example, it might be important in differentiation processes, then aberrant methylation of this gene may favor the transformation process by interfering with tissue-specific differentiation pathways. In a recent

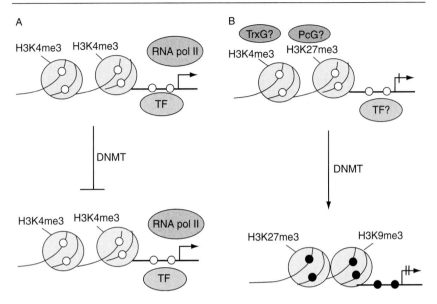

Figure 10.1. Targeted methylation or "passenger methylation" events. (A) Many genes are resistant to promoter methylation in cancer. Among the causes for this resistance can be association of a gene promoter with specific transcription factors or transcription factor complexes or presence of the activating histone mark H3K4me3, which interferes with DNA methylation. (B) Genes are targeted for methylation by a specific chromatin environment, for example, lack of transcription and presence of the repressive Polycomb complex and the histone mark, H3K27me3. These genes may initially be associated with "bivalent" chromatin characterized by both activating marks (H3K4me3 catalyzed by the Trithorax complex, TrxG) and repressive marks (H3K27me3 catalyzed by the Polycomb complex, PcG). In cancer tissue, these genes undergo DNA methylation and may be associated with other repressive marks (H3K9me3 and/or H3K27me3). The ball-shaped symbols represent nucleosome core particles over which the DNA is wrapped. Histone tail modifications are indicated and the open and closed circles represent unmethylated and methylated CpG sites, respectively.

study on astrocytomas, we observed that many of the tumor-specific methylated genes had roles in neuronal differentiation supporting a model in which methylation of these genes in neural stem cells favors proliferation versus differentiation and may contribute to initiation of the malignancy (Wu *et al.*, 2010). One key step in deciphering the role of methylation of Polycomb target genes in cancer will be the characterization of the expression and chromatin structure of these genes in adult somatic tissue stem cells.

There is evidence that methylation of genes within well-defined cellular pathways can contribute to tumorigenesis. For example, a large body of literature has established aberrant activation of Wingless-type (Wnt) signaling in various cancers such as colorectal cancer (Suzuki *et al.*, 2004; Yue *et al.*, 2008), head and neck carcinoma (Rhee *et al.*, 2002), melanoma (Weeraratna *et al.*, 2002), gastric cancer (Nojima *et al.*, 2007), hepatocellular carcinoma (Shih *et al.*, 2007; Takagi *et al.*, 2008), bladder cancer (Urakami *et al.*, 2006), and leukemia (Lu *et al.*, 2004). The secreted frizzled-related proteins function as negative regulators of Wnt signaling and have important roles in tumorigenesis. Notably, methylation of Wnt pathway inhibitory genes, such as secreted frizzled-related protein 1 and 2 (SFRP1 and SFRP2), whose inactivation enhances Wnt signaling, was observed in very early lesions of colon carcinogenesis, aberrant crypt foci (Suzuki *et al.*, 2004). Aberrant methylation of SFRP promoters and activation of the Wnt signaling pathway with excessive accumulation of beta-catenin in the nucleus was prominent in colorectal cancer (Suzuki *et al.*, 2004) and gastric cancers (Nojima *et al.*, 2007). Hypermethylation of SFRP genes, except for SFRP4, is frequent in hepatocellular carcinomas (HCCs). Reactivation of SFRP1 function by overexpressing SFRP1 in HCC cell lines blocked Wnt signaling and decreased abnormal accumulation of beta-catenin in the nucleus leading to arrest of cell growth. Overexpressed SFRPs downregulated T cell factor/lymphocyte enhancer factor (TCF/LEF) transcriptional activity in HCCs (Takagi *et al.*, 2008). siRNA-mediated downregulation of SFRP1 in beta-catenin-deficient cell lines promotes cell growth by activating Wnt signaling (Shih *et al.*, 2007). SFRP domain similarity with WNT-receptor frizzled proteins allows SFRPs to inhibit WNT receptor binding to consequently influence downstream pathway signaling during cell proliferation. Methylation silencing of other Wnt antagonists including Dickkopf 1 (DKK1) and Wnt inhibitory factor-1 (WIF-1) are also observed in different malignancies (Aguilera *et al.*, 2006; Taniguchi *et al.*, 2005).

SHP1 negatively regulates the Janus kinase/signal transducer and activator of transcription (Jak/STAT) signaling pathway (Chim *et al.*, 2004a,b). SHP1 in myeloma showed hypermethylation with constitutive STAT3 phosphorylation. Demethylated myeloma samples restored SHP1 expression with parallel downregulation of phosphorylated STAT3 (Chim *et al.*, 2004a). SHP1 methylation leading to epigenetic activation of the Jak/STAT pathway might have a tentative role in the pathogenesis of myeloma. Similarly, frequent methylation of SHP1 was observed in mantle cell and follicular lymphoma (Chim *et al.*, 2004c) and also in acute myeloid leukemia (Chim *et al.*, 2004b). Hypermethylation of SHP1-mediated activation of the Jak/STAT signaling pathway along with upregulation of cyclin D1 and BCL2 could be the basis for tumorigenicity in follicular lymphoma (Chim *et al.*, 2004c).

One emerging cellular growth control pathway is the Hippo pathway, a proapoptotic and antiproliferation signaling pathway initially identified in *Drosophila*. The pathway consists of a kinase cascade including the Drosophila Hippo kinase orthologues MST1 and MST2 and the LATS/WARTS serine/threonine kinases as well as several adapter proteins including RASSF proteins, SAV1, and

MOB1 (Guo et al., 2007; Harvey and Tapon, 2007). The kinase cascade functions to inactivate the gene product of the YAP oncogene, a transcriptional
coactivator of antiapoptotic and proproliferative genes, by phosphorylation and
cytoplasmic retention. Although YAP is overexpressed in some tumors, mutations in other components of the pathway are rare. However, several of the
Hippo pathway genes including RASSF1A, MST1, MST2, and LATS1 are
frequently methylated in human cancers leading to inactivation or partial
dysfunction of the pathway and tumorigenesis (Dammann et al., 2005a; Seidel
et al., 2007; Takahashi et al., 2005). This scenario is supported by several mouse
models, in which gene targeting of Rassf1a, Mst1 and Mst2, or Lats1, leads to
tumorigenesis (St John et al., 1999; Tommasi et al., 2005; Zhou et al., 2009).

      The preceding paragraphs illustrate some of the potential tumor-driving
mechanisms that are connected to aberrant methylation of genes within specific
growth control pathways. However, recent evidence also implicates growth-
signaling pathways in aberrant DNA methylation patterns. One example is the
significant correlation that has been reported between mutated BRAF kinase and
the CpG island methylator phenotype (CIMP) in colorectal cancer (Kambara
et al., 2004; Weisenberger et al., 2006). The exact mechanistic basis of this link is
unclear, but DNA hypermethylation of unknown targets might create a favorable
context for the acquisition of mutated BRAF(V600E) in CIMP-positive colorectal cancer (Hinoue et al., 2009). These findings raise the important and unan-
swered question of whether genetic lesions are driving DNA methylation, or
whether DNA methylation promotes or favors the selection of genetic lesions.

VII. METHYLATION OF NONCODING RNA PROMOTERS IN CANCER

The emerging field of noncoding RNA adds to the complexity of cellular
mechanisms that maintain normal cellular integrity. MicroRNAs (miRNAs)
represent a class of small noncoding RNAs that play important roles in carcinogenesis. Aberrant miRNA expression by promoter methylation has been illustrated for several human malignancies, and tumor suppressor functions have
been recognized for this new class of small regulatory RNAs. Recently, miRNAs
have been shown to play a role as targets in gene hypermethylation and silencing
in malignant cells. Similar to protein-coding genes, an aberrant pattern of
methylation of CpG islands near or within miRNAs genes could result in
alterations in the expression of miRNAs that could lead to tumorigenesis
(Davalos and Esteller, 2010).

      Studies with myeloid leukemia showed that overexpression of miR-29b
in acute myeloid leukemia cells caused significant downregulation of DNA
methyltransferases DNMT1, DNMT3A, and DNMT3B at both RNA and

protein levels (Garzon *et al.*, 2009). This decrease in expression of DNA methyl-transferases resulted in a decrease of global DNA methylation and reactivation of several genes via promoter DNA hypomethylation. Yet, this miR-29b down-regulation of DNA methyltransferase was indirect by targeting Sp1, a transacti-vator of the DNMT1 gene (Garzon *et al.*, 2009). A study in bacteria-induced gastric cancers highlighted the involvement of three miRNAs (miR-124a-1, miR-124a-2, and miR-124a-3). Silencing of these miRNAs due to aberrant methylation at their promoters, in addition to that of protein-coding genes, favors gastric tumorigenesis (Ando *et al.*, 2009).

A recent study analyzing miRNAs that are aberrantly expressed in ovarian cancer identified a number of hypomethylated miRNAs genes (including miR-21, miR-203, and miR-205) with the encoded miRNAs displaying upmo-dulated expression (Iorio *et al.*, 2007). Colon cancer cell line studies revealed hypermethylation of the CpG island of miR-124a in the cancer cell line but not in the normal tissue (Lujambio *et al.*, 2007). Subsequent studies proved that miR-124a is also frequently methylated in other colon, breast, and lung carcinoma cell lines, as well as in leukemias and lymphomas. Further studies showed that silencing of miR-124a by hypermethylation in cancer cells could result in increased expression of its target CDK6, an important regulator of the Rb protein. Another recent study of methylation of miRNA genes has shown hypermethylation of miR-9-1, miR-124a3, miR-148a, miR-152, and miR-663 in most of the human breast cancers tested (Lehmann *et al.*, 2008).

Hypermethylation of CpG islands near miR-34b, miR-137, miR-193a, and miR-203 miRNA genes in oral squamous cell carcinomas resulted in silen-cing of these miRNAs in the cancer setting (Kozaki *et al.*, 2008). miR-137 and miR-193a were found consistently hypermethylated in tumors, and normal expression of these two miRNAs was responsible for reduction in cell growth and proliferation factors, suggesting tumor suppressor characteristics for these miRNAs that are silenced during oral cancer progression.

Hypermethylation of the miR-34b/c CpG islands was commonly ob-served in colorectal cancer cell lines and in primary colorectal tumors (Toyota *et al.*, 2008). Both miR-34b and miR-34c are part of the p53 pathway. Reintro-ducing miR-34b or miR-34c into colorectal cancer cells induced changes in gene expression that overlapped (Toyota *et al.*, 2008). Hypermethylation of miR-34b/c along with miR-148a and miR-9 was observed in lymph node metastatic cancer (Lujambio *et al.*, 2008). The reintroduction of miR-148a and miR-34b/c into cancer cells resulted in inhibition of their motility, reduced tumor growth, and inhibited metastasis formation in xenograft models, with an associated down-regulation of the miRNA target genes, such as C-MYC, E2F3, CDK6, and TGIF2 (Lujambio *et al.*, 2008).

VIII. EXPERIMENTS THAT COULD DISTINGUISH BETWEEN "DRIVER METHYLATION" AND "PASSENGER METHYLATION" IN CANCER

In order to identify driver methylation events and to set them apart from passenger methylation events, it is important to know how to define driver methylation. In simple terms, a methylation event, which promotes tumorigenesis, can be considered as driver methylation. If a methylation change is a tumor-driving or initiating event, it is more likely to occur during early stages of tumorigenesis. Using mouse models or early stage human tumor specimens and premalignant lesions, one can observe the timing of methylation changes from early, preneoplastic tissues to late malignant disease in established cancer models. This approach has been used successfully by Chen et al. (2009). Early hypermethylation of a transcription factor gene, Foxd3, influenced methylation of genes later in disease progression. Early changes are more likely to be driving the cancer phenotype, whereas later changes may simply reflect the transformed phenotype.

Driver methylation is not only represented by inactivation of tumor suppressor genes, but may also be equivalent to activation of oncogenes either directly or indirectly. Moreover, methylation-silenced genes may be placed into any pathway that is represented by the hallmarks of cancer (Fig. 10.2; Hanahan and Weinberg, 2000). Thus, methylation silencing may enable the cell to acquire properties of increased cell growth and higher potential to invade tissue, to metastasize or to initiate angiogenesis, or to evade apoptosis. In addition, there are methylation-associated events that do not immediately lead to altered cellular phenotypes; for example, silencing of a DNA repair or cell cycle checkpoint gene will lead to enhanced genomic instability and that event might have secondary consequences such as promoting mutations in genes that counteract any of the hallmarks of cancer. Strong support for a cancer-driving role of a promoter methylation event will come from cancer genome sequencing data that show that the gene of interest is also mutated in a significant fraction of the tumors.

When genome-wide screens or other discovery platforms have provided information about which genes are methylated in specific cancers, the first question to ask is whether the methylation event indeed is associated with gene silencing. Such a test is of importance inasmuch as many of the methylation events in cancer occur at genes that are already silenced in the corresponding normal tissue in which the tumor originates (Hahn et al., 2008; Song et al., 2002; Takeshima et al., 2009). Methylation-associated gene inactivation will generally be more likely for genes that are methylated in the 5' promoter region. Expression studies in tumor and normal tissue and 5-azacytidine reactivation experiments with cancer cell lines are standard approaches to address this point. Next, gene ontology analysis can be useful to determine if the gene of interest functions

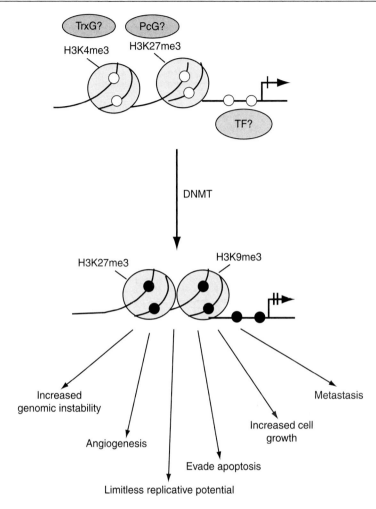

Figure 10.2. "Driver methylation" events. In order to promote tumorigenesis, a driver methylation event should occur in pathways important to prevent the emergence of any of the hallmarks of cancer, for example, increased cell growth, angiogenesis, metastasis and tissue invasion, evasion of apoptosis, unlimited replicative potential, or increased genomic instability.

in a particular relevant pathway. Pathways of interest will be those potentially associated with the hallmarks of cancer or with other important cellular processes, for example, defense against genome instability. If all these criteria are met, then experiments to test gene function are called for.

Usually, forced overexpression of methylation-suppressed genes in cancer cells or siRNA-mediated downregulation of the same genes in normal cells might give the first hints to differentiate between driver and passenger methylation. Forced expression of some of the frequently promoter-methylated genes in various cancer cell lines often inhibits cancer cell growth and colony formation and suppresses *in vivo* tumor growth in mice emphasizing the potential importance of these genes in tumorigenesis. Based on these observations, it is logical to propose that genes that are frequently methylated in various types of cancer could be driver methylation events in tumorigenesis. However, to provide more definitive evidence for the causative role of a methylation-silenced gene in cancer, *in vivo* mouse models employing gene targeting are necessary. This proof of principal has been accomplished for a few genes that are frequently methylated in human tumors, for example, HIC1, RASSF1A, SOCS3, and WIF-1 (Chen *et al.*, 2003; Kansara *et al.*, 2009; Ogata *et al.*, 2006; Tommasi *et al.*, 2005). More studies of this type should be conducted to assess the overall importance of hypermethylation of gene promoters in cancer.

In summary, genome-wide methylation-profiling studies have begun to portray a much more complete picture of all the methylation changes that occur in different types of malignancies. The published data have expanded our view of the extent of these changes, and it has now become a challenge to identify the critical, that is, driver methylation events that contribute to the transformed phenotype. In analogy to the cancer genome sequencing projects, the identification of driver versus passenger events will be of importance for our understanding of cancer etiology, development of tumor-relevant biomarkers, and eventually therapeutic approaches to target the driver methylation event.

Acknowledgment

Work of the authors was supported by NIH grant CA084469.

References

Abate-Shen, C. (2002). Deregulated homeobox gene expression in cancer: Cause or consequence? *Nat. Rev. Cancer* **2**, 777–785.

Aguilera, O., Fraga, M. F., Ballestar, E., Paz, M. F., Herranz, M., Espada, J., Garcia, J. M., Munoz, A., Esteller, M., and Gonzalez-Sancho, J. M. (2006). Epigenetic inactivation of the Wnt antagonist DICKKOPF-1 (DKK-1) gene in human colorectal cancer. *Oncogene* **25**, 4116–4121.

Ando, T., Yoshida, T., Enomoto, S., Asada, K., Tatematsu, M., Ichinose, M., Sugiyama, T., and Ushijima, T. (2009). DNA methylation of microRNA genes in gastric mucosae of gastric cancer patients: its possible involvement in the formation of epigenetic field defect. *Int. J. Cancer* **124**, 2367–2374.

Arai, E., Ushijima, S., Gotoh, M., Ojima, H., Kosuge, T., Hosoda, F., Shibata, T., Kondo, T., Yokoi, S., Imoto, I., *et al.* (2009). Genome-wide DNA methylation profiles in liver tissue at the precancerous stage and in hepatocellular carcinoma. *Int. J. Cancer* **125**, 2854–2862.

Bachman, K. E., Park, B. H., Rhee, I., Rajagopalan, H., Herman, J. G., Baylin, S. B., Kinzler, K. W., and Vogelstein, B. (2003). Histone modifications and silencing prior to DNA methylation of a tumor suppressor gene. *Cancer Cell* **3**, 89–95.

Ball, M. P., Li, J. B., Gao, Y., Lee, J. H., LeProust, E. M., Park, I. H., Xie, B., Daley, G. Q., and Church, G. M. (2009). Targeted and genome-scale strategies reveal gene-body methylation signatures in human cells. *Nat. Biotechnol.* **27**, 361–368.

Barrera, L. O., Li, Z., Smith, A. D., Arden, K. C., Cavenee, W. K., Zhang, M. Q., Green, R. D., and Ren, B. (2008). Genome-wide mapping and analysis of active promoters in mouse embryonic stem cells and adult organs. *Genome Res.* **18**, 46–59.

Bennett, K. L., Karpenko, M., Lin, M. T., Claus, R., Arab, K., Dyckhoff, G., Plinkert, P., Herpel, E., Smiraglia, D., and Plass, C. (2008). Frequently methylated tumor suppressor genes in head and neck squamous cell carcinoma. *Cancer Res.* **68**, 4494–4499.

Bennett, L. B., Schnabel, J. L., Kelchen, J. M., Taylor, K. H., Guo, J., Arthur, G. L., Papageorgio, C. N., Shi, H., and Caldwell, C. W. (2009). DNA hypermethylation accompanied by transcriptional repression in follicular lymphoma. *Genes Chromosom. Cancer* **48**, 828–841.

Bernstein, B. E., Mikkelsen, T. S., Xie, X., Kamal, M., Huebert, D. J., Cuff, J., Fry, B., Meissner, A., Wernig, M., Plath, K., *et al.* (2006). A bivalent chromatin structure marks key developmental genes in embryonic stem cells. *Cell* **125**, 315–326.

Bibikova, M., Lin, Z., Zhou, L., Chudin, E., Garcia, E. W., Wu, B., Doucet, D., Thomas, N. J., Wang, Y., Vollmer, E., *et al.* (2006). High-throughput DNA methylation profiling using universal bead arrays. *Genome Res.* **16**, 383–393.

Bird, A. P. (1986). CpG-rich islands and the function of DNA methylation. *Nature* **321**, 209–213.

Bird, A. P., and Southern, E. M. (1978). Use of restriction enzymes to study eukaryotic DNA methylation: I The methylation pattern in ribosomal DNA from Xenopus laevis. *J. Mol. Biol.* **118**, 27–47.

Bowman, R. V., Yang, I. A., Semmler, A. B., and Fong, K. M. (2006). Epigenetics of lung cancer. *Respirology* **11**, 355–365.

Bracken, A. P., Dietrich, N., Pasini, D., Hansen, K. H., and Helin, K. (2006). Genome-wide mapping of Polycomb target genes unravels their roles in cell fate transitions. *Genes Dev.* **20**, 1123–1136.

Bracken, A. P., Pasini, D., Capra, M., Prosperini, E., Colli, E., and Helin, K. (2003). EZH2 is downstream of the pRB-E2F pathway, essential for proliferation and amplified in cancer. *EMBO J.* **22**, 5323–5335.

Carter, H., Chen, S., Isik, L., Tyekucheva, S., Velculescu, V. E., Kinzler, K. W., Vogelstein, B., and Karchin, R. (2009). Cancer-specific high-throughput annotation of somatic mutations: computational prediction of driver missense mutations. *Cancer Res.* **69**, 6660–6667.

Chen, S. S., Raval, A., Johnson, A. J., Hertlein, E., Liu, T. H., Jin, V. X., Sherman, M. H., Liu, S. J., Dawson, D. W., Williams, K. E., *et al.* (2009). Epigenetic changes during disease progression in a murine model of human chronic lymphocytic leukemia. *Proc. Natl. Acad. Sci. USA* **106**, 13433–13438.

Chen, W. Y., Zeng, X., Carter, M. G., Morrell, C. N., Chiu Yen, R. W., Esteller, M., Watkins, D. N., Herman, J. G., Mankowski, J. L., and Baylin, S. B. (2003). Heterozygous disruption of Hic1 predisposes mice to a gender-dependent spectrum of malignant tumors. *Nat. Genet.* **33**, 197–202.

Cheung, H. H., Lee, T. L., Davis, A. J., Taft, D. H., Rennert, O. M., and Chan, W. Y. (2010). Genome-wide DNA methylation profiling reveals novel epigenetically regulated genes and non-coding RNAs in human testicular cancer. *Br. J. Cancer* **102**, 419–427.

Chim, C. S., Fung, T. K., Cheung, W. C., Liang, R., and Kwong, Y. L. (2004a). SOCS1 and SHP1 hypermethylation in multiple myeloma: implications for epigenetic activation of the Jak/STAT pathway. *Blood* **103**, 4630–4635.

Chim, C. S., Wong, A. S., and Kwong, Y. L. (2004b). Epigenetic dysregulation of the Jak/STAT pathway by frequent aberrant methylation of SHP1 but not SOCS1 in acute leukaemias. *Ann. Hematol.* **83,** 527–532.

Chim, C. S., Wong, K. Y., Loong, F., and Srivastava, G. (2004c). SOCS1 and SHP1 hypermethylation in mantle cell lymphoma and follicular lymphoma: implications for epigenetic activation of the Jak/STAT pathway. *Leukemia* **18,** 356–358.

Coletta, R. D., Jedlicka, P., Gutierrez-Hartmann, A., and Ford, H. L. (2004). Transcriptional control of the cell cycle in mammary gland development and tumorigenesis. *J. Mammary Gland Biol. Neoplasia* **9,** 39–53.

Costello, J. F., Fruhwald, M. C., Smiraglia, D. J., Rush, L. J., Robertson, G. P., Gao, X., Wright, F. A., Feramisco, J. D., Peltomaki, P., Lang, J. C., *et al.* (2000). Aberrant CpG-island methylation has non-random and tumour-type-specific patterns. *Nat. Genet.* **24,** 132–138.

Coussens, L. M., and Werb, Z. (2002). Inflammation and cancer. *Nature* **420,** 860–867.

Cross, S. H., Charlton, J. A., Nan, X., and Bird, A. P. (1994). Purification of CpG islands using a methylated DNA binding column. *Nat. Genet.* **6,** 236–244.

Dammann, R., Li, C., Yoon, J. H., Chin, P. L., Bates, S., and Pfeifer, G. P. (2000). Epigenetic inactivation of a RAS association domain family protein from the lung tumour suppressor locus 3p21.3. *Nat. Genet.* **25,** 315–319.

Dammann, R., Schagdarsurengin, U., Seidel, C., Strunnikova, M., Rastetter, M., Baier, K., and Pfeifer, G. P. (2005a). The tumor suppressor RASSF1A in human carcinogenesis: an update. *Histol. Histopathol.* **20,** 645–663.

Dammann, R., Strunnikova, M., Schagdarsurengin, U., Rastetter, M., Papritz, M., Hattenhorst, U. E., Hofmann, H. S., Silber, R. E., Burdach, S., and Hansen, G. (2005b). CpG island methylation and expression of tumour-associated genes in lung carcinoma. *Eur. J. Cancer* **41,** 1223–1236.

Davalos, V., and Esteller, M. (2010). MicroRNAs and cancer epigenetics: a macrorevolution. *Curr. Opin. Oncol.* **22,** 35–45.

Davies, H., Bignell, G. R., Cox, C., Stephens, P., Edkins, S., Clegg, S., Teague, J., Woffendin, H., Garnett, M. J., Bottomley, W., *et al.* (2002). Mutations of the BRAF gene in human cancer. *Nature* **417,** 949–954.

Deneberg, S., Grovdal, M., Karimi, M., Jansson, M., Nahi, H., Corbacioglu, A., Gaidzik, V., Dohner, K., Paul, C., Ekstrom, T. J., *et al.* (2010). Gene-specific and global methylation patterns predict outcome in patients with acute myeloid leukemia. *Leukemia* **24,** 932–941.

Ding, L., and Kleer, C. G. (2006). Enhancer of Zeste 2 as a marker of preneoplastic progression in the breast. *Cancer Res.* **66,** 9352–9355.

Dudley, K. J., Revill, K., Whitby, P., Clayton, R. N., and Farrell, W. E. (2008). Genome-wide analysis in a murine Dnmt1 knockdown model identifies epigenetically silenced genes in primary human pituitary tumors. *Mol. Cancer Res.* **6,** 1567–1574.

Dunwell, T., Hesson, L., Rauch, T. A., Wang, L., Clark, R. E., Dallol, A., Gentle, D., Catchpoole, D., Maher, E. R., Pfeifer, G. P., and Latif, F. (2010). A genome-wide screen identifies frequently methylated genes in haematological and epithelial cancers. *Mol. Cancer* **9,** 44.

Eden, E., Lipson, D., Yogev, S., and Yakhini, Z. (2007). Discovering motifs in ranked lists of DNA sequences. *PLoS Comput. Biol.* **3,** e39.

Ehrlich, M. (2002). DNA methylation in cancer: too much, but also too little. *Oncogene* **21,** 5400–5413.

Estecio, M. R., Yan, P. S., Ibrahim, A. E., Tellez, C. S., Shen, L., Huang, T. H., and Issa, J. P. (2007). High-throughput methylation profiling by MCA coupled to CpG island microarray. *Genome Res.* **17,** 1529–1536.

Esteller, M. (2007). Cancer epigenomics: DNA methylomes and histone-modification maps. *Nat. Rev. Genet.* **8,** 286–298.

Feltus, F. A., Lee, E. K., Costello, J. F., Plass, C., and Vertino, P. M. (2003). Predicting aberrant CpG island methylation. *Proc. Natl. Acad. Sci. USA* **100**, 12253–12258.

Figueroa, M. E., Melnick, A., and Greally, J. M. (2009). Genome-wide determination of DNA methylation by Hpa II tiny fragment enrichment by ligation-mediated PCR (HELP) for the study of acute leukemias. *Methods Mol. Biol.* **538**, 395–407.

Fraga, M. F., Ballestar, E., Montoya, G., Taysavang, P., Wade, P. A., and Esteller, M. (2003). The affinity of different MBD proteins for a specific methylated locus depends on their intrinsic binding properties. *Nucleic Acids Res.* **31**, 1765–1774.

Franklin, W. A. (2004). Premalignant evolution of lung cancer: Gilles F Filley lecture. *Chest* **125**, 90S–94S.

Frommer, M., McDonald, L. E., Millar, D. S., Collis, C. M., Watt, F., Grigg, G. W., Molloy, P. L., and Paul, C. L. (1992). A genomic sequencing protocol that yields a positive display of 5-methylcytosine residues in individual DNA strands. *Proc. Natl. Acad. Sci. USA* **89**, 1827–1831.

Furuta, J., Nobeyama, Y., Umebayashi, Y., Otsuka, F., Kikuchi, K., and Ushijima, T. (2006). Silencing of Peroxiredoxin 2 and aberrant methylation of 33 CpG islands in putative promoter regions in human malignant melanomas. *Cancer Res.* **66**, 6080–6086.

Garzon, R., Liu, S., Fabbri, M., Liu, Z., Heaphy, C. E., Callegari, E., Schwind, S., Pang, J., Yu, J., Muthusamy, N., et al. (2009). MicroRNA-29b induces global DNA hypomethylation and tumor suppressor gene reexpression in acute myeloid leukemia by targeting directly DNMT3A and 3B and indirectly DNMT1. *Blood* **113**, 6411–6418.

Gaudet, F., Hodgson, J. G., Eden, A., Jackson-Grusby, L., Dausman, J., Gray, J. W., Leonhardt, H., and Jaenisch, R. (2003). Induction of tumors in mice by genomic hypomethylation. *Science* **300**, 489–492.

Gebhard, C., Benner, C., Ehrich, M., Schwarzfischer, L., Schilling, E., Klug, M., Dietmaier, W., Thiede, C., Holler, E., Andreesen, R., and Rehli, M. (2010). General transcription factor binding at CpG islands in normal cells correlates with resistance to de novo DNA methylation in cancer cells. *Cancer Res.* **70**, 1398–1407.

Geiman, T. M., and Robertson, K. D. (2002). Chromatin remodeling, histone modifications, and DNA methylation-how does it all fit together? *J. Cell. Biochem.* **87**, 117–125.

Gitan, R. S., Shi, H., Chen, C. M., Yan, P. S., and Huang, T. H. (2002). Methylation-specific oligonucleotide microarray: a new potential for high-throughput methylation analysis. *Genome Res.* **12**, 158–164.

Goto, Y., Shinjo, K., Kondo, Y., Shen, L., Toyota, M., Suzuki, H., Gao, W., An, B., Fujii, M., Murakami, H., et al. (2009). Epigenetic profiles distinguish malignant pleural mesothelioma from lung adenocarcinoma. *Cancer Res.* **69**, 9073–9082.

Gu, H., Bock, C., Mikkelsen, T. S., Jager, N., Smith, Z. D., Tomazou, E., Gnirke, A., Lander, E. S., and Meissner, A. (2010). Genome-scale DNA methylation mapping of clinical samples at single-nucleotide resolution. *Nat. Methods* **7**, 133–136.

Guo, C., Tommasi, S., Liu, L., Yee, J. K., Dammann, R., and Pfeifer, G. P. (2007). RASSF1A is part of a complex similar to the Drosophila Hippo/Salvador/Lats tumor-suppressor network. *Curr. Biol.* **17**, 700–705.

Hahn, M. A., Hahn, T., Lee, D. H., Esworthy, R. S., Kim, B. W., Riggs, A. D., Chu, F. F., and Pfeifer, G. P. (2008). Methylation of polycomb target genes in intestinal cancer is mediated by inflammation. *Cancer Res.* **68**, 10280–10289.

Hanahan, D., and Weinberg, R. A. (2000). The hallmarks of cancer. *Cell* **100**, 57–70.

Harvey, K., and Tapon, N. (2007). The Salvador-Warts-Hippo pathway - an emerging tumour-suppressor network. *Nat. Rev. Cancer* **7**, 182–191.

Hatada, I., Hayashizaki, Y., Hirotsune, S., Komatsubara, H., and Mukai, T. (1991). A genomic scanning method for higher organisms using restriction sites as landmarks. *Proc. Natl. Acad. Sci. USA* **88**, 9523–9527.

Hendrich, B., and Bird, A. (1998). Identification and characterization of a family of mammalian methyl-CpG binding proteins. *Mol. Cell. Biol.* **18,** 6538–6547.

Herman, J. G., and Baylin, S. B. (2003). Gene silencing in cancer in association with promoter hypermethylation. *N. Engl. J. Med.* **349,** 2042–2054.

Hinoue, T., Weisenberger, D. J., Pan, F., Campan, M., Kim, M., Young, J., Whitehall, V. L., Leggett, B. A., and Laird, P. W. (2009). Analysis of the association between CIMP and BRAF in colorectal cancer by DNA methylation profiling. *PLoS ONE* **4,** e8357.

Hoon, D. S., Spugnardi, M., Kuo, C., Huang, S. K., Morton, D. L., and Taback, B. (2004). Profiling epigenetic inactivation of tumor suppressor genes in tumors and plasma from cutaneous melanoma patients. *Oncogene* **23,** 4014–4022.

Hou, P., Shen, J. Y., Ji, M. J., He, N. Y., and Lu, Z. H. (2004). Microarray-based method for detecting methylation changes of p16(Ink4a) gene 5'-CpG islands in gastric carcinomas. *World J. Gastroenterol.* **10,** 3553–3558.

Huang, T. H., Perry, M. R., and Laux, D. E. (1999). Methylation profiling of CpG islands in human breast cancer cells. *Hum. Mol. Genet.* **8,** 459–470.

Iorio, M. V., Visone, R., Di Leva, G., Donati, V., Petrocca, F., Casalini, P., Taccioli, C., Volinia, S., Liu, C. G., Alder, H., *et al.* (2007). MicroRNA signatures in human ovarian cancer. *Cancer Res.* **67,** 8699–8707.

Irizarry, R. A., Ladd-Acosta, C., Carvalho, B., Wu, H., Brandenburg, S. A., Jeddeloh, J. A., Wen, B., and Feinberg, A. P. (2008). Comprehensive high-throughput arrays for relative methylation (CHARM). *Genome Res.* **18,** 780–790.

Issa, J. P. (1999). Aging, DNA methylation and cancer. *Crit. Rev. Oncol. Hematol.* **32,** 31–43.

Jia, D., Jurkowska, R. Z., Zhang, X., Jeltsch, A., and Cheng, X. (2007). Structure of Dnmt3a bound to Dnmt3L suggests a model for de novo DNA methylation. *Nature* **449,** 248–251.

Jiang, C. L., Jin, S. G., and Pfeifer, G. P. (2004). MBD3L1 is a transcriptional repressor that interacts with methyl-CpG-binding protein 2 (MBD2) and components of the NuRD complex. *J. Biol. Chem.* **279,** 52456–52464.

Jin, M., Kawakami, K., Fukui, Y., Tsukioka, S., Oda, M., Watanabe, G., Takechi, T., Oka, T., and Minamoto, T. (2009). Different histological types of non-small cell lung cancer have distinct folate and DNA methylation levels. *Cancer Sci.* **100,** 2325–2330.

Jones, P. A., and Baylin, S. B. (2002). The fundamental role of epigenetic events in cancer. *Nat. Rev. Genet.* **3,** 415–428.

Kambara, T., Simms, L. A., Whitehall, V. L., Spring, K. J., Wynter, C. V., Walsh, M. D., Barker, M. A., Arnold, S., McGivern, A., Matsubara, N., *et al.* (2004). BRAF mutation is associated with DNA methylation in serrated polyps and cancers of the colorectum. *Gut* **53,** 1137–1144.

Kanduri, M., Cahill, N., Goransson, H., Enstrom, C., Ryan, F., Isaksson, A., and Rosenquist, R. (2010). Differential genome-wide array-based methylation profiles in prognostic subsets of chronic lymphocytic leukemia. *Blood* **115,** 296–305.

Kansara, M., Tsang, M., Kodjabachian, L., Sims, N. A., Trivett, M. K., Ehrich, M., Dobrovic, A., Slavin, J., Choong, P. F., Simmons, P. J., *et al.* (2009). Wnt inhibitory factor 1 is epigenetically silenced in human osteosarcoma, and targeted disruption accelerates osteosarcomagenesis in mice. *J. Clin. Invest.* **119,** 837–851.

Keshet, I., Schlesinger, Y., Farkash, S., Rand, E., Hecht, M., Segal, E., Pikarski, E., Young, R. A., Niveleau, A., Cedar, H., and Simon, I. (2006). Evidence for an instructive mechanism of de novo methylation in cancer cells. *Nat. Genet.* **38,** 149–153.

Khulan, B., Thompson, R. F., Ye, K., Fazzari, M. J., Suzuki, M., Stasiek, E., Figueroa, M. E., Glass, J. L., Chen, Q., Montagna, C., *et al.* (2006). Comparative isoschizomer profiling of cytosine methylation: the HELP assay. *Genome Res.* **16,** 1046–1055.

Kim, Y. T., Park, S. J., Lee, S. H., Kang, H. J., Hahn, S., Kang, C. H., Sung, S. W., and Kim, J. H. (2005). Prognostic implication of aberrant promoter hypermethylation of CpG islands in adenocarcinoma of the lung. *J. Thorac. Cardiovasc. Surg.* **130,** 1378.

Kirmizis, A., Bartley, S. M., and Farnham, P. J. (2003). Identification of the polycomb group protein SU(Z)12 as a potential molecular target for human cancer therapy. *Mol. Cancer Ther.* **2,** 113–121.

Koga, Y., Pelizzola, M., Cheng, E., Krauthammer, M., Sznol, M., Ariyan, S., Narayan, D., Molinaro, A. M., Halaban, R., and Weissman, S. M. (2009). Genome-wide screen of promoter methylation identifies novel markers in melanoma. *Genome Res.* **19,** 1462–1470.

Kondo, Y., Shen, L., Cheng, A. S., Ahmed, S., Boumber, Y., Charo, C., Yamochi, T., Urano, T., Furukawa, K., Kwabi-Addo, B., *et al.* (2008). Gene silencing in cancer by histone H3 lysine 27 trimethylation independent of promoter DNA methylation. *Nat. Genet.* **40,** 741–750.

Kozaki, K., Imoto, I., Mogi, S., Omura, K., and Inazawa, J. (2008). Exploration of tumor-suppressive microRNAs silenced by DNA hypermethylation in oral cancer. *Cancer Res.* **68,** 2094–2105.

Kron, K., Pethe, V., Briollais, L., Sadikovic, B., Ozcelik, H., Sunderji, A., Venkateswaran, V., Pinthus, J., Fleshner, N., van der Kwast, T., and Bapat, B. (2009). Discovery of novel hypermethylated genes in prostate cancer using genomic CpG island microarrays. *PLoS ONE* **4,** e4830.

Kuang, S. Q., Tong, W. G., Yang, H., Lin, W., Lee, M. K., Fang, Z. H., Wei, Y., Jelinek, J., Issa, J. P., and Garcia-Manero, G. (2008). Genome-wide identification of aberrantly methylated promoter associated CpG islands in acute lymphocytic leukemia. *Leukemia* **22,** 1529–1538.

Kuzmichev, A., Margueron, R., Vaquero, A., Preissner, T. S., Scher, M., Kirmizis, A., Ouyang, X., Brockdorff, N., Abate-Shen, C., Farnham, P., and Reinberg, D. (2005). Composition and histone substrates of polycomb repressive group complexes change during cellular differentiation. *Proc. Natl. Acad. Sci. USA* **102,** 1859–1864.

Laird, P. W. (2010). Principles and challenges of genome-wide DNA methylation analysis. *Nat. Rev. Genet.* **11,** 191–203.

Laird, P. W., Jackson-Grusby, L., Fazeli, A., Dickinson, S. L., Jung, W. E., Li, E., Weinberg, R. A., and Jaenisch, R. (1995). Suppression of intestinal neoplasia by DNA hypomethylation. *Cell* **81,** 197–205.

Lee, T. I., Jenner, R. G., Boyer, L. A., Guenther, M. G., Levine, S. S., Kumar, R. M., Chevalier, B., Johnstone, S. E., Cole, M. F., Isono, K., *et al.* (2006). Control of developmental regulators by Polycomb in human embryonic stem cells. *Cell* **125,** 301–313.

Lehmann, U., Hasemeier, B., Christgen, M., Muller, M., Romermann, D., Langer, F., and Kreipe, H. (2008). Epigenetic inactivation of microRNA gene hsa-mir-9-1 in human breast cancer. *J. Pathol.* **214,** 17–24.

Leshchenko, V.V., Kuo, P.Y., Shaknovich, R., Yang, D.T., Gellen, T., Petrich, A., Yu, Y., Remache, Y., Weniger, M.A., Rafiq, S., et al. (2010). Genome wide DNA methylation analysis reveals novel targets for drug development in mantle cell lymphoma. Blood, in press

Lippman, Z., Gendrel, A. V., Colot, V., and Martienssen, R. (2005). Profiling DNA methylation patterns using genomic tiling microarrays. *Nat. Methods* **2,** 219–224.

Lister, R., Pelizzola, M., Dowen, R. H., Hawkins, R. D., Hon, G., Tonti-Filippini, J., Nery, J. R., Lee, L., Ye, Z., Ngo, Q. M., *et al.* (2009). Human DNA methylomes at base resolution show widespread epigenomic differences. *Nature* **462,** 315–322.

Lu, D., Zhao, Y., Tawatao, R., Cottam, H. B., Sen, M., Leoni, L. M., Kipps, T. J., Corr, M., and Carson, D. A. (2004). Activation of the Wnt signaling pathway in chronic lymphocytic leukemia. *Proc. Natl. Acad. Sci. USA* **101,** 3118–3123.

Lu, H., Ouyang, W., and Huang, C. (2006). Inflammation, a key event in cancer development. *Mol. Cancer Res.* **4,** 221–233.

Lujambio, A., Calin, G. A., Villanueva, A., Ropero, S., Sanchez-Cespedes, M., Blanco, D., Montuenga, L. M., Rossi, S., Nicoloso, M. S., Faller, W. J., et al. (2008). A microRNA DNA methylation signature for human cancer metastasis. *Proc. Natl. Acad. Sci. USA* **105**, 13556–13561.

Lujambio, A., Ropero, S., Ballestar, E., Fraga, M. F., Cerrato, C., Setien, F., Casado, S., Suarez-Gauthier, A., Sanchez-Cespedes, M., Git, A., et al. (2007). Genetic unmasking of an epigenetically silenced microRNA in human cancer cells. *Cancer Res.* **67**, 1424–1429.

Maegawa, S., Hinkal, G., Kim, H. S., Shen, L., Zhang, L., Zhang, J., Zhang, N., Liang, S., Donehower, L. A., and Issa, J. P. (2010). Widespread and tissue specific age-related DNA methylation changes in mice. *Genome Res.* **20**, 332–340.

Martin-Subero, J. I., Ammerpohl, O., Bibikova, M., Wickham-Garcia, E., Agirre, X., Alvarez, S., Bruggemann, M., Bug, S., Calasanz, M. J., Deckert, M., et al. (2009). A comprehensive microarray-based DNA methylation study of 367 hematological neoplasms. *PLoS ONE* **4**, e6986.

Martinez, R., Martin-Subero, J. I., Rohde, V., Kirsch, M., Alaminos, M., Fernandez, A. F., Ropero, S., Schackert, G., and Esteller, M. (2009). A microarray-based DNA methylation study of glioblastoma multiforme. *Epigenetics* **4**, 255–264.

McCabe, M. T., Lee, E. K., and Vertino, P. M. (2009). A multifactorial signature of DNA sequence and polycomb binding predicts aberrant CpG island methylation. *Cancer Res.* **69**, 282–291.

McGarvey, K. M., Greene, E., Fahrner, J. A., Jenuwein, T., and Baylin, S. B. (2007). DNA methylation and complete transcriptional silencing of cancer genes persist after depletion of EZH2. *Cancer Res.* **67**, 5097–5102.

Meissner, A., Mikkelsen, T. S., Gu, H., Wernig, M., Hanna, J., Sivachenko, A., Zhang, X., Bernstein, B. E., Nusbaum, C., Jaffe, D. B., et al. (2008). Genome-scale DNA methylation maps of pluripotent and differentiated cells. *Nature* **454**, 766–770.

Nephew, K. P., and Huang, T. H. (2003). Epigenetic gene silencing in cancer initiation and progression. *Cancer Lett.* **190**, 125–133.

Nishiyama, N., Arai, E., Chihara, Y., Fujimoto, H., Hosoda, F., Shibata, T., Kondo, T., Tsukamoto, T., Yokoi, S., Imoto, I., et al. (2010). Genome-wide DNA methylation profiles in urothelial carcinomas and urothelia at the precancerous stage. *Cancer Sci.* **101**, 231–240.

Nojima, M., Suzuki, H., Toyota, M., Watanabe, Y., Maruyama, R., Sasaki, S., Sasaki, Y., Mita, H., Nishikawa, N., Yamaguchi, K., et al. (2007). Frequent epigenetic inactivation of SFRP genes and constitutive activation of Wnt signaling in gastric cancer. *Oncogene* **26**, 4699–4713.

Nouzova, M., Holtan, N., Oshiro, M. M., Isett, R. B., Munoz-Rodriguez, J. L., List, A. F., Narro, M. L., Miller, S. J., Merchant, N. C., and Futscher, B. W. (2004). Epigenomic changes during leukemia cell differentiation: analysis of histone acetylation and cytosine methylation using CpG island microarrays. *J. Pharmacol. Exp. Ther.* **311**, 968–981.

Novak, P., Jensen, T., Oshiro, M. M., Watts, G. S., Kim, C. J., and Futscher, B. W. (2008). Agglomerative epigenetic aberrations are a common event in human breast cancer. *Cancer Res.* **68**, 8616–8625.

Ogata, H., Kobayashi, T., Chinen, T., Takaki, H., Sanada, T., Minoda, Y., Koga, K., Takaeu, G., Maehara, Y., Iida, M., and Yoshimura, A. (2006). Deletion of the SOCS3 gene in liver parenchymal cells promotes hepatitis-induced hepatocarcinogenesis. *Gastroenterology* **131**, 179–193.

Ohm, J. E., McGarvey, K. M., Yu, X., Cheng, L., Schuebel, K. E., Cope, L., Mohammad, H. P., Chen, W., Daniel, V. C., Yu, W., et al. (2007). A stem cell-like chromatin pattern may predispose tumor suppressor genes to DNA hypermethylation and heritable silencing. *Nat. Genet.* **39**, 237–242.

Omura, N., Li, C. P., Li, A., Hong, S. M., Walter, K., Jimeno, A., Hidalgo, M., and Goggins, M. (2008). Genome-wide profiling of methylated promoters in pancreatic adenocarcinoma. *Cancer Biol. Ther.* **7**, 1146–1156.

Ooi, S. K., Qiu, C., Bernstein, E., Li, K., Jia, D., Yang, Z., Erdjument-Bromage, H., Tempst, P., Lin, S. P., Allis, C. D., et al. (2007). DNMT3L connects unmethylated lysine 4 of histone H3 to de novo methylation of DNA. Nature **448**, 714–717.

Pardal, R., Molofsky, A. V., He, S., and Morrison, S. J. (2005). Stem cell self-renewal and cancer cell proliferation are regulated by common networks that balance the activation of proto-oncogenes and tumor suppressors. Cold Spring Harb. Symp. Quant. Biol. **70**, 177–185.

Parsons, D. W., Jones, S., Zhang, X., Lin, J. C., Leary, R. J., Angenendt, P., Mankoo, P., Carter, H., Siu, I. M., Gallia, G. L., et al. (2008). An integrated genomic analysis of human glioblastoma multiforme. Science **321**, 1807–1812.

Pasini, D., Bracken, A. P., and Helin, K. (2004). Polycomb group proteins in cell cycle progression and cancer. Cell Cycle **3**, 396–400.

Pfeifer, G. P. (2006). Mutagenesis at methylated CpG sequences. Curr. Top. Microbiol. Immunol. **301**, 259–281.

Pfeifer, G. P., and Besaratinia, A. (2009). Mutational spectra of human cancer. Hum. Genet. **125**, 493–506.

Raaphorst, F. M. (2005b). Deregulated expression of Polycomb-group oncogenes in human malignant lymphomas and epithelial tumors. Hum Mol Genet **14**(Spec No 1), R93–R100.

Raaphorst, F. M. (2005a). Of mice, flies, and man: the emerging role of polycomb-group genes in human malignant lymphomas. Int. J. Hematol. **81**, 281–287.

Rauch, T., Li, H., Wu, X., and Pfeifer, G. P. (2006). MIRA-assisted microarray analysis, a new technology for the determination of DNA methylation patterns, identifies frequent methylation of homeodomain-containing genes in lung cancer cells. Cancer Res. **66**, 7939–7947.

Rauch, T., and Pfeifer, G. P. (2005). Methylated-CpG island recovery assay: a new technique for the rapid detection of methylated-CpG islands in cancer. Lab. Invest. **85**, 1172–1180.

Rauch, T., Wang, Z., Zhang, X., Zhong, X., Wu, X., Lau, S. K., Kernstine, K. H., Riggs, A. D., and Pfeifer, G. P. (2007). Homeobox gene methylation in lung cancer studied by genome-wide analysis with a microarray-based methylated CpG island recovery assay. Proc. Natl. Acad. Sci. USA **104**, 5527–5532.

Rauch, T. A., Zhong, X., Wu, X., Wang, M., Kernstine, K. H., Wang, Z., Riggs, A. D., and Pfeifer, G. P. (2008). High-resolution mapping of DNA hypermethylation and hypomethylation in lung cancer. Proc. Natl. Acad. Sci. USA **105**, 252–257.

Rhee, C. S., Sen, M., Lu, D., Wu, C., Leoni, L., Rubin, J., Corr, M., and Carson, D. A. (2002). Wnt and frizzled receptors as potential targets for immunotherapy in head and neck squamous cell carcinomas. Oncogene **21**, 6598–6605.

Ringrose, L. (2007). Polycomb comes of age: genome-wide profiling of target sites. Curr. Opin. Cell Biol. **19**, 290–297.

Ruike, Y., Imanaka, Y., Sato, F., Shimizu, K., and Tsujimoto, G. (2010). Genome-wide analysis of aberrant methylation in human breast cancer cells using methyl-DNA immunoprecipitation combined with high-throughput sequencing. BMC Genomics **11**, 137.

Schlesinger, Y., Straussman, R., Keshet, I., Farkash, S., Hecht, M., Zimmerman, J., Eden, E., Yakhini, Z., Ben-Shushan, E., Reubinoff, B. E., et al. (2007). Polycomb-mediated methylation on Lys27 of histone H3 pre-marks genes for de novo methylation in cancer. Nat. Genet. **39**, 232–236.

Seidel, C., Schagdarsurengin, U., Blumke, K., Wurl, P., Pfeifer, G. P., Hauptmann, S., Taubert, H., and Dammann, R. (2007). Frequent hypermethylation of MST1 and MST2 in soft tissue sarcoma. Mol. Carcinog. **46**, 865–871.

Sellers, W. R., and Loda, M. (2002). The EZH2 polycomb transcriptional repressor—a marker or mover of metastatic prostate cancer? Cancer Cell **2**, 349–350.

Sharpless, N. E., and DePinho, R. A. (2005). Cancer: crime and punishment. Nature **436**, 636–637.

Shih, Y. L., Hsieh, C. B., Lai, H. C., Yan, M. D., Hsieh, T. Y., Chao, Y. C., and Lin, Y. W. (2007). SFRP1 suppressed hepatoma cells growth through Wnt canonical signaling pathway. *Int. J. Cancer* **121**, 1028–1035.

Shiraishi, M., Sekiguchi, A., Terry, M. J., Oates, A. J., Miyamoto, Y., Chuu, Y. H., Munakata, M., and Sekiya, T. (2002). A comprehensive catalog of CpG islands methylated in human lung adenocarcinomas for the identification of tumor suppressor genes. *Oncogene* **21**, 3804–3813.

Smith, R. J., and Kelsey, G. (2001). Identification of imprinted loci by methylation: use of methylation-sensitive representational difference analysis (Me-RDA). *Methods Mol. Biol.* **181**, 113–132.

Song, J. Z., Stirzaker, C., Harrison, J., Melki, J. R., and Clark, S. J. (2002). Hypermethylation trigger of the glutathione-S-transferase gene (GSTP1) in prostate cancer cells. *Oncogene* **21**, 1048–1061.

Spugnardi, M., Tommasi, S., Dammann, R., Pfeifer, G. P., and Hoon, D. S. (2003). Epigenetic inactivation of RAS association domain family protein 1 (RASSF1A) in malignant cutaneous melanoma. *Cancer Res.* **63**, 1639–1643.

Squazzo, S. L., O'Geen, H., Komashko, V. M., Krig, S. R., Jin, V. X., Jang, S. W., Margueron, R., Reinberg, D., Green, R., and Farnham, P. J. (2006). Suz12 binds to silenced regions of the genome in a cell-type-specific manner. *Genome Res.* **16**, 890–900.

St John, M. A., Tao, W., Fei, X., Fukumoto, R., Carcangiu, M. L., Brownstein, D. G., Parlow, A. F., McGrath, J., and Xu, T. (1999). Mice deficient of Lats1 develop soft-tissue sarcomas, ovarian tumours and pituitary dysfunction. *Nat. Genet.* **21**, 182–186.

Suzuki, H., Watkins, D. N., Jair, K. W., Schuebel, K. E., Markowitz, S. D., Chen, W. D., Pretlow, T. P., Yang, B., Akiyama, Y., Van Engeland, M., *et al.* (2004). Epigenetic inactivation of SFRP genes allows constitutive WNT signaling in colorectal cancer. *Nat. Genet.* **36**, 417–422.

Takagi, H., Sasaki, S., Suzuki, H., Toyota, M., Maruyama, R., Nojima, M., Yamamoto, H., Omata, M., Tokino, T., Imai, K., and Shinomura, Y. (2008). Frequent epigenetic inactivation of SFRP genes in hepatocellular carcinoma. *J. Gastroenterol.* **43**, 378–389.

Takahashi, Y., Miyoshi, Y., Takahata, C., Irahara, N., Taguchi, T., Tamaki, Y., and Noguchi, S. (2005). Down-regulation of LATS1 and LATS2 mRNA expression by promoter hypermethylation and its association with biologically aggressive phenotype in human breast cancers. *Clin. Cancer Res.* **11**, 1380–1385.

Takeshima, H., Yamashita, S., Shimazu, T., Niwa, T., and Ushijima, T. (2009). The presence of RNA polymerase II, active or stalled, predicts epigenetic fate of promoter CpG islands. *Genome Res.* **19**, 1974–1982.

Tan, J., Yang, X., Zhuang, L., Jiang, X., Chen, W., Lee, P. L., Karuturi, R. K., Tan, P. B., Liu, E. T., and Yu, Q. (2007). Pharmacologic disruption of Polycomb-repressive complex 2-mediated gene repression selectively induces apoptosis in cancer cells. *Genes Dev.* **21**, 1050–1063.

Taniguchi, H., Yamamoto, H., Hirata, T., Miyamoto, N., Oki, M., Nosho, K., Adachi, Y., Endo, T., Imai, K., and Shinomura, Y. (2005). Frequent epigenetic inactivation of Wnt inhibitory factor-1 in human gastrointestinal cancers. *Oncogene* **24**, 7946–7952.

Teschendorff, A. E., Menon, U., Gentry-Maharaj, A., Ramus, S. J., Gayther, S. A., Apostolidou, S., Jones, A., Lechner, M., Beck, S., Jacobs, I. J., and Widschwendter, M. (2009). An epigenetic signature in peripheral blood predicts active ovarian cancer. *PLoS ONE* **4**, e8274.

Tommasi, S., Dammann, R., Zhang, Z., Wang, Y., Liu, L., Tsark, W. M., Wilczynski, S. P., Li, J., You, M., and Pfeifer, G. P. (2005). Tumor susceptibility of Rassf1a knockout mice. *Cancer Res.* **65**, 92–98.

Tommasi, S., Karm, D. L., Wu, X., Yen, Y., and Pfeifer, G. P. (2009). Methylation of homeobox genes is a frequent and early epigenetic event in breast cancer. *Breast Cancer Res.* **11**, R14.

Topaloglu, O., Hoque, M. O., Tokumaru, Y., Lee, J., Ratovitski, E., Sidransky, D., and Moon, C. S. (2004). Detection of promoter hypermethylation of multiple genes in the tumor and bronchoalveolar lavage of patients with lung cancer. *Clin. Cancer Res.* **10**, 2284–2288.

Toyooka, S., Maruyama, R., Toyooka, K. O., McLerran, D., Feng, Z., Fukuyama, Y., Virmani, A. K., Zochbauer-Muller, S., Tsukuda, K., Sugio, K., et al. (2003). Smoke exposure, histologic type and geography-related differences in the methylation profiles of non-small cell lung cancer. Int. J. Cancer 103, 153–160.

Toyota, M., Ho, C., Ahuja, N., Jair, K. W., Li, Q., Ohe-Toyota, M., Baylin, S. B., and Issa, J. P. (1999). Identification of differentially methylated sequences in colorectal cancer by methylated CpG island amplification. Cancer Res. 59, 2307–2312.

Toyota, M., Suzuki, H., Sasaki, Y., Maruyama, R., Imai, K., Shinomura, Y., and Tokino, T. (2008). Epigenetic silencing of microRNA-34b/c and B-cell translocation gene 4 is associated with CpG island methylation in colorectal cancer. Cancer Res. 68, 4123–4132.

Urakami, S., Shiina, H., Enokida, H., Kawakami, T., Tokizane, T., Ogishima, T., Tanaka, Y., Li, L. C., Ribeiro-Filho, L. A., Terashima, M., et al. (2006). Epigenetic inactivation of Wnt inhibitory factor-1 plays an important role in bladder cancer through aberrant canonical Wnt/ beta-catenin signaling pathway. Clin. Cancer Res. 12, 383–391.

Ushijima, T. (2005). Detection and interpretation of altered methylation patterns in cancer cells. Nat. Rev. Cancer 5, 223–231.

Ushijima, T., Morimura, K., Hosoya, Y., Okonogi, H., Tatematsu, M., Sugimura, T., and Nagao, M. (1997). Establishment of methylation-sensitive-representational difference analysis and isolation of hypo- and hypermethylated genomic fragments in mouse liver tumors. Proc. Natl. Acad. Sci. USA 94, 2284–2289.

Ushijima, T., and Yamashita, S. (2009). Methylation-sensitive representational difference analysis (MS-RDA). Methods Mol. Biol. 507, 117–130.

Valk-Lingbeek, M. E., Bruggeman, S. W., and van Lohuizen, M. (2004). Stem cells and cancer; the polycomb connection. Cell 118, 409–418.

Varambally, S., Dhanasekaran, S. M., Zhou, M., Barrette, T. R., Kumar-Sinha, C., Sanda, M. G., Ghosh, D., Pienta, K. J., Sewalt, R. G., Otte, A. P., et al. (2002). The polycomb group protein EZH2 is involved in progression of prostate cancer. Nature 419, 624–629.

Vire, E., Brenner, C., Deplus, R., Blanchon, L., Fraga, M., Didelot, C., Morey, L., Van Eynde, A., Bernard, D., Vanderwinden, J. M., et al. (2006). The Polycomb group protein EZH2 directly controls DNA methylation. Nature 439, 871–874.

Weber, M., Davies, J. J., Wittig, D., Oakeley, E. J., Haase, M., Lam, W. L., and Schubeler, D. (2005). Chromosome-wide and promoter-specific analyses identify sites of differential DNA methylation in normal and transformed human cells. Nat. Genet. 37, 853–862.

Weeraratna, A. T., Jiang, Y., Hostetter, G., Rosenblatt, K., Duray, P., Bittner, M., and Trent, J. M. (2002). Wnt5a signaling directly affects cell motility and invasion of metastatic melanoma. Cancer Cell 1, 279–288.

Wei, S. H., Chen, C. M., Strathdee, G., Harnsomburana, J., Shyu, C. R., Rahmatpanah, F., Shi, H., Ng, S. W., Yan, P. S., Nephew, K. P., et al. (2002). Methylation microarray analysis of late-stage ovarian carcinomas distinguishes progression-free survival in patients and identifies candidate epigenetic markers. Clin. Cancer Res. 8, 2246–2252.

Weisenberger, D. J., Siegmund, K. D., Campan, M., Young, J., Long, T. I., Faasse, M. A., Kang, G. H., Widschwendter, M., Weener, D., Buchanan, D., et al. (2006). CpG island methylator phenotype underlies sporadic microsatellite instability and is tightly associated with BRAF mutation in colorectal cancer. Nat. Genet. 38, 787–793.

Widschwendter, M., Fiegl, H., Egle, D., Mueller-Holzner, E., Spizzo, G., Marth, C., Weisenberger, D. J., Campan, M., Young, J., Jacobs, I., and Laird, P. W. (2007). Epigenetic stem cell signature in cancer. Nat. Genet. 39, 157–158.

Wu, X., Rauch, T. A., Zhong, X., Latif, F., Krex, D., and Pfeifer, G. P. (2010). CpG island hypermethylation in human astrocytomas. Cancer Res. 70, 2718–2727.

Yagi, K., Akagi, K., Hayashi, H., Nagae, G., Tsuji, S., Isagawa, T., Midorikawa, Y., Nishimura, Y., Sakamoto, H., Seto, Y., *et al.* (2010). Three DNA methylation epigenotypes in human colorectal cancer. *Clin. Cancer Res.* **16,** 21–33.

Yamada, Y., Jackson-Grusby, L., Linhart, H., Meissner, A., Eden, A., Lin, H., and Jaenisch, R. (2005). Opposing effects of DNA hypomethylation on intestinal and liver carcinogenesis. *Proc. Natl. Acad. Sci. USA* **102,** 13580–13585.

Yan, P. S., Chen, C. M., Shi, H., Rahmatpanah, F., Wei, S. H., Caldwell, C. W., and Huang, T. H. (2001). Dissecting complex epigenetic alterations in breast cancer using CpG island microarrays. *Cancer Res.* **61,** 8375–8380.

Yanagawa, N., Tamura, G., Oizumi, H., Takahashi, N., Shimazaki, Y., and Motoyama, T. (2003). Promoter hypermethylation of tumor suppressor and tumor-related genes in non-small cell lung cancers. *Cancer Sci.* **94,** 589–592.

Yang, H. H., Hu, N., Wang, C., Ding, T., Dunn, B. K., Goldstein, A. M., Taylor, P. R., and Lee, M. P. (2010). Influence of genetic background and tissue types on global DNA methylation patterns. *PLoS ONE* **5,** e9355.

Yue, W., Sun, Q., Dacic, S., Landreneau, R. J., Siegfried, J. M., Yu, J., and Zhang, L. (2008). Downregulation of Dkk3 activates beta-catenin/TCF-4 signaling in lung cancer. *Carcinogenesis* **29,** 84–92.

Zhou, D., Conrad, C., Xia, F., Park, J. S., Payer, B., Yin, Y., Lauwers, G. Y., Thasler, W., Lee, J. T., Avruch, J., and Bardeesy, N. (2009). Mst1 and Mst2 maintain hepatocyte quiescence and suppress hepatocellular carcinoma development through inactivation of the Yap1 oncogene. *Cancer Cell* **16,** 425–438.

Zochbauer-Muller, S., Fong, K. M., Virmani, A. K., Geradts, J., Gazdar, A. F., and Minna, J. D. (2001). Aberrant promoter methylation of multiple genes in non-small cell lung cancers. *Cancer Res.* **61,** 249–255.

11

Epigenetic Drivers of Genetic Alterations

Minoru Toyota* and Hiromu Suzuki*,†

*Department of Biochemistry, Sapporo Medical University, Sapporo, Japan
†First Department of Internal Medicine, Sapporo Medical University, Sapporo, Japan

Advances in Genetics, Vol. 70
Copyright 2010, Elsevier Inc. All rights reserved.

0065-2660/10 $35.00
DOI: 10.1016/S0065-2660(10)70011-0

ABSTRACT

DNA methylation plays a key role in the silencing of cancer-related genes, thereby affecting numerous cellular processes, including the cell cycle checkpoint, apoptosis, signal transduction, cell adhesion, and angiogenesis. DNA methylation also affects the expression of genes involved in maintaining the integrity of the genome through DNA repair and detoxification of reactive oxygen species. Here, we discuss how epigenetic changes lead to genetic alterations, including microsatellite instability and nucleotide and chromosomal alterations. Epigenetic inactivation of hMLH1 is a major cause of microsatellite instability in sporadic colorectal cancers, and germline epimutation of hMLH1 and hMSH2 is a cause of hereditary nonpolyposis colorectal cancers, which do not show mutation of mismatch repair genes. Epigenetic inactivation of MGMT is often associated with G:C-to-A:T mutations in K-ras and p53, while epigenetic inactivation of BRCA1, WRN, FANCF, and CHFR impairs the machinery involved in maintaining genomic integrity. Epigenetic alteration of the genes involved in the induction of senescence is often associated with cancers showing mutations in the Ras signaling pathway. In addition to regional hypermethylation, global hypomethylation is also a common feature of tumors. Hypomethylation of short and long interspersed repetitive elements has been reported, and hypomethylation affecting the integrity of the genome has been observed in ICF syndrome and various cancers. Dissection of the epigenetic drivers of genetic instability may be important for the development of novel approaches to the treatment of cancer.   © 2010, Elsevier Inc.

I. INTRODUCTION

Changes in DNA methylation play a critical role in tumorigenesis. The first identified tumor-associated change was global DNA hypomethylation, and subsequently DNA methylation of 5′ CpG islands was shown to be involved in silencing tumor suppressor genes. Three methyltransferases, DNMT1, DNMT3A, and DNMT3B, catalyze methylation of cytosines at CpG sites. In addition, DNA methylation of promoter regions is associated with modification of repressive histone marks, including deacetylation of lysine residues and tri-methylation of histones H3K9 and H3K27. DNA methylation is also involved in the maintenance of the tumor phenotype through modulation of cell cycle checkpoints, apoptosis, signal transduction, cell adhesion, angiogenesis, and immune responses. We previously showed that epigenetic inactivation of the SFRP and DKK gene families, together with APC/β-catenin mutations, plays a critical role in the activation of WNT signaling (Nojima et al., 2007; Sato et al., 2007; Suzuki et al., 2004). DNA methylation of certain genes also influences the sensitivity to chemotherapeutic drugs and can be a molecular marker predicting the response to drugs (Toyota *et al.*, 2009). In addition, subsets of cancers show simultaneous methylation of multiple genes, which is indicative of the CpG island methylator phenotype, or CIMP (Toyota *et al.*, 1999a). In gastric cancer, for example, infection with Epstein–Barr virus is associated with CIMP (Kusano *et al.*, 2006). The molecular mechanisms underlying DNA methylation remain unclear; however, recent studies indicate that inflammation and various pathogens are involved in the process.

Genomic integrity is maintained through several mechanisms, which are often impaired in cancer cells. For example, in patients with a hereditary cancer syndrome such as hereditary nonpolyposis colorectal cancer (HNPCC), the mismatch repair genes contain germline mutations (Kinzler and Vogelstein, 1996), while in patients with familial breast cancer, mutations are found in the genes involved in the response to DNA damage (Futreal *et al.*, 1994). By contrast, sporadic cancers with genetic instability show few mutations in genes involved in DNA repair (Merajver *et al.*, 1995; Moslein *et al.*, 1996). Notably in that regard, it is a unique feature of epigenetic changes that they can drive genetic alterations by attenuating maintenance of the integrity of the genome. In this review, we discuss the role of epigenetic inactivation of genes involved in DNA repair, the mitotic checkpoint, and carcinogen detoxification as a cause of genetic alterations in tumors (Table 11.1).

II. GENETIC INSTABILITY AND CANCER

In eukaryotic cells, the genome is under continuous attack from a variety of endogenous and exogenous agents that damage the DNA. The damage caused by these agents includes single-strand breaks, double-strand breaks, base mismatches, and chemical modification of the bases (Fig. 11.1). To avoid the accumulation of DNA damage, multiple DNA repair pathways have evolved,

Table 11.1. The Role of Epigenetic Drivers in Tumorigenesis

Genes	Function	Roles in tumorigenesis
hMLH1	Mismatch repair	Cause of microsatellite instability
MGMT	DNA repair	Increased mutation of G to A
BRCA1	DNA repair	Checkpoint defect
FANCF	DNA repair	Associated with sensitivity to cisplatin
WRN	DNA repair	Associated with sensitivity to topoisomerase inhibitor
CHFR	Mitotic checkpoint	Mitotic checkpoint defect
GSTp1	Detoxification	Increased hormone-related mutation caused

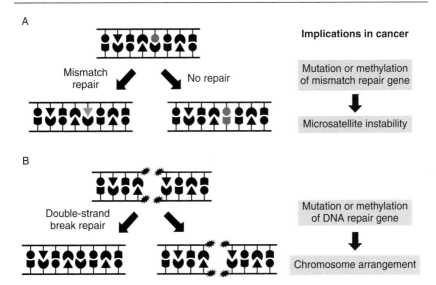

Figure 11.1. DNA repair system and cancer. (A) Impairment of mismatch repair leads to microsatellite instability. Mutated nucleotides are shown in red. Repaired nucleotides is shown in green. (B) Impairment of DNA repair of double-strand breaks leads to chromosome rearrangement. (For interpretation of the references to color in this figure legend, the reader is referred to the Web version of this chapter.)

including base excision repair (BER), nucleotide excision repair (NER), homologous recombination (HR), nonhomologous end joining (NHEJ), and mismatch repair (MMR), each of which is associated with a specific class of lesions. Moreover, mutation of the genes involved in these DNA repair pathways is often the cause of a cancer prone syndrome. Genetic instability can also occur at the level of the chromosome, leading to chromosomal instability or aneuploidy.

III. EPIGENETIC DRIVERS OF GENETIC ALTERATIONS

A. Epigenetic defects in mismatch repair genes

HNPCC is an autosomal dominant cancer syndrome characterized by an early age of onset, multiplicity of tumors, and mutation of microsatellite sequences. Defects in the mismatch repair system caused by germline mutation of *hMSH2* and *hMLH1* account for about 60% of HNPCCs (Lynch and de la Chapelle, 2003). Microsatellite instability caused by a mismatch repair defects can lead to frame shift mutations of genes such as *hMSH6*, *BAX*, *TGFβ-RII*, *IGFIIR*, and *MBD4* (Duval and Hamelin, 2002).

Among sporadic colorectal cancers, about 10–15% of tumors show microsatellite instability, though mutation of mismatch repair genes is not frequently found. On the other hand, Kane *et al.* reported that in two colon cancer cell lines showing microsatellite instability, *hMLH1* was methylated and its expression was silenced (Kane *et al.*, 1997). Similarly, they found that *hMLH1* was methylated in four primary colorectal cancers exhibiting microsatellite instability, and follow-up studies have shown that epigenetic inactivation of *hMLH1* is a common mode of mismatch repair gene inactivation in sporadic colorectal cancers (Cunningham *et al.*, 1998; Herman *et al.*, 1998). Moreover, treating colorectal cancer cell lines in which *hMLH1* expression is silenced due to DNA methylation with 5′-aza-deoxycytidine not only restores expression of *hMLH1*, but also restores the capacity for mismatch repair, confirming the functional consequences of DNA methylation, leading to mismatch repair deficiency (Herman *et al.*, 1998). Epigenetic inactivation of *hMLH1* is also found in gastric, endometrial, and ovarian tumors, which are frequently observed in HNPCC patients (Esteller *et al.*, 1999a; Leung *et al.*, 1999; Strathdee *et al.*, 1999; Toyota *et al.*, 1999b). Induction of microsatellite instability through epigenetic inactivation of a mismatch repair gene is a good example of an epigenetic change driving genetic alterations.

About one-third of HNPCC patients do not show mismatch repair gene mutations. Recently, germline epimutation of *hMLH1* and *hMSH2*, in which methylation of a single allele is associated with transcriptional silencing, was implicated in the pathogenesis of colorectal cancers with microsatellite instability. In these cases, *hMLH1* is inactivated by epimutation not only in cancer cells, but also in germ cells and somatic cells (Suter *et al.*, 2004). Colorectal cancers in these patients exhibit microsatellite instability due to absence of hMLH1 protein caused by a combination of epigenetic inactivation and loss of the wild-type allele. Germline epimutation has also been reported for *hMSH2*, and several familial HNPCC cases in which *hMSH2* expression was lost without mutation of the gene have been reported (Chan *et al.*, 2006). Deletion of *TACSTD1/EPCAM* gene is associated with a read-through of aberrant mRNA, which appears to be the cause of the *hMSH2* promoter methylation in those cases (Kovacs *et al.*, 2009; Ligtenberg *et al.*, 2009; Niessen *et al.*, 2009).

B. An epigenetic defect in a DNA repair gene is associated with a specific mutation profile in cancer

Point mutations play a critical role in the activation of oncogenes and inactivation of tumor suppressor genes. Although there are twice as many possible transversions, transition mutations occur much more frequently during tumorigenesis. O^6-methylguanine is often a cause of G:C-to-A:T transition mutations. Failure of O^6-methylguanine repair results in the conversion G:C to A:T due to the pairing of O^6-methylguanine and thymine during replication. *MGMT* encodes a DNA repair enzyme that removes the mutagenic adduct from O^6-methylguanine (Pegg, 1990). Methylation of *MGMT* is observed in wide variety of tumors, including colorectal, lung, and brain tumors (Costello *et al.*, 1994; Esteller *et al.*, 1999b). Codon 12 of *K-ras* (GGT) is often mutated in cancer, and G:C-to-A:T mutations are particularly common. There is thus a significant association between G:C-to-A:T mutations and the methylation of *MGMT* (Esteller *et al.*, 2000). G:C-to-A:T transition is also a common mutation in *p53*, and G-to-A transitions at nonCpG sites of *p53* correlate with methylation of *MGMT* (Esteller *et al.*, 2000).

C. Epigenetic inactivation of genes involved in the repair of double-strand breaks

BRCA1 is a DNA repair protein that contains an N-terminal RING domain and a C-terminal BRCT domain. When there is damage to the DNA in the form of a double-strand break, BRCA1 colocalizes at the appropriate nuclear foci, along with RAD51, and contributes to DNA repair through homologous recombination (Scully *et al.*, 1997). BRCA1 also plays a role in the G2/M cell cycle checkpoint through activation of CHK1 (Yarden *et al.*, 2002). Germline mutation of *BRCA1* is often found in hereditary breast and ovarian cancers. Tumors with *BRCA1* mutations show a high degree of genomic imbalance, which is indicative of the importance of *BRCA1* to the integrity of the genome (Tirkkonen *et al.*, 1997). In sporadic breast and ovarian cancers, somatic mutations of *BRCA1* are rare, but methylation of its promoter is observed in about 10–15% of cases. There have been few studies examining the role of *BRCA1* methylation in the impairment of DNA repair or checkpoint function, and the functional consequences of *BRCA1* methylation remain to be determined. Hedenfalk *et al.* showed that the biological effect of the loss of *BRCA1* due to methylation is similar to that of a coding region mutation (Hedenfalk *et al.*, 2001), and methylation of *BRCA1* is associated with a poor prognosis in breast cancer patients (Buller *et al.*, 2002; Chiang *et al.*, 2006).

Fanconi anemia is a rare autosomal recessive disease characterized by bone marrow failure, developmental abnormalities, and susceptibility to cancer (D'Andrea and Grompe, 2003). To date, seven Fanconi anemia genes have been identified, and five Fanconi anemia proteins are known to interact in the nucleus. Epigenetic inactivation of *FANCF* has been detected in 30% of cervical cancers, 15% of head and neck squamous cell cancers, and 14% of nonsmall cell lung cancers (Marsit *et al.*, 2004; Narayan *et al.*, 2004; Taniguchi *et al.*, 2003). In addition, monoubiquitination of FANCD2, which plays an important role in the protein's localization at DNA lesions, and in the interaction between FANCD2 and BRCA1, is impaired in cancer cells that lack FANCF, and is restored by treating cancer cells with a demethylating agent (Taniguchi *et al.*, 2003). Defects in the FA-BRCA pathway are also associated with genomic instability and increased sensitivity to DNA damaging agents such as mitomycin C and cisplatin, and there is a significant correlation between *FANCF* methylation and cisplatin sensitivity in ovarian cancer cell lines (Taniguchi *et al.*, 2003). Epigenetic inactivation of *FANCC* and *FANCL* has also been reported in acute myeloid leukemia (Hess *et al.*, 2008).

RecQ-like helicases also contribute to the maintenance of genomic integrity. Germline mutations that disrupt these genes cause chromosome breakage syndromes, such as Bloom syndrome, Rothmund–Thomson syndrome, and Werner syndrome, the last of which is an inherited disorder characterized by the premature onset of aging and susceptibility to various types of cancer. Epigenetic inactivation of *WRN* has been reported in colorectal cancers and in colorectal cancer cell lines, and *WRN* methylation is associated with sensitivity to the topoisomerase inhibitor camptothecin and to the interstrand crosslinker mitomycin C (Agrelo *et al.*, 2006). Clinically, colorectal cancers exhibiting *WRN* methylation respond well to the topoisomerase inhibitor irinotecan. Hypermethylation of *WRN* in colorectal tumors could thus be a useful predictor of a robust clinical response to a topoisomerase inhibitor.

D. Epigenetic defects in mitotic checkpoint genes

Genomic integrity is also maintained by mitotic checkpoint machinery, which controls chromosome separation. Failure of the mitotic checkpoint leads to aneuploidy, which is a hallmark of human tumors, and it has been shown that genes involved in this process are mutated in a subset of human tumors. The affected genes include *BUB1*, *BUB1R*, and *hCDC4/FBXW7*. By themselves, however, mutations of mitotic checkpoint genes cannot explain the chromosomal instability seen in some tumors. CHFR encodes a protein that contains both a fork head-associated domain and a RING domain, and plays a key role in the early mitotic checkpoint (Fig. 11.2A) (Scolnick and Halazonetis, 2000). *CHFR* is inactivated by DNA methylation in various tumors, including colorectal, gastric, lung, and breast (Ogi *et al.*, 2005; Satoh *et al.*, 2003; Toyota *et al.*, 2003). Cancer cells lacking *CHFR* expression are sensitive to microtubule inhibitors such as docetaxel (Fig. 11.2B) (Satoh *et al.*, 2003). CHFR has E3 ubiquitin ligase activity and controls Aurora A protein levels through ubiquitination; consequently, overexpression of Aurora A in cancer may be explained in part by inactivation of *CHFR*. *CHFR* knockout mice develop solid tumors and lymphomas, and embryonic fibroblasts from *CHFR* knockout mice show a mitotic checkpoint defect that causes chromosomal instability (Yu *et al.*, 2005). Although CHFR plays a predominant role in the mitotic checkpoint, there also appears to be novel link between CHFR and microsatellite instability, as colorectal cancers with

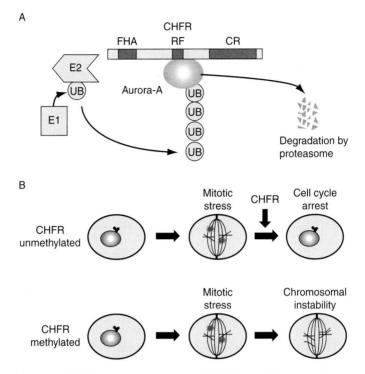

Figure 11.2. Role of CHFR in mitotic checkpoint. (A) CHFR is an E3 ubiquitin ligase regulating Aurora A. (B) In cancer cells lacking CHFR, impairment of the mitotic checkpoint leads to chromosomal instability.

microsatellite instability frequently show epigenetic inactivation of *CHFR* (Brandes *et al.*, 2005). In addition, mice deficient in both *Chfr* and *Mlh1* develop colon tumors, including intestinal tumors and lymphomas, again suggesting a link between microsatellite instability and the mitotic checkpoint (Fu *et al.*, 2009). Alternatively, hMLH1 may be involved in the sensor that detects mitotic defects. If so, treating cancers positive for microsatellite instability with a microtubule inhibitor may be a useful new approach to therapy.

E. Carcinogen detoxifying genes and increased mutability of methylcytosines

Oxidative stress because of reactive oxygen species is a major cause of DNA damage leading to cancer. *Glutathione S-transferase* (*GST*) and *glutathione peroxidase* (*GPX*) family genes encode enzymes that catalyze the conjugation of glutathione to electrophilic compounds, with the reduction of H_2O_2 and lipid peroxides. These enzymes also reduce environmental mutagens through detoxification of endogenous and exogenous agents. Methylation of $GST\pi1$ ($GSTP1$) has been demonstrated in several tumors, including prostate, breast, renal, and hepatocellular cancers (Esteller et al., 1998; Lee et al., 1994; Zhong et al., 2002). In addition, a link between $GSTP1$ methylation and hormone-related tumors may be explained by the role of $GSTP1$ in the detoxification of estrogen metabolites. Epigenetic inactivation of $GSTP1$ may thus increase the accumulation of endogenous adduct, which may in turn induce gene mutation (Esteller, 2000). Moreover, the detection of GPX3, GPX7, and GSTM2 methylation in esophageal adenocarcinomas suggests that the silencing of *mu-class* GST and GPX may also contribute to tumorigenesis (Peng et al., 2009).

F. Global DNA hypomethylation and chromosomal instability

The global loss of methylcytosine is one of the earliest epigenetic changes in cancer (Feinberg and Vogelstein, 1983). Consistent with that finding, embryonic stem cells from *DNMT1* knockout mice show increases in chromosomal alterations (Chen et al., 1998). Hypomethylation of satellite DNA in chromosomes 1 and 16 has been detected in patients with ICF syndrome, which is caused by mutation of *DNMT3B* and is characterized by immunodeficiency, centromere instability, and facial anomalies (Hansen et al., 1999; Jeanpierre et al., 1993; Okano et al., 1999). In addition, repetitive sequences such as short and long interspersed repeat elements (SINEs and LINEs) are often hypomethylated in cancers, and genome-wide methylation analysis revealed that subtelomeric CpG islands composed of short direct or indirect repeat sequences are demethylated in lung cancer (Rauch et al., 2008). Although the functional consequence of hypomethylation is not fully understood, it appears to lead to chromosomal instability (Eden et al., 2003). Hypomethylation of LINEs is also detected in preoneoplastic conditions such as enlarged fold gastritis, which is a strong risk factor for gastric cancer (Yamamoto et al., 2008).

IV. CONCURRENT GENETIC AND EPIGENETIC ALTERATIONS ASSOCIATED WITH CANCER CELL SURVIVAL

DNA methylation of multiple genes is observed in several tumor types. This phenomenon has been termed the CpG island methylator phenotype or CIMP (Toyota et al., 1999a). Interestingly, CIMP-positive colorectal cancers often show mutations in the Ras signaling pathway. For example, we showed that there is a high rate of *K-ras* mutation in colorectal cancers with CIMP (Toyota et al., 2000), while Weisenberger et al. reported a strong association between

BRAF mutations and CIMP in colorectal cancer (Weisenberger *et al.*, 2006). However, it remains to be determined why these tumors have concurrent alterations in their Ras signaling pathway and CIMP. It has been reported that activation of Ras and RAF induces senescence (Michaloglou *et al.*, 2005). One plausible explanation for the association between CIMP and altered Ras signaling is that CIMP leads to the silencing of genes involved in senescence. Recently, IGFBP7 was shown to induce senescence in cells expressing mutant *BRAF* (Wajapeyee *et al.*, 2008). For example, expression of *IGFBP7* is necessary and sufficient to induce senescence and apoptosis in mutant BRAF-activated melanocytes, and inactivation of *IGFBP7* through DNA methylation has been detected in melanomas. Likewise, *IGFBP7* is frequently silenced by DNA methylation in colorectal cancer with CIMP (Hinoue *et al.*, 2009; Suzuki *et al.* 2010), suggesting CIMP-related silencing of *IGFBP7* may enable cells expressing mutant *BRAF* to avoid senescence (Fig. 11.3). The regulatory function of *IGFBP7* expression is largely unknown, but p53, which is also known to induce senescence, appears to control *IGFBP7* in cellular stress-dependent manner (Suzuki *et al.*, 2010). Thus, p53 may be involved in mediating the senescence induced in

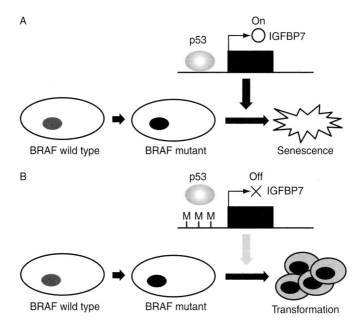

Figure 11.3. Role of IGFBP7 in transformation induced by BRAF mutation. (A) p53-induced IGFBP7 expression suppresses transformation of cells expressing mutant BRAF by inducing senescence. (B) Epigenetic silencing of IGFBP7 expression allows cells expressing mutant BRAF to escape senescence, which leads to their transformation.

cells expressing mutant *BRAF* through activation of *IGFBP7*, and epigenetic inactivation of *IGFBP7* may enable these cells to avoid senescence. Similar genetic and epigenetic interactions have been seen in colorectal cancers with *RASSF2* methylation, which also show high rates of *K-ras* and *BRAF* mutations (Akino *et al.*, 2005).

V. CONCLUSIONS

In this review, we summarized the role of epigenetic alterations in the induction of mismatch repair deficiency, altered DNA repair, and loss of chromosome integrity. DNA repair defects are often used as a target for cancer therapy and may enable more personalized therapy in the future. Genome-wide analysis using next-generation sequencers will give us a more complete epigenomic information about the genes involved in maintaining the integrity of the genome.

Acknowledgments

The authors thank Dr. William F. Goldman for editing the manuscript. This study was supported in part by Grants-in-Aid for Scientific Research on Priority Areas from the Ministry of Education, Culture, Sports, Science, and Technology (M. T.), a Grant-in-Aid for the Third-term Comprehensive 10-year Strategy for Cancer Control, and Grant-in-Aid for Cancer Research from the Ministry of Health, Labor, and Welfare, Japan (M. T.).

References

Agrelo, R., Cheng, W. H., Setien, F., Ropero, S., Espada, J., Fraga, M. F., Herranz, M., Paz, M. F., Sanchez-Cespedes, M., Artiga, M. J., *et al.* (2006). Epigenetic inactivation of the premature aging Werner syndrome gene in human cancer. *Proc. Natl. Acad. Sci. USA* **103**, 8822–8827.

Akino, K., Toyota, M., Suzuki, H., Mita, H., Sasaki, Y., Ohe-Toyota, M., Issa, J. P., Hinoda, Y., Imai, K., and Tokino, T. (2005). The Ras effector RASSF2 is a novel tumor-suppressor gene in human colorectal cancer. *Gastroenterology* **129**, 156–169.

Brandes, J. C., van Engeland, M., Wouters, K. A., Weijenberg, M. P., and Herman, J. G. (2005). CHFR promoter hypermethylation in colon cancer correlates with the microsatellite instability phenotype. *Carcinogenesis* **26**, 1152–1156.

Buller, R. E., Shahin, M. S., Geisler, J. P., Zogg, M., De Young, B. R., and Davis, C. S. (2002). Failure of BRCA1 dysfunction to alter ovarian cancer survival. *Clin. Cancer Res.* **8**, 1196–1202.

Chan, T. L., Yuen, S. T., Kong, C. K., Chan, Y. W., Chan, A. S., Ng, W. F., Tsui, W. Y., Lo, M. W., Tam, W. Y., Li, V. S., *et al.* (2006). Heritable germline epimutation of MSH2 in a family with hereditary nonpolyposis colorectal cancer. *Nat. Genet.* **38**, 1178–1183.

Chen, R. Z., Pettersson, U., Beard, C., Jackson-Grusby, L., and Jaenisch, R. (1998). DNA hypomethylation leads to elevated mutation rates. *Nature* **395**, 89–93.

Chiang, J. W., Karlan, B. Y., Cass, L., and Baldwin, R. L. (2006). BRCA1 promoter methylation predicts adverse ovarian cancer prognosis. *Gynecol. Oncol.* **101**, 403–410.

Costello, J. F., Futscher, B. W., Kroes, R. A., and Pieper, R. O. (1994). Methylation-related chromatin structure is associated with exclusion of transcription factors from and suppressed expression of the O-6-methylguanine DNA methyltransferase gene in human glioma cell lines. *Mol. Cell. Biol.* **14,** 6515–6521.

Cunningham, J. M., Christensen, E. R., Tester, D. J., Kim, C. Y., Roche, P. C., Burgart, L. J., and Thibodeau, S. N. (1998). Hypermethylation of the hMLH1 promoter in colon cancer with microsatellite instability. *Cancer Res.* **58,** 3455–3460.

D'Andrea, A. D., and Grompe, M. (2003). The Fanconi anaemia/BRCA pathway. *Nat. Rev. Cancer* **3,** 23–34.

Duval, A., and Hamelin, R. (2002). Mutations at coding repeat sequences in mismatch repair-deficient human cancers: Toward a new concept of target genes for instability. *Cancer Res.* **62,** 2447–2454.

Eden, A., Gaudet, F., Waghmare, A., and Jaenisch, R. (2003). Chromosomal instability and tumors promoted by DNA hypomethylation. *Science* **300,** 455.

Esteller, M. (2000). Epigenetic lesions causing genetic lesions in human cancer: Promoter hypermethylation of DNA repair genes. *Eur. J. Cancer* **36,** 2294–2300.

Esteller, M., Corn, P. G., Urena, J. M., Gabrielson, E., Baylin, S. B., and Herman, J. G. (1998). Inactivation of glutathione S-transferase P1 gene by promoter hypermethylation in human neoplasia. *Cancer Res.* **58,** 4515–4518.

Esteller, M., Catasus, L., Matias-Guiu, X., Mutter, G. L., Prat, J., Baylin, S. B., and Herman, J. G. (1999a). hMLH1 promoter hypermethylation is an early event in human endometrial tumorigenesis. *Am. J. Pathol.* **155,** 1767–1772.

Esteller, M., Hamilton, S. R., Burger, P. C., Baylin, S. B., and Herman, J. G. (1999b). Inactivation of the DNA repair gene O6-methylguanine-DNA methyltransferase by promoter hypermethylation is a common event in primary human neoplasia. *Cancer Res.* **59,** 793–797.

Esteller, M., Toyota, M., Sanchez-Cespedes, M., Capella, G., Peinado, M. A., Watkins, D. N., Issa, J. P., Sidransky, D., Baylin, S. B., and Herman, J. G. (2000). Inactivation of the DNA repair gene O6-methylguanine-DNA methyltransferase by promoter hypermethylation is associated with G to A mutations in K-ras in colorectal tumorigenesis. *Cancer Res.* **60,** 2368–2371.

Feinberg, A. P., and Vogelstein, B. (1983). Hypomethylation distinguishes genes of some human cancers from their normal counterparts. *Nature* **301,** 89–92.

Fu, Z., Regan, K., Zhang, L., Muders, M. H., Thibodeau, S. N., French, A., Wu, Y., Kaufmann, S. H., Lingle, W. L., Chen, J., *et al.* (2009). Deficiencies in Chfr and Mlh1 synergistically enhance tumor susceptibility in mice. *J. Clin. Invest.* **119,** 2714–2724.

Futreal, P. A., Liu, Q., Shattuck-Eidens, D., Cochran, C., Harshman, K., Tavtigian, S., Bennett, L. M., Haugen-Strano, A., Swensen, J., Miki, Y., *et al.* (1994). BRCA1 mutations in primary breast and ovarian carcinomas. *Science* **266,** 120–122.

Hansen, R. S., Wijmenga, C., Luo, P., Stanek, A. M., Canfield, T. K., Weemaes, C. M., and Gartler, S. M. (1999). The DNMT3B DNA methyltransferase gene is mutated in the ICF immunodeficiency syndrome. *Proc. Natl. Acad. Sci. USA* **96,** 14412–14417.

Hedenfalk, I., Duggan, D., Chen, Y., Radmacher, M., Bittner, M., Simon, R., Meltzer, P., Gusterson, B., Esteller, M., Kallioniemi, O. P., *et al.* (2001). Gene-expression profiles in hereditary breast cancer. *N. Engl. J. Med.* **344,** 539–548.

Herman, J. G., Umar, A., Polyak, K., Graff, J. R., Ahuja, N., Issa, J. P., Markowitz, S., Willson, J. K., Hamilton, S. R., Kinzler, K. W., *et al.* (1998). Incidence and functional consequences of hMLH1 promoter hypermethylation in colorectal carcinoma. *Proc. Natl. Acad. Sci. USA* **95,** 6870–6875.

Hess, C. J., Ameziane, N., Schuurhuis, G. J., Errami, A., Denkers, F., Kaspers, G. J., Cloos, J., Joenje, H., Reinhardt, D., Ossenkoppele, G. J., *et al.* (2008). Hypermethylation of the FANCC and FANCL promoter regions in sporadic acute leukaemia. *Cell. Oncol.* **30,** 299–306.

Hinoue, T., Weisenberger, D. J., Pan, F., Campan, M., Kim, M., Young, J., Whitehall, V. L., Leggett, B. A., and Laird, P. W. (2009). Analysis of the association between CIMP and BRAF in colorectal cancer by DNA methylation profiling. *PLoS ONE* **4**, e8357.

Jeanpierre, M., Turleau, C., Aurias, A., Prieur, M., Ledeist, F., Fischer, A., and Viegas-Pequignot, E. (1993). An embryonic-like methylation pattern of classical satellite DNA is observed in ICF syndrome. *Hum. Mol. Genet.* **2**, 731–735.

Kane, M. F., Loda, M., Gaida, G. M., Lipman, J., Mishra, R., Goldman, H., Jessup, J. M., and Kolodner, R. (1997). Methylation of the hMLH1 promoter correlates with lack of expression of hMLH1 in sporadic colon tumors and mismatch repair-defective human tumor cell lines. *Cancer Res.* **57**, 808–811.

Kinzler, K. W., and Vogelstein, B. (1996). Lessons from hereditary colorectal cancer. *Cell* **87**, 159–170.

Kovacs, M. E., Papp, J., Szentirmay, Z., Otto, S., and Olah, E. (2009). Deletions removing the last exon of TACSTD1 constitute a distinct class of mutations predisposing to Lynch syndrome. *Hum. Mutat.* **30**, 197–203.

Kusano, M., Toyota, M., Suzuki, H., Akino, K., Aoki, F., Fujita, M., Hosokawa, M., Shinomura, Y., Imai, K., and Tokino, T. (2006). Genetic, epigenetic, and clinicopathologic features of gastric carcinomas with the CpG island methylator phenotype and an association with Epstein-Barr virus. *Cancer* **106**, 1467–1479.

Lee, W. H., Morton, R. A., Epstein, J. I., Brooks, J. D., Campbell, P. A., Bova, G. S., Hsieh, W. S., Isaacs, W. B., and Nelson, W. G. (1994). Cytidine methylation of regulatory sequences near the pi-class glutathione S-transferase gene accompanies human prostatic carcinogenesis. *Proc. Natl. Acad. Sci. USA* **91**, 11733–11737.

Leung, S. Y., Yuen, S. T., Chung, L. P., Chu, K. M., Chan, A. S., and Ho, J. C. (1999). hMLH1 promoter methylation and lack of hMLH1 expression in sporadic gastric carcinomas with high-frequency microsatellite instability. *Cancer Res.* **59**, 159–164.

Ligtenberg, M. J., Kuiper, R. P., Chan, T. L., Goossens, M., Hebeda, K. M., Voorendt, M., Lee, T. Y., Bodmer, D., Hoenselaar, E., Hendriks-Cornelissen, S. J., et al. (2009). Heritable somatic methylation and inactivation of MSH2 in families with Lynch syndrome due to deletion of the 3' exons of TACSTD1. *Nat. Genet.* **41**, 112–117.

Lynch, H. T., and de la Chapelle, A. (2003). Hereditary colorectal cancer. *N. Engl. J. Med.* **348**, 919–932.

Marsit, C. J., Liu, M., Nelson, H. H., Posner, M., Suzuki, M., and Kelsey, K. T. (2004). Inactivation of the Fanconi anemia/BRCA pathway in lung and oral cancers: Implications for treatment and survival. *Oncogene* **23**, 1000–1004.

Merajver, S. D., Pham, T. M., Caduff, R. F., Chen, M., Poy, E. L., Cooney, K. A., Weber, B. L., Collins, F. S., Johnston, C., and Frank, T. S. (1995). Somatic mutations in the BRCA1 gene in sporadic ovarian tumours. *Nat. Genet.* **9**, 439–443.

Michaloglou, C., Vredeveld, L. C., Soengas, M. S., Denoyelle, C., Kuilman, T., van der Horst, C. M., Majoor, D. M., Shay, J. W., Mooi, W. J., and Peeper, D. S. (2005). BRAFE600-associated senescence-like cell cycle arrest of human naevi. *Nature* **436**, 720–724.

Moslein, G., Tester, D. J., Lindor, N. M., Honchel, R., Cunningham, J. M., French, A. J., Halling, K. C., Schwab, M., Goretzki, P., and Thibodeau, S. N. (1996). Microsatellite instability and mutation analysis of hMSH2 and hMLH1 in patients with sporadic, familial and hereditary colorectal cancer. *Hum. Mol. Genet.* **5**, 1245–1252.

Narayan, G., Arias-Pulido, H., Nandula, S. V., Basso, K., Sugirtharaj, D. D., Vargas, H., Mansukhani, M., Villella, J., Meyer, L., Schneider, A., et al. (2004). Promoter hypermethylation of FANCF: Disruption of Fanconi Anemia-BRCA pathway in cervical cancer. *Cancer Res.* **64**, 2994–2997.

Niessen, R. C., Hofstra, R. M., Westers, H., Ligtenberg, M. J., Kooi, K., Jager, P. O., de Groote, M. L., Dijkhuizen, T., Olderode-Berends, M. J., Hollema, H., et al. (2009). Germline hypermethylation of MLH1 and EPCAM deletions are a frequent cause of Lynch syndrome. Genes Chromosom. Cancer 48, 737–744.

Nojima, M., Suzuki, H., Toyota, M., Watanabe, Y., Maruyama, R., Sasaki, S., Sasaki, Y., Mita, H., Nishikawa, N., Yamaguchi, K., et al. (2007). Frequent epigenetic inactivation of SFRP genes and constitutive activation of Wnt signaling in gastric cancer. Oncogene 26, 4699–4713.

Ogi, K., Toyota, M., Mita, H., Satoh, A., Kashima, L., Sasaki, Y., Suzuki, H., Akino, K., Nishikawa, N., Noguchi, M., et al. (2005). Small interfering RNA-induced CHFR silencing sensitizes oral squamous cell cancer cells to microtubule inhibitors. Cancer Biol. Ther. 4, 773–780.

Okano, M., Bell, D. W., Haber, D. A., and Li, E. (1999). DNA methyltransferases Dnmt3a and Dnmt3b are essential for de novo methylation and mammalian development. Cell 99, 247–257.

Pegg, A. E. (1990). Mammalian O6-alkylguanine-DNA alkyltransferase: Regulation and importance in response to alkylating carcinogenic and therapeutic agents. Cancer Res. 50, 6119–6129.

Peng, D. F., Razvi, M., Chen, H., Washington, K., Roessner, A., Schneider-Stock, R., and El-Rifai, W. (2009). DNA hypermethylation regulates the expression of members of the Mu-class glutathione S-transferases and glutathione peroxidases in Barrett's adenocarcinoma. Gut 58, 5–15.

Rauch, T. A., Zhong, X., Wu, X., Wang, M., Kernstine, K. H., Wang, Z., Riggs, A. D., and Pfeifer, G. P. (2008). High-resolution mapping of DNA hypermethylation and hypomethylation in lung cancer. Proc. Natl. Acad. Sci. USA 105, 252–257.

Sato, H., Suzuki, H., Toyota, M., Nojima, M., Maruyama, R., Sasaki, S., Takagi, H., Sogabe, Y., Sasaki, Y., Idogawa, M., et al. (2007). Frequent epigenetic inactivation of DICKKOPF family genes in human gastrointestinal tumors. Carcinogenesis 28, 2459–2466.

Satoh, A., Toyota, M., Itoh, F., Sasaki, Y., Suzuki, H., Ogi, K., Kikuchi, T., Mita, H., Yamashita, T., Kojima, T., et al. (2003). Epigenetic inactivation of CHFR and sensitivity to microtubule inhibitors in gastric cancer. Cancer Res. 63, 8606–8613.

Scolnick, D. M., and Halazonetis, T. D. (2000). Chfr defines a mitotic stress checkpoint that delays entry into metaphase. Nature 406, 430–435.

Scully, R., Chen, J., Plug, A., Xiao, Y., Weaver, D., Feunteun, J., Ashley, T., and Livingston, D. M. (1997). Association of BRCA1 with Rad51 in mitotic and meiotic cells. Cell 88, 265–275.

Strathdee, G., MacKean, M. J., Illand, M., and Brown, R. (1999). A role for methylation of the hMLH1 promoter in loss of hMLH1 expression and drug resistance in ovarian cancer. Oncogene 18, 2335–2341.

Suter, C. M., Martin, D. I., and Ward, R. L. (2004). Germline epimutation of MLH1 in individuals with multiple cancers. Nat. Genet. 36, 497–501.

Suzuki, H., Watkins, D. N., Jair, K. W., Schuebel, K. E., Markowitz, S. D., Chen, W. D., Pretlow, T. P., Yang, B., Akiyama, Y., Van Engeland, M., et al. (2004). Epigenetic inactivation of SFRP genes allows constitutive WNT signaling in colorectal cancer. Nat. Genet. 36, 417–422.

Suzuki, H., Igarashi, S., Nojima, M., Maruyama, R., Yamamoto, E., Kai, M., Akashi, H., Watanabe, Y., Yamamoto, H., Sasaki, Y., et al. (2010). IGFBP7 is a p53-responsive gene specifically silenced in colorectal cancer with CpG island methylator phenotype. Carcinogenesis 31, 342–349.

Taniguchi, T., Tischkowitz, M., Ameziane, N., Hodgson, S. V., Mathew, C. G., Joenje, H., Mok, S. C., and D'Andrea, A. D. (2003). Disruption of the Fanconi anemia-BRCA pathway in cisplatin-sensitive ovarian tumors. Nat. Med. 9, 568–574.

Tirkkonen, M., Johannsson, O., Agnarsson, B. A., Olsson, H., Ingvarsson, S., Karhu, R., Tanner, M., Isola, J., Barkardottir, R. B., Borg, A., et al. (1997). Distinct somatic genetic changes associated with tumor progression in carriers of BRCA1 and BRCA2 germ-line mutations. Cancer Res. 57, 1222–1227.

Toyota, M., Ahuja, N., Ohe-Toyota, M., Herman, J. G., Baylin, S. B., and Issa, J. P. (1999a). CpG island methylator phenotype in colorectal cancer. *Proc. Natl. Acad. Sci. USA* **96,** 8681–8686.

Toyota, M., Ahuja, N., Suzuki, H., Itoh, F., Ohe-Toyota, M., Imai, K., Baylin, S. B., and Issa, J. P. (1999b). Aberrant methylation in gastric cancer associated with the CpG island methylator phenotype. *Cancer Res.* **59,** 5438–5442.

Toyota, M., Ohe-Toyota, M., Ahuja, N., and Issa, J. P. (2000). Distinct genetic profiles in colorectal tumors with or without the CpG island methylator phenotype. *Proc. Natl. Acad. Sci. USA* **97,** 710–715.

Toyota, M., Sasaki, Y., Satoh, A., Ogi, K., Kikuchi, T., Suzuki, H., Mita, H., Tanaka, N., Itoh, F., Issa, J. P., *et al.* (2003). Epigenetic inactivation of CHFR in human tumors. *Proc. Natl. Acad. Sci. USA* **100,** 7818–7823.

Toyota, M., Suzuki, H., Yamashita, T., Hirata, K., Imai, K., Tokino, T., and Shinomura, Y. (2009). Cancer epigenomics: implications of DNA methylation in personalized cancer therapy. *Cancer Sci.* **100,** 787–791.

Wajapeyee, N., Serra, R. W., Zhu, X., Mahalingam, M., and Green, M. R. (2008). Oncogenic BRAF induces senescence and apoptosis through pathways mediated by the secreted protein IGFBP7. *Cell* **132,** 363–374.

Weisenberger, D. J., Siegmund, K. D., Campan, M., Young, J., Long, T. I., Faasse, M. A., Kang, G. H., Widschwendter, M., Weener, D., Buchanan, D., *et al.* (2006). CpG island methylator phenotype underlies sporadic microsatellite instability and is tightly associated with BRAF mutation in colorectal cancer. *Nat. Genet.* **38,** 787–793.

Yamamoto, E., Toyota, M., Suzuki, H., Kondo, Y., Sanomura, T., Murayama, Y., Ohe-Toyota, M., Maruyama, R., Nojima, M., Ashida, M., *et al.* (2008). LINE-1 hypomethylation is associated with increased CpG island methylation in Helicobacter pylori-related enlarged-fold gastritis. *Cancer Epidemiol. Biomarkers Prev.* **17,** 2555–2564.

Yarden, R. I., Pardo-Reoyo, S., Sgagias, M., Cowan, K. H., and Brody, L. C. (2002). BRCA1 regulates the G2/M checkpoint by activating Chk1 kinase upon DNA damage. *Nat. Genet.* **30,** 285–289.

Yu, X., Minter-Dykhouse, K., Malureanu, L., Zhao, W. M., Zhang, D., Merkle, C. J., Ward, I. M., Saya, H., Fang, G., van Deursen, J., *et al.* (2005). Chfr is required for tumor suppression and Aurora A regulation. *Nat. Genet.* **37,** 401–406.

Zhong, S., Tang, M. W., Yeo, W., Liu, C., Lo, Y. M., and Johnson, P. J. (2002). Silencing of GSTP1 gene by CpG island DNA hypermethylation in HBV-associated hepatocellular carcinomas. *Clin. Cancer Res.* **8,** 1087–1092.

Epigenetic Therapy and Epigenetic Drugs

12

DNA Demethylating Agents and Epigenetic Therapy of Cancer

Samson Mani and Zdenko Herceg
Epigenetics Group, International Agency for Research on Cancer (IARC), Lyon, France

ABSTRACT

Epigenetic events have been associated with virtually every step of tumor development and progression, and epigenetic alterations are believed to occur early in tumor development and may precede the malignant process. In contrast to genetic changes, epigenetic alterations arise in a gradual manner, leading to a progressive silencing of specific genes. An important distinction between epigenetic and genetic alterations is intrinsic reversibility of the former, making cancer-associated changes in DNA methylation, histone modifications, and expression of noncoding RNAs attractive targets for therapeutic intervention. This realization has triggered an impressive quest for the development of "epigenetic drugs" and epigenetic therapies. A number of agents have been subjected to an intensive investigation, many of which have been found capable of altering

Advances in Genetics, Vol. 70
Copyright 2010, Elsevier Inc. All rights reserved.
0065-2660/10 $35.00
DOI: 10.1016/S0065-2660(10)70012-2

epigenetic states, including DNA methylation patterns and histone modification states. Many of these agents are currently being tested in clinical trials, while several of them are already used in clinics. This review will focus on the recent advances in the development of epigenetic drugs based on the inhibition of DNA methylation. Combinatorial therapies that couple DNA demethylating agents with histone deacetylase inhibitors will also be discussed. © 2010, Elsevier Inc.

I. INTRODUCTION

It is now widely accepted that epigenetic inheritance is an essential mechanism that allows a stable propagation of gene activity states over many cell generations and that epigenetic deregulation is intimately linked to cancer development and progression. Recent mechanistic studies provided evidence that different epigenetic mechanisms (DNA methylation, noncoding RNAs, and histone modifications) are interconnected and may work together to establish and maintain specific gene activity states in normal cells (Jenuwein and Allis, 2001; Johnson et al., 2002; Jones and Baylin, 2002, 2007; Lehnertz et al., 2003; Soppe et al., 2002; Strahl and Allis, 2000). Importantly, all three distinct classes of epigenetic mechanisms have been found profoundly altered in a wide range of human cancers.

Although deregulation of epigenetic mechanisms is primarily studied in the context of cancer, epigenetic changes have been also implicated in several developmental syndromes (Egger et al., 2004; Feinberg and Tycko, 2004; Jiang et al., 2004), diabetes, obesity, cardiovascular diseases (Maier and Olek, 2002; McKinsey and Olson, 2004), and neurological disorders (Urdinguio et al., 2009). Epigenetic events have been associated with virtually every step of tumor development and progression (Feinberg and Tycko, 2004; Feinberg et al., 2006; Jones and Baylin, 2002). Importantly, epigenetic alterations are believed to occur early in tumor development and may precede the malignant process (Belinsky, 2004; Laird, 2003; Nephew and Huang, 2003). Therefore, epigenetic deregulation can be exploited as a powerful tool in the clinic and as a novel approach to early diagnosis, prediction of clinical outcome, and risk assessment (Belinsky, 2004; Egger et al., 2004; Fraga et al., 2005; Laird, 2003; Seligson et al., 2005). An important distinction between epigenetic and genetic alterations is intrinsic reversibility of the former, making cancer-associated changes in DNA methylation, histone modifications, and expression of noncoding RNAs particularly, attractive targets for the epigenetic therapy (Egger et al., 2004; Feinberg and Tycko, 2004; Jones and Baylin, 2007). Another distinguishing feature of epigenetic changes is that they arise in a gradual manner, leading to a progressive silencing of specific genes (such are tumor suppressor genes and DNA repair genes). This represents an exciting opportunity that can also be exploited in the development of novel strategies aiming to modulate the susceptibility to cancer and to prevent disease.

The ubiquity of epigenetic changes in many malignancies and other significant human diseases has triggered an impressive quest for the development of "epigenetic drugs" and epigenetic therapies. A number of agents have been subjected to an intensive investigation, many of which have been found capable of altering epigenetic states, including DNA methylation patterns and histone modification states (Table 12.1). Many of these molecules are currently being tested in clinical trials, while several of them are already used in clinics. A number of excellent reviews on epigenetics and epigenomics have been recently published (Belinsky, 2004; Bird, 2002; Callinan and Feinberg, 2006; Egger *et al.*, 2004; Esteller, 2007; Feinberg and Tycko, 2004; Feinberg *et al.*, 2006; Jaenisch and Bird, 2003; Jenuwein and Allis, 2001; Jones and Baylin, 2002, 2007;

Table 12.1. Classification of Epigenetic Drugs According to Their Potential Therapeutic Application

DNMT inhibitors:		
5-Azacytidine	MDS	Laird (2005), Sekeres *et al.* (in press)
	Prostrate cancer	Gravina *et al.* (2010)
	Solid tumors	Rudek *et al.* (2005)
	Leukemia	Rudek *et al.* (2005)
Decitabine	MDS	Steensma (2009), van den Bosch *et al.* (2004)
	Leukemia	Yang *et al.* (2005), Cashen *et al.* (2010)
Zebularine	Bladder cancer	Foubister (2003)
Procainamide	Prostate cancer	Lin *et al.* (2001)
Procaine	Breast cancer	Villar-Garea *et al.* (2003)
EGCG	Cervical cancer	Noguchi *et al.* (2006)
DNMT1 ASO	Solid tumors	Davis *et al.* (2003)
HDAC Inhibitors:		
Trichostatin A	Cervical cancer	Lin *et al.* (2009)
	Breast cancer	Joung *et al.* (2004)
	Ovarian cancer	Reid *et al.* (2005)
SAHA	Colorectal cancer	Kim *et al.* (2009)
	Liver cancer	Zhou *et al.* (2010)
	Leukemia	Garcia-Manero *et al.* (2008)
Apicidin	Leukemia	Park *et al.* (2009), Kim *et al.* (2004)
	Breast/ovarian cancer	Reid *et al.* (2005)
Depsipeptide	Leukemia	Oki and Issa (2006)
	Melanoma	Klisovic *et al.* (2005)
	Colon cancer	Xiao *et al.* (2005)
Phenyl butyrate	MDS, Leukemia	Maslak *et al.* (2006)

Laird, 2003, 2010; Lund and van Lohuizen, 2004; Ushijima, 2005). This review
will focus on recent advances in the development of epigenetic drugs and
therapies as well as on challenges and opportunities in targeting epigenetic
changes in cancer therapy and prevention (Fig. 12.1).

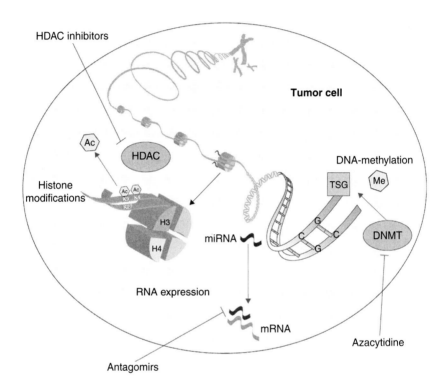

Figure 12.1. Epigenetic changes in cancer as a target for epigenetic therapy. Main epigenetic
mechanisms are illustrated as 1 = DNA methylation, 2 = histone modifications, and
3 = microRNA (miRNA) expression. DNA methyl-transferases (DNMTs) are
involved in the addition of CpG methyl marks (Me), while histone deacetylases
(HDACs) participate in removing histone acetyl marks (Ac). Therapeutical targeting
of HDACs, DNMTs, and miRNAs is depicted in the figure. TSG = tumor suppressor
genes. K = histone lysine residues. Tumor cells are characterized by an epigenetic
deregulation at different levels (1 = DNA methylation, 2 = histone modifications,
and 3 = microRNA expression). This deregulation may create a "signature" that can
be identified with modern screening tools. This information can be useful in clinical
settings and in the development of strategies for cancer treatment and prevention.
(*Adapted from* Thomson Reuters (Scientific) Ltd ISSN 2040-3445 *and Lima SCS,
Hernandez-Vargas H Herceg Z*: Epigenetic signatures in cancer: Implications for the
control of cancer in the clinic. Current Opinion in Molecular Therapeutics 2010 12
(3) © 2010 Thomson Reuters (Scientific) Ltd).

II. EPIGENETIC DEREGULATION IN CANCER

Epigenetic states are maintained through three distinct self-reinforcing mechanisms: DNA methylation, RNA-mediated gene silencing, and histone modifications (Feinberg and Tycko, 2004; Jones and Baylin, 2002; Sawan et al., 2008; Vaissiere et al., 2008). DNA methylation is the best understood and most extensively studied epigenetic mechanism which consists of the attachment of a methyl group to the 5-carbon (C^5) position of cytosine bases that are located 5′ to a guanosine base in a CpG dinucleotide (so-called CpG dinucleotide). This process is mediated by two main categories of DNA methyltransferases (DNMTs): enzymes involved in de novo DNA methylation (DNMT3A and DNMT3B) and in the maintenance of DNA methylation (DNMT1). This covalent modification of DNA plays an important role in cellular processes such as regulation of gene expression and cellular defense against foreign DNA sequences (such as viruses). Noncoding RNAs found in the form of small RNAs (microRNAs) or long noncoding RNAs (lncRNAs) are the most recently discovered epigenetic mechanisms. They play an important role in the regulation of the gene transcription and gene silencing. Deregulation of noncoding RNA expression has been associated with human diseases, including cancer. Histone modifications include a variety of posttranslational modifications of histone (specialized proteins associated with genomic DNA form the chromatin, a DNA–protein complex). Histone modifications comprise acetylation, phosphorylation, ubiquitination, and methylation of histone proteins at specific amino-acid residues. Based on previous studies, it was suggested that different histone marks may act in a combinatorial fashion to regulate cellular processes and consistently dictate cellular outcome, a concept known as the "histone code." For a detailed review on histone modifications, readers are directed to Chapter 3.

Previous genetic studies demonstrated that DNA methylation is essential in embryonic development and for faithful propagation of gene activity states during cell proliferation. Although epigenetic deregulation in cancer has been reviewed in detail in other chapters, we will give a brief description of DNA methylation changes in cancer. Aberrant DNA methylation can be found in two forms: gene promoter-associated DNA hypermethylation and global DNA hypomethylation, and both types of aberrant DNA methylation are found in most, if not all, types of human cancer. Hypermethylation typically occurs at CpG islands and is associated with gene inactivation (Feinberg and Tycko, 2004; Jones and Baylin, 2002). A large number of cancer-related genes, including those involved in tumor suppression, DNA repair, and carcinogen detoxification, has been found silenced by aberrant promoter hypermethylation (Jones and Baylin, 2002). p16 gene (*CDKN2A*), MLH1, RASSF1A, retinoblastima (*RB*) gene, VHL, E-cadherin, CDH1, LKB1, GSTP1, and MGMT are examples of cancer-associated genes that are found frequently hypermethylated in a wide range of human malignancies (Esteller, 2007; Jones and Baylin, 2002). However, the list of genes that are targeted by aberrant hypermethylation is likely to grow rapidly with the advent of high-throughput and genome-wide epigenomic tools (Esteller, 2007; Jones and Martienssen, 2005; Laird, 2010).

A global loss of DNA methylation (DNA hypomethylation) has also been consistently found in virtually all types of human cancers (Feinberg and Tycko, 2004). It is believed that global hypomethylation may promote tumor development through activation of cellular proto-oncogenes and induction of chromosomal instability (Feinberg and Tycko, 2004). However, further studies are needed to establish the precise underlying mechanism.

III. EPIGENETIC THERAPIES

Since there is now compelling evidence that deregulation of epigenetic mechanisms is involved in the development of most human malignancies, many efforts and resources have been mobilized in the development of different therapeutic approaches that are known as "epigenetic therapies." These approaches are directed to modify DNA methylation profiles and histone modification states in cancer cells and are based on specific properties of various chemical agents affecting the activity of molecules (enzymes) involved in the establishment and maintenance of epigenetic marking. Among these agents, demethylating agents (inhibitors of DNA methyltransferases) are the most extensively studied epigenetic agents. These include 5-azacytidine (5-aza-CR) and 5-aza-2-deoxycytidine (5-aza-CdR), both of which efficiently inhibit DNA methyltransferases and lower DNA methylation levels in a variety of cancer cell lines, leading to reactivation of gene expression (Egger et al., 2004). 5-aza-CR and 5-aza-CdR are nucleoside analogs of cytosine and the mechanism by which they inhibit DNA methylation involves incorporation of these molecules at the position of cytosine during DNA replication. This event results in trapping and inactivation of DNA methyltransferases. Therefore, event transient treatment of cells with demethylating agents can lead to a long-lasting demethylation effect (Egger et al., 2004). In a similar manner, pseudoisocytosine (also known as zebularine) can induce efficient demethylation (Marquez et al., 2005). Numerous studies showed that these agents can efficiently reactivate expression of aberrantly silenced genes in a variety of cancer cells.

Another class of epigenetic therapies is based on inhibition of histone deacetylases (HDACs). The currently available HDAC inhibitors (HDACi) are thought to inhibit the class I and II of HDACs (Egger et al., 2004; Marks et al., 2004). Treatment of cells with HDACi can cause the induction of differentiation, growth arrest, and/or apoptosis in a broad spectrum of transformed cells in culture and tumors in animals (Altucci and Minucci, 2009; Marks et al., 2004).

The anticancer effects of these drugs are thought to be due to the accumulation of acetylated histones, leading to the modulation of the transcription of specific genes whose expression causes inhibition of cancer cell growth (Marks *et al.*, 2004). The progress in therapeutic approached targeting histone modifications has been comprehensively reviewed in Chapter 13.

IV. EPIGENETIC DRUGS

A plethora of mechanistic and preclinical studies generated over the last decade provided a compelling evidence that different chemical agents may be exploited as potent modulators of epigenetic states (epigenetic drugs) in clinics and cancer control (Table 12.1). Four epigenetic drugs have been approved by the US Food and Drug Administration (FDA) and are currently available for clinical use. These include two DNA demethylating agents, 5-azacytidine and decitabine, and two histone deacetylase (HDAC) inhibitors, vorinostat and valproic acid. 5-azacytidine and decitabine inhibit DNA methyltransferase enzymes. The latter two drugs work by blocking HDACs, enzymes that remove acetyl groups from histone tails. Both classes of epigenetic drugs are thought to function, at least in part, by derepressing genes, such as tumor suppressor genes, aberrantly silenced through epigenetic mechanisms in diseased tissues.

A. DNA demethylating agents

The DNA demethylating agents inhibit DNA methyltransferase enzymes, causing reduced overall levels of DNA methylation. There are two classes of DNA demethylating agents: nucleoside DNMT inhibitors and nonnucleoside DNMT inhibitors (Table 12.1).

   1. *Nucleoside inhibitors of DNA methyltransferases*. The first nucleoside DNMT inhibitor (5-azacytidine) was discovered in the early 1960s (Sorm *et al.*, 1964), and this cytidine ribose nucleoside analog was only subsequently identified as an inhibitor of DNA methyltransferases. The deoxyribose analog of 5-azacytidine (5-aza-2′-deoxycytidine, also known as decitabine), which can be rather directly incorporated into DNA, was developed later.

   Aberrant promoter hypermethylation can be successfully targeted by inhibitors of DNMT enzymes. In cells, 5-azacytidine undergoes chemical modification to deoxyribonucleoside triphosphate that can be incorporated into DNA, where it is methylated by DNMTs. 5-azacytidine, 5-aza-20-deoxycytidine (5-AZA-CdR), 5-fluoro-20-deoxycytidine, 5,6-dihydro-5-azacytidine, and zebularine have been successfully used as epigenetic drugs. All these drugs are phosphorylated to the deoxynucleotide triphosphate and then incorporated instead of cytosine into replicating DNA. Hence, these are S-phase-specific drugs, acting as potent inhibitors of DNMTs (Egger et al., 2004). However, these drugs were used at relatively high doses, and more recent studies using significantly lower doses showed encouraging results, especially considering the relatively mild cytotoxicity of these doses. The disadvantage of 5-aza-CR and decitabine is that these molecules must be administered parenterally and are myelotoxic, resulting in cytopenia (Gilbert et al., 2004). The myelotoxicity of these drugs is thought to be due to their incorporation into DNA and not related to their DNA hypomethylation effects (Herman and Baylin, 2003). Zebularine is a newer cytosine analog that is less toxic and can be administered orally (Foubister, 2003). However, it has the disadvantage of being much less potent than 5-azacytidine and decitabine and needs to be administered in high doses (Foubister, 2003).

2. *Nonnucleoside inhibitors of DNMT enzymes.* This class comprises different nonnucleoside agents that are capable of inhibiting DNMT activity (Flynn et al., 2003; Ting et al., 2004). Antisense oligonucleotides designed to target DNMT enzymes (such as MG98) is an example of a nonnucleoside DNMT inhibitor that has been already tested in different models. Because these agents are not incorporated into DNA, they may be less toxic and better tolerated in clinical settings. However, the clinical trials with MG98 on a specific cancer type failed to provide encouraging results, mainly due to the lack of efficacy. Therefore, further studies are needed to test the efficacy and toxicity of different nonnucleoside agents.

B. Histone deacetylase inhibitors

Histone deacetylase inhibitors (HDACis) represent a novel class of targeted drugs that alter the acetylation status of several cellular proteins (Altucci and Minucci, 2009). Effects of HDAC inhibitors on gene expression may be highly selective, leading to transcriptional activation of certain genes such as the cyclin-dependent kinase inhibitor p21WAF1/CIP1 and repression of others (Vigushin and Coombes, 2002). Increased HDAC expression and activity often correlates with pathological gene repression and neoplastic transformation. Therefore, intense efforts have been made to find small molecule inhibitors of HDACs (HDACis). Drugs with HDAC inhibitory effects have been described and many are currently under clinical trials (Mai and Altucci, 2009). Wide spectrums of drugs inhibiting HDAC enzymes are designed to interfere with the catalytic domain and thereby block substrate recognition and induce gene expression (Table 12.1). Since aberrant expression of different HDAC isoforms has been associated with different malignancies (Bicaku et al., 2008; Oehme et al., 2009), it is often of interest to design isoform-specific HDAC inhibitors. With respect to the possible mechanisms of action of these drugs, it is important to consider that the response to HDACi appears also to depend, at least in part, on the nature of the HDACi, on the concentration and time of exposure, and on the cellular context. HDACi are able, for example, to induce cell growth arrest both in normal and cancer cells (Marks and Dokmanovic, 2005; Marks et al., 2000; Ungerstedt et al., 2005). For a more comprehensive overview of the progress in therapeutic approaches targeting histone modifications, readers are directed to Chapter 13.

V. EPIGENETIC COUPLING THERAPIES

In addition to the development and application of epigenetic drugs in the clinic as monotherapies, recent studies indicated that different epigenetic agents can be used in combination. This idea originates from the fact that an intimate cross talk exists between different epigenetic mechanisms. For example, there is evidence of a self-reinforcing link between DNA methylation and histone modifications in the maintenance of gene activity states in normal cells and pathological silencing of a wide range of genes in cancer cells. This encouraged a search for combinatorial therapies that couple DNA methylation inhibitors with HDAC inhibitors. In vitro experiments in colorectal carcinoma cell lines have shown that transcriptional reactivation of four hypermethylated genes was possible in the presence of low doses of DNMT inhibitor decitabine and HDAC inhibitor trichostatin A (Cameron et al., 1999). In addition, it has also been shown that a combination of decitabine and the HDAC inhibitor sodium phenylbutyrate has a synergistic effect in preventing tobacco carcinogen-induced lung cancer in mice (Belinsky et al., 2003). Several clinical trials are underway to investigate combinations of different epigenetic drugs in specific human malignancies. Finally, epigenetic drugs can be exploited in combination with other (conventional, nonepigenetic) anticancer drugs (Bhalla, 2005; Bolden et al., 2006), although these studies are at an early stage.

VI. CONCLUDING REMARKS

Interest in the epigenetics of cancer is considerably augmented by the realization that epigenetic alterations can be exploited as an attractive target in the clinic and cancer control. One important distinction between an epigenetic change and genetic alteration is that the former is, in turn, reversible. Therefore, aberrant levels and patterns of DNA methylation, histone modifications, and noncoding RNAs represent an enormous potential for the epigenetic therapy. Studies in transgenic and knockout mice have given fundamental insights into the regulation and biological functions of epigenetic therapy and drugs used for cancer treatment. As epigenetic mechanisms for cancer and other human disorders are identified, aberrant epigenetic states proved to be attractive molecular targets for therapeutic intervention and prevention. Indeed, the race is on to find efficient drugs and therapeutic strategies that can reverse epigenetic changes and unscheduled gene silencing. Several epigenetic drugs are already used for specific human malignancies, while many other epigenome-modifying substances are currently in clinical trials. However, despite the promise of epigenetic therapy, several concerns need to be addressed before it can be fully exploited in clinics. This includes the lack of systematic research into the function of HDAC family members in different cancer types or their effects on specific tumor suppressor genes as well as on other nonhistone deacetylation processes related to carcinogenesis. There is also a need to develop target-specific DNMT inhibitors and isotype-selective HDAC inhibitors to minimize the toxicity associated with these drugs. Early clinical trials with demethylating agents showed a relatively strong cytotoxic effect and were not well tolerated. However, these drugs were used at relatively high doses, and more recent studies with significantly lower doses showed encouraging results with relatively mild cytotoxicity. Therefore, combinatorial therapies that target different epigenetic mechanisms such as DNA methylation inhibitors with HDAC inhibitors may prove particularly efficient. Ongoing and future studies and clinical trials should give an answer on important questions regarding dosing schedules, routes of administration, and combination regimens. Furthermore, epigenetic drugs, including DNA demethylating agent inhibitors and HDAC inhibitors, exhibited a marked efficacy in the treatment of hematological malignancies; however, their potential in solid tumors remains less studied (Graham *et al.*, 2009). Epigenetic drugs that could target noncoding RNAs are also eagerly awaited. Together, with the advent of a wide range of drugs that target-specific molecules involved in epigenetic mechanisms, the discovery and characterization of epigenetic targets may prove particularly attractive in devising novel strategies for chemotherapy and chemoprevention of human cancer.

References

Altucci, L., and Minucci, S. (2009). Epigenetic therapies in haematological malignancies: Searching for true targets. *Eur. J. Cancer* **45,** 1137–1145.

Belinsky, S. A. (2004). Gene-promoter hypermethylation as a biomarker in lung cancer. *Nat. Rev. Cancer* **4,** 707–717.

Belinsky, S. A., Klinge, D. M., Stidley, C. A., Issa, J. P., Herman, J. G., March, T. H., and Baylin, S. B. (2003). Inhibition of DNA methylation and histone deacetylation prevents murine lung cancer. *Cancer Res.* **63,** 7089–7093.

Bhalla, K. N. (2005). Epigenetic and chromatin modifiers as targeted therapy of hematologic malignancies. *J. Clin. Oncol.* **23,** 3971–3993.

Bicaku, E., Marchion, D. C., Schmitt, M. L., and Munster, P. N. (2008). Selective inhibition of histone deacetylase 2 silences progesterone receptor-mediated signaling. *Cancer Res.* **68,** 1513–1519.

Bird, A. (2002). DNA methylation patterns and epigenetic memory. *Genes Dev.* **16,** 6–21.

Bolden, J. E., Peart, M. J., and Johnstone, R. W. (2006). Anticancer activities of histone deacetylase inhibitors. *Nat. Rev. Drug Discov.* **5,** 769–784.

Callinan, P. A., and Feinberg, A. P. (2006). The emerging science of epigenomics. *Hum. Mol. Genet.* **15**(Special No 1), R95–R101.

Cameron, E. E., Bachman, K. E., Myohanen, S., Herman, J. G., and Baylin, S. B. (1999). Synergy of demethylation and histone deacetylase inhibition in the re-expression of genes silenced in cancer. *Nat. Genet.* **21,** 103–107.

Cashen, A. F., Schiller, G. J., O'Donnell, M. R., and DiPersio, J. F. (2010). Multicenter, phase II study of decitabine for the first-line treatment of older patients with acute myeloid leukemia. *J. Clin. Oncol.* **28,** 556–561.

Davis, A. J., Gelmon, K. A., Siu, L. L., Moore, M. J., Britten, C. D., Mistry, N., Klamut, H., D'Aloisio, S., MacLean, M., Wainman, N., Ayers, D., Firby, P., *et al.* (2003). Phase I and pharmacologic study of the human DNA methyltransferase antisense oligodeoxynucleotide MG98 given as a 21-day continuous infusion every 4 weeks. *Invest. New Drugs* **21,** 85–97.

Egger, G., Liang, G., Aparicio, A., and Jones, P. A. (2004). Epigenetics in human disease and prospects for epigenetic therapy. *Nature* **429,** 457–463.

Esteller, M. (2007). Cancer epigenomics: DNA methylomes and histone-modification maps. *Nat. Rev. Genet.* **8,** 286–298.

Feinberg, A. P., and Tycko, B. (2004). The history of cancer epigenetics. *Nat. Rev. Cancer* **4,** 143–153.

Feinberg, A. P., Ohlsson, R., and Henikoff, S. (2006). The epigenetic progenitor origin of human cancer. *Nat. Rev. Genet.* **7,** 21–33.

Flynn, J., Fang, J. Y., Mikovits, J. A., and Reich, N. O. (2003). A potent cell-active allosteric inhibitor of murine DNA cytosine C5 methyltransferase. *J. Biol. Chem.* **278,** 8238–8243.

Foubister, V. (2003). Drug reactivates genes to inhibit cancer. *Drug Discov. Today* **8,** 430–431.

Fraga, M. F., Ballestar, E., Villar-Garea, A., Boix-Chornet, M., Espada, J., Schotta, G., Bonaldi, T., Haydon, C., Ropero, S., Petrie, K., Iyer, N. G., Perez-Rosado, A., *et al.* (2005). Loss of acetylation at Lys16 and trimethylation at Lys20 of histone H4 is a common hallmark of human cancer. *Nat. Genet.* **37,** 391–400.

Garcia-Manero, G., Yang, H., Bueso-Ramos, C., Ferrajoli, A., Cortes, J., Wierda, W. G., Faderl, S., Koller, C., Morris, G., Rosner, G., Loboda, A., Fantin, V. R., *et al.* (2008). Phase 1 study of the histone deacetylase inhibitor vorinostat (suberoylanilide hydroxamic acid [SAHA]) in patients with advanced leukemias and myelodysplastic syndromes. *Blood* **111,** 1060–1066.

Gilbert, J., Gore, S. D., Herman, J. G., and Carducci, M. A. (2004). The clinical application of targeting cancer through histone acetylation and hypomethylation. *Clin. Cancer Res.* **10,** 4589–4596.

Graham, J. S., Kaye, S. B., and Brown, R. (2009). The promises and pitfalls of epigenetic therapies in solid tumours. *Eur. J. Cancer* **45,** 1129–1136.

Gravina, G. L., Marampon, F., Di Staso, M., Bonfili, P., Vitturini, A., Jannini, E. A., Pestell, R. G., Tombolini, V., and Festuccia, C. (2010). 5-Azacitidine restores and amplifies the bicalutamide response on preclinical models of androgen receptor expressing or deficient prostate tumors. *Prostate.*

Herman, J. G., and Baylin, S. B. (2003). Gene silencing in cancer in association with promoter hypermethylation. *N. Engl. J. Med.* **349,** 2042–2054.

Jaenisch, R., and Bird, A. (2003). Epigenetic regulation of gene expression: How the genome integrates intrinsic and environmental signals. *Nat. Genet.* **33**(Suppl.), 245–254.

Jenuwein, T., and Allis, C. D. (2001). Translating the histone code. *Science* **293,** 1074–1080.

Jiang, Y. H., Bressler, J., and Beaudet, A. L. (2004). Epigenetics and human disease. *Annu. Rev. Genomics Hum. Genet.* **5,** 479–510.

Johnson, L., Cao, X., and Jacobsen, S. (2002). Interplay between two epigenetic marks. DNA methylation and histone H3 lysine 9 methylation. *Curr. Biol.* **12,** 1360–1367.

Jones, P. A., and Baylin, S. B. (2002). The fundamental role of epigenetic events in cancer. *Nat. Rev. Genet.* **3,** 415–428.

Jones, P. A., and Baylin, S. B. (2007). The epigenomics of cancer. *Cell* **128,** 683–692.

Jones, P. A., and Martienssen, R. (2005). A blueprint for a human epigenome project: The AACR human epigenome workshop. *Cancer Res.* **65,** 11241–11246.

Joung, K. E., Kim, D. K., and Sheen, Y. Y. (2004). Antiproliferative effect of trichostatin A and HC-toxin in T47D human breast cancer cells. *Arch. Pharm. Res.* **27,** 640–645.

Kim, J. S., Jeung, H. K., Cheong, J. W., Maeng, H., Lee, S. T., Hahn, J. S., Ko, Y. W., and Min, Y. H. (2004). Apicidin potentiates the imatinib-induced apoptosis of Bcr-Abl-positive human leukaemia cells by enhancing the activation of mitochondria-dependent caspase cascades. *Br. J. Haematol.* **124,** 166–178.

Kim, J. C., Shin, E. S., Kim, C. W., Roh, S. A., Cho, D. H., Na, Y. S., Kim, T. W., Kim, M. B., Hyun, Y. L., Ro, S., Kim, S. Y., and Kim, Y. S. (2009). In vitro evaluation of histone deacetylase inhibitors as combination agents for colorectal cancer. *Anticancer Res.* **29,** 3027–3034.

Klisovic, D. D., Klisovic, M. I., Effron, D., Liu, S., Marcucci, G., and Katz, S. E. (2005). Depsipeptide inhibits migration of primary and metastatic uveal melanoma cell lines in vitro: A potential strategy for uveal melanoma. *Melanoma Res.* **15,** 147–153.

Laird, P. W. (2003). The power and the promise of DNA methylation markers. *Nat. Rev. Cancer* **3,** 253–266.

Laird, P. W. (2005). Cancer epigenetics. *Hum. Mol. Genet.* **14**(Spec No 1), R65–R76.

Laird, P. W. (2010). Principles and challenges of genome-wide DNA methylation analysis. *Nat. Rev. Genet.* **11,** 191–203.

Lehnertz, B., Ueda, Y., Derijck, A. A., Braunschweig, U., Perez-Burgos, L., Kubicek, S., Chen, T., Li, E., Jenuwein, T., and Peters, A. H. (2003). Suv39h-mediated histone H3 lysine 9 methylation directs DNA methylation to major satellite repeats at pericentric heterochromatin. *Curr. Biol.* **13,** 1192–1200.

Lin, X., Asgari, K., Putzi, M. J., Gage, W. R., Yu, X., Cornblatt, B. S., Kumar, A., Piantadosi, S., DeWeese, T. L., De Marzo, A. M., and Nelson, W. G. (2001). Reversal of GSTP1 CpG island hypermethylation and reactivation of pi-class glutathione S-transferase (GSTP1) expression in human prostate cancer cells by treatment with procainamide. *Cancer Res.* **61,** 8611–8616.

Lin, Z., Bazzaro, M., Wang, M. C., Chan, K. C., Peng, S., and Roden, R. B. (2009). Combination of proteasome and HDAC inhibitors for uterine cervical cancer treatment. *Clin. Cancer Res.* **15,** 570–577.

Lund, A. H., and van Lohuizen, M. (2004). Epigenetics and cancer. *Genes Dev.* **18,** 2315–2335.

Mai, A., and Altucci, L. (2009). Epi-drugs to fight cancer: From chemistry to cancer treatment, the road ahead. *Int. J. Biochem. Cell Biol.* **41,** 199–213.

Maier, S., and Olek, A. (2002). Diabetes: A candidate disease for efficient DNA methylation profiling. *J. Nutr.* **132,** 2440S–2443S.

Marks, P. A., and Dokmanovic, M. (2005). Histone deacetylase inhibitors: Discovery and development as anticancer agents. *Expert Opin. Investig. Drugs* **14,** 1497–1511.

Marks, P. A., Richon, V. M., and Rifkind, R. A. (2000). Histone deacetylase inhibitors: Inducers of differentiation or apoptosis of transformed cells. *J. Natl. Cancer Inst.* **92,** 1210–1216.

Marks, P. A., Richon, V. M., Kelly, W. K., Chiao, J. H., and Miller, T. (2004). Histone deacetylase inhibitors: Development as cancer therapy. *Novartis Found Symp.* **259,** 269–281.

Marquez, V. E., Kelley, J. A., Agbaria, R., Ben-Kasus, T., Cheng, J. C., Yoo, C. B., and Jones, P. A. (2005). Zebularine: A unique molecule for an epigenetically based strategy in cancer chemotherapy. *Ann. NY Acad. Sci.* **1058,** 246–254.

Maslak, P., Chanel, S., Camacho, L. H., Soignet, S., Pandolfi, P. P., Guernah, I., Warrell, R., and Nimer, S. (2006). Pilot study of combination transcriptional modulation therapy with sodium phenylbutyrate and 5-azacytidine in patients with acute myeloid leukemia or myelodysplastic syndrome. *Leukemia* **20**, 212–217.

McKinsey, T. A., and Olson, E. N. (2004). Cardiac histone acetylation–therapeutic opportunities abound. *Trends Genet.* **20**, 206–213.

Nephew, K. P., and Huang, T. H. (2003). Epigenetic gene silencing in cancer initiation and progression. *Cancer Lett.* **190**, 125–133.

Noguchi, M., Yokoyama, M., Watanabe, S., Uchiyama, M., Nakao, Y., Hara, K., and Iwasaka, T. (2006). Inhibitory effect of the tea polyphenol, (−)-epigallocatechin gallate, on growth of cervical adenocarcinoma cell lines. *Cancer Lett.* **234**, 135–142.

Oehme, I., Deubzer, H. E., Wegener, D., Pickert, D., Linke, J. P., Hero, B., Kopp-Schneider, A., Westermann, F., Ulrich, S. M., von Deimling, A., Fischer, M., and Witt, O. (2009). Histone deacetylase 8 in neuroblastoma tumorigenesis. *Clin. Cancer Res.* **15**, 91–99.

Oki, Y., and Issa, J. P. (2006). Review: Recent clinical trials in epigenetic therapy. *Rev. Recent Clin. Trials* **1**, 169–182.

Park, S. J., Kim, M. J., Kim, H. B., Sohn, H. Y., Bae, J. H., Kang, C. D., and Kim, S. H. (2009). Cotreatment with apicidin overcomes TRAIL resistance via inhibition of Bcr-Abl signaling pathway in K562 leukemia cells. *Exp. Cell Res.* **315**, 1809–1818.

Reid, G., Metivier, R., Lin, C. Y., Denger, S., Ibberson, D., Ivacevic, T., Brand, H., Benes, V., Liu, E. T., and Gannon, F. (2005). Multiple mechanisms induce transcriptional silencing of a subset of genes, including oestrogen receptor alpha, in response to deacetylase inhibition by valproic acid and trichostatin A. *Oncogene* **24**, 4894–4907.

Rudek, M. A., Zhao, M., He, P., Hartke, C., Gilbert, J., Gore, S. D., Carducci, M. A., and Baker, S. D. (2005). Pharmacokinetics of 5-azacitidine administered with phenylbutyrate in patients with refractory solid tumors or hematologic malignancies. *J. Clin. Oncol.* **23**, 3906–3911.

Sawan, C., Vaissiere, T., Murr, R., and Herceg, Z. (2008). Epigenetic drivers and genetic passengers on the road to cancer. *Mutat. Res.* **642**, 1–13.

Sekeres, M. A., List, A. F., Cuthbertson, D., Paquette, R., Ganetsky, R., Latham, D., Paulic, K., Afable, M., Saba, H. I., Loughran, T. P., Jr., and Maciejewski, J. P. (in press). Phase I combination trial of lenalidomide and azacitidine in patients with higher-risk myelodysplastic syndromes. *J. Clin. Oncol.*

Seligson, D. B., Horvath, S., Shi, T., Yu, H., Tze, S., Grunstein, M., and Kurdistani, S. K. (2005). Global histone modification patterns predict risk of prostate cancer recurrence. *Nature* **435**, 1262–1266.

Soppe, W. J., Jasencakova, Z., Houben, A., Kakutani, T., Meister, A., Huang, M. S., Jacobsen, S. E., Schubert, I., and Fransz, P. F. (2002). DNA methylation controls histone H3 lysine 9 methylation and heterochromatin assembly in Arabidopsis. *EMBO J.* **21**, 6549–6559.

Sorm, F., Piskala, A., Cihak, A., and Vesely, J. (1964). 5-Azacytidine, a new, highly effective cancerostatic. *Experientia* **20**, 202–203.

Steensma, D. P. (2009). Decitabine treatment of patients with higher-risk myelodysplastic syndromes. *Leuk. Res.* **33**(Suppl. 2), S12–S17.

Strahl, B. D., and Allis, C. D. (2000). The language of covalent histone modifications. *Nature* **403**, 41–45.

Ting, A. H., Jair, K. W., Suzuki, H., Yen, R. W., Baylin, S. B., and Schuebel, K. E. (2004). CpG island hypermethylation is maintained in human colorectal cancer cells after RNAi-mediated depletion of DNMT1. *Nat. Genet.* **36**, 582–584.

Ungerstedt, J. S., Sowa, Y., Xu, W. S., Shao, Y., Dokmanovic, M., Perez, G., Ngo, L., Holmgren, A., Jiang, X., and Marks, P. A. (2005). Role of thioredoxin in the response of normal and transformed cells to histone deacetylase inhibitors. *Proc. Natl. Acad. Sci. USA* **102**, 673–678.

Urdinguio, R. G., Sanchez-Mut, J. V., and Esteller, M. (2009). Epigenetic mechanisms in neurological diseases: Genes, syndromes, and therapies. *Lancet Neurol.* **8,** 1056–1072.

Ushijima, T. (2005). Detection and interpretation of altered methylation patterns in cancer cells. *Nat. Rev. Cancer* **5,** 223–231.

Vaissiere, T., Sawan, C., and Herceg, Z. (2008). Epigenetic interplay between histone modifications and DNA methylation in gene silencing. *Mutat. Res.* **659,** 40–48.

van den Bosch, J., Lubbert, M., Verhoef, G., and Wijermans, P. W. (2004). The effects of 5-aza-2'-deoxycytidine (Decitabine) on the platelet count in patients with intermediate and high-risk myelodysplastic syndromes. *Leuk. Res.* **28,** 785–790.

Vigushin, D. M., and Coombes, R. C. (2002). Histone deacetylase inhibitors in cancer treatment. *Anticancer Drugs* **13,** 1–13.

Villar-Garea, A., Fraga, M. F., Espada, J., and Esteller, M. (2003). Procaine is a DNA-demethylating agent with growth-inhibitory effects in human cancer cells. *Cancer Res.* **63,** 4984–4989.

Xiao, J. J., Foraker, A. B., Swaan, P. W., Liu, S., Huang, Y., Dai, Z., Chen, J., Sadee, W., Byrd, J., Marcucci, G., and Chan, K. K. (2005). Efflux of depsipeptide FK228 (FR901228, NSC-630176) is mediated by P-glycoprotein and multidrug resistance-associated protein 1. *J. Pharmacol. Exp. Ther.* **313,** 268–276.

Yang, H., Hoshino, K., Sanchez-Gonzalez, B., Kantarjian, H., and Garcia-Manero, G. (2005). Antileukemia activity of the combination of 5-aza-2'-deoxycytidine with valproic acid. *Leuk. Res.* **29,** 739–748.

Zhou, X., Yang, X. Y., and Popescu, N. C. (2010). Synergistic antineoplastic effect of DLC1 tumor suppressor protein and histone deacetylase inhibitor, suberoylanilide hydroxamic acid (SAHA), on prostate and liver cancer cells: Perspectives for therapeutics. *Int. J. Oncol.* **36,** 999–1005.

13

Histone Modification Therapy of Cancer

Chiara Biancotto,*,1 Gianmaria Frigè,*,1 and Saverio Minucci*,†
*Department of Experimental Oncology, European Institute of Oncology,
Via Adamello 16, Milan, Italy
†Department of Biomolecular Sciences and Biotechnologies, University of
Milan, Via Celoria 26, Milan, Italy

I. Introduction
II. Histone Acetylation
   A. Histone acetyltransferases
   B. Alteration of HATs in cancer
   C. HAT inhibitors
   D. Histone deacetylases
   E. Alteration of HDACs in cancer
   F. HDAC inhibitors and the clinical validation of histone
      modification therapies of cancer
III. Histone Methylation
   A. Lysine and arginine methyltransferases and PR-domain
      containing proteins
   B. Histone demethylases
   C. Alteration of histone methylation in cancer
   D. Histone methyltransferases and demethylase inhibitors
IV. Perspectives
   Acknowledgments
   References

1These authors contributed equally to this work.

Advances in Genetics, Vol. 70   0065-2660/10 $35.00
Copyright 2010, Elsevier Inc. All rights reserved.   DOI: 10.1016/S0065-2660(10)70013-4

ABSTRACT

The state of modification of histone tails plays an important role in defining the accessibility of DNA for the transcription machinery and other regulatory factors. It has been extensively demonstrated that the posttranslational modifications of the histone tails, as well as modifications within the nucleosome domain, regulate the level of chromatin condensation and are therefore important in regulating gene expression and other nuclear events. Together with DNA methylation, they constitute the most relevant level of epigenetic regulation of cell functions.

Histone modifications are carried out by a multipart network of macromolecular complexes endowed with enzymatic, regulatory, and recognition domains. Not surprisingly, epigenetic alterations caused by aberrant activity of these enzymes are linked to the establishment and maintenance of the cancer phenotype and, importantly, are potentially reversible, since they do not involve genetic mutations in the underlying DNA sequence. Histone modification therapy of cancer is based on the generation of drugs able to interfere with the activity of enzymes involved in histone modifications: new drugs have recently been approved for use in cancer patients, clinically validating this strategy. Unfortunately, however, clinical responses are not always consistent and do not parallel closely the results observed in preclinical models.

Here, we present a brief overview of the deregulation of chromatin-associated enzymatic activities in cancer cells and of the main results achieved by histone modification therapeutic approaches.   © 2010, Elsevier Inc.

I. INTRODUCTION

The basic building block of chromatin is the nucleosome, formed by 147 base pairs of DNA wrapped around an octamer of histone proteins. There are five forms of histone proteins: histones H1, H2A, H2B, H3, and H4 (Finch *et al.*, 1977), as well as several histone variants, which are found in a subfraction of the nucleosomes and are involved in specific functions, such as DNA repair and gene activation (Sarma and Reinberg, 2005). The octamer structure of the nucleosome is composed of a H3-H4 tetramer, flanked on either side by a H2A-H2B dimer. The N-terminal tails of these "core" histones are extensively modified by methylation (Jenuwein, 2001), acetylation (Wade *et al.*, 1997), phosphorylation, sumoylation (Shiio and Eisenman, 2003), and ubiquitination (Shilatifard, 2006). Together with DNA methylation, histone posttranslational modifications are the main effectors of the epigenetic control of cell function, and they have been shown to have a central part in regulating gene expression, DNA replication and repair, cell division, and apoptosis.

The state of modification of histone tails plays an important role in defining the accessibility of DNA for transcription machinery and other regulatory factors. It has been widely demonstrated that the posttranslational modifications of the histone amino-terminal tails, as well as modifications within the nucleosome domain, regulate the level of chromatin condensation and are therefore important in regulating gene expression (Jenuwein and Allis, 2001). The correlation between regulation of transcription and chromatin acetylation is well characterized: acetylated histones are associated with a transcriptionally active state of the chromatin (Lehrmann et al., 2002). These regions are more accessible to regulatory factors, as experimentally indicated by the increased sensitivity of their DNA sequence to digestion with exogenous nucleases.

Histone modifications are carried out by an intricate network of macromolecular complexes, endowed with enzymatic, regulatory, and recognition domains. The acetylation status of the lysine residues at the N-terminus of histones is regulated by the opposing actions of histone acetyltransferases (HATs) and histone deacetylases (HDACs) (Roth et al., 2001; Thiagalingam et al., 2003), while histone methylation is occurring on lysine and arginine residues, and is due to the action of histone lysine methyltransferases (HMT), protein arginine methyltransferases (PRMT), and demethylating enzymes.

Interestingly, though most of these enzymatic activities have been initially described as acting mainly, if not exclusively, on histones, it is now clear that they should be considered to have much broader function in posttranslational modification of cell proteomes: indeed, a few recent studies have described hundreds of proteins regulated by reversible acetylation, that are most likely regulated by the same HATs and HDACs (Choudhary et al., 2009). Less information is available for nonhistone substrates of lysine and arginine methylases and demethylases, but there are no reasons to believe that the picture is different. This situation complicates enormously the interpretation of pharmacological outcomes of treatment aimed at inhibiting the activity of histone modifying enzymes. In a sense, the title of this chapter is—at least in part—erroneous, since we cannot selectively modulate histone modifications, without an equivalent impact on the large set of cellular proteins modified by the same enzymes.

It is perhaps not surprising to find that alterations in DNA methylation and histone modification patterns are present universally in cancer cells. These alterations can be mechanistically linked to the aberrant activity of activated oncogenes: as an example, we have previously shown the part played by acute myeloid leukemia (AML) fusion proteins, such as PML-RAR in acute promyelocytic leukemia (APL) (Minucci and Pelicci, 2006), in silencing differentiation genes through the recruitment of HDACs and other chromatin modifiers. A few reports have suggested a more complex set of chromatin modifications occurring in cancer cells. Global loss of monoacetylation and trimethylation of histone H4 (the latter regulated by SET-8 histone methyltransferase and JMJD2 histone demethylase) has been proposed to be a common hallmark of human tumor cells (Fraga *et al.*, 2005). Importantly, these global chromatin changes may be predictive of clinical outcome (Seligson *et al.*, 2005). Another important issue, that we will not further describe for lack of space, is the possible interplay between DNA methylation and histone modifications in regulating epigenetic silencing in mammalian cells. This process is normally orchestrated by polycomb-based histone H3 lysine 27 trimethylation (H3K27me3), histone H3 lysine 9 dimethylation (H3K9me2), and DNA methylation (Jenuwein and Allis, 2001). Whole-genome approaches for assessing the status of DNA methylation and histones modifications in normal and in cancer cells pave the way for identifying the link between them. This topic, however, is still controversial. In fact, while it has been reported that in cancer cells a number of CpG islands undergo *de novo* DNA methylation when previously marked by Polycomb-directed H3K27me3 (Schlesinger *et al.*, 2007; Vire *et al.*, 2006), a more recent work excluded the relationship between these two silencing pathways (Kondo *et al.*, 2008b). A more detailed analysis has been presented elsewhere (Cedar and Bergman, 2009).

Irrespective of these issues, it is clear that epigenetic alterations are related to the establishment and maintenance of the cancer phenotype. Importantly, those alterations are potentially reversible, since they do not involve genetic mutations in the underlying DNA sequence. Epigenetic therapies are based on the identification of drugs able to interfere with the activity of enzymes responsible for the epigenetic alterations occurring in tumor cells. For the reasons mentioned above, these histone modification therapies of cancer cannot be considered as "purely epigenetic," as they are bound to have an impact on nonhistone substrates. Targeting epigenetic mechanisms, however, is the main difference that they present with other targeted therapies currently in use or being developed, and the reason for a particular interest existing for this approach.

New drugs have recently been approved for use in cancer patients, validating clinically this strategy (reviewed in Mai and Altucci, 2009). Paradoxically, different classes of enzymes with opposing activities (histone acetylases and deacetylases; histone methyltransferases and demethylases) are being tested as antitumor therapies. In fact, the paradox is only apparent, the reason being that these enzymes may impinge on several concurrent mechanisms and opposing enzymatic activities may act at different stages of the same process (transcription, DNA repair, other nuclear events), providing therefore multiple starting points for therapeutic intervention.

Unfortunately, clinical responses following the use of epigenetic drugs are not always consistent and do not parallel closely the results observed in preclinical models.

The main explanations for this failure are (a) the use of epigenetic drugs in a modality of treatment similar to traditional chemotherapy, not based on a targeted approach; (b) the lack of biomarkers for directing therapeutic choices, at least due to the lack of appropriate technologies for epigenome profiling of cancer samples, that are just beginning to emerge; (c) the use of inappropriate preclinical models, mainly based on cell lines where epigenetic drift has occurred as a consequence of long-term *in vitro* culture.

A potential key to optimize this approach is based on the postulate that epigenetic therapies have to be directed against specific functional epigenetic alterations present in cancer cells: this is a strong theoretical rationale for searching drugs able to interfere with all chromatin modifying activities, since in different tumor cells different alterations are observed, driven by the aberrant activity of different enzymes.

Herein we will summarize essential findings on the epigenetic drugs developed in recent years in order to revert the aberrant epigenetic status of cancer cells.

II. HISTONE ACETYLATION

A. Histone acetyltransferases

HATs catalyze the transfer of acetyl groups from acetyl-coA to the ε-amino groups of conserved histone lysine residues, thus reducing the net positive charge on these proteins. In addition to histones, several nonhistone substrates are known: transcription factors (p53, myoD, EKLF), chromatin-associated proteins (HMGB1, 2, 4 17) (Sterner and Berger, 2000), nuclear import factors, and others (Bannister *et al.*, 2000). For DNA binding proteins, acetylation has been found to modulate their DNA binding affinity. HATs are classified in three major families based on sequence similarity: GNAT (Gcn5-related *N*-acetyl transferases), MYST (MOZ, Ybf2/Sas3, Sas2, and Tip60), and p300/CREB binding protein (CBP) (Fig. 13.1).

1. GNAT family

This family is conserved throughout eukaryotes, and the main members are Gcn5 and PCAF. GNAT family members are characterized by the presence of three conserved domains: an acetyltransferase domain called PCAF homology domain, an interaction domain for ADA2 (a protein found in GNAT containing complexes), and a bromodomain (domain for recognition of acetyl-lysine), believed to interact with acetyl-lysine residues at the C-terminus (Roth *et al.*, 2001).

Gcn5 was initially identified as a transcriptional regulator in yeast, and due to its homology with the *Tetrahymena* enzyme p55 HAT was characterized as a HAT (Brownell *et al.*, 1996). PCAF shares high homology with Gcn5 and was identified as a protein that interacts with p300/CBP and competes for their

GNAT family

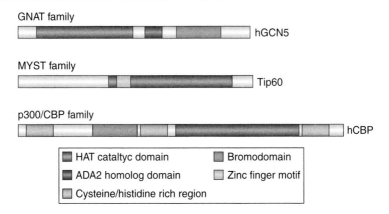

hGCN5

MYST family

Tip60

p300/CBP family

hCBP

■ HAT cataltyc domain	■ Bromodomain
■ ADA2 homolog domain	□ Zinc finger motif
□ Cysteine/histidine rich region	

Figure 13.1. Prototype members of the various families of histone acetyltransferases (HATs). Conserved structural features are indicated according to the legend showed at the bottom.

binding with the adenoviral E1A oncoprotein (Yang *et al.*, 1996). PCAF HAT activity shows a strong preference for lysine 14 of histone H3 and, to a less extent, for lysine 8 of histone H4 (Schiltz *et al.*, 1999).

2. MYST family

The MYST family is named after its four originating members. In human, the family currently includes: Tip60, MOZ, MORF, HBO, and MOF.

MYST HATs share a highly conserved MYST domain composed of an acetyl-coA binding motif involved in enzymatic activity and a zinc finger domain required for substrate interaction (Yan *et al.*, 2000). They are catalytic components of different multiprotein complexes.

Tip60 (Tat interacting protein 60 kDa), the first discovered human HAT, was originally described as an HIV Tat interacting protein and represents the catalytic subunit of the coactivator complex NuA4 (nucleosome acetyltransferase of histone H4), which selectively acetylates histone H4 and H2A tails (Ikura *et al.*, 2000; Kamine *et al.*, 1996). Tip60 acts as a coregulator of many transcription factors including androgen receptor, Myc, STAT3, NF-kB, E2F1, and p53 (Avvakumov and Cote, 2007). Due to its activities on histones and nonhistone substrates, it has been shown to play also an important role in DNA damage response pathways and apoptosis (Sun *et al.*, 2005).

HBO1 was originally identified in a complex with ORC1, a protein that binds to DNA replication origins (Iizuka and Stillman, 1999). This HAT is responsible for the bulk of histone H4 acetylation, leading to a more accessible chromatin conformation (Doyon et al., 2006). It plays an active role in DNA replication, positively regulating the assembly of the prereplication complex (preRC) and licensing of replication (Miotto and Struhl, 2008).

The MOZ acetyltransferase (monocytic leukemia zinc finger) was first identified in a study that examined the t(8;16)(p11;p13) chromosomal translocation, common in AML, as a fusion partner of CBP (Borrow et al., 1996). The MOZ-related factor (MORF) was identified based on its structural and functional homology to MOZ (Champagne et al., 1999).

MOF was discovered in Drosophila melanogaster as one of the key components of the male specific lethal (MSL) complex, required for dose compensation in flies (Hilfiker et al., 1997). Studies on the human MSL complex show that the interacting partners are evolutionary conserved and the fly and human proteins possess the same enzymatic activity.

Both MOZ and MOF proteins have HAT activity that is specific for lysine 16 of histone H4. Knockdown experiments for hMOF result in severely reduced or complete loss of H4K16 acetylation (H4K16Ac), suggesting that any process mediated through H4K16Ac can potentially be interpreted as being mediated by hMOF (Smith et al., 2005; Taipale et al., 2005).

3. p300/CBP family

p300/CBP is probably the most widely studied HAT family. p300 was first identified as a protein interacting with the adenovirus E1A, while CBP was proposed to act as a transcriptional coactivator of cAMP-responsive transcription factor CREB (Arany et al., 1994).

p300 and CBP are highly homologous and contain many common structural domains (bromodomain, HAT domain, and the cysteine/histidine (C/H)-rich domain). They are ubiquitously expressed and considered as global transcriptional coactivators, playing a critical role in cell cycle regulation, cellular differentiation, and apoptosis (Chan and La Thangue, 2001).

They interact with components of the basal transcriptional machinery (TBP and RNA polymerase II complex) (Chan and La Thangue, 2001), with several transcription factors (such as c-jun, c-fos, p53, c-myb, and several others) and with other HATs, including PCAF (Shiama, 1997). They can acetylate multiple lysines on all histones within a nucleosomal context and additionally they efficiently catalyze the acetylation of many nonhistone substrates.

B. Alteration of HATs in cancer

Alterations of HAT function induced by mutations, translocations, or simple overexpression have been shown to be relevant in several diseases, including cancer. Interestingly, HATs have versatile roles in cancer biology, since they can act as either tumor suppressors or as oncogenes, depending upon the genetic context.

Global loss of histone H4 lysine 16 acetylation has been shown to be a very common alteration present in several cancer cell types. In mammalian cells, hMOF specifically acetylates H4K16 and its depletion leads to global reduction of H4K16Ac *in vitro* (Taipale *et al.*, 2005). It appears therefore as if hMOF function is very commonly altered in cancer cells, not necessarily due to a unique mechanism.

Based on genetic studies in mice, p300/CBP have been considered as tumor suppressor genes; moreover, missense and deletion mutations in p300 are present in colorectal, gastric, and epithelial cancer samples (Gayther *et al.*, 2000; Muraoka *et al.*, 1996). Eighty percent of glioblastoma cases have been associated with loss of heterozygosity of p300 (Phillips and Vousden, 2000). In confirmation of these findings, patients suffering from the Rubinstein–Taybi syndrome, due to loss of one allele of CBP, have an increased risk of tumor formation (Kalkhoven *et al.*, 2003).

HATs can act also as oncogenes. Downregulation of p300 activity resulted in growth inhibition and activation of a senescence checkpoint in human melanocytes (Bandyopadhyay *et al.*, 2002). More in general, p300/CBP HAT activity is required for the G1/S transition (Ait-Si-Ali *et al.*, 2000). CBP/p300 contribute to the action of many cellular (*c-Fos, c-Jun*, and *c-Myb*) and viral oncogenes, that is, interact with the viral oncoprotein E1A, an SV40 large T antigen, leading to cell transformation (Miura *et al.*, 2007).

In AML, CBP can be found translocated and fused to either the HAT monocytic leukemia zinc finger (MOZ) gene, or to MLL (mixed lineage leukemia, a homeotic regulator with the ability to methylate histones) (Chaffanet *et al.*, 2000; Satake *et al.*, 1997); in both cases, the HAT activity remains unaffected. A translocation between CBP and a member of the MYST family (MORF), resulting in a fusion protein with two functional HAT domains, has also been reported linked to AML (Panagopoulos *et al.*, 2001). In these cases, animal studies have shown that the HAT activity is invariably required for leukemogenesis, thus suggesting that the altered enzymatic activity found in the AML-associated fusion proteins is indeed a potential target for therapy.

C. HAT inhibitors

Taken together, the data described above offer contrasting views about the potential use of drugs acting against HATs, suggesting that the outcome can be different, depending on cancer type and specific genetic/epigenetic alterations.

Considerable effort has been made in the recent years to develop HAT modulators to be used either as chemical tools for mechanistic studies, or as potential anticancer drugs. The HAT inhibitors developed so far can be classified as natural or synthetic compounds (Fig. 13.2).

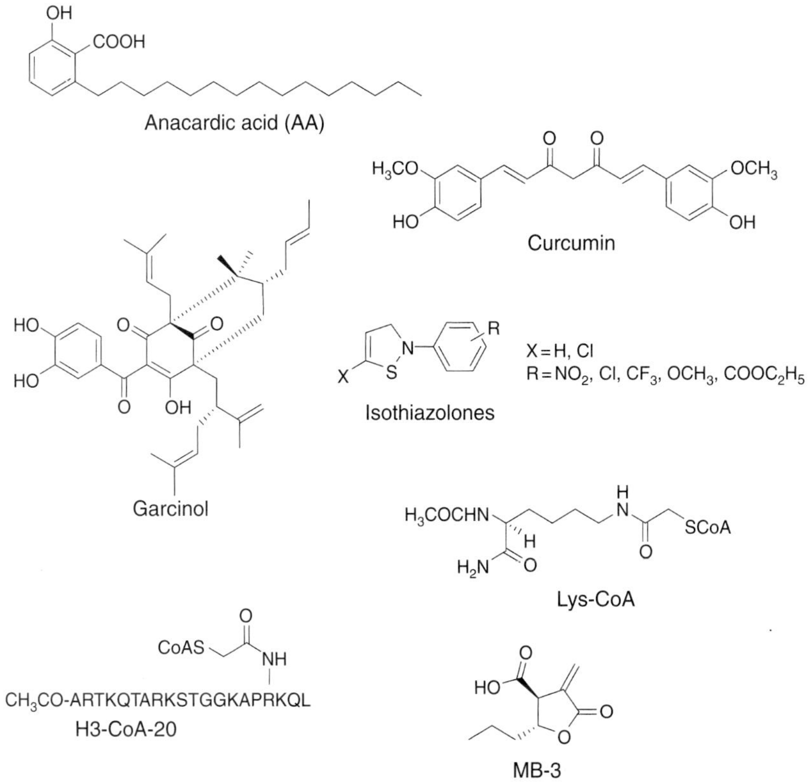

Figure 13.2. Chemical structure of histone acetyltransferase inhibitors (HATi): anacardic acid, curcumin, and garcinol are derived from natural compounds; Lys-CoA and H3-CoA-20 are synthetic peptides based on CoA; MB-3 is a small synthetic molecule; isothiazolones are represented with the basic structure and functional groups used for their modification.

Anacardic acid, garcinol, and curcumin are the most important HAT inhibitors obtained from plants. Anacardic acid, isolated from cashew nut shell liquid, and garcinol, isolated from *Garcinia indica* fruit rind, are reported to be potent inhibitors of both p300 and PCAF (Ayroldi *et al.*, 1992; Balasubramanyam *et al.*, 2004a). Their anticancer properties have also been investigated: anacardic acid was shown to sensitize human tumor cells to the cytotoxic effects of ionizing radiation, providing a novel therapeutic approach for enhancing the efficacy of radiation therapy in clinic (Sun *et al.*, 2006). Mechanistically, it is not clear how HAT inhibition results in enhanced radiosensitivity. Garcinol and its derivatives showed significant growth suppression due to apoptosis mediated by the activation of caspase-3 in four human leukemic cell lines (Pan *et al.*, 2001). Treatment of

HeLa cells with garcinol was shown to inhibit histone acetylation, to induce apoptosis, and to downregulate expression of proto-oncogenes (due to a decrease in histone acetylation) (Balasubramanyam et al., 2004a).

Among natural compounds, the most studied is curcumin, derived from the rhizome of the *Curcuma longa* plant, that has shown efficacy in the prevention and treatment of various cancers. Several studies have demonstrated that curcumin can affect numerous molecular and biochemical cascades, interacting and modulating different targets including transcription factors (i.e., NF-kB, AP-1), growth factors, and their receptors, and various protein kinases that contribute to malignant transformation and metastasis (Goel et al., 2008). Curcumin is also reported to be a specific inhibitor of the p300/CREB-binding protein (CBP), a transcriptional coactivator frequently mutated and deregulated in cancer (Balasubramanyam et al., 2004b). Curcumin has also been shown to inhibit the expression of genes and pathways involved in apoptosis, cell invasion, and adhesion. Given the wide variety of mechanisms proposed, treatment with curcumin shows a promise as an agent to use in therapy, and indeed curcumin is currently being investigated in human clinical trials for a variety of tumors, including multiple myeloma, pancreatic cancer, myelodysplastic syndrome, and colon cancer. In these studies, clinical effects have been weak, perhaps due to the poor bioavailability of the compound (Bar-Sela et al., 2010). A note of concern in interpreting the studies on HAT inhibitors derived from natural sources is that the molecules identified (as most natural compounds) are highly complex, and most likely act through multiple mechanisms. Genetic studies (knockdown) should be performed to confirm that the data obtained with drugs are indeed due to the postulated HAT inhibitory effect.

Synthetic peptide-CoA-based compounds were the first class of potent and selective HAT inhibitors described for p300 and PCAF acetyltransferases. Coenzyme A conjugated with the lysine amino acid (lys-CoA) specifically inhibits p300, whereas a 20 amino acid long peptide (H3-CoA-20) targets PCAF (Lau et al., 2000). Unfortunately, these synthetic inhibitors showed cell permeability problems and high metabolic instability, which limit their uses *in vivo*. To address this problem, novel inhibitors have been synthesized attaching cell-permeabilizing peptide sequences (Lys-CoA-Tat and H3-CoA-20-Tat). Lys-CoA-Tat has been shown to block p300/CBP-mediated acetylation of the promyelocytic zinc finger protein (PLZF), which interrupts its function as a transcriptional repressor (Guidez et al., 2005). These molecules are unlikely to be used differently than as chemical tools; however, they are of great interest since they seem to act rather selectively on HATs.

In addition to peptides, a few small molecules have been described as synthetic HAT inhibitors. The γ-butyrolactone MB-3 has been discovered as a cell-permeable GCN5 inhibitor (Biel et al., 2004). A group of isothiazolones, which act as p300/PCAF inhibitors, display an interesting anticancer property against different human colon and ovarian cancer cell lines. Moreover, the selective inhibition of p300 HAT by semisynthetic derivatives of garcinol (i.e., LTK-14) and anacardic acid was also described (Mantelingu et al., 2007). In all these instances, studies are limited to cell assays, and more advanced compounds able to work in animal models are badly needed to further validate HATs as a relevant target.

D. Histone deacetylases

In mammals, the HDAC family comprises 18 members that can be grouped in four classes based on their sequence homology to yeast proteins Fig. 13.3. Class I members (HDAC 1, 2, 3, 8, and 11) share a single deacetylase domain at the N-terminus, are homologous to the yeast Rpd3, and are localized to the nucleus (Yang and Seto, 2008). Class II members show greater similarity with yeast

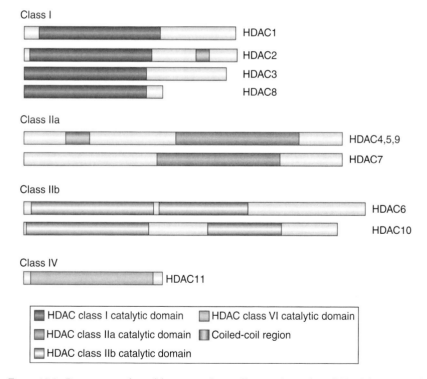

Figure 13.3. Prototype members of the various classes of histone deacetylases (HDACs). Conserved structural features are indicated according to the legend showed at the bottom.

Hda1, shuttle between nucleus and cytoplasm, and can be further subdivided into class IIa (HDAC 4, 5, 7, and 9) and class IIb (HDAC6c and 10) (Verdin *et al*., 2003). HDAC6, mainly located in the cytoplasm, contains two catalytic domains, whereas HDAC10 possesses an N-terminal catalytic domain and a C-terminal incomplete catalytic domain. Class III HDACs (SIRT1–SIRT7) called Sirtuins are homologs of yeast Sir2, share a conserved catalytic domain structurally and mechanistically different from that of other HDACs described above (Zn-dependent HDACs), and need NAD + coenzyme to deacetylate histone and nonhistone substrates (Michan and Sinclair, 2007). Though class III HDACs have been also involved in the pathogenesis of cancer, there are much less data available when compared with the other classes of HDACs, and therefore they will not be further discussed here. HDAC11 represents its own class (class IV): its catalytic domain shares sequence homology with both class I and class II domain HDACs (Gao *et al*., 2002).

HDACs are found associated with a large number of transcription- and chromatin-related factors, and in some cases they are found in multiprotein complexes including multiple HDACs and other enzymes: for example, DNA methyltransferase 1 (DNMT1) interacts with HDAC1 and together constitute a complex that contains also the transcription factor E2F and the tumor suppressor Rb (Robertson *et al*., 2000). Since HDACs do not possess DNA-binding activity, the interactions with complexes containing sequence-specific DNA-binding proteins facilitate the access to DNA and the recruitment to specific chromosomal regions. These corepression complexes (including YY1, Sin3A, N-CoR, and SMRT) direct the HDAC activity to certain promoters to regulate transcription (Yang and Seto, 2008). Interestingly, recent results show conclusively that HDACs may not only be involved in repressing transcription but are also found at transcriptionally active genes, being required for their proper expression (Wang *et al*., 2009b).

As for HATs, the activity of the different HDACs is not exclusively targeted at histones but also at nonhistone proteins including transcription factors and many other proteins involved in DNA repair and replication, metabolism, cytoskeletal dynamics, apoptosis, nuclear import, protein folding, and cellular signaling (Glozak *et al*., 2005). Therefore, HDACs not only contribute to the proper coordination of gene expression through modulating the acetylation at chromatin level but also control the many aspects of biological processes such as protein stability, protein translocation, enzymatic activity, protein–protein interaction, and DNA binding affinity via acetylation of nonhistone proteins.

E. Alteration of HDACs in cancer

There is growing evidence that aberrant HDAC activity is strongly associated with the development of cancer. There are several studies showing altered expression of HDACs in tumor samples. For example, prostate cancer cells overexpress HDAC1 (Halkidou *et al.*, 2004). In addition, HDAC3 protein levels have been found increased in some colon cancers (Spurling *et al.*, 2008). HDAC2 has been found overexpressed in gastric carcinomas, colorectal carcinomas, cervical dysplasias, and endometrial stromal sarcomas as compared to their normal counterparts (Hrzenjak *et al.*, 2006; Huang *et al.*, 2005; Song *et al.*, 2005). Reduced expression levels of HDAC1 in gastric and HDAC5 and HDAC10 in lung cancers have been linked with poor prognosis (Osada *et al.*, 2004).

Class III enzyme SIRT1 has been found upregulated in mouse lung carcinomas, lymphomas, and prostate cancer, and in human AMLs, glioblastoma, colorectal, prostate, and skin cancer (Bradbury *et al.*, 2005; Chen *et al.*, 2005; Huffman *et al.*, 2007; Liu *et al.*, 2006; Ozdag *et al.*, 2006).

SIRT1 is responsible for the reduction of the monoacetylated-lysine 16 of histone H4 and acts mainly at gene promoter regions (Pruitt *et al.*, 2006). The levels of H4K16 and H3K9 (another known histone target of SIRT1) have been found altered in different types of tumors. Consistent loss of H4K16Ac and H4K20 trimethylation (H4K20me3 is a marker of constitutive heterochromatin) has been observed in various cancer cell lines and primary tumors, suggesting that the level of this H4 "cancer signature" is a common hallmark of human cancer, and a potential role of SIRT1 in this process (Fraga *et al.*, 2005).

Aberrant transcriptional repression by the recruitment of HDAC containing complexes is commonly observed in leukemia. APL was the first hematological malignancy in which a role for HDAC has been demonstrated in the pathogenesis of the disease (Minucci *et al.*, 2001).

APL is caused by the oncoprotein resulting from the fusion between retinoic acid receptor (RAR) with either the promyelocytic leukemia protein (PML) (majority of the cases, 95%) or the promyelocytic leukemia zinc finger protein (PLZF) (5%). PML/RAR has an increased ability to associate with complexes containing HDAC and chromatin modifiers (DNMTs, HMTs, MBDs) at RAR target genes (Minucci and Pelicci, 2006). This aberrant recruitment of corepressor complexes blocks the differentiation process at the promyelocytic stage. Since the fusion protein is still able to bind the ligand retinoic acid (RA), treatment with pharmacological doses of RA induces the dissociation of PML/RAR/HDAC complex and the degradation of the fusion protein, restoring the cell differentiation program that results in apoptotic cell death (Warrell *et al.*, 1991). In APL patients expressing the fusion protein PLZF/RAR, however, HDAC/corepressor complexes bind to both RAR and PLZF moieties of the fusion protein. Upon RA treatment, corepressors are released only from RAR, and they can still induce transcriptional repression through the PLZF moiety (Grignani *et al.*, 1998; He *et al.*, 1998). In this case, terminal differentiation is possible only by combining RA treatment with HDAC inhibitors (HDACi) (He *et al.*, 2001).

In the M2 subtype of AML, the *AML1* gene is disrupted by a translocation that results in a fusion to the ETO protein. Similar to the case of APL, the
fusion protein AML1-ETO works as a potent transcriptional repressor through
the interaction with the complex N-CoR/Sin3/HDAC1 (Wang *et al.*, 1998).

HDAC involvement in leukemia is not only confined to myeloid
malignances but is also found in lymphomas. In the B cell-derived non-Hodgkin
lymphomas (NHLs), the transcriptional repressor BCL6 is overexpressed and
aberrantly recruits HDACs complexes to mediate transcriptional repression
(Dhordain *et al.*, 1998; Lemercier *et al.*, 2002; Shaffer *et al.*, 2002).

The role of HDACs in cancer is not confined only to their contribution
to histone deacetylation but also to their role in deacetylation of nonhistone
proteins. In fact, many proteins clearly involved in oncogenesis such as p53,
HSP90, E2F, pRB, and BCL6 have been identified as substrates of HDACs
(Glozak *et al.*, 2005).

These results (and several others that for space limits cannot be summarized) clearly show that HDACs play an active role in tumor onset and progression,
and that makes them attractive targets for anticancer drugs and therapies.

F. HDAC inhibitors and the clinical validation of histone modification therapies of cancer

Both natural molecules and synthesized compounds with HDAC inhibitory
activity have been used extensively in several settings, opening new perspectives
for cancer therapy (Fig. 13.4). Interfering with HDAC activity resulted mainly in
the reexpression of genes with tumor suppressor function and thereby in the
reversion of the aberrant epigenetic status of tumor cells (Finnin *et al.*, 1999),
cell cycle arrest, induction of differentiation, or apoptosis (Mai *et al.*, 2005;
Nebbioso *et al.*, 2005). HDACi appear to target preferentially cancer cells,
while they do not induce apoptosis of normal cells if not at much higher doses
(Ungerstedt *et al.*, 2005).

The first HDACi have been initially discovered for their differentiating
and antiproliferative activities of natural compounds such as trichostatin A
(TSA, Selleck Chemicals LLC, US) that was able to induce differentiation of
erythroleukemia cells (Tsuji *et al.*, 1976; Ungerstedt *et al.*, 2005). It was then
demonstrated that these molecules increase histone acetylation as a direct
consequence of HDAC inhibition: currently, the structural interaction of
HDAC–HDACi complexes has been solved in a few cases, using a panel of
HDACi and a few HDACs (that are not easily amenable to structural studies)
(Somoza *et al.*, 2004).

The growth-inhibitory effect of HDACi was subsequently tested in
many transformed cell types deriving from both hematological and solid tumors.
Biochemical and structural information accompanied with the observations

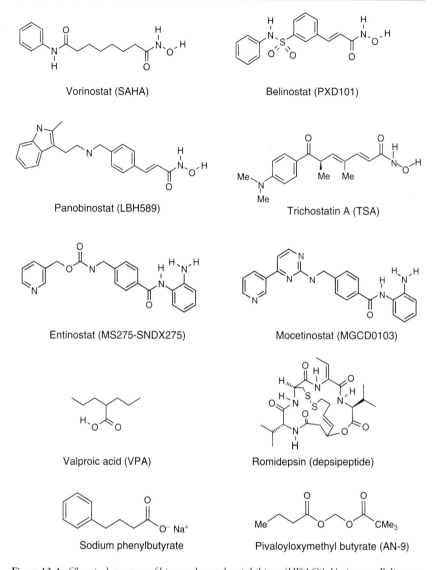

Figure 13.4. Chemical structure of histone deacetylase inhibitors (HDACi): Vorinostat, Belinostat, Panobinostat, Trichostatin are Hydroxamic acid derivatives; Entinostat, Mocetinostat are benzamides; Valproic Acid, Sodium Phenilbutyrate, and Pivaloyloxymethyl butyrate are short chain fatty acids; and Romidepsin is a cyclic peptide.

made in cancer cell lines contributed to the design of novel molecules, trying to minimize side- and dose-related secondary effects (Butler *et al.*, 2000; Richon *et al.*, 1998).

HDACi showed also potent antitumor effects *in vivo* (Marks *et al.*, 2000, 2001): TSA (Yoshida *et al.*, 1990), CHAP1 and CHAP31, SAHA (Butler *et al.*, 2000), pyroxamide (Butler *et al.*, 2001), CBHA (Coffey *et al.*, 2001), and oxamflatin (Kim *et al.*, 1999) displayed a reduction of tumor growth in xenograft models of breast, prostate, lung, and stomach cancer; neuroblastoma; and leukemia showing overall low toxicity.

The majority of preclinical studies show that HDACi affect preferentially hematologic malignancies as single agents, while they have been used in association with other chemotherapeutic agents in a variety of malignancies (Camphausen and Tofilon, 2007). The observation of HDACi anticancer effects in preclinical models allowed their introduction into clinical trials, and Vorinostat (known also as SAHA—Merck & co.) has been the first HDACi approved by US Food and Drug Administration for the treatment of cutaneous T cell lymphoma (CTCL) in patients who have progressive or recurrent disease on or following two systemic therapies (Marks, 2007). More recently, Romidepsin (see below) has also been approved for the same indication (Kim *et al.*, 2008). At the cellular level, Vorinostat increases the acetylation of histones in tumors as observed in bone marrow and peripheral blood cells and increases the expression of p21, an inhibitor of cellular proliferation, while decreases the expression of phospho-STAT6 that has been correlated with clinical response (Fantin *et al.*, 2008). Nevertheless, multiple processes can concurrently occur following treatment with this and other epigenetic drugs, suggesting that the mechanisms underlying the cell response to HDACi are pleiotropic and context dependent.

Looking back at the clinical studies with Vorinostat, the first promising results were obtained in different hematological malignancies showing a good safety profile and antitumoral activity; moreover, only for relapsed CTCL the treatment results led to registration studies (Mann *et al.*, 2007). In particular, two independent phase II studies reported similar results: Vorinostat was safe and effective at an oral dose of 400 mg/day with an overall response rate of 30% of the enrolled patients; however, nausea, diarrhea, thrombocytopenia, and anemia were the observed side effects (Duvic and Vu, 2007; Olsen *et al.*, 2007).

Vorinostat was clinically tested in other hematological malignancies such as NHL, in which tumor regression and symptomatic improvement were observed at doses that have no clinical toxicity (Kirschbaum *et al.*, 2007). Other advanced refractory leukemia patients were enrolled in clinical trials with Vorinostat, but clinical responses were observed only for the therapy of AML and, independently from clinical efficacy, an hyperacetylation of histone H3 was observed at all dose levels (Garcia-Manero *et al.*, 2008b).

A phase II study in patients with glioblastoma multiforme (GBM) indicated that vorinostat mono-therapy has an activity also in solid tumors. The increased acetylation levels of histones H2B, H4, and histone H3 following the treatment and changes in the expression pattern of genes regulated by vorinostat, such as upregulation of E-cadherin, suggests that the treatment had biological effect on the glioblastoma tumor and affects critical target pathways in GBM (Galanis *et al.*, 2009).

Being one of the most extensively investigated epigenetic drugs in clinical studies, Vorinostat exemplifies perhaps what has been discussed at the beginning of this chapter: it is a striking case of a drug that when tested in preclinical models showed a tremendous promise, which has been since validated clinically as possessing antitumor efficacy leading to its approval by FDA. Nonetheless, it has fallen short of its promises, giving a very limited clinical response in most cancer patients, irrespective of the extent of target modulation observed and leaving us with an almost complete lack of understanding as to the criteria needed for the stratification and selection of responder patients. We will further discuss this conundrum, and how to find directions for its solution at the end.

Unlike Vorinostat (that similarly to TSA contains a hydroxamic acid moiety required for enzyme inhibition), Romidepsin (also known as FK228 or Depsipeptide—Gloucester Pharmaceuticals, Cambridge, MA) is a natural cyclic peptide which is atypically metabolized to an active, reduced form by cellular reducing activities (Itoh *et al.*, 2008; Konstantinopoulos *et al.*, 2006). This prodrug once administered to the cells induces cell cycle arrest and apoptosis in a variety of cancer cell lines, and a potent *in vivo* antitumor activity against human tumor xenografts (Furumai *et al.*, 2002; Piekarz *et al.*, 2001). Romidepsin has been extensively investigated in several clinical trials mainly for hematological malignancies such as chronic myeloid leukemia (CML), AML, and CTCL showing a limited antitumoral activity, and in a single case complete remission has been reported for an AML patient (Byrd *et al.*, 2005; Klimek *et al.*, 2008). Nonetheless, the response observed in CTCL patients in two phase II clinical trials encouraged the FDA to approve the treatment with Romidepsin for CTCL patients who have received at least one previous systemic therapy (Bates *et al.*, 2008; Kim *et al.*, 2008). Recent phase II trials conducted in AML patients highlight a more selective antileukemic effect of Romidepsin, since it was restricted to patients with specific chromosomal translocations (Odenike *et al.*, 2008).

Clinical trials on patients with solid tumors have been carried out, but Romidepsin is ineffective in most cases: further clinical trials combining this inhibitor with chemotherapeutic or demethylating agents will be required (Stadler *et al.*, 2006; Whitehead *et al.*, 2009).

Among hydroxamic acid derivatives in clinical trials, LBH 589 (Panobinostat—Novartis Pharmaceuticals, East Hanover, NJ) is one of the most potent HDACi available tested for CML, refractory CTCL, and multiple myeloma (Bayes *et al.*, 2007; Khan *et al.*, 2008) while PDX-101 (Belinostat) has entered phase II clinical trials for hematological malignancies and for many solid tumors (Khan *et al.*, 2008; Qian *et al.*, 2008). The results with these second generation (rationally designed with improved properties and specificity)

HDACi are of clear interest because they overcome in part some limitations in pharmacokinetic/pharmacodynamic properties of the first generation HDACi (the first showing clinical benefits such as Vorinostat and Romidepsin), and therefore they may provide an explanation (if showing better clinical profiles) for the partially disappointing clinical results obtained so far.

Valproic acid (VPA) is a short chain fatty acid used as an antiepileptic and mood stabilizer (Johannessen and Johannessen, 2003). VPA has been observed to affect the growth of malignant cells *in vitro* and to prolong the G1 phase of the cell cycle (Bacon *et al.*, 2002; Cinatl *et al.*, 2002). Subsequently, it has been administered to patients with myelodysplastic syndrome (MDS) and AML, alone or in combination with RA as previous *in vitro* studies indicated a synergism of the two agents in inducing cellular differentiation and apoptosis in leukemia cell lines (Kuendgen and Gattermann, 2007). Despite the *in vitro* results, the clinical outcome did not recapitulate the improvement observed using both agents, and overall the clinical responses in phase I and II clinical trials were not frequent (Bug *et al.*, 2005; Cimino *et al.*, 2006). More promising results were obtained using VPA in combination with DNA demethylating agents such as azacytidine (5-azacytidine; Vidaza, Celgene, Corp.) and decitabine (5-aza-2′-deoxycytidine; Dacogen, MGI Pharma, Inc.) (Garcia-Manero *et al.*, 2006).

More recently, a phase I/II study was conducted in patients with advanced leukemia ($N = 46$ AML, $N = 6$ MDS) combining VPA and decitabine (Garcia-Manero *et al.*, 2006). Twenty-two percent of the patients had an objective response, including 19% with complete remissions (CRs) and 3% of CRs with incomplete platelet recovery (CRp). Transient DNA hypomethylation and global histone H3 and H4 acetylation were observed suggesting that the combination of epigenetic therapies rather than single compounds might be more efficient to revert aberrant epigenetic marks.

Short chain fatty acids (SCFA) comprise also sodium butyrate, a natural compound derived from anaerobic bacterial fermentation of dietary fibers. Sodium butyrate has been extensively characterized by *in vitro* studies where it was shown to induce cell cycle arrest and cellular differentiation (Marks *et al.*, 2000). Despite the fact that effectiveness has also been reported for colon, prostate, endometrial, and cervical carcinomas (Emenaker *et al.*, 2001; Hinnebusch *et al.*, 2002), the clinical employment of sodium butyrate has been hampered by the high (millimolar) concentrations required for antitumor activity. In order to achieve the same effect using clinically manageable doses, some derivaties of sodium butyrate have been synthesized such as phenylbutyrate (or Tributyrate–Ucyclyd Pharma, Scottsdale, AZ) and AN-9 (also known as pivaloyloxymethyl butyrate or Pivanex).

Although SCFA display pleiotropic effects, they are still widely studied as these compounds are already available in clinic for other medical conditions and can be orally administered (Melchior *et al.*, 1999). Phenylbutyrate is metabolized in the liver and kidney in phenyl acetate which is able to penetrate the central nervous system (Carducci et al., 1996; Piscitelli et al., 1995). This compound was initially introduced in clinical use to induce fetal erythropoiesis in the treatment of thalassemia, and later it has been demonstrated to induce differentiation and to inhibit the growth of primary leukemic cells in vitro. Clinical trials were reported in MDS and AML, but no complete or partial responses were observed (DiGiuseppe et al., 1999; Gore et al., 2001).

AN-9 (Titan, South San Francisco, CA) is a prodrug metabolized to butyric acid and aldehyde which inhibits growth, proliferation, and clonogenicity of a wide range of cancer cells achieving the same effect of butyric acid at considerably lower concentrations (Batova et al., 2002; Zimra et al., 2000). It has been demonstrated that AN-9 works in combination with doxorubicin to kill multidrug resistant cancer cells (Engel et al., 2006) and that its combination with radiation therapy significantly increased mortality of glioma cells (Entin-Meer et al., 2007). Phase I clinical trials in patients with advanced solid tumors revealed low toxicity and in phase II clinical trials AN-9 increased survival of non-small-cell lung carcinoma patients (Reid et al., 2004).

MS-275 (Syndax Pharmaceuticals, Waltham, MA) is a benzamide derivative described to preferentially target class I HDACs (Hess-Stumpp et al., 2007). It has been shown to inhibit the growth of leukemia cell lines, of primary blasts, and of prostate cancer cells (Lucas et al., 2004; Qian et al., 2007). Moreover, inhibitory effects were observed in seven out of eight tumor lines implanted into nude mice. Clinical studies in refractory leukemia patients (in particular AML) did not give positive results in terms of malignancy relapse, notwithstanding there were clear signs of target modulation (Gojo et al., 2007).

In a phase I study in patients with advanced and refractory solid tumors, stabilization was observed only in highly pretreated subjects (Ryan et al., 2005). This drug is currently undergoing phase II clinical trials, in the majority of cases in combination with other agents.

MGCD0103 (Mocetinostat–MethylGene, Inc., Saint Laurent, Quebec) inhibits specifically HDACs of classes I and IV (Bonfils et al., 2008; Fournel et al., 2008; Kell, 2007). The clinical studies showed a good efficacy of the compound in relapsed or refractory leukemia and in MDS: 3/23 AML patients achieved a complete bone marrow remission, and the inhibition of HDAC activity was found to be dose dependent, confirmed by a second clinical trial where the drug administrated at lower doses stabilized the disease (Garcia-Manero et al., 2008a; Stimson et al., 2009). Clinical development for this compound has been put on hold for some time, due to pericardial effusion observed in patients during clinical trials (most likely a side effect specific for this kind of molecule because it has not been observed with other HDACi): very recently, trials have been allowed to continue.

MS-275, VPA, and Mocetinostat are examples of so-called selective HDACi, since they target only a subset of HDACs, while the other HDACi described are thought to act as pan-HDACi. Technically, it is rather difficult to extrapolate the spectrum of selectivity of a given HDAC inhibitor in cells from biochemical assays *in vitro*, where HDACs are present in complexes and can be modulated by posttranslational modifications. In any case, at this stage it is unclear whether selective inhibition of specific HDACs is an advantage in cancer therapy, since it remains to be conclusively established (studying animal models and patient samples) (a) whether specific HDACs are indeed selectively involved in tumor maintenance and (b) whether side effects due to HDAC inhibition can be avoided by sparing HDACs that are neutral in the cancer process.

III. HISTONE METHYLATION

The methylation of histone tail residues is a heritable modification which influences the state of chromatin. In normal cells, these epigenetic changes regulate the expression level of genes in different tissues and other aspects of cell biology, such as chromosome X inactivation and imprinting. Among histone modifications, methylation and especially demethylation are poorly understood. The difficulty lies mainly in the fact that a particular modification is not unequivocally associated with a precise outcome in terms of gene regulation; instead, different types of modification act in concert and are context dependent. Indeed, histone methylation can be either an activating or repressing mark (Kouzarides, 2002; Morgunkova and Barlev, 2006; Qian and Zhou, 2006; Rice and Allis, 2001; Trievel, 2004), depending on the site and degree of methylation. New hints in understanding the complexity of chromatin organization come from genome-wide studies in which distinct localization and combinatorial patterns of histone marks in the genome were identified showing how these diverse modifications act in a cooperative manner to regulate globally gene expression (Barski *et al.*, 2007; Wang *et al.*, 2008).
      Histone methylation occurs on lysine and arginine residues and is mediated by histone lysine methyltransferases (HMTs) and protein arginine methyltransferases (PRMTs), respectively, which transfer a methyl group to their substrates from the cofactor S-adenosyl-L-methionine. Abnormal activity of these enzymes has been correlated with a variety of human disease states, including cancer (Glozak and Seto, 2007).

A. Lysine and arginine methyltransferases and PR-domain containing proteins

1. Lysine methyltransferases

HMTs consist of two main classes, the SET [Su(var)3-9, enhancer-of-zeste, Trithorax] domain containing family and the DOT1 family (Allis *et al.*, 2007; Letunic *et al.*, 2004). The first class carries out mono-, di-, or trimethylation of its target lysine residue, while members of the latter methylate lysine-79 within the core domain of H3 only in nucleosomal substrates and not in free histones (Feng *et al.*, 2002). In general, lysine methylation at histone H3 K9 and K27 and histone H4 K20 is associated with gene silencing, whereas methylation of histone H3 in K4, K36, and K79 is associated with gene activation (Martin and Zhang, 2005). However, it is now clear that these correlations are far from absolute and the presence of other histone modifications as well as the mono-, di-, tristate of methylation can alter the functional consequences of these modifications.

The major group of lysine methyltransferases contains a conserved methyltransferase domain termed SET that was originally identified as a suppressor of position effect variegation in *Drosophila*. The homologues in mice (Suv39h1 and Suv39h2) have catalytic activity methylating lysine 9 of histone H3 (Rea *et al.*, 2000). It has also been demonstrated that Suv39h1 modulates chromatin organization; in fact by imposing the histone mark, it creates a scaffold for the heterochromatin-associated HP1 protein, that in turn propagates H3K9 methylation for the establishment of repressive heterochromatin (Bannister *et al.*, 2001).

Lysine methyltransferases comprise also the human polycomb proteins, enhancer of zeste 1 and 2 (EZH1, EZH2), and the human trithorax proteins MLL1-3 (Sparmann and van Lohuizen, 2006). Both classes of enzymes were described in *Drosophila* for the first time: the polycomb (PcG) as repressors of homeotic (hox) gene activity, whereas the trithorax (trxG) group as activators of hox gene activity beyond mid-embryogenesis (Fig. 13.5).

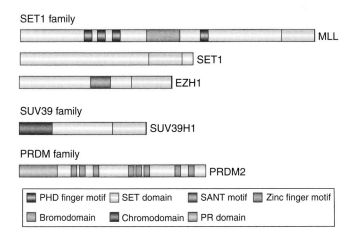

Figure 13.5. Prototype members of the various families of histone methyltransferases (HMTs). Conserved structural features are indicated according to the legend showed at the bottom.

2. Arginine methyltransferases

Arginine can be either mono- or dimethylated, with the latter in symmetric or asymmetric configurations. Arginine methyltransferases form two classes: type I and type II. Type II (PRMT 5, 7, and 9) catalyze the formation of symmetric methylation of the histone H4 tail, whereas the formation of asymmetric tails is maintained by type I (PRMT 1–4, 6, and 8) enzymes.

3. PR-domain containing proteins

Recently, a new class of candidate histone modifying proteins has been characterized, consisting of 17 members that share the N-terminal PR-domain (PR-domain containing proteins—PRDMs) among the protein family (Fig. 13.5). From the structural point of view, PRDM proteins are DNA binding factors, owing to the carboxy-terminal Krüppel-type zinc fingers (C2H2 zinc fingers) subdomain (Davis *et al.*, 2006), which is present in nearly all of the members of the family. In addition, PRDM proteins harbor a PR-domain toward the N-terminus, which is a motif with 20–30% of sequence identity to the SET domain of histone lysine methyltransferases, and are structurally related to it. The enzymatic activity of PRDM proteins remains to be fully elucidated, and it has already been demonstrated only for a few members of the family. In particular, PRDM2 (RIZ1) methylates histone H3 on lysine 9 (Kim *et al.*, 2003) and PRDM6 methylates histone H4 on lysine 20 (Funabiki *et al.*, 1994). PRDM1/Blimp-1, a direct DNA binding protein, is capable of acting as a scaffold for the recruitment of multiple chromatin modifying enzymes, much like a corepressor complex does. For example, it has been demonstrated that PRDM1 assembles silent chromatin over the interferon-β (IFN-β) promoter in U2OS cell line through the recruitment of histone H3 lysine methyltransferase G9a (Calame *et al.*, 2003).

B. Histone demethylases

Quite recently, it has been discovered that histone methylation can be reversibly modified by histone demethylases (Fig. 13.7) (Shi *et al.*, 2004). Lysine-specific demethylase-1 (LSD1) is an amine oxidase type protein that works in concert with methyltransferases to maintain global histone methylation patterns (Culhane and Cole, 2007). LSD1 associates with transcriptional repression, and in fact it was originally identified as a histone H3 K4 demethylase repressing its own targets. Due to the mechanisms of the oxidative reaction leading to demethylation (that also leads to release of potential signaling molecules: formaldehyde and hydrogen peroxide), LSD1 shows specificity toward mono- and dimethylated H3K4 while it is not able to turn over the trimethylated mark (Shi *et al.*, 2004). Although it has also been proposed to act as a transcriptional activator removing H3K9 methylation in order to activate androgen receptor target genes, this latter mechanism has not been demonstrated, and it appears to be due to associated demethylases of the JmjC family (see below) (Metzger *et al.*, 2005).

The search for other histone demethylases is an area of intensive investigation, since it is expected that different residues and different degrees of methylation are target of diverse enzymes. Following the identification of LSD1, Jumonji (Jmj) domain proteins have been discovered, a group of enzymes that target trimethylated lysine residues through an oxidative reaction different than the one carried out by LSD1 (Tsukada *et al.*, 2006).

Five JmjC domain subfamilies have been shown so far to carry out histone demethylation: JHDM1, JHDM2, JMJD2, JMJD3, and JARID1. Together these enzymes represent a key addition to the panel of chromatin modifying enzymes and are the subject of active research.

C. Alteration of histone methylation in cancer

1. Histone methyltransferases

The genes codifying for histone methyltransferases and histone demethylases can be targeted by mutations and expression changes in human cancers. As a consequence of alterations in the activity of histone methylating enzymes, the loss of trimethylation at lysine 20 of histone H4 has been described as a common hallmark of cancer (Fraga *et al.*, 2005). Additionally, alterations occurring in histone H3 K9 and K27 methylation patterns have been associated with aberrant gene expression in various forms of cancer (Okada *et al.*, 2005).

In genetic animal models, double knockout of Suv39h1/h2 in mice compromises genomic stability as a consequence of a greatly reduced level of histone H3 K9 levels (Peters *et al.*, 2001); animals develop late-onset B cell lymphomas with a similar phenotype of those in human slow-progressing NHLs. Consistently, human SUV39H1/2 play a role as regulators of cycline E gene expression, interacting *in vivo* with the protein Rb for the G1- to S-phase transition in the cell cycle. The failure of this interaction may be involved in cancer progression when Rb is mutated (Nielsen *et al.*, 2001).

The most frequently mutated members of the SET1 subfamily (human polycomb and trithorax proteins) in cancer belong to the trxG group: MLL is a frequent target of chromosomal rearrangements found in human leukemias, yielding multiple aberrant fusion proteins (Ziemin-van der Poel *et al.*, 1991). Approximately 30 different chromosomal fusions interest the chromosomal band in which MLL1 gene is located (11q23), identifying it as a fragile site predisposed to chromosomal rearrangements (So *et al.*, 2004). The carboxy-terminal SET domain of wild-type MLL, which associates with three highly conserved core components (RbBP5, Ash2L, and WDR5) and mediates H3K4 trimethylation

(Lund and van Lohuizen, 2004) is replaced by sequences from fusion partner genes (Cozzio *et al.*, 2003). Notably, all the fusion products encompass the N-terminal 1400 amino acids of MLL1 and lack the SET domain (Zeleznik-Le *et al.*, 1994). Although deletion of the SET domain would indicate a loss of transactivation function, the resulting MLL fusion proteins act as dominant-positive transcriptional regulators that can bind DNA and aberrantly activate, via unknown mechanisms, the expression of MLL downstream targets such as Hox genes, a group of transcription factors involved in embryonic development and hematopoietic cell differentiation (Zeleznik-Le *et al.*, 1994).

As previously mentioned, the SET-domain proteins include also the human polycomb proteins EZH1 and EZH2. In particular, EZH2 is part of the polycomb repressive complex 2 (PRC2), which also includes EED, SUZ12, RbAp48, and is involved in silencing (Sparmann and van Lohuizen, 2006). EZH2 contains the highly conserved SET domain and exerts its silencing function through methylation of H3K27. The study of genes silenced by PcG shows a strong preference for genes involved in cell fate decisions, including genes belonging to Hox, Notch, Hedgehog, Wnt, and TGF signaling pathways. EZH2 overexpression has been observed in prostate, breast, endometrium, and bladder cancer. In breast cancer, its overexpression is associated with poor prognosis. Although the mechanism of action of EZH2 in cancer is not yet clear, it appears to play a role in the regulation of pRB-E2F growth control pathway and of genes involved in homologous recombination pathway of DNA repair (Zeidler *et al.*, 2005). A recent study has shown that EZH2 is also overexpressed in preneoplastic breast lesions, as well as morphologically normal breast epithelium adjacent to the preinvasive and invasive lesions, and may therefore mark epithelium at higher risk for neoplastic transformation (Ding *et al.*, 2006). Surprisingly, EZH2 has been found mutated in B cell lymphomas and these mutations decrease its enzymatic activity, leading to the conclusion that alterations in either sense of the activity of this enzyme can exert prooncogenic functions (Martinez-Garcia and Licht, 2010; Morin *et al.*, 2010).

Increased levels of G9a (H3K9 KMT) are described in human liver cancers and have been implicated in maintenance of the malignant phenotype (Kondo *et al.*, 2007); the knockdown of G9a in cancer cells inhibited cell growth and led to morphologically senescent cells with telomere abnormalities (Kondo *et al.*, 2008a).

Although alterations of arginine methyltransferases have been correlated with diverse type of tumors, no mutations were identified to affect them. Knockdown experiments of PRMT1 supported its oncogenic role in MLL-mediated transformation (Cheung *et al.*, 2007). PRMT4 (also known as CARM1) is overexpressed in hormone-dependent tumors (cancer and prostate) and its knockdown impairs androgen receptor signaling (Majumder *et al.*, 2006). PRMT5 negatively regulates the expression of tumor suppressor genes in fibroblast cells and participates in the maintenance of the tumorigenic state (Pal *et al.*, 2004).

2. PR-domain containing proteins

Deregulation of members of the PR-domain containing proteins (PRDM) family occurs in various types of cancers, including multiple myeloma and lymphomas. PRDM proteins are often associated with transcriptional repression and are believed to act as tumor suppressors. Remarkably, both transcriptional repression ability and tumor suppressor role of various PRDM proteins rely on an intact PR/SET domain (Fears et al., 1996; Morishita et al., 1988; Steele-Perkins et al., 2001) and, in human tumors, naturally occurring truncated isoforms of three of the PRDM gene family proteins have been identified so far, in which the PR-domain is disrupted (Gyory et al., 2003; Liu et al., 1997).

For example, the overexpression of the PR-lacking (PR-) isoform of PRDM16 (sPRDM16) induces AML in mice. Notably, PRDM16 was firstly identified as involved in genomic translocation in patients (Lahortiga et al., 2004).

The human PRDM1 gene contains an alternative promoter capable of generating a PR-domain deleted protein (called PRDM1-β), which lacks 101 amino acids comprising most of the regulatory domain. The function of this isoform is currently unknown as it is found abundantly expressed in human myelomas (Gyory et al., 2003) but it has not been described in normal mouse development (apparently there is not a PRDM1β counterpart in mouse). In multiple myeloma cell lines (U266 and NCI-H929), PRDM1β has a significantly impaired transcription repressor function on multiple target genes (Gyory et al., 2003).

Deregulation of PRDMs' expression by epigenetic inactivation has been identified to be associated with cancer progression: the methylation of the PRDM2 (or RIZ1) promoter has been detected in several cancer cell lines and in liver and breast tumors whereas PRDM5 is most frequently silenced in colorectal and gastric cancer cell lines and also in some primary cancers.

3. Histone demethylases

Several observations support a role for LSD1 in tumorigenesis: (a) It is over-expressed in many solid tumors. LSD1 overexpression has been correlated with an adverse clinical outcome in neuroblastoma, prostate, and colon cancers, suggesting a tumor-promoting role for LSD1 (Kahl et al., 2006). In addition, inhibition of LSD1 suppresses colon cancer and neuroblastoma growth in vitro and in xenograft mouse models (Huang et al., 2007; Schulte et al., 2009). A recent study showed that LSD1 is highly expressed in breast cancer and correlates with tumor grades and poor clinical outcome (Lim et al., 2010). (b) LSD1 is actively engaged in key differentiation processes. Crucial is the recent discovery that LSD1 can tightly associate to the metastasis tumor antigen (MTA) proteins, which have been repeatedly reported to be overexpressed in a wide range of human cancers and directly implicated in metastasis. In particular, LSD1/MTA is recruited by the NuRD nucleosome-remodeling complex to regulate pathways that critically control cell proliferation (Wang *et al.*, 2009a). (c) Work done in our group demonstrated that LSD1/2 inhibitors can strongly potentiate the effect of RA in the treatment of APL (Binda *et al.*, 2010; Huang *et al.*, 2009).

Among JmjC family members, UTX and JARID1C somatic mutations have been identified, strongly supporting a direct role for these enzymes in tumorigenesis (Dalgliesh *et al.*, 2010; van Haaften *et al.*, 2009). JARID1B and JMJD2C are overexpressed in breast, testis cancer, and esophageal squamous carcinoma (Barrett *et al.*, 2002; Yamane *et al.*, 2007). Knockdown of JMJD2C resulted in the inhibition of cell proliferation (Cloos *et al.*, 2006). Additional studies on the biology of JmjC proteins and their role in cancer are urgently needed, and indeed these proteins are subject to intense investigation.

D. Histone methyltransferases and demethylase inhibitors

From the results presented so far, it is clear that the use of specific inhibitors blocking histone methyltransferases (Fig. 13.6) or demethylases (Fig. 13.8) is of great interest, but of uncertain outcome, given the dual properties of these enzymes. The search for histone methyltransferase inhibitors is still in its infancy. There are several major obstacles in transferring biochemical studies to the design of compounds. First of all, there is the lack of structural information for enzyme–inhibitor complex needed for the creation of molecular modeling and docking studies in order to understand the structural requirements for inhibitor binding. In the second instance, these compounds often lack specificity, indiscriminately targeting all enzymes that use S-adenosyl-L-methionine as cofactor.

The available rat PRMT1 and PRMT3 crystal structures have been used to provide a structural explanation for the inhibitory activity of a set of synthesized analogs of AMI-1. AMI-1 was found to inhibit substrate *in vitro* methylation by Hmt1p, the yeast homolog of PRMT1 (Cheng *et al.*, 2004) and docking studies have proposed its binding at the substrate and SAM cofactor binding pocket (Spannhoff *et al.*, 2007). The same screening led to the identification of nine AMI compounds able to inhibit all arginine methyltransferases: two of them (AMI-1 and 6) show specific inhibition of arginine methyltransferases. Following structural and biochemical studies, AMIs were used to model the design of new inhibitors. For example, AMI-5 analogs have been reported as both lysine and arginine methyltransferase inhibitors (Mai *et al.*, 2007) and for structural similarities they were also tested against other enzymes (SIRT1/2 and p300/CBP) demonstrating that some of them behave as multiple epigenetic ligands (epi-MLs) (Mai *et al.*, 2008). Recently, a new model utilizes PRMT1 as a template for the virtual screening of a compound library (Cheng

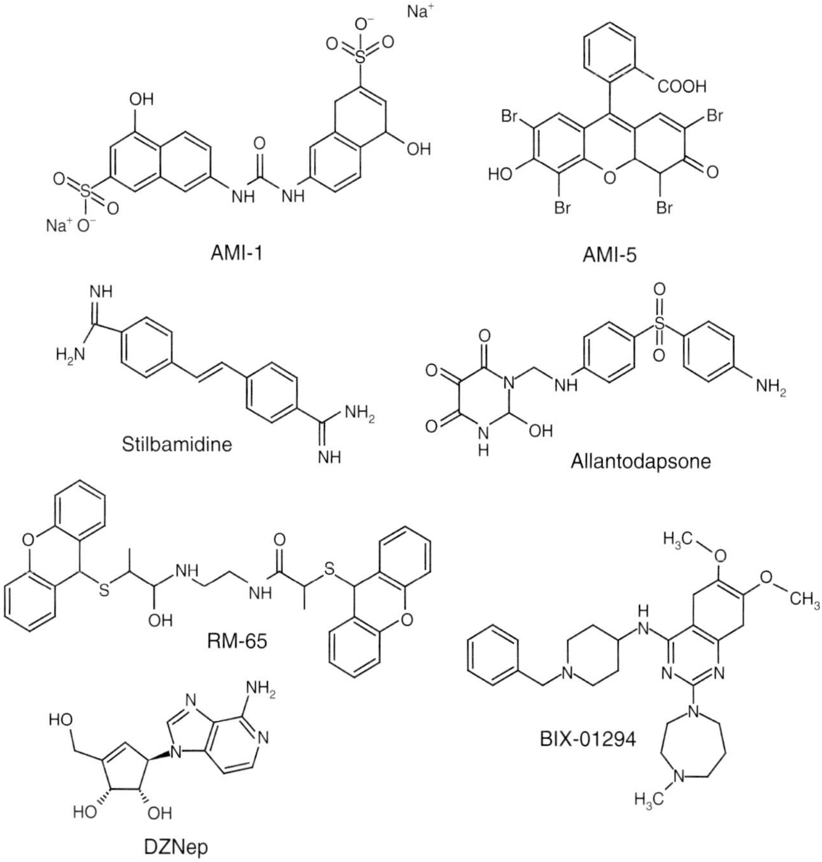

Figure 13.6. Chemical structure of histone methyltransferase inhibitors: AMIs, stilbamidine, and allantodapsone are potent inhibitors of arginine methyltransferases. RM-65 was selected to inhibit RmtA/PRMT1. DZNep and BIX-01294 are inhibitors of lysine methyltransferases.

et al., 2004). These molecules that show competition with the substrate were then assayed for inhibitory effects in cell culture and two of them (stilbamidine and allantodapsone) displayed a strong hypomethylating effect (Paik and Kim, 1973, 1974). The inhibitory effect predicted by ligand docking is due to the interaction of a basic or polar group with the acidic residue Glu 152. The basic moiety mimics the guanidine side of the substrate peptide. Other fragment-based virtual screenings gave rise to a novel lead structure against Rtma/PRMT1. The lead structure, an α-methylthioglycolic amide, displays chemical instability while a new compound (RM65) possessing the same substructure inhibits both RmtA (the fungal enzyme) and PRMT1 with equal strength (Spannhoff *et al.*, 2007).

To find small molecules that affect HMT function, a small library of compounds was screened for their ability to inhibit the activity of recombinant *Drosophila melanogaster* SU(VAR)3-9 protein. One of the strongest inhibitors was the fungal mycotoxin chaetocin that was initially isolated from the fermentation broth of *Chaetomium minutum*. The compound did not inhibit E(Z) or SET7/9 at low concentrations (below 90 μM); however, an inhibitory effect toward G9a (2.5 μM) was observed, and it showed cytotoxic effects in cultured *Drosophila melanogaster* cells (Greiner *et al.*, 2005). For the histone methyltransferase G9a inhibition, the BIX-01294 compound was selected for its specificity without affecting SUV39H1 or PRMT1. *In vitro* tests demonstrate that cell lines treated with 2.7 μM BIX-01294 decrease the content of H3K9 dimethylation without affecting the mono- or trimethyl states. A similar readout was obtained in G9a null cells confirming the specificity of the inhibitor (Chang *et al.*, 2009).

Recently, 3-deazaneplanocin A (DZNep) was shown to selectively inhibit H3K27me3 and H4K20me3 as well as to induce apoptosis in cancer cells. However, DZNep is a known S-adenosylhomocysteine (AdoHcy) hydrolase inhibitor, which leads to the indirect inhibition of S-adenosyl-methionine (AdoMet)-dependent reactions, including those carried out by many methyltransferases (Tan *et al.*, 2007).

Inhibitors of histone demethylases have also been synthesized. Due to the similarity of LDS1 with monoamine oxidase (MAO) A and B enzymes, MAO inhibitors (used clinically for central nervous system disorders) have been a starting point for the search of LSD1 inhibitors (Kanazawa, 1994). Both MAO and LSD1 are flavin-dependent enzymes; they remove amine groups in a similar oxidative manner and they show homology in the catalytic site. For these reasons known MAO inhibitors were tested against LSD1 and some compounds showed an inhibitory effect on bulk histones. However, only phenelzine and tranylcypromine show a high potency at the lowest dose (21 μM) (Schmidt and McCafferty, 2007). Tranylcypromine forms a covalent binding with FAD cofactor leading to the formation of an irreversible inhibitor–cofactor intermediate. More recently, tranylcypromine derivatives with a much higher potency against LSD1/2 and reduced activity against MAO have been synthesized and show to induce differentiation and apoptosis of APL cells, alone or in combination with RA (Binda *et al.*, 2010).

Taken together, the attempts of inhibition of histone methylating and demethylating enzymes are still at the beginning as compared, for example, to HDACi: we expect that in the forthcoming 2–3 years a burst of novel tools/candidate drugs with improved potency and drug-like properties will emerge and will allow more advanced studies (including *in vivo* studies) to fully validate this modification as a target for cancer therapy.

IV. PERSPECTIVES

At the end of this survey, it is clear that the concept of histone modification therapy of cancer is an "in progress" story, with, hopefully, several parallel lines of investigations finally beginning to converge. Clearly more time is needed to definitively claim this approach as a success or a failure.

It is remarkable how it is now largely expected for epigenetic therapies to work. Looking at the "arsenal of weapons" in our hands, it should not be forgotten that many of them were not born by design, but rather became available by chance. Indeed, to think that inhibiting an entire class of enzymes (such as HDACi do, including approved drugs) would lead to a clinical benefit appears a "quasi-heretical" view, and probably without the fortuitous discovery that compounds with interesting biological properties (TSA) were pan-HDACi, the entire field would have followed different routes. However, the drawback of this jump-start is that too many expectations have been casted on the possibility to immediately reach optimal use of epigenetic drugs. In turn, this has led to disappointment after the clinical results with HDACi have shown that—as for all anticancer therapies—the way to follow is a long one.

On purpose, we have treated so far histone modifications and histone modifying enzymes as separate areas, but the idea of an "epigenetic therapy" without combining drugs able to modulate the multiple blocks of epigenetic information is a nonsense. Regulatory and practical reasons limit the use of combination therapies with epigenetic drugs for now, but there is no doubt that the future of this approach lies in the possibility to carefully modulate the epigenome with a combination of drugs (Fig. 13.9).

A few examples are already available. In APL, treatment with HDACi and RA is extremely effective in derepressing RA target genes and inducing cellular responses to RA both *in vitro* and *in vivo*. Another approach is to combine hypomethylating agents with HDACi. The rational for this combination emerged from studies demonstrating synergistic effects in the reactivation of epigenetically silenced genes. The treatment, mainly explored in preclinical trials, has been investigated also in clinical trials in AML and MDS patients using valproic acid and phenylbutirate in one case and in a phase I/II study using vorinostat and 5-azacitidine, with extremely encouraging results. Once other inhibitors become available for both animal and clinical studies (histone methylation modulators first and foremost), we may expect a quantum leap in efficacy of these approaches (Figs. 13.7–13.9).

As repeatedly stated in the text, the classification of these approaches as "epigenetic therapies" is an oversimplification, since they combine epigenetic and nonepigenetic effects through their effect on nonhistone substrates. We cannot avoid coping with this "natural imperfection"; however, more subtle ways to design epigenetic drugs may be found. As an example, three-dimensional

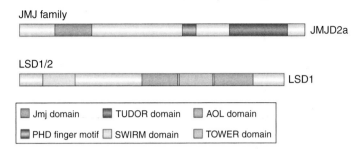

JMJ family

JMJD2a

LSD1/2

LSD1

| | Jmj domain | | TUDOR domain | | AOL domain |
| | PHD finger motif | | SWIRM domain | | TOWER domain |

Figure 13.7. Prototype members of the various families of histone demethylases (HDMs). Conserved structural features are indicated according to the legend showed at the bottom.

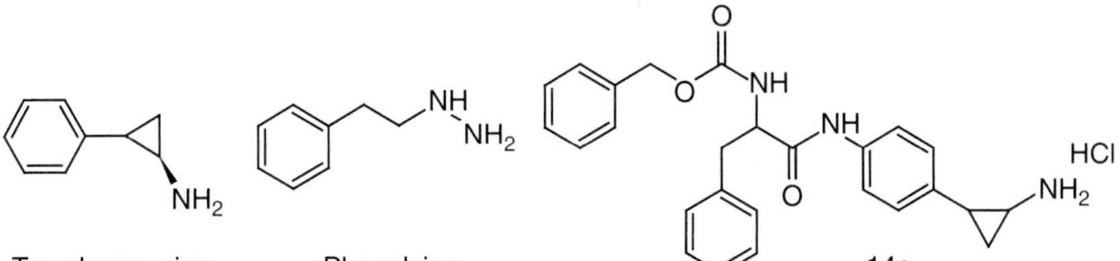

Tranylcypromine   Phenelzine   14e

Figure 13.8. Chemical structure of histone demethylase inhibitors. See text for a more detailed description of the compounds.

structures of the bromodomains have been reported. The bromodomain is part of chromatin-associated proteins and HATs known to bind acetyl-lysine motifs. The solved structure of bromodomain reveals a poor overall sequence similarity among the family, but the distinct structural fold conformation is quite maintained with a left-handed four-helix bundle in which two interhelical loops constitute a hydrophobic pocket recognizing the acetyl-lysine. The two loops create a hydrophobic cavity that stabilizes the structure and that interacts with the acetyl-lysine. This pocket was used to model molecules blocking PCAF and CBP bromodomain that respectively bind to acetyl-lysine 50 of HIV Tat protein and to acetyl-lysine 382 of p53 tumor suppressor. The synthesis of these compounds represents a potent research instrument and the proof of principle that manipulating chromatin recognition motifs may be a more selective approach to the epigenome (Sanchez and Zhou, 2009).

Other than the acquisition of genome mutations, very recent works highlight the possibility of different routes to drug resistance in tumor cells, including dynamic chromatin modifications based on the phenomena of drug

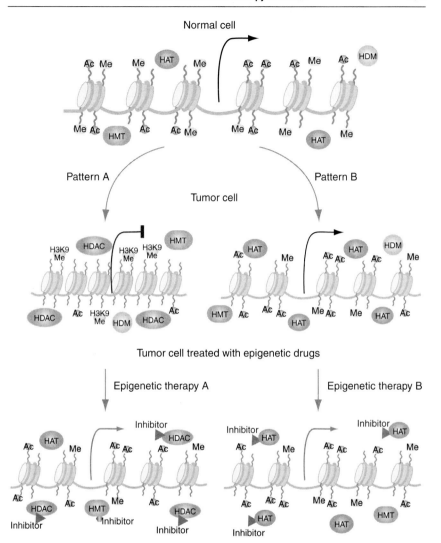

Figure 13.9. Chromatin alterations occurring in cancer, and the emerging potential of epigenetic therapy. The degree of histone modifications influences transcriptional processes in mammalian cells. The aberrant activity of chromatin modifying enzymes occurring in cancer cells is often associated with silencing of tumor suppressor genes (pattern A), activation of oncogenes (pattern B), and in general with altered gene expression. The pharmacologically reversible nature of these modifications makes them a promising target for the development of epigenetic therapy.

resistance. A study focuses on the drug resistance of PC9 non-small-cell lung cancer cell line carrying an activating mutation in the epidermal growth factor receptor (EGFR) not responsive to after the exposure to erlotinib that abrogates EGFR signaling. The resistance of a small fraction of cells—so-called persisters (0.3%)—is reported as a different chromatin configuration compared with the sensible cells, and with the expression of stem cell-like markers (CD133 and CD24). Supporting this, the persisters after expansion were cultured in the absence of erlotinib and gave rise to drug-tolerant cells. After gene expression profiling of the different cell populations, the authors focused the attention on the upregulation of the KDM5A histone demethylase observed in drug-tolerant and expanded persisters with respect to the parental ones. The fact that the yeast ortholog of KDM5A is known to reduce H3K14 acetylation combined with the observed H3K14 decrease in drug-tolerant cells and expanded persisters gave the rational for the treatment with different HDACi that surprisingly caused a rapid death of drug-tolerant and expanded persisters without affecting the parental cell line. Epigenetic treatments, therefore, can also be used to expand and potentiate the effects of traditional or nonepigenetic targeted drugs (Sharma et al., 2010).

These results imply that the search for a responder hypothesis (i.e., to identify specific epigenetic alterations underlying sensitivity to specific epigenetic treatments) should be the leading principle for the next generation of studies in this field: new technologies are becoming rapidly available for the study of the epigenome of patient samples (Gargiulo and Minucci, 2009; Gargiulo et al., 2009), that should make this a feasible approach for the clinical treatment of cancer as a complex "epigenetic" disease. Last but not least, so far most epigenetic approaches to therapy have neglected the possibility that distinct subpopulations of cancer cells (such as cancer stem cells) may respond differently to treatment with epigenetic drugs: this will surely be another critical issue needed to be addressed aggressively, to optimize clinical strategies.

Acknowledgments

Research in SM laboratory is supported by AIRC (Associazione Italiana Ricerca sul Cancro), EEC (Epitron, EU), MIUR (Ministero dell'Istruzione, Universita' e Ricerca), MIS, Cariplo.

We thank Alicja Gruszka for constructive criticisms and reviewing the text.

C. B., G. F., and S. M. are members of the IFOM-IEO Campus for Oncogenomics, Via Adamello 16, Milano.

References

Ait-Si-Ali, S., Polesskaya, A., Filleur, S., Ferreira, R., Duquet, A., Robin, P., Vervish, A., Trouche, D., Cabon, F., and Harel-Bellan, A. (2000). CBP/p300 histone acetyl-transferase activity is important for the G1/S transition. *Oncogene* **19**(20), 2430–2437.

Allis, C. D., Berger, S. L., Cote, J., Dent, S., Jenuwien, T., Kouzarides, T., Pillus, L., Reinberg, D., Shi, Y., Shiekhattar, R., *et al.* (2007). New nomenclature for chromatin-modifying enzymes. *Cell* **131**(4), 633–636.

Arany, Z., Sellers, W. R., Livingston, D. M., and Eckner, R. (1994). E1A-associated p300 and CREB-associated CBP belong to a conserved family of coactivators. *Cell* **77**(6), 799–800.

Avvakumov, N., and Cote, J. (2007). The MYST family of histone acetyltransferases and their intimate links to cancer. *Oncogene* **26**(37), 5395–5407.

Ayroldi, E., Blasi, E., Varesio, L., and Wiltrout, R. H. (1992). Inhibition of proliferation of retrovirus-immortalized macrophages by LPS and IFN-gamma: Possible autocrine down-regulation of cell growth by induction of IL1 and TNF. *Biotherapy* **4**(4), 267–276.

Bacon, C. L., Gallagher, H. C., Haughey, J. C., and Regan, C. M. (2002). Antiproliferative action of valproate is associated with aberrant expression and nuclear translocation of cyclin D3 during the C6 glioma G1 phase. *J. Neurochem.* **83**(1), 12–19.

Balasubramanyam, K., Altaf, M., Varier, R. A., Swaminathan, V., Ravindran, A., Sadhale, P. P., and Kundu, T. K. (2004a). Polyisoprenylated benzophenone, garcinol, a natural histone acetyltransferase inhibitor, represses chromatin transcription and alters global gene expression. *J. Biol. Chem.* **279**(32), 33716–33726.

Balasubramanyam, K., Varier, R. A., Altaf, M., Swaminathan, V., Siddappa, N. B., Ranga, U., and Kundu, T. K. (2004b). Curcumin, a novel p300/CREB-binding protein-specific inhibitor of acetyltransferase, represses the acetylation of histone/nonhistone proteins and histone acetyltransferase-dependent chromatin transcription. *J. Biol. Chem.* **279**(49), 51163–51171.

Bandyopadhyay, D., Okan, N. A., Bales, E., Nascimento, L., Cole, P. A., and Medrano, E. E. (2002). Down-regulation of p300/CBP histone acetyltransferase activates a senescence checkpoint in human melanocytes. *Cancer Res.* **62**(21), 6231–6239.

Bannister, A. J., Miska, E. A., Gorlich, D., and Kouzarides, T. (2000). Acetylation of importin-alpha nuclear import factors by CBP/p300. *Curr. Biol.* **10**(8), 467–470.

Bannister, A. J., Zegerman, P., Partridge, J. F., Miska, E. A., Thomas, J. O., Allshire, R. C., and Kouzarides, T. (2001). Selective recognition of methylated lysine 9 on histone H3 by the HP1 chromo domain. *Nature* **410**(6824), 120–124.

Barrett, A., Madsen, B., Copier, J., Lu, P. J., Cooper, L., Scibetta, A. G., Burchell, J., and Taylor-Papadimitriou, J. (2002). PLU-1 nuclear protein, which is upregulated in breast cancer, shows restricted expression in normal human adult tissues: A new cancer/testis antigen? *Int. J. Cancer* **101**(6), 581–588.

Bar-Sela, G., Epelbaum, R., and Schaffer, M. (2010). Curcumin as an Anti-Cancer Agent: Review of the Gap between Basic and Clinical Applications. *Curr. Med. Chem.* **17**(3), 190–197.

Barski, A., Cuddapah, S., Cui, K., Roh, T. Y., Schones, D. E., Wang, Z., Wei, G., Chepelev, I., and Zhao, K. (2007). High-resolution profiling of histone methylations in the human genome. *Cell* **129**(4), 823–837.

Batova, A., Shao, L. E., Diccianni, M. B., Yu, A. L., Tanaka, T., Rephaeli, A., Nudelman, A., and Yu, J. (2002). The histone deacetylase inhibitor AN-9 has selective toxicity to acute leukemia and drug-resistant primary leukemia and cancer cell lines. *Blood* **100**(9), 3319–3324.

Bayes, M., Rabasseda, X., and Prous, J. R. (2007). Gateways to clinical trials. *Methods Find Exp. Clin. Pharmacol.* **29**(7), 467–509.

Biel, M., Kretsovali, A., Karatzali, E., Papamatheakis, J., and Giannis, A. (2004). Design, synthesis, and biological evaluation of a small-molecule inhibitor of the histone acetyltransferase Gcn5. *Angew. Chem. Int. Ed Engl.* **43**(30), 3974–3976.

Binda, C., Valente, S., Romanenghi, M., Pilotto, S., Cirilli, R., Karytinos, A., Ciossani, G., Botrugno, O. A., Forneris, F., Tardugno, M., *et al.* (2010). Biochemical, structural, and biological evaluation of tranylcypromine derivatives as inhibitors of histone demethylases LSD1 and LSD2. *J. Am. Chem. Soc.* **132**(19), 6827–6833.

Bonfils, C., Kalita, A., Dubay, M., Siu, L. L., Carducci, M. A., Reid, G., Martell, R. E., Besterman, J. M., and Li, Z. (2008). Evaluation of the pharmacodynamic effects of MGCD0103 from preclinical models to human using a novel HDAC enzyme assay. *Clin. Cancer Res.* **14**(11), 3441–3449.

Borrow, J., Stanton, V. P., Jr., Andresen, J. M., Becher, R., Behm, F. G., Chaganti, R. S., Civin, C. I., Disteche, C., Dube, I., Frischauf, A. M., *et al.* (1996). The translocation t(8;16)(p11;p13) of acute myeloid leukaemia fuses a putative acetyltransferase to the CREB-binding protein. *Nat. Genet.* **14**(1), 33–41.

Bradbury, C. A., Khanim, F. L., Hayden, R., Bunce, C. M., White, D. A., Drayson, M. T., Craddock, C., and Turner, B. M. (2005). Histone deacetylases in acute myeloid leukaemia show a distinctive pattern of expression that changes selectively in response to deacetylase inhibitors. *Leukemia* **19**(10), 1751–1759.

Brownell, J. E., Zhou, J., Ranalli, T., Kobayashi, R., Edmondson, D. G., Roth, S. Y., and Allis, C. D. (1996). Tetrahymena histone acetyltransferase A: A homolog to yeast Gcn5p linking histone acetylation to gene activation. *Cell* **84**(6), 843–851.

Bug, G., Ritter, M., Wassmann, B., Schoch, C., Heinzel, T., Schwarz, K., Romanski, A., Kramer, O. H., Kampfmann, M., Hoelzer, D., *et al.* (2005). Clinical trial of valproic acid and all-trans retinoic acid in patients with poor-risk acute myeloid leukemia. *Cancer* **104**(12), 2717–2725.

Butler, L. M., Agus, D. B., Scher, H. I., Higgins, B., Rose, A., Cordon-Cardo, C., Thaler, H. T., Rifkind, R. A., Marks, P. A., and Richon, V. M. (2000). Suberoylanilide hydroxamic acid, an inhibitor of histone deacetylase, suppresses the growth of prostate cancer cells in vitro and in vivo. *Cancer Res.* **60**(18), 5165–5170.

Butler, L. M., Webb, Y., Agus, D. B., Higgins, B., Tolentino, T. R., Kutko, M. C., LaQuaglia, M. P., Drobnjak, M., Cordon-Cardo, C., Scher, H. I., *et al.* (2001). Inhibition of transformed cell growth and induction of cellular differentiation by pyroxamide, an inhibitor of histone deacetylase. *Clin. Cancer Res.* **7**(4), 962–970.

Byrd, J. C., Marcucci, G., Parthun, M. R., Xiao, J. J., Klisovic, R. B., Moran, M., Lin, T. S., Liu, S., Sklenar, A. R., Davis, M. E., *et al.* (2005). A phase 1 and pharmacodynamic study of depsipeptide (FK228) in chronic lymphocytic leukemia and acute myeloid leukemia. *Blood* **105**(3), 959–967.

Calame, K. L., Lin, K. I., and Tunyaplin, C. (2003). Regulatory mechanisms that determine the development and function of plasma cells. *Annu. Rev. Immunol.* **21**, 205–230.

Camphausen, K., and Tofilon, P. J. (2007). Inhibition of histone deacetylation: A strategy for tumor radiosensitization. *J. Clin. Oncol.* **25**(26), 4051–4056.

Carducci, M. A., Nelson, J. B., Chan-Tack, K. M., Ayyagari, S. R., Sweatt, W. H., Campbell, P. A., Nelson, W. G., and Simons, J. W. (1996). Phenylbutyrate induces apoptosis in human prostate cancer and is more potent than phenylacetate. *Clin. Cancer Res.* **2**(2), 379–387.

Cedar, H., and Bergman, Y. (2009). Linking DNA methylation and histone modification: Patterns and paradigms. *Nat. Rev. Genet.* **10**(5), 295–304.

Chaffanet, M., Gressin, L., Preudhomme, C., Soenen-Cornu, V., Birnbaum, D., and Pebusque, M. J. (2000). MOZ is fused to p300 in an acute monocytic leukemia with t(8;22). *Genes Chromosom. Cancer* **28**(2), 138–144.

Champagne, N., Bertos, N. R., Pelletier, N., Wang, A. H., Vezmar, M., Yang, Y., Heng, H. H., and Yang, X. J. (1999). Identification of a human histone acetyltransferase related to monocytic leukemia zinc finger protein. *J. Biol. Chem.* **274**(40), 28528–28536.

Chan, H. M., and La Thangue, N. B. (2001). p300/CBP proteins: HATs for transcriptional bridges and scaffolds. *J. Cell Sci.* **114**(Pt. 13), 2363–2373.

Chang, Y., Zhang, X., Horton, J. R., Upadhyay, A. K., Spannhoff, A., Liu, J., Snyder, J. P., Bedford, M. T., and Cheng, X. (2009). Structural basis for G9a-like protein lysine methyltransferase inhibition by BIX-01294. *Nat. Struct. Mol. Biol.* **16**(3), 312–317.

Chen, W. Y., Wang, D. H., Yen, R. C., Luo, J., Gu, W., and Baylin, S. B. (2005). Tumor suppressor HIC1 directly regulates SIRT1 to modulate p53-dependent DNA-damage responses. *Cell* **123**(3), 437–448.

Cheng, D., Yadav, N., King, R. W., Swanson, M. S., Weinstein, E. J., and Bedford, M. T. (2004). Small molecule regulators of protein arginine methyltransferases. *J. Biol. Chem.* **279**(23), 23892–23899.

Cheung, N., Chan, L. C., Thompson, A., Cleary, M. L., and So, C. W. (2007). Protein arginine-methyltransferase-dependent oncogenesis. *Nat. Cell Biol.* **9**(10), 1208–1215.

Choudhary, C., Kumar, C., Gnad, F., Nielsen, M. L., Rehman, M., Walther, T. C., Olsen, J. V., and Mann, M. (2009). Lysine acetylation targets protein complexes and co-regulates major cellular functions. *Science* **325**(5942), 834–840.

Cimino, G., Lo-Coco, F., Fenu, S., Travaglini, L., Finolezzi, E., Mancini, M., Nanni, M., Careddu, A., Fazi, F., Padula, F., *et al.* (2006). Sequential valproic acid/all-trans retinoic acid treatment reprograms differentiation in refractory and high-risk acute myeloid leukemia. *Cancer Res.* **66**(17), 8903–8911.

Cinatl, J., Jr., Kotchetkov, R., Blaheta, R., Driever, P. H., Vogel, J. U., and Cinatl, J. (2002). Induction of differentiation and suppression of malignant phenotype of human neuroblastoma BE(2)-C cells by valproic acid: Enhancement by combination with interferon-alpha. *Int. J. Oncol.* **20**(1), 97–106.

Cloos, P. A., Christensen, J., Agger, K., Maiolica, A., Rappsilber, J., Antal, T., Hansen, K. H., and Helin, K. (2006). The putative oncogene GASC1 demethylates tri- and dimethylated lysine 9 on histone H3. *Nature* **442**(7100), 307–311.

Coffey, D. C., Kutko, M. C., Glick, R. D., Butler, L. M., Heller, G., Rifkind, R. A., Marks, P. A., Richon, V. M., and La Quaglia, M. P. (2001). The histone deacetylase inhibitor, CBHA, inhibits growth of human neuroblastoma xenografts in vivo, alone and synergistically with all-trans retinoic acid. *Cancer Res.* **61**(9), 3591–3594.

Cozzio, A., Passegue, E., Ayton, P. M., Karsunky, H., Cleary, M. L., and Weissman, I. L. (2003). Similar MLL-associated leukemias arising from self-renewing stem cells and short-lived myeloid progenitors. *Genes Dev.* **17**(24), 3029–3035.

Culhane, J. C., and Cole, P. A. (2007). LSD1 and the chemistry of histone demethylation. *Curr. Opin. Chem. Biol.* **11**(5), 561–568.

Dalgliesh, G. L., Furge, K., Greenman, C., Chen, L., Bignell, G., Butler, A., Davies, H., Edkins, S., Hardy, C., Latimer, C., *et al.* (2010). Systematic sequencing of renal carcinoma reveals inactivation of histone modifying genes. *Nature* **463**(7279), 360–363.

Davis, C. A., Haberland, M., Arnold, M. A., Sutherland, L. B., McDonald, O. G., Richardson, J. A., Childs, G., Harris, S., Owens, G. K., and Olson, E. N. (2006). PRISM/PRDM6, a transcriptional repressor that promotes the proliferative gene program in smooth muscle cells. *Mol. Cell. Biol.* **26**(7), 2626–2636.

Dhordain, P., Lin, R. J., Quief, S., Lantoine, D., Kerckaert, J. P., Evans, R. M., and Albagli, O. (1998). The LAZ3(BCL-6) oncoprotein recruits a SMRT/mSIN3A/histone deacetylase containing complex to mediate transcriptional repression. *Nucleic Acids Res.* **26**(20), 4645–4651.

DiGiuseppe, J. A., Weng, L. J., Yu, K. H., Fu, S., Kastan, M. B., Samid, D., and Gore, S. D. (1999). Phenylbutyrate-induced G1 arrest and apoptosis in myeloid leukemia cells: Structure-function analysis. *Leukemia* **13**(8), 1243–1253.

Ding, L., Erdmann, C., Chinnaiyan, A. M., Merajver, S. D., and Kleer, C. G. (2006). Identification of EZH2 as a molecular marker for a precancerous state in morphologically normal breast tissues. *Cancer Res.* **66**(8), 4095–4099.

Doyon, Y., Cayrou, C., Ullah, M., Landry, A. J., Cote, V., Selleck, W., Lane, W. S., Tan, S., Yang, X. J., and Cote, J. (2006). ING tumor suppressor proteins are critical regulators of chromatin acetylation required for genome expression and perpetuation. *Mol. Cell* **21**(1), 51–64.

Duvic, M., and Vu, J. (2007). Vorinostat in cutaneous T-cell lymphoma. *Drugs Today* **43**(9), 585–599.

Emenaker, N. J., Calaf, G. M., Cox, D., Basson, M. D., and Qureshi, N. (2001). Short-chain fatty acids inhibit invasive human colon cancer by modulating uPA, TIMP-1, TIMP-2, mutant p53, Bcl-2, Bax, p21 and PCNA protein expression in an in vitro cell culture model. *J. Nutr.* **131**(11 Suppl.), 3041S–3046S.

Engel, D., Nudelman, A., Levovich, I., Gruss-Fischer, T., Entin-Meer, M., Phillips, D. R., Cutts, S. M., and Rephaeli, A. (2006). Mode of interaction between butyroyloxymethyl-diethyl phosphate (AN-7) and doxorubicin in MCF-7 and resistant MCF-7/Dx cell lines. *J. Cancer Res. Clin. Oncol.* **132**(10), 673–683.

Entin-Meer, M., Yang, X., VandenBerg, S. R., Lamborn, K. R., Nudelman, A., Rephaeli, A., and Haas-Kogan, D. A. (2007). In vivo efficacy of a novel histone deacetylase inhibitor in combination with radiation for the treatment of gliomas. *Neuro-oncology* **9**(2), 82–88.

Fantin, V. R., Loboda, A., Paweletz, C. P., Hendrickson, R. C., Pierce, J. W., Roth, J. A., Li, L., Gooden, F., Korenchuk, S., Hou, X. S., *et al.* (2008). Constitutive activation of signal transducers and activators of transcription predicts vorinostat resistance in cutaneous T-cell lymphoma. *Cancer Res.* **68**(10), 3785–3794.

Fears, S., Mathieu, C., Zeleznik-Le, N., Huang, S., Rowley, J. D., and Nucifora, G. (1996). Intergenic splicing of MDS1 and EVI1 occurs in normal tissues as well as in myeloid leukemia and produces a new member of the PR domain family. *Proc. Natl Acad. Sci. USA* **93**(4), 1642–1647.

Feng, Q., Wang, H., Ng, H. H., Erdjument-Bromage, H., Tempst, P., Struhl, K., and Zhang, Y. (2002). Methylation of H3-lysine 79 is mediated by a new family of HMTases without a SET domain. *Curr. Biol.* **12**(12), 1052–1058.

Finch, J. T., Lutter, L. C., Rhodes, D., Brown, R. S., Rushton, B., Levitt, M., and Klug, A. (1977). Structure of nucleosome core particles of chromatin. *Nature* **269**(5623), 29–36.

Finnin, M. S., Donigian, J. R., Cohen, A., Richon, V. M., Rifkind, R. A., Marks, P. A., Breslow, R., and Pavletich, N. P. (1999). Structures of a histone deacetylase homologue bound to the TSA and SAHA inhibitors. *Nature* **401**(6749), 188–193.

Fournel, M., Bonfils, C., Hou, Y., Yan, P. T., Trachy-Bourget, M. C., Kalita, A., Liu, J., Lu, A. H., Zhou, N. Z., Robert, M. F., *et al.* (2008). MGCD0103, a novel isotype-selective histone deacetylase inhibitor, has broad spectrum antitumor activity in vitro and in vivo. *Mol. Cancer Ther.* **7**(4), 759–768.

Fraga, M. F., Ballestar, E., Villar-Garea, A., Boix-Chornet, M., Espada, J., Schotta, G., Bonaldi, T., Haydon, C., Ropero, S., Petrie, K., *et al.* (2005). Loss of acetylation at Lys16 and trimethylation at Lys20 of histone H4 is a common hallmark of human cancer. *Nat. Genet.* **37**(4), 391–400.

Funabiki, T., Kreider, B. L., and Ihle, J. N. (1994). The carboxyl domain of zinc fingers of the Evi-1 myeloid transforming gene binds a consensus sequence of GAAGATGAG. *Oncogene* **9**(6), 1575–1581.

Furumai, R., Matsuyama, A., Kobashi, N., Lee, K. H., Nishiyama, M., Nakajima, H., Tanaka, A., Komatsu, Y., Nishino, N., Yoshida, M., *et al.* (2002). FK228 (depsipeptide) as a natural prodrug that inhibits class I histone deacetylases. *Cancer Res.* **62**(17), 4916–4921.

Galanis, E., Jaeckle, K. A., Maurer, M. J., Reid, J. M., Ames, M. M., Hardwick, J. S., Reilly, J. F., Loboda, A., Nebozhyn, M., Fantin, V. R., *et al.* (2009). Phase II trial of vorinostat in recurrent glioblastoma multiforme: A north central cancer treatment group study. *J. Clin. Oncol.* **27**(12), 2052–2058.

Gao, L., Cueto, M. A., Asselbergs, F., and Atadja, P. (2002). Cloning and functional characterization of HDAC11, a novel member of the human histone deacetylase family. *J. Biol. Chem.* **277**(28), 25748–25755.

Garcia-Manero, G., Kantarjian, H. M., Sanchez-Gonzalez, B., Yang, H., Rosner, G., Verstovsek, S., Rytting, M., Wierda, W. G., Ravandi, F., Koller, C., et al. (2006). Phase 1/2 study of the combination of 5-aza-2′-deoxycytidine with valproic acid in patients with leukemia. *Blood* **108** (10), 3271–3279.

Garcia-Manero, G., Assouline, S., Cortes, J., Estrov, Z., Kantarjian, H., Yang, H., Newsome, W. M., Miller, W. H., Jr., Rousseau, C., Kalita, A., et al. (2008a). Phase 1 study of the oral isotype specific histone deacetylase inhibitor MGCD0103 in leukemia. *Blood* **112**(4), 981–989.

Garcia-Manero, G., Yang, H., Bueso-Ramos, C., Ferrajoli, A., Cortes, J., Wierda, W. G., Faderl, S., Koller, C., Morris, G., Rosner, G., et al. (2008b). Phase 1 study of the histone deacetylase inhibitor vorinostat (suberoylanilide hydroxamic acid [SAHA]) in patients with advanced leukemias and myelodysplastic syndromes. *Blood* **111**(3), 1060–1066.

Gargiulo, G., and Minucci, S. (2009). Epigenomic profiling of cancer cells. *Int. J. Biochem. Cell Biol.* **41**(1), 127–135.

Gargiulo, G., Levy, S., Bucci, G., Romanenghi, M., Fornasari, L., Beeson, K. Y., Goldberg, S. M., Cesaroni, M., Ballarini, M., Santoro, F., et al. (2009). NA-Seq: A discovery tool for the analysis of chromatin structure and dynamics during differentiation. *Dev. Cell* **16**(3), 466–481.

Gayther, S. A., Batley, S. J., Linger, L., Bannister, A., Thorpe, K., Chin, S. F., Daigo, Y., Russell, P., Wilson, A., Sowter, H. M., et al. (2000). Mutations truncating the EP300 acetylase in human cancers. *Nat. Genet.* **24**(3), 300–303.

Glozak, M. A., and Seto, E. (2007). Histone deacetylases and cancer. *Oncogene* **26**(37), 5420–5432.

Glozak, M. A., Sengupta, N., Zhang, X., and Seto, E. (2005). Acetylation and deacetylation of non-histone proteins. *Gene* **363**, 15–23.

Goel, A., Kunnumakkara, A. B., and Aggarwal, B. B. (2008). Curcumin as "Curecumin": From kitchen to clinic. *Biochem. Pharmacol.* **75**(4), 787–809.

Gojo, I., Jiemjit, A., Trepel, J. B., Sparreboom, A., Figg, W. D., Rollins, S., Tidwell, M. L., Greer, J., Chung, E. J., Lee, M. J., et al. (2007). Phase 1 and pharmacologic study of MS-275, a histone deacetylase inhibitor, in adults with refractory and relapsed acute leukemias. *Blood* **109**(7), 2781–2790.

Gore, S. D., Weng, L. J., Zhai, S., Figg, W. D., Donehower, R. C., Dover, G. J., Grever, M., Griffin, C. A., Grochow, L. B., Rowinsky, E. K., et al. (2001). Impact of the putative differentiating agent sodium phenylbutyrate on myelodysplastic syndromes and acute myeloid leukemia. *Clin. Cancer Res.* **7**(8), 2330–2339.

Greiner, D., Bonaldi, T., Eskeland, R., Roemer, E., and Imhof, A. (2005). Identification of a specific inhibitor of the histone methyltransferase SU(VAR)3-9. *Nat. Chem. Biol.* **1**(3), 143–145.

Grignani, F., De Matteis, S., Nervi, C., Tomassoni, L., Gelmetti, V., Cioce, M., Fanelli, M., Ruthardt, M., Ferrara, F. F., Zamir, I., et al. (1998). Fusion proteins of the retinoic acid receptor-alpha recruit histone deacetylase in promyelocytic leukaemia. *Nature* **391**(6669), 815–818.

Guidez, F., Howell, L., Isalan, M., Cebrat, M., Alani, R. M., Ivins, S., Hormaeche, I., McConnell, M. J., Pierce, S., Cole, P. A., et al. (2005). Histone acetyltransferase activity of p300 is required for transcriptional repression by the promyelocytic leukemia zinc finger protein. *Mol. Cell. Biol.* **25**(13), 5552–5566.

Gyory, I., Fejer, G., Ghosh, N., Seto, E., and Wright, K. L. (2003). Identification of a functionally impaired positive regulatory domain I binding factor 1 transcription repressor in myeloma cell lines. *J. Immunol.* **170**(6), 3125–3133.

Halkidou, K., Gaughan, L., Cook, S., Leung, H. Y., Neal, D. E., and Robson, C. N. (2004). Upregulation and nuclear recruitment of HDAC1 in hormone refractory prostate cancer. *Prostate* **59**(2), 177–189.

He, L. Z., Guidez, F., Triboli, C., Peruzzi, D., Ruthardt, M., Zelent, A., and Pandolfi, P. P. (1998). Distinct interactions of PML-RARalpha and PLZF-RARalpha with co-repressors determine differential responses to RA in APL. *Nat. Genet.* **18**(2), 126–135.

He, L. Z., Tolentino, T., Grayson, P., Zhong, S., Warrell, R. P., Jr., Rifkind, R. A., Marks, P. A., Richon, V. M., and Pandolfi, P. P. (2001). Histone deacetylase inhibitors induce remission in transgenic models of therapy-resistant acute promyelocytic leukemia. *J. Clin. Invest.* **108**(9), 1321–1330.

Hess-Stumpp, H., Bracker, T. U., Henderson, D., and Politz, O. (2007). MS-275, a potent orally available inhibitor of histone deacetylases–the development of an anticancer agent. *Int. J. Biochem. Cell Biol.* **39**(7–8), 1388–1405.

Hilfiker, A., Hilfiker-Kleiner, D., Pannuti, A., and Lucchesi, J. C. (1997). mof, a putative acetyl transferase gene related to the Tip60 and MOZ human genes and to the SAS genes of yeast, is required for dosage compensation in Drosophila. *EMBO J.* **16**(8), 2054–2060.

Hinnebusch, B. F., Meng, S., Wu, J. T., Archer, S. Y., and Hodin, R. A. (2002). The effects of short-chain fatty acids on human colon cancer cell phenotype are associated with histone hyperacetylation. *J. Nutr.* **132**(5), 1012–1017.

Hrzenjak, A., Moinfar, F., Kremser, M. L., Strohmeier, B., Staber, P. B., Zatloukal, K., and Denk, H. (2006). Valproate inhibition of histone deacetylase 2 affects differentiation and decreases proliferation of endometrial stromal sarcoma cells. *Mol. Cancer Ther.* **5**(9), 2203–2210.

Huang, B. H., Laban, M., Leung, C. H., Lee, L., Lee, C. K., Salto-Tellez, M., Raju, G. C., and Hooi, S. C. (2005). Inhibition of histone deacetylase 2 increases apoptosis and p21Cip1/WAF1 expression, independent of histone deacetylase 1. *Cell Death Differ.* **12**(4), 395–404.

Huang, J., Sengupta, R., Espejo, A. B., Lee, M. G., Dorsey, J. A., Richter, M., Opravil, S., Shiekhattar, R., Bedford, M. T., Jenuwein, T., et al. (2007). p53 is regulated by the lysine demethylase LSD1. *Nature* **449**(7158), 105–108.

Huang, Y., Stewart, T. M., Wu, Y., Baylin, S. B., Marton, L. J., Perkins, B., Jones, R. J., Woster, P. M., and Casero, R. A., Jr. (2009). Novel oligoamine analogues inhibit lysine-specific demethylase 1 and induce reexpression of epigenetically silenced genes. *Clin. Cancer Res.* **15**(23), 7217–7228.

Huffman, D. M., Grizzle, W. E., Bamman, M. M., Kim, J. S., Eltoum, I. A., Elgavish, A., and Nagy, T. R. (2007). SIRT1 is significantly elevated in mouse and human prostate cancer. *Cancer Res.* **67**(14), 6612–6618.

Iizuka, M., and Stillman, B. (1999). Histone acetyltransferase HBO1 interacts with the ORC1 subunit of the human initiator protein. *J. Biol. Chem.* **274**(33), 23027–23034.

Ikura, T., Ogryzko, V. V., Grigoriev, M., Groisman, R., Wang, J., Horikoshi, M., Scully, R., Qin, J., and Nakatani, Y. (2000). Involvement of the TIP60 histone acetylase complex in DNA repair and apoptosis. *Cell* **102**(4), 463–473.

Itoh, Y., Suzuki, T., and Miyata, N. (2008). Isoform-selective histone deacetylase inhibitors. *Curr. Pharm. Des.* **14**(6), 529–544.

Jenuwein, T. (2001). Re-SET-ting heterochromatin by histone methyltransferases. *Trends Cell Biol.* **11**(6), 266–273.

Jenuwein, T., and Allis, C. D. (2001). Translating the histone code. *Science* **293**(5532), 1074–1080.

Johannessen, C. U., and Johannessen, S. I. (2003). Valproate: Past, present, and future. *CNS Drug Rev.* **9**(2), 199–216.

Kahl, P., Gullotti, L., Heukamp, L. C., Wolf, S., Friedrichs, N., Vorreuther, R., Solleder, G., Bastian, P. J., Ellinger, J., Metzger, E., et al. (2006). Androgen receptor coactivators lysine-specific histone demethylase 1 and four and a half LIM domain protein 2 predict risk of prostate cancer recurrence. *Cancer Res.* **66**(23), 11341–11347.

Kalkhoven, E., Roelfsema, J. H., Teunissen, H., den Boer, A., Ariyurek, Y., Zantema, A., Breuning, M. H., Hennekam, R. C., and Peters, D. J. (2003). Loss of CBP acetyltransferase activity by PHD finger mutations in Rubinstein-Taybi syndrome. *Hum. Mol. Genet.* **12**(4), 441–450.

Kamine, J., Elangovan, B., Subramanian, T., Coleman, D., and Chinnadurai, G. (1996). Identification of a cellular protein that specifically interacts with the essential cysteine region of the HIV-1 Tat transactivator. *Virology* **216**(2), 357–366.

Kanazawa, I. (1994). Short review on monoamine oxidase and its inhibitors. *Eur. Neurol.* **34**(Suppl. 3), 36–39.

Kell, J. (2007). Drug evaluation: MGCD-0103, a histone deacetylase inhibitor for the treatment of cancer. *Curr. Opin. Investig. Drugs* **8**(6), 485–492.

Khan, N., Jeffers, M., Kumar, S., Hackett, C., Boldog, F., Khramtsov, N., Qian, X., Mills, E., Berghs, S. C., Carey, N., *et al.* (2008). Determination of the class and isoform selectivity of small-molecule histone deacetylase inhibitors. *Biochem. J.* **409**(2), 581–589.

Kim, Y. B., Lee, K. H., Sugita, K., Yoshida, M., and Horinouchi, S. (1999). Oxamflatin is a novel antitumor compound that inhibits mammalian histone deacetylase. *Oncogene* **18**(15), 2461–2470.

Kim, K. C., Geng, L., and Huang, S. (2003). Inactivation of a histone methyltransferase by mutations in human cancers. *Cancer Res.* **63**(22), 7619–7623.

Kim, Y. H., Whittaker, S., Demierre, M. F., Rook, A. H., Lerner, A., Duvic, M., Reddy, S., Kim, E. J., Robak, T., Becker, J. C., *et al.* (2008). Clinically significant responses achieved with romidepsin in treatment-refractory cutaneous T-cell lymphoma: final results from a Phase 2B, International, multicenter, registration study. *ASH Annual Meeting Abstracts.*

Kirschbaum, M., Zain, J., Popplewell, L., Pullarkat, V., Obadike, N., Frankel, P., Zwiebel, J., Forman, S., Newman, E., and Gandara, D. (2007). Phase 2 study of suberoylanilide hydroxamic acid (SAHA) in relapsed or refractory indolent non-Hodgkin lymphoma: A California Cancer Consortium study. *J. Clin. Oncol. (Meeting Abstracts)* **25**, 18515.

Klimek, V. M., Fircanis, S., Maslak, P., Guernah, I., Baum, M., Wu, N., Pangeas, K., Wright, J. J., Pandolfi, P. P., and Nimer, S. D. (2008). Tolerability, pharmacodynamics, and pharmacokinetics studies of depsipeptide (romidepsin) in patients with acute myelogenous leukemia or advanced myelodysplastic syndromes. *Clin. Cancer Res.* **14**(3), 826–832.

Kondo, Y., Shen, L., Suzuki, S., Kurokawa, T., Masuko, K., Tanaka, Y., Kato, H., Mizuno, Y., Yokoe, M., Sugauchi, F., *et al.* (2007). Alterations of DNA methylation and histone modifications contribute to gene silencing in hepatocellular carcinomas. *Hepatol. Res.* **37**(11), 974–983.

Kondo, Y., Shen, L., Ahmed, S., Boumber, Y., Sekido, Y., Haddad, B. R., and Issa, J. P. (2008a). Downregulation of histone H3 lysine 9 methyltransferase G9a induces centrosome disruption and chromosome instability in cancer cells. *PLoS ONE* **3**(4), e2037.

Kondo, Y., Shen, L., Cheng, A. S., Ahmed, S., Boumber, Y., Charo, C., Yamochi, T., Urano, T., Furukawa, K., Kwabi-Addo, B., *et al.* (2008b). Gene silencing in cancer by histone H3 lysine 27 trimethylation independent of promoter DNA methylation. *Nat. Genet.* **40**(6), 741–750.

Konstantinopoulos, P. A., Vandoros, G. P., and Papavassiliou, A. G. (2006). FK228 (depsipeptide): A HDAC inhibitor with pleiotropic antitumor activities. *Cancer Chemother. Pharmacol.* **58**(5), 711–715.

Kouzarides, T. (2002). Histone methylation in transcriptional control. *Curr. Opin. Genet. Dev.* **12**(2), 198–209.

Kuendgen, A., and Gattermann, N. (2007). Valproic acid for the treatment of myeloid malignancies. *Cancer* **110**(5), 943–954.

Lahortiga, I., Agirre, X., Belloni, E., Vazquez, I., Larrayoz, M. J., Gasparini, P., Lo Coco, F., Pelicci, P. G., Calasanz, M. J., and Odero, M. D. (2004). Molecular characterization of a t(1;3) (p36;q21) in a patient with MDS. MEL1 is widely expressed in normal tissues, including bone marrow, and it is not overexpressed in the t(1;3) cells. *Oncogene* **23**(1), 311–316.

Lau, O. D., Kundu, T. K., Soccio, R. E., Ait-Si-Ali, S., Khalil, E. M., Vassilev, A., Wolffe, A. P., Nakatani, Y., Roeder, R. G., and Cole, P. A. (2000). HATs off: Selective synthetic inhibitors of the histone acetyltransferases p300 and PCAF. *Mol. Cell* **5**(3), 589–595.

Lehrmann, H., Pritchard, L. L., and Harel-Bellan, A. (2002). Histone acetyltransferases and deacetylases in the control of cell proliferation and differentiation. *Adv. Cancer Res.* **86,** 41–65.

Lemercier, C., Brocard, M. P., Puvion-Dutilleul, F., Kao, H. Y., Albagli, O., and Khochbin, S. (2002). Class II histone deacetylases are directly recruited by BCL6 transcriptional repressor. *J. Biol. Chem.* **277**(24), 22045–22052.

Letunic, I., Copley, R. R., Schmidt, S., Ciccarelli, F. D., Doerks, T., Schultz, J., Ponting, C. P., and Bork, P. (2004). SMART 4.0: Towards genomic data integration. *Nucleic Acids Res.* **32**(Database issue), D142–144.

Lim, S., Janzer, A., Becker, A., Zimmer, A., Schule, R., Buettner, R., and Kirfel, J. (2010). Lysine-specific demethylase 1 (LSD1) is highly expressed in ER-negative breast cancers and a biomarker predicting aggressive biology. *Carcinogenesis* **31**(3), 512–520.

Liu, L., Shao, G., Steele-Perkins, G., and Huang, S. (1997). The retinoblastoma interacting zinc finger gene RIZ produces a PR domain-lacking product through an internal promoter. *J. Biol. Chem.* **272**(5), 2984–2991.

Liu, G., Yuan, X., Zeng, Z., Tunici, P., Ng, H., Abdulkadir, I. R., Lu, L., Irvin, D., Black, K. L., and Yu, J. S. (2006). Analysis of gene expression and chemoresistance of CD133+ cancer stem cells in glioblastoma. *Mol. Cancer* **5**, 67.

Lucas, D. M., Davis, M. E., Parthun, M. R., Mone, A. P., Kitada, S., Cunningham, K. D., Flax, E. L., Wickham, J., Reed, J. C., Byrd, J. C., *et al.* (2004). The histone deacetylase inhibitor MS-275 induces caspase-dependent apoptosis in B-cell chronic lymphocytic leukemia cells. *Leukemia* **18**(7), 1207–1214.

Lund, A. H., and van Lohuizen, M. (2004). Epigenetics and cancer. *Genes Dev.* **18**(19), 2315–2335.

Mai, A., and Altucci, L. (2009). Epi-drugs to fight cancer: From chemistry to cancer treatment, the road ahead. *Int. J. Biochem. Cell Biol.* **41**(1), 199–213.

Mai, A., Massa, S., Rotili, D., Cerbara, I., Valente, S., Pezzi, R., Simeoni, S., and Ragno, R. (2005). Histone deacetylation in epigenetics: An attractive target for anticancer therapy. *Med. Res. Rev.* **25**(3), 261–309.

Mai, A., Valente, S., Cheng, D., Perrone, A., Ragno, R., Simeoni, S., Sbardella, G., Brosch, G., Nebbioso, A., Conte, M., *et al.* (2007). Synthesis and biological validation of novel synthetic histone/protein methyltransferase inhibitors. *ChemMedChem* **2**(7), 987–991.

Mai, A., Cheng, D., Bedford, M. T., Valente, S., Nebbioso, A., Perrone, A., Brosch, G., Sbardella, G., De Bellis, F., Miceli, M., *et al.* (2008). epigenetic multiple ligands: Mixed histone/protein methyltransferase, acetyltransferase, and class III deacetylase (sirtuin) inhibitors. *J. Med. Chem.* **51**(7), 2279–2290.

Majumder, S., Liu, Y., Ford, O. H., III, Mohler, J. L., and Whang, Y. E. (2006). Involvement of arginine methyltransferase CARM1 in androgen receptor function and prostate cancer cell viability. *Prostate* **66**(12), 1292–1301.

Mann, B. S., Johnson, J. R., Cohen, M. H., Justice, R., and Pazdur, R. (2007). FDA approval summary: Vorinostat for treatment of advanced primary cutaneous T-cell lymphoma. *Oncologist* **12**(10), 1247–1252.

Mantelingu, K., Reddy, B. A., Swaminathan, V., Kishore, A. H., Siddappa, N. B., Kumar, G. V., Nagashankar, G., Natesh, N., Roy, S., Sadhale, P. P., *et al.* (2007). Specific inhibition of p300-HAT alters global gene expression and represses HIV replication. *Chem. Biol.* **14**(6), 645–657.

Marks, P. A. (2007). Discovery and development of SAHA as an anticancer agent. *Oncogene* **26**(9), 1351–1356.

Marks, P. A., Richon, V. M., and Rifkind, R. A. (2000). Histone deacetylase inhibitors: Inducers of differentiation or apoptosis of transformed cells. *J. Natl Cancer Inst.* **92**(15), 1210–1216.

Marks, P. A., Richon, V. M., Breslow, R., and Rifkind, R. A. (2001). Histone deacetylase inhibitors as new cancer drugs. *Curr. Opin. Oncol.* **13**(6), 477–483.

Martin, C., and Zhang, Y. (2005). The diverse functions of histone lysine methylation. *Nat. Rev. Mol. Cell Biol.* **6**(11), 838–849.

Martinez-Garcia, E., and Licht, J. D. (2010). Deregulation of H3K27 methylation in cancer. *Nat. Genet.* **42**(2), 100–101.

Melchior, S. W., Brown, L. G., Figg, W. D., Quinn, J. E., Santucci, R. A., Brunner, J., Thuroff, J. W., Lange, P. H., and Vessella, R. L. (1999). Effects of phenylbutyrate on proliferation and apoptosis in human prostate cancer cells in vitro and in vivo. *Int. J. Oncol.* **14**(3), 501–508.

Metzger, E., Wissmann, M., Yin, N., Muller, J. M., Schneider, R., Peters, A. H., Gunther, T., Buettner, R., and Schule, R. (2005). LSD1 demethylates repressive histone marks to promote androgen-receptor-dependent transcription. *Nature* **437**(7057), 436–439.

Michan, S., and Sinclair, D. (2007). Sirtuins in mammals: Insights into their biological function. *Biochem. J.* **404**(1), 1–13.

Minucci, S., and Pelicci, P. G. (2006). Histone deacetylase inhibitors and the promise of epigenetic (and more) treatments for cancer. *Nat. Rev. Cancer* **6**(1), 38–51.

Minucci, S., Nervi, C., Lo Coco, F., and Pelicci, P. G. (2001). Histone deacetylases: A common molecular target for differentiation treatment of acute myeloid leukemias? *Oncogene* **20**(24), 3110–3115.

Miotto, B., and Struhl, K. (2008). HBO1 histone acetylase is a coactivator of the replication licensing factor Cdt1. *Genes Dev.* **22**(19), 2633–2638.

Miura, T. A., Cook, J. L., Potter, T. A., Ryan, S., and Routes, J. M. (2007). The interaction of adenovirus E1A with p300 family members modulates cellular gene expression to reduce tumorigenicity. *J. Cell. Biochem.* **100**(4), 929–940.

Morgunkova, A., and Barlev, N. A. (2006). Lysine methylation goes global. *Cell Cycle* **5**(12), 1308–1312.

Morin, R. D., Johnson, N. A., Severson, T. M., Mungall, A. J., An, J., Goya, R., Paul, J. E., Boyle, M., Woolcock, B. W., Kuchenbauer, F., *et al.* (2010). Somatic mutations altering EZH2 (Tyr641) in follicular and diffuse large B-cell lymphomas of germinal-center origin. *Nat. Genet.* **42**(2), 181–185.

Morishita, K., Parker, D. S., Mucenski, M. L., Jenkins, N. A., Copeland, N. G., and Ihle, J. N. (1988). Retroviral activation of a novel gene encoding a zinc finger protein in IL-3-dependent myeloid leukemia cell lines. *Cell* **54**(6), 831–840.

Muraoka, M., Konishi, M., Kikuchi-Yanoshita, R., Tanaka, K., Shitara, N., Chong, J. M., Iwama, T., and Miyaki, M. (1996). p300 gene alterations in colorectal and gastric carcinomas. *Oncogene* **12**(7), 1565–1569.

Nebbioso, A., Clarke, N., Voltz, E., Germain, E., Ambrosino, C., Bontempo, P., Alvarez, R., Schiavone, E. M., Ferrara, F., Bresciani, F., *et al.* (2005). Tumor-selective action of HDAC inhibitors involves TRAIL induction in acute myeloid leukemia cells. *Nat. Med.* **11**(1), 77–84.

Nielsen, S. J., Schneider, R., Bauer, U. M., Bannister, A. J., Morrison, A., O'Carroll, D., Firestein, R., Cleary, M., Jenuwein, T., Herrera, R. E., *et al.* (2001). Rb targets histone H3 methylation and HP1 to promoters. *Nature* **412**(6846), 561–565.

Odenike, O. M., Alkan, S., Sher, D., Godwin, J. E., Huo, D., Brandt, S. J., Green, M., Xie, J., Zhang, Y., Vesole, D. H., *et al.* (2008). Histone deacetylase inhibitor romidepsin has differential activity in core binding factor acute myeloid leukemia. *Clin. Cancer Res.* **14**(21), 7095–7101.

Okada, Y., Takeda, S., Tanaka, Y., Belmonte, J. C., and Hirokawa, N. (2005). Mechanism of nodal flow: A conserved symmetry breaking event in left-right axis determination. *Cell* **121**(4), 633–644.

Olsen, E. A., Kim, Y. H., Kuzel, T. M., Pacheco, T. R., Foss, F. M., Parker, S., Frankel, S. R., Chen, C., Ricker, J. L., Arduino, J. M., *et al.* (2007). Phase IIb multicenter trial of vorinostat in patients with persistent, progressive, or treatment refractory cutaneous T-cell lymphoma. *J. Clin. Oncol.* **25**(21), 3109–3115.

Osada, H., Tatematsu, Y., Saito, H., Yatabe, Y., Mitsudomi, T., and Takahashi, T. (2004). Reduced expression of class II histone deacetylase genes is associated with poor prognosis in lung cancer patients. *Int. J. Cancer* **112**(1), 26–32.

Ozdag, H., Teschendorff, A. E., Ahmed, A. A., Hyland, S. J., Blenkiron, C., Bobrow, L., Veerakumarasivam, A., Burtt, G., Subkhankulova, T., Arends, M. J., *et al.* (2006). Differential expression of selected histone modifier genes in human solid cancers. *BMC Genomics* **7**, 90.

Paik, W. K., and Kim, S. (1973). Enzymatic demethylation of calf thymus histones. *Biochem. Biophys. Res. Commun.* **51**(3), 781–788.

Paik, W. K., and Kim, S. (1974). Epsilon-alkyllysinase. New assay method, purification, and biological significance. *Arch. Biochem. Biophys.* **165**(1), 369–378.

Pal, S., Vishwanath, S. N., Erdjument-Bromage, H., Tempst, P., and Sif, S. (2004). Human SWI/SNF-associated PRMT5 methylates histone H3 arginine 8 and negatively regulates expression of ST7 and NM23 tumor suppressor genes. *Mol. Cell. Biol.* **24**(21), 9630–9645.

Pan, M. H., Chang, W. L., Lin-Shiau, S. Y., Ho, C. T., and Lin, J. K. (2001). Induction of apoptosis by garcinol and curcumin through cytochrome c release and activation of caspases in human leukemia HL-60 cells. *J. Agric. Food Chem.* **49**(3), 1464–1474.

Panagopoulos, I., Fioretos, T., Isaksson, M., Samuelsson, U., Billstrom, R., Strombeck, B., Mitelman, F., and Johansson, B. (2001). Fusion of the MORF and CBP genes in acute myeloid leukemia with the t(10;16)(q22;p13). *Hum. Mol. Genet.* **10**(4), 395–404.

Peters, A. H., O'Carroll, D., Scherthan, H., Mechtler, K., Sauer, S., Schofer, C., Weipoltshammer, K., Pagani, M., Lachner, M., Kohlmaier, A., *et al.* (2001). Loss of the Suv39h histone methyltransferases impairs mammalian heterochromatin and genome stability. *Cell* **107**(3), 323–337.

Phillips, A. C., and Vousden, K. H. (2000). Acetyltransferases and tumour suppression. *Breast Cancer Res.* **2**(4), 244–246.

Piekarz, R. L., Robey, R., Sandor, V., Bakke, S., Wilson, W. H., Dahmoush, L., Kingma, D. M., Turner, M. L., Altemus, R., and Bates, S. E. (2001). Inhibitor of histone deacetylation, depsipeptide (FR901228), in the treatment of peripheral and cutaneous T-cell lymphoma: A case report. *Blood* **98**(9), 2865–2868.

Piscitelli, S. C., Thibault, A., Figg, W. D., Tompkins, A., Headlee, D., Lieberman, R., Samid, D., and Myers, C. E. (1995). Disposition of phenylbutyrate and its metabolites, phenylacetate and phenylacetylglutamine. *J. Clin. Pharmacol.* **35**(4), 368–373.

Pruitt, K., Zinn, R. L., Ohm, J. E., McGarvey, K. M., Kang, S. H., Watkins, D. N., Herman, J. G., and Baylin, S. B. (2006). Inhibition of SIRT1 reactivates silenced cancer genes without loss of promoter DNA hypermethylation. *PLoS Genet.* **2**(3), e40.

Qian, C., and Zhou, M. M. (2006). SET domain protein lysine methyltransferases: Structure, specificity and catalysis. *Cell. Mol. Life Sci.* **63**(23), 2755–2763.

Qian, D. Z., Wei, Y. F., Wang, X., Kato, Y., Cheng, L., and Pili, R. (2007). Antitumor activity of the histone deacetylase inhibitor MS-275 in prostate cancer models. *Prostate* **67**(11), 1182–1193.

Qian, X., Ara, G., Mills, E., LaRochelle, W. J., Lichenstein, H. S., and Jeffers, M. (2008). Activity of the histone deacetylase inhibitor belinostat (PXD101) in preclinical models of prostate cancer. *Int. J. Cancer* **122**(6), 1400–1410.

Rea, S., Eisenhaber, F., O'Carroll, D., Strahl, B. D., Sun, Z. W., Schmid, M., Opravil, S., Mechtler, K., Ponting, C. P., Allis, C. D., *et al.* (2000). Regulation of chromatin structure by site-specific histone H3 methyltransferases. *Nature* **406**(6796), 593–599.

Reid, T., Valone, F., Lipera, W., Irwin, D., Paroly, W., Natale, R., Sreedharan, S., Keer, H., Lum, B., Scappaticci, F., *et al.* (2004). Phase II trial of the histone deacetylase inhibitor pivaloyloxymethyl butyrate (Pivanex, AN-9) in advanced non-small cell lung cancer. *Lung Cancer* **45**(3), 381–386.

Rice, J. C., and Allis, C. D. (2001). Histone methylation versus histone acetylation: New insights into epigenetic regulation. *Curr. Opin. Cell Biol.* **13**(3), 263–273.

Richon, V. M., Emiliani, S., Verdin, E., Webb, Y., Breslow, R., Rifkind, R. A., and Marks, P. A. (1998). A class of hybrid polar inducers of transformed cell differentiation inhibits histone deacetylases. *Proc. Natl Acad. Sci. USA* **95**(6), 3003–3007.

Robertson, K. D., Ait-Si-Ali, S., Yokochi, T., Wade, P. A., Jones, P. L., and Wolffe, A. P. (2000). DNMT1 forms a complex with Rb, E2F1 and HDAC1 and represses transcription from E2F-responsive promoters. *Nat. Genet.* **25**(3), 338–342.

Roth, S. Y., Denu, J. M., and Allis, C. D. (2001). Histone acetyltransferases. *Annu. Rev. Biochem.* **70**, 81–120.

Ryan, Q. C., Headlee, D., Acharya, M., Sparreboom, A., Trepel, J. B., Ye, J., Figg, W. D., Hwang, K., Chung, E. J., Murgo, A., *et al.* (2005). Phase I and pharmacokinetic study of MS-275, a histone deacetylase inhibitor, in patients with advanced and refractory solid tumors or lymphoma. *J. Clin. Oncol.* **23**(17), 3912–3922.

Sanchez, R., and Zhou, M. M. (2009). The role of human bromodomains in chromatin biology and gene transcription. *Curr. Opin Drug Discov. Devel.* **12**(5), 659–665.

Sarma, K., and Reinberg, D. (2005). Histone variants meet their match. *Nat. Rev. Mol. Cell Biol.* **6**(2), 139–149.

Satake, N., Ishida, Y., Otoh, Y., Hinohara, S., Kobayashi, H., Sakashita, A., Maseki, N., and Kaneko, Y. (1997). Novel MLL-CBP fusion transcript in therapy-related chronic myelomonocytic leukemia with a t(11;16)(q23;p13) chromosome translocation. *Genes Chromosom. Cancer* **20**(1), 60–63.

Schiltz, R. L., Mizzen, C. A., Vassilev, A., Cook, R. G., Allis, C. D., and Nakatani, Y. (1999). Overlapping but distinct patterns of histone acetylation by the human coactivators p300 and PCAF within nucleosomal substrates. *J. Biol. Chem.* **274**(3), 1189–1192.

Schlesinger, Y., Straussman, R., Keshet, I., Farkash, S., Hecht, M., Zimmerman, J., Eden, E., Yakhini, Z., Ben-Shushan, E., Reubinoff, B. E., *et al.* (2007). Polycomb-mediated methylation on Lys27 of histone H3 pre-marks genes for de novo methylation in cancer. *Nat. Genet.* **39**(2), 232–236.

Schmidt, D. M., and McCafferty, D. G. (2007). *trans*-2-Phenylcyclopropylamine is a mechanism-based inactivator of the histone demethylase LSD1. *Biochemistry* **46**(14), 4408–4416.

Schulte, J. H., Lim, S., Schramm, A., Friedrichs, N., Koster, J., Versteeg, R., Ora, I., Pajtler, K., Klein-Hitpass, L., Kuhfittig-Kulle, S., *et al.* (2009). Lysine-specific demethylase 1 is strongly expressed in poorly differentiated neuroblastoma: Implications for therapy. *Cancer Res.* **69**(5), 2065–2071.

Seligson, D. B., Horvath, S., Shi, T., Yu, H., Tze, S., Grunstein, M., and Kurdistani, S. K. (2005). Global histone modification patterns predict risk of prostate cancer recurrence. *Nature* **435** (7046), 1262–1266.

Shaffer, A. L., Rosenwald, A., and Staudt, L. M. (2002). Lymphoid malignancies: The dark side of B-cell differentiation. *Nat. Rev. Immunol.* **2**(12), 920–932.

Sharma, S. V., Lee, D. Y., Li, B., Quinlan, M. P., Takahashi, F., Maheswaran, S., McDermott, U., Azizian, N., Zou, L., Fischbach, M. A., *et al.* (2010). A chromatin-mediated reversible drug-tolerant state in cancer cell subpopulations. *Cell* **141**(1), 69–80.

Shi, Y., Lan, F., Matson, C., Mulligan, P., Whetstine, J. R., Cole, P. A., and Casero, R. A. (2004). Histone demethylation mediated by the nuclear amine oxidase homolog LSD1. *Cell* **119**(7), 941–953.

Shiama, N. (1997). The p300/CBP family: Integrating signals with transcription factors and chromatin. *Trends Cell Biol.* **7**(6), 230–236.

Shiio, Y., and Eisenman, R. N. (2003). Histone sumoylation is associated with transcriptional repression. *Proc. Natl Acad. Sci. USA* **100**(23), 13225–13230.

Shilatifard, A. (2006). Chromatin modifications by methylation and ubiquitination: Implications in the regulation of gene expression. *Annu. Rev. Biochem.* **75**, 243–269.

Smith, E. R., Cayrou, C., Huang, R., Lane, W. S., Cote, J., and Lucchesi, J. C. (2005). A human protein complex homologous to the Drosophila MSL complex is responsible for the majority of histone H4 acetylation at lysine 16. *Mol. Cell. Biol.* **25**(21), 9175–9188.

So, C. W., Karsunky, H., Wong, P., Weissman, I. L., and Cleary, M. L. (2004). Leukemic transformation of hematopoietic progenitors by MLL-GAS7 in the absence of Hoxa7 or Hoxa9. *Blood* **103**(8), 3192–3199.

Somoza, J. R., Skene, R. J., Katz, B. A., Mol, C., Ho, J. D., Jennings, A. J., Luong, C., Arvai, A., Buggy, J. J., Chi, E., *et al.* (2004). Structural snapshots of human HDAC8 provide insights into the class I histone deacetylases. *Structure* **12**(7), 1325–1334.

Song, J., Noh, J. H., Lee, J. H., Eun, J. W., Ahn, Y. M., Kim, S. Y., Lee, S. H., Park, W. S., Yoo, N. J., Lee, J. Y., *et al.* (2005). Increased expression of histone deacetylase 2 is found in human gastric cancer. *APMIS* **113**(4), 264–268.

Spannhoff, A., Machmur, R., Heinke, R., Trojer, P., Bauer, I., Brosch, G., Schule, R., Hanefeld, W., Sippl, W., and Jung, M. (2007). A novel arginine methyltransferase inhibitor with cellular activity. *Bioorg. Med. Chem. Lett.* **17**(15), 4150–4153.

Sparmann, A., and van Lohuizen, M. (2006). Polycomb silencers control cell fate, development and cancer. *Nat. Rev. Cancer* **6**(11), 846–856.

Spurling, C. C., Godman, C. A., Noonan, E. J., Rasmussen, T. P., Rosenberg, D. W., and Giardina, C. (2008). HDAC3 overexpression and colon cancer cell proliferation and differentiation. *Mol. Carcinog.* **47**(2), 137–147.

Stadler, W. M., Margolin, K., Ferber, S., McCulloch, W., and Thompson, J. A. (2006). A phase II study of depsipeptide in refractory metastatic renal cell cancer. *Clin. Genitourin. Cancer* **5**(1), 57–60.

Steele-Perkins, G., Fang, W., Yang, X. H., Van Gele, M., Carling, T., Gu, J., Buyse, I. M., Fletcher, J. A., Liu, J., Bronson, R., *et al.* (2001). Tumor formation and inactivation of RIZ1, an Rb-binding member of a nuclear protein-methyltransferase superfamily. *Genes Dev.* **15**(17), 2250–2262.

Sterner, D. E., and Berger, S. L. (2000). Acetylation of histones and transcription-related factors. *Microbiol. Mol. Biol. Rev.* **64**(2), 435–459.

Stimson, L., Wood, V., Khan, O., Fotheringham, S., and La Thangue, N. B. (2009). HDAC inhibitor-based therapies and haematological malignancy. *Ann. Oncol.* **20**(8), 1293–1302.

Sun, Y., Jiang, X., Chen, S., Fernandes, N., and Price, B. D. (2005). A role for the Tip60 histone acetyltransferase in the acetylation and activation of ATM. *Proc. Natl Acad. Sci. USA* **102**(37), 13182–13187.

Sun, Y., Jiang, X., Chen, S., and Price, B. D. (2006). Inhibition of histone acetyltransferase activity by anacardic acid sensitizes tumor cells to ionizing radiation. *FEBS Lett.* **580**(18), 4353–4356.

Taipale, M., Rea, S., Richter, K., Vilar, A., Lichter, P., Imhof, A., and Akhtar, A. (2005). hMOF histone acetyltransferase is required for histone H4 lysine 16 acetylation in mammalian cells. *Mol. Cell. Biol.* **25**(15), 6798–6810.

Tan, J., Yang, X., Zhuang, L., Jiang, X., Chen, W., Lee, P. L., Karuturi, R. K., Tan, P. B., Liu, E. T., and Yu, Q. (2007). Pharmacologic disruption of Polycomb-repressive complex 2-mediated gene repression selectively induces apoptosis in cancer cells. *Genes Dev.* **21**(9), 1050–1063.

Thiagalingam, S., Cheng, K. H., Lee, H. J., Mineva, N., Thiagalingam, A., and Ponte, J. F. (2003). Histone deacetylases: Unique players in shaping the epigenetic histone code. *Ann. NY Acad. Sci.* **983**, 84–100.

Trievel, R. C. (2004). Structure and function of histone methyltransferases. *Crit. Rev. Eukaryot. Gene Expr.* **14**(3), 147–169.

Tsuji, N., Kobayashi, M., Nagashima, K., Wakisaka, Y., and Koizumi, K. (1976). A new antifungal antibiotic, trichostatin. *J. Antibiot. (Tokyo)* **29**(1), 1–6.

Tsukada, Y., Fang, J., Erdjument-Bromage, H., Warren, M. E., Borchers, C. H., Tempst, P., and Zhang, Y. (2006). Histone demethylation by a family of JmjC domain-containing proteins. *Nature* **439**(7078), 811–816.

Ungerstedt, J. S., Sowa, Y., Xu, W. S., Shao, Y., Dokmanovic, M., Perez, G., Ngo, L., Holmgren, A., Jiang, X., and Marks, P. A. (2005). Role of thioredoxin in the response of normal and transformed cells to histone deacetylase inhibitors. *Proc. Natl Acad. Sci. USA* **102**(3), 673–678.

van Haaften, G., Dalgliesh, G. L., Davies, H., Chen, L., Bignell, G., Greenman, C., Edkins, S., Hardy, C., O'Meara, S., Teague, J., *et al.* (2009). Somatic mutations of the histone H3K27 demethylase gene UTX in human cancer. *Nat. Genet.* **41**(5), 521–523.

Verdin, E., Dequiedt, F., and Kasler, H. G. (2003). Class II histone deacetylases: Versatile regulators. *Trends Genet.* **19**(5), 286–293.

Vire, E., Brenner, C., Deplus, R., Blanchon, L., Fraga, M., Didelot, C., Morey, L., Van Eynde, A., Bernard, D., Vanderwinden, J. M., *et al.* (2006). The Polycomb group protein EZH2 directly controls DNA methylation. *Nature* **439**(7078), 871–874.

Wade, P. A., Pruss, D., and Wolffe, A. P. (1997). Histone acetylation: Chromatin in action. *Trends Biochem. Sci.* **22**(4), 128–132.

Wang, J., Hoshino, T., Redner, R. L., Kajigaya, S., and Liu, J. M. (1998). ETO, fusion partner in t(8;21) acute myeloid leukemia, represses transcription by interaction with the human N-CoR/mSin3/HDAC1 complex. *Proc. Natl Acad. Sci. USA* **95**(18), 10860–10865.

Wang, Z., Zang, C., Rosenfeld, J. A., Schones, D. E., Barski, A., Cuddapah, S., Cui, K., Roh, T. Y., Peng, W., Zhang, M. Q., *et al.* (2008). Combinatorial patterns of histone acetylations and methylations in the human genome. *Nat. Genet.* **40**(7), 897–903.

Wang, Y., Zhang, H., Chen, Y., Sun, Y., Yang, F., Yu, W., Liang, J., Sun, L., Yang, X., Shi, L., *et al.* (2009a). LSD1 is a subunit of the NuRD complex and targets the metastasis programs in breast cancer. *Cell* **138**(4), 660–672.

Wang, Z., Zang, C., Cui, K., Schones, D. E., Barski, A., Peng, W., and Zhao, K. (2009b). Genomewide mapping of HATs and HDACs reveals distinct functions in active and inactive genes. *Cell* **138**(5), 1019–1031.

Warrell, R. P., Jr., Frankel, S. R., Miller, W. H., Jr., Scheinberg, D. A., Itri, L. M., Hittelman, W. N., Vyas, R., Andreeff, M., Tafuri, A., Jakubowski, A., *et al.* (1991). Differentiation therapy of acute promyelocytic leukemia with tretinoin (all-*trans*-retinoic acid). *N Engl J. Med.* **324**(20), 1385–1393.

Whitehead, R. P., Rankin, C., Hoff, P. M., Gold, P. J., Billingsley, K. G., Chapman, R. A., Wong, L., Ward, J. H., Abbruzzese, J. L., and Blanke, C. D. (2009). Phase II trial of romidepsin (NSC-630176) in previously treated colorectal cancer patients with advanced disease: A Southwest Oncology Group study (S0336). *Invest New Drugs* **27**(5), 469–475.

Yamane, K., Tateishi, K., Klose, R. J., Fang, J., Fabrizio, L. A., Erdjument-Bromage, H., Taylor-Papadimitriou, J., Tempst, P., and Zhang, Y. (2007). PLU-1 is an H3K4 demethylase involved in transcriptional repression and breast cancer cell proliferation. *Mol. Cell* **25**(6), 801–812.

Yan, Y., Barlev, N. A., Haley, R. H., Berger, S. L., and Marmorstein, R. (2000). Crystal structure of yeast Esa1 suggests a unified mechanism for catalysis and substrate binding by histone acetyltransferases. *Mol. Cell* **6**(5), 1195–1205.

Yang, X. J., and Seto, E. (2008). The Rpd3/Hda1 family of lysine deacetylases: From bacteria and yeast to mice and men. *Nat. Rev. Mol. Cell Biol.* **9**(3), 206–218.

Yang, X. J., Ogryzko, V. V., Nishikawa, J., Howard, B. H., and Nakatani, Y. (1996). A p300/CBP-associated factor that competes with the adenoviral oncoprotein E1A. *Nature* **382**(6589), 319–324.

Yoshida, M., Hoshikawa, Y., Koseki, K., Mori, K., and Beppu, T. (1990). Structural specificity for biological activity of trichostatin A, a specific inhibitor of mammalian cell cycle with potent differentiation-inducing activity in Friend leukemia cells. *J. Antibiot. (Tokyo)* **43**(9), 1101–1106.

Zeidler, M., Varambally, S., Cao, Q., Chinnaiyan, A. M., Ferguson, D. O., Merajver, S. D., and Kleer, C. G. (2005). The Polycomb group protein EZH2 impairs DNA repair in breast epithelial cells. *Neoplasia* **7**(11), 1011–1019.

Zeleznik-Le, N. J., Harden, A. M., and Rowley, J. D. (1994). 11q23 translocations split the "AT-hook" cruciform DNA-binding region and the transcriptional repression domain from the activation domain of the mixed-lineage leukemia (MLL) gene. *Proc. Natl Acad. Sci. USA* **91**(22), 10610–10614.

Ziemin-van der Poel, S., McCabe, N. R., Gill, H. J., Espinosa, R., III, Patel, Y., Harden, A., Rubinelli, P., Smith, S. D., LeBeau, M. M., Rowley, J. D., *et al.* (1991). Identification of a gene, MLL, that spans the breakpoint in 11q23 translocations associated with human leukemias. *Proc. Natl. Acad. Sci. USA* **88**(23), 10735–10739.

Zimra, Y., Nudelman, A., Zhuk, R., Rabizadeh, E., Shaklai, M., Aviram, A., and Rephaeli, A. (2000). Uptake of pivaloyloxymethyl butyrate into leukemic cells and its intracellular esterase-catalyzed hydrolysis. *J. Cancer Res. Clin. Oncol.* **126**(12), 693–698.

Index